Lecture Notes in Networks and Systems 872

The series "Lecture Notes in Networks and Systems" publishes the latest developments in Networks and Systems—quickly, informally and with high quality. Original research reported in proceedings and post-proceedings represents the core of LNNS.

Volumes published in LNNS embrace all aspects and subfields of, as well as new challenges in, Networks and Systems.

The series contains proceedings and edited volumes in systems and networks, spanning the areas of Cyber-Physical Systems, Autonomous Systems, Sensor Networks, Control Systems, Energy Systems, Automotive Systems, Biological Systems, Vehicular Networking and Connected Vehicles, Aerospace Systems, Automation, Manufacturing, Smart Grids, Nonlinear Systems, Power Systems, Robotics, Social Systems, Economic Systems and other. Of particular value to both the contributors and the readership are the short publication timeframe and the worldwide distribution and exposure which enable both a wide and rapid dissemination of research output.

The series covers the theory, applications, and perspectives on the state of the art and future developments relevant to systems and networks, decision making, control, complex processes and related areas, as embedded in the fields of interdisciplinary and applied sciences, engineering, computer science, physics, economics, social, and life sciences, as well as the paradigms and methodologies behind them.

Indexed by SCOPUS, INSPEC, WTI Frankfurt eG, zbMATH, SCImago.

All books published in the series are submitted for consideration in Web of Science.

For proposals from Asia please contact Aninda Bose (aninda.bose@springer.com).

Miroslav Trajanovic · Nenad Filipovic ·
Milan Zdravkovic
Editors

Disruptive Information Technologies for a Smart Society

Proceedings of the 13th International
Conference on Information Society and
Technology (ICIST)

Editors
Miroslav Trajanovic
Faculty of Mechanical Engineering
University of Nis
Nis, Serbia

Nenad Filipovic
Faculty of Engineering
University of Kragujevac
Kragujevac, Serbia

Milan Zdravkovic
Faculty of Mechanical Engineering
University of Nis
Nis, Serbia

ISSN 2367-3370   ISSN 2367-3389 (electronic)
Lecture Notes in Networks and Systems
ISBN 978-3-031-50754-0   ISBN 978-3-031-50755-7 (eBook)
https://doi.org/10.1007/978-3-031-50755-7

This Springer imprint is published by the registered company Springer Nature Switzerland AG
The registered company address is: Gewerbestrasse 11, 6330 Cham, Switzerland

Paper in this product is recyclable.

Preface

This book includes selected high-quality peer-reviewed research papers presented at 13th International Conference on Information Society and Technology held on Kopaonik Mountain, Serbia, on Mar 12–15, 2023. In an era where technology disrupts many facets of our lives, the papers included in this issue exemplify the remarkable ways in which information technologies are reshaping our world, driving innovation, and paving the path toward a smarter, more efficient society.

The selected papers represent a diverse range of topics, all connected by the common commitment to explore information technologies as tools for positive transformation. From e-government requirements specification to advanced machine learning algorithms for river water quality management and from the application of artificial intelligence in healthcare to the analysis of financial markets using social media data, these contributions collectively illuminate the profound impact of information technologies on various domains.

One of the several focal points of this special issue is the intersection of artificial intelligence and public administration. Some papers delve into this area, addressing topics such as the role of AI in public administration and business sectors and the adoption of e-contracting and smart contracts for legally enforceable conformance checking in collaborative production. These papers underscore the potential of information technologies to enhance governance, streamline processes, and promote transparency in the public sector.

Another significant theme explored in this issue is the application of disruptive technologies in healthcare. Whether it's the prediction of coronary plaque progression using data mining and artificial neural networks or the risk stratification of patients with hypertrophic cardiomyopathy through genetic and clinical data features, these studies demonstrate how advanced information technologies are contributing to the diagnosis, treatment, and overall well-being of individuals. The use of technology in healthcare is becoming increasingly indispensable, and the papers presented here showcase the latest trends in this field.

Furthermore, this special issue highlights the importance of sustainability and environmental consciousness in our technologically driven society. From estimating solar power potential for rooftops to optimizing wind production forecasting and analyzing hydropower system resilience, these papers underscore the critical role of information technologies in promoting eco-friendly practices and renewable energy solutions.

The paper review process, organized as a single blind, had two stages. In the first stage, the papers were reviewed to be accepted for presentation at the 13th International Conference on Information Society and Technology. A total of 80 papers were accepted for presentation at the conference. The authors had the opportunity to prepare an improved version of the manuscript for these proceedings. After the second stage of review, 48 papers were accepted for publication in the proceedings.

In conclusion, "Disruptive Information Technologies for a Smart Society" represents a collective effort to explore the transformative power of information technologies in diverse domains. We extend our heartfelt gratitude to all the authors who have contributed their valuable research. Also, we would like to thank the reviewers who, with their expertise and comments, contributed to significantly improving the quality of the selected papers. We believe that the insights and findings presented in these papers will not only advance our understanding of the potential impact digitalization may have but also inspire further research and innovation in the pursuit of a smarter, more connected, and sustainable society.

November 2023

Miroslav Trajanovic
Nenad Filipovic
Milan Zdravkovic

Organization

General Chair

Miodrag Ivkovic — Information Society of Serbia, Serbia

Program Committee Chairs

Miroslav Trajanovic — Ministry of Science, TD and Innovation, Serbia
Milan Zdravkovic — University of Nis, Serbia
Nenad Filipovic — University of Kragujevac, Serbia
Zora Konjovic — Singidunum University, Serbia

Steering Committee

Hervé Panetto — Université de Lorraine, Nancy, France
Georg Weichhart — Primetals Technologies, Austria
Anton Kos — University of Ljubljana, Slovenia
Osiris Canciglieri Jr. — Pontifical Catholic University of Parana, Brazil
Nikola Milivojevic — Jaroslav Cerni Institute, Serbia
Boban Stojanovic — University of Kragujevac, Serbia
Milan Stojkovic — The Institute for Artificial Intelligence R&D of Serbia

Program Committee

Thomas Ahmad — University of Minho, Portugal
Alexis Aubry — University of Lorraine, France
Marija Boban — University of Split, Croatia
Zorica Bogdanovic — University of Belgrade, Serbia
Miloš Bogdanović — University of Nis, Serbia
Ivan Ćirić — University of Nis, Serbia
Zarko Cojbasic — University of Nis, Serbia
Dubravko Culibrk — Institute for Artificial Intelligence R&D of Serbia
Michele Dassisti — Politecnico di Bari, Italy
Igor Dejanovic — University of Novi Sad, Serbia

Additional Reviewers

Ivan Ilic
Milos Ivanovic
Lidija Korunovic
Nikola Korunovic
Srdjan Kostic
Brankica Majkic-Dursun
Jelena Mitic

Vladimir Mitrovic
Jelena Rakocevic
Mirko Stojiljkovic
Dejan Tanikic
Vesna Tripkovic
Nikola Vitkovic

Contents

Applied Artificial Intelligence

Digital Water

Computer Based Learning

Digitalization in Energy Sector

Advanced Information Systems

ICT for Health, Well-Being and Sports

Prediction of Coronary Plaque Progression Using Data Mining and Artificial Neural Networks

Lemana Spahić[1][✉] ⓘ, Leo Benolić[1] ⓘ, Safi Ur Rehman Qamar[1] ⓘ,
Vladimir Simic[1,2] ⓘ, Bogdan Miličević[1,3] ⓘ, Miljan Milošević[1,2,4] ⓘ,
Tijana Geroski[1,3] ⓘ, and Nenad Filipović[1,3] ⓘ

[1] Research and Development Center for Bioengineering, BioIRC, Prvoslava Stojanovica, 34000 Kragujevac, Serbia
lemanaspahic@gmail.com
[2] Institute for Information Technologies, Jovana Cvijića, 34000 Kragujevac, Serbia
[3] Faculty of Engineering, University of Kragujevac, Sestre Janjić, 34000 Kragujevac, Serbia
[4] Metropolitan University, Tadeuša Košćuška 63, 11158 Belgrade, Serbia

Abstract. Coronary artery disease represents one of the most significant health burdens worldwide. As the onset of the disease is multifactorial in nature, physicians are often struggling with determining the rate of progression of the arterial narrowing caused by buildup of plaque. Computational models have brought upon a significant shift in the paradigm and the advent of Big Data and machine learning has enabled far better understanding of disease dynamics. This study is based on a cohort of patients recruited through SMARTool project for whom an extensive monitoring system was set up. In order to select for the most influential parameters on the progression of coronary atherosclerosis, different feature selection algorithms were used. The dataset used for development of the system for prediction of coronary plaque progression consisted of demographic data, data considering comorbidities and different blood cholesterol parameters. The developed artificial neural network showed significant strength in diagnosing progression of coronary arterial plaque, and features selected within this study indicate the high potential of machine learning to be used in clinical practice as well as that specific types of cholesterol are important markers impacting plaque progression.

Keywords: coronary artery disease · coronary atherosclerosis progression · machine learning · prediction · feature selection

1 Introduction

Coronary artery disease (CAD), more specifically coronary atherosclerosis (CATS), is one of the leading causes of death worldwide, accounting for approximately 17.9 million deaths annually (Bhatt et al. 2010). Coronary atherosclerosis is a condition marked by the accumulation of plaque on the artery wall, which is made up of fat, cholesterol, calcium, and other components. This causes arteries to gradually narrow, eventually occluding

M. Trajanovic et al. (Eds.): ICIST 2023, LNNS 872, pp. 3–13, 2024.
https://doi.org/10.1007/978-3-031-50755-7_1

and preventing blood flow (Libby 2021, Weber and Noels 2011) The most prevalent signs and symptoms of CAD are chest pain and discomfort, which are medically known as angina (Ong et al. 2018). Excessive plaque buildup in the arteries, which obstructs blood flow to the heart and the rest of the body, causes angina. Reduced oxygen and nutrition delivery as a result of this insufficient blood flow runs the risk of causing tissue damage and, in extreme circumstances, even death (Taqueti and Di Carli 2018).

Obesity, physical inactivity, an unhealthy diet, smoking, a family history of CAD or heart disease, and comorbidities such as diabetes, high blood pressure, and elevated blood cholesterol levels are all risk factors contributing to coronary artery disease (Yusuf et al. 2020, Khera and Kathiresan 2017). The significance of early detection and prevention techniques is emphasized by the fact that many of these characteristics can be altered by alterations in lifestyle and medical treatment (Arnett et al., 2019).

Aside from causing partial or total blockage of arteries, plaque can separate from the artery wall and flow into the bloodstream, resulting in an acute thrombotic event (Falk et al. 2013). This can lead to a heart attack or a stroke, which both have high morbidity and death rates (Benjamin et al. 2018).

It is essential to comprehend the relevance of factors influencing the evolution of atherosclerotic lesions in order to properly treat and prevent future cardiac events. Inflammation, endothelial dysfunction, and oxidative stress are a few of the mechanisms that have been linked to the development of atherosclerosis in studies (Hansson 2005). It has been demonstrated that pharmaceutical therapies that target these processes, such as statins and antihypertensive drugs, lower the incidence of CAD-related events (Trialists 2012). In addition, crucial elements of CAD management and prevention include stress management, regular physical activity, a heart-healthy diet, and quitting smoking (Eckel et al. 2014). These adjustments can enhance cardiovascular health overall, lower the risk of future cardiac events, and slow the development of atherosclerosis. Successful treatment and prevention of coronary artery disease depend on an understanding of the variables influencing the development of atherosclerotic plaques. It is possible to lessen the overall burden of CAD and enhance patient outcomes by focusing on modifiable risk factors and the underlying processes of atherosclerosis.

It is well known that atherosclerosis occurs because of an interplay of a variety of factors. The correlations of these factors to atherosclerosis is explored computationally in order to aid physicians in treating the exact cause of CATS, however research has found that most commonly several factors influence characteristics and hence optimal treatment strategy in the case of arterial plaque (Obermeyer and Emanuel 2016). For this reason, it is crucial to apply a multiscale approach to analysis of risk factors leading to CATS, starting from cells that make up the coronary arteries, through tissues to the entire organism and its environment (Tran et al. 2017). Pinpointing the most significant combination of risk factors for CATS development and treatment prognosis would enable physicians to target the disease with optimal treatment strategy and enable better patient outcomes.

Numerous computational methods, from basic statistics to more complex machine learning (ML) techniques, have been used to analyze the mechanism of coronary atherosclerosis (CATS) progression (Krittanawong et al. 2017). Data analytics has increased in capability with the development of big data and artificial intelligence (AI), allowing machines to learn from data without human input. Machine learning, a subfield of artificial intelligence, refers to a collection of techniques that may learn from data, i.e., find hidden patterns in databases with the intention of using the learned information to forecast future system outputs (LeCun et al. 2015). For descriptive or prescriptive purposes, a variety of mathematical models are used to approximate complex relationships in data (Hastie et al. 2009). As a result, ML has emerged as a valuable approach for various domains where collections of relevant real data could not be explored deterministically due to the presence of uncertainty or noise and nonlinear dependencies among features (Motwani et al. 2017).

Machine learning has been used to examine enormous datasets of patient medical records, genetic data, and imaging data in the context of CATS progression with the goal of revealing fresh insights and creating predictive models (Ambale-Venkatesh et al. 2017, Kwon et al. 2019). For instance, ML algorithms have been applied to discover novel biomarkers linked to CATS progression, forecast patient outcomes, and evaluate the effectiveness of treatments (Coelho-Filho et al. 2011).

These methods have the potential to advance clinical decision-making and increase understanding of CATS. Researchers and clinicians can create individualized treatment plans for patients at risk by utilizing the capacity of ML and AI to better understand the intricate interplay of factors that contribute to CATS progression (Al'Aref et al. 2019, Johnson et al. 2018). Improved patient care and management have resulted from the use of machine learning in the study of CATS progression, which has permitted the creation of more precise predictive models and sophisticated risk stratification tools (Shameer et al. 2018). As an illustration, ML algorithms have been used to forecast the probability of plaque rupture, a major contributor to acute coronary syndromes (Narula et al. 2013).

In order to better characterize the composition and form of atherosclerotic plaques, machine learning has also been used to analyze medical imaging, including coronary computed tomography angiography (CCTA) and intravascular ultrasound (IVUS) (Kolossváry et al. 2017, Lee et al. 2016). ML-guided imaging makes it possible to more precisely identify high-risk plaques, which helps in selecting the best therapies and tracking treatment response.

In the past few years, there has been a substantial increase in the use of machine learning (ML) approaches in the study of coronary artery disease (CAD) and coronary atherosclerosis (CATS), which has had a positive impact on risk assessment, diagnosis, prognosis, treatment, and management. The growing body of writing on the subject has reflected this tendency. The 2017 review "Artificial Intelligence in Precision Cardiovascular Medicine" by Krittanawong et al., is a significant instance. The review, examines the potential impact of machine learning and artificial intelligence on cardiovascular medicine, particularly in the areas of precision medicine, risk prediction, and treatment optimization (Krittanawong et al. 2017). A study by Motwani et al., entitled "Machine learning for prediction of all-cause mortality in patients with suspected coronary artery disease: a 5-year multicentre prospective registry analysis," was released in the same year. In it, the authors showed how clinical, demographic, and computed tomography angiography (CTA) data can be used to predict all-cause mortality in patients with suspected CAD (Motwani et al. 2017). The Multi-Ethnic Study of Atherosclerosis (MESA) cohort is used to predict cardiovascular events in a second compelling research article titled "Cardiovascular Event Prediction by Machine Learning: The Multi-Ethnic Study of Atherosclerosis," which was published in Circulation Research (Ambale-Venkatesh et al. 2017). Another outstanding review, "Clinical applications of machine learning in cardiovascular disease and its relevance to cardiac imaging," provides an in-depth analysis of ML applications in cardiovascular disease with a focus on how ML can enhance cardiac imaging's diagnostic precision, risk assessment, and treatment planning (Al'Aref et al. 2019). The potential uses and difficulties of AI in cardiology were examined in a study of "Artificial Intelligence in Cardiology," which also examined ML methods for CAD and CATS as well as other cardiovascular illnesses (Johnson et al. 2018).

2 Methodology

The dataset used in this study is derived from SMARTool project. It is comprised of patient's data from two-time moments, at the beginning of their monitoring and after certain timepoints with various data domains carefully collected to obtain the comprehensive description of disease and its mechanisms. Data about pharmacological treatment in combination with other parameters, represented by this dataset, could reveal significant medication factors that affect physician's prescription decision and a plenty of research initiatives in the recent period is directed towards the application of data science approach for this purpose.

There is a standardized process to perform classification tasks and discover the potential the database holds, a concept known as Knowledge Discovery in Databases (KDD) (Frawley et al. 1992). Prior to starting with the application of ANN it is necessary to perform in depth analysis of the dataset and to mitigate all potential sources of error arising from ambiguities in the dataset. In general, the methodological approach taken in this research is:

– Data elucidation – defining the clinical significance to prognosis of CATS of individual parameters
– Database preparation – raw data preprocessing by dealing with missing values, different feature types, unbalanced dataset in terms of classes etc.
– Data division into training and testing dataset for ANN
– Training of ANN-based prediction model using a defined subset of data (training dataset)
– Validation of prediction model using a defined subset of data (testing dataset)
– Evaluation of the results

This is a generalized approach used in most studies concerning application of ML on real-world data. However, due to the complexity of the obtained dataset, an additional step was applied prior to implementation of ANN.

Missing values problem must be mitigated at first. The main justification for addressing this issue is the fact that ANNs cannot manage data with missing values. Additionally, if a feature value is absent for some patients, the algorithm may underestimate or exaggerate its significance for the discrimination issue. In some circumstances, it is possible to generate accurate guesses for the missing values by utilizing a feature's unique characteristics and its correlation with other features. There are solutions that resolve this issue effectively:

- completely remove the missing value-related features. However, in case of high significance of a certain feature, this cannot be done.
- substitute particular values for missing ones. These values can either be the mean or median of the current numerical values, or for a nominal characteristic, the mode, which is the nominal value that occurs most frequently (this can also be used when the numerical values are discontinuous and typically few in number).

For further analysis, the features with more than 10% missing values were disregarded. The latter method was used for the remaining features.

Secondly, class imbalance must be addressed. Imbalanced class distribution can have a detrimental impact on ML models, and misclassifying the minor class can result in lower decision-making capacities. Performance measures that consider class distribution should be used for model evaluation. The majority of solutions rely on ensemble methods, while cost-sensitive learning is based on algorithm tweaks, and resampling techniques for data transformation. Resampling techniques are the most popular way for handling uneven data, according to (Haixiang et al. 2017). When a data collection has a limited number of minority cases, the SMOTE algorithm (Chawla et al. 2002) is demonstrated to be a good option. This method is based on the k-NN methodology, and it oversamples instances from minority classes by producing artificial minority class examples that are comparable to the ones that are already available. The key result is the identification of more focused decision areas for the minority class in the feature space, without encroaching on the region for the majority class.

Feature selection represents identification of subsets of significant features. This step is important for decreasing the dimensionality of the data and increasing the overall quality of data by disregarding irrelevant features. For the purpose of this study, the following feature selection methods were used: ReliefF, MRMR and wrapper technique with genetic algorithm. The relief (Chikhi and Benhammada 2009) estimates the contextual usefulness of the data in accordance to differences and similarities between the values for neigboring instances belonging to the different or same classes respectively. According to literature, the algorithm is able to correctly estimate the quality of attributes in problems with strong intra-attribute dependencies. mRMR (minimum redundancy-maximum relevance) feature selection algorithm (Parmar et al. 2007)selects the subset of features with the highest relevance in terms of their correlation to the output and least correlation to other input features. Lastly, wrapper algorithm in conjunction with genetic algorithm (Huang et al. 2007) is a feature selection algorithm evaluates the selected subsets of features by estimating the "area under the curve" (AUC) parameters using -fold cross validation.

As the aim of this study is to determine the risk and pace of progression of CATS, lipid-species, anti-thrombotic drugs, clinical data, risk factors and general biomarkers were selected as parameters of interest from the database. As the database consisted of 242 samples in total and an overall 112 parameters, with a significant class imbalance towards the minority class (patients with no progression or insignificant progression of CATS) it was important to decrease the number of used parameters to optimize the performance of the neural network. Upon applying the reliefF feature extraction algorithm, the following parameters were selected for use: No of plaques, CAD score, Max stenosis class, Gender, Hypertension, Dyslipidemia, Uric Acid, Statins, LDL, TG(52:2)

TG(18:1/18:1/16:0), CE(16:0), TG(52:3) TG(18:1/18:1/16:1), CE(20:3), PC(38:4), CE(18:2), CE(22,6), CE(18:1), PC(38:6), PC(40:6), TG(50:2)TG(16:0/16:1/18:1). Main statistical characteristics of the selected parameters in terms of mean, median and standard deviation are presented in Table 1. Along with correlation and p-value evaluation for each parameter.

Table 1. Summary statistics for the selected parameters.

Variable	Mean class 0	Median class 0	StdDev class 0	Mean class 1	Median class 1	StdDev class 1	Correlation	p-value
Gender	0.257	0.000	0.437	0.343	0.000	0.475	−0.150	0.391
Plaque count	3.886	2.000	3.115	5.314	5.000	2.806	0.125	0.474
CAD score	11.767	10.000	8.223	15.472	14.800	6.797	0.026	0.880
Max stenosis classes scan 2 (0 = no stenosis, 1 = < 30%, 2 = 30–50%, 3 = 50–70%, 4 = > 70%)	2.057	2.000	1.120	2.171	2.000	1.000	−0.034	0.845
Dyslipidemia	0.800	1.000	0.400	0.743	1.000	0.437	−0.294	0.086
Hypertension	0.857	1.000	0.350	0.743	1.000	0.437	0.133	0.445
Uric acid	5.874	5.800	1.241	5.494	5.200	1.061	0.049	0.778
LDL	89.789	84.000	31.172	72.859	68.200	31.361	−0.046	0.794
TG(52:2) TG(18:1/18:1/16:0) M	502.277	209.993	505.725	541.737	224.298	411.222	−0.024	0.890
TG(52:3) TG(18:1/18:1/16:1) M	111.813	107.451	24.585	113.250	114.270	24.804	0.151	0.388
TG(50:2)TG(16:0/16:1/18:1) M	104.865	99.764	38.872	126.073	119.885	55.284	−0.100	0.569
CE(16:0)	501.783	465.631	216.459	397.497	352.312	246.851	−0.126	0.469
CE(18:2)	1932.305	1897.609	491.412	1757.615	1665.873	373.051	−0.079	0.654

When observing the individual correlations, taking into account that 1 is the highest correlation while 0 is the lowest, it can be concluded that none of the parameters individually have significant impact on the overall classification. In order to account the difference between the classes, p-value estimation was calculated. The cut-off p-value, indicating statistical significance of intergroup variability is $p < 0.05$. As it can be concluded from Table 1, there is no statistical significance observed for individual parameters.

Upon feature selection and data preprocessing, a total of 204 samples remained in the database with 79% of them belonging to the majority class (patients with significant progression of CATS). The dataset was divided in an 80%–20% manner for training and testing respectively.

3   Results and Discussion

As artificial neural networks are the most robust algorithms for classification tasks and most prominently used when complex datasets with different types of data are used, they were chosen as the classification algorithm for prediction of CATS progression. During the development of the ANN, several different combinations of activation functions and the number of neurons in the hidden layers were tested in order to improve the overall performance of the system. The measure of performance for this binary classification task was mean squared error.

In the field of machine learning, hyperparameters are parameters that are defined before the learning process begins, which distinguishes them from other parameters learned during model training. These can include things like the learning rate, the number of hidden layers in the neural network, the number of neurons in each layer, and the type of optimizer used, among other things. The goal of hyperparameter optimization is to search the hyperparameter space for an optimal set of hyperparameters that will produce a model with the best performance, usually evaluated through a defined metric. Predetermined, such as precision, mean error, or a combination of several metrics. This is an important step in ANN development because the selected hyperparameters can significantly affect the performance of the model. For example, a high learning rate may cause the model to converge rapidly but miss the global minimum, while a low learning rate may cause slow convergence or even stagnation at a suboptimal point.. Likewise, the number of hidden layers and neurons in each layer can significantly affect the model's ability to capture complex patterns in the data, potentially leading to missing or overfitting pages. Level. In the present study, we used the Keras Tuner library to optimize the hyperparameters for our ANN model. We used a variety of techniques, including stochastic search and Bayesian optimization, to tune the model's hyperparameters, such as the learning rate, the number of hidden layers, and the number of neurons in each layer.. Through rigorous optimization, we can significantly improve the performance of the model, highlighting the important role of hyperparameter optimization in ANN development.

A closer examination of the training process revealed a discrepancy between training and validation metrics (Fig. 1). Specifically, while the Mean Squared Error (MSE) and accuracy on the training set showed a consistent improvement throughout the training iterations, the validation MSE and accuracy did not follow the same trend. Instead, they fluctuated and did not converge with the training metrics, indicating a potential issue with overfitting.

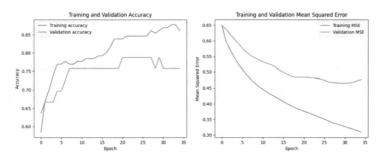

Fig. 1. Training and validation accuracy and MSE before regularization

Overfitting is a common problem in machine learning and occurs when a model learns to perform exceptionally well on training data but fails to generalize to unseen data. In our case, the continuous improvement of the training indices suggests that our ANN may have become too specialized in capturing the nuances and noise of the training set, which is not necessarily the case. It is necessary to apply to the authentication training set or to the general public. Several factors can contribute to this overfitting phenomenon. One possibility is the complexity of our ANN. The model, equipped with many layers and a large number of neurons, may have too much capacity, thus learning the training data too well. Another factor could be the lack of regularization techniques, such as dropout or L1/L2 regularization, which are often used to avoid overlearning by adding constraints to the learning process. To solve these problems and improve the performance of ANNs, we have studied several strategies. First, we tried to simplify the model by reducing the number of layers or neurons. Second, we explored the use of regularization techniques to limit model learning and avoid overfitting. For this purpose, dropout layers were used. After many iterations of the system, the overall performance has improved significantly (Fig. 2), eventhough the overfitting problem persisted.

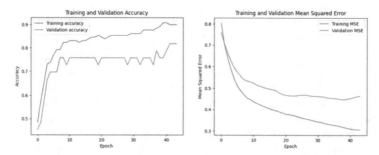

Fig. 2. Training and validation accuracy after regularization

In spite of overfitting during training, the final validation of the system performed on an independent testing set showed an accuracy of 81.81% for the developed system (Fig. 3).

The specificity of 37.5% indicates that the ANN does not generalize well for the insignificant plaque progression samples. This problem persisted across all iterations of the ANN. Considering significant class imbalance of the dataset where only 22% of the data corresponded to the minority class.

Future perspectives of this research should include expanding the dataset with more samples of minority class to decrease the class bias as this would significantly improve the overall accuracy of the system and potentially render it suitable for practical application in form of a decision support system for CATS diagnosis.

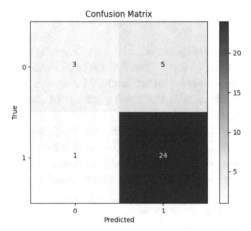

Fig. 3. Confusion matrix for the developed system (accuracy 81.81%, sensitivity 96% and specificity 37.5%)

4 Conclusion

This study demonstrates the significant potential of machine learning, especially artificial neural networks (ANNs), in the study of coronary artery disease (CAD) and coronary atherosclerosis (CATS). The application of ANNs in this study highlights the ability to process complex multidimensional data and create models suitable for risk assessment, diagnosis and prognosis.

In the course of this work, an ANN model with 19 input parameters representing binary classifiers of two output classes was developed. Application of hyperparameter optimization techniques, by means of the Keras Tuner library, was key to fine-tuning the model's architecture to improve performance.

By employing an early-stopping regularization strategy, the overfitting problem was effectively resolved, adding an additional layer of complexity to the training process. These strategies help prevent the model from learning noise in the training data, resulting in a better ability to generalize to unseen data. The final model achieved an accuracy of 81.81%, a significant improvement over the original model. This improved model demonstrates the importance of careful hyperparameter tuning and good regularization techniques in developing robust and reliable ANN models.

However, the model had a specificity of 37.5% and a sensitivity of 96%. High sensitivity indicates that the model is very good at identifying truly positive outcomes. Conversely, lower specificity suggests that the model may have more difficulty identifying true negatives. This disparity underscores the need for balanced datasets in training and highlights the importance of considering various performance metrics beyond mere accuracy when evaluating a model's performance.

Despite the challenges faced during model development, this study highlights the potential of ANNs and machine learning in biomedical research. As the understanding and application of these advanced computational techniques continue to advance, they will undoubtedly play a key role in advancing precision medicine, improving patient outcomes, and expanding the understanding of complex diseases such as CAD and CATS.

It would be interesting to explore other machine learning techniques, and even different model combinations, to see if specificity can be further improved without significantly compromising sensitivity in the future. Moreover, applying these methods to larger and more diverse datasets will lead to a more comprehensive and more informed understanding of the potential in this area.

Overall, the results of this study suggest that ANNs, if properly tuned and regulated, have great potential in analyzing CAD and CATS. It is hoped that these findings will contribute to the ongoing debate in this field and stimulate further research to realize the full potential of machine learning in cardiovascular medicine.

Acknowledgements. This paper is supported by the DECODE project (www.decodeitn.eu) that has received funding from the European Union's Horizon 2020 research and innovation programme under grant agreement No 956470. This article reflects only the author's view. The Commission is not responsible for any use that may be made of the information it contains.

References

Al'aref, S.J., et al.: Clinical applications of machine learning in cardiovascular disease and its relevance to cardiac imaging. Eur. Heart J. **40**, 1975–1986 (2019)

Ambale-Venkatesh, B., et al.: Cardiovascular event prediction by machine learning: the multi-ethnic study of atherosclerosis. Circ. Res. **121**, 1092–1101 (2017)

Arnett, D.K., et al.: 2019 ACC/AHA guideline on the primary prevention of cardiovascular disease: a report of the American College of Cardiology/American Heart Association Task Force on Clinical Practice Guidelines. Circulation **140**, e596–e646 (2019)

Benjamin, E.J., et al.: Heart disease and stroke statistics—2018 update: a report from the American Heart Association. Circulation **137**, e67–e492 (2018)

Bhatt, D.L., et al.: Comparative determinants of 4-year cardiovascular event rates in stable outpatients at risk of or with atherothrombosis. JAMA **304**, 1350–1357 (2010)

Chawla, N.V., Bowyer, K.W., Hall, L.O., Kegelmeyer, W.P.: SMOTE: synthetic minority over-sampling technique. J. Artif. Intell. Res. **16**, 321–357 (2002)

Chikhi, S., Benhammada, S.: ReliefMSS: a variation on a feature ranking ReliefF algorithm. Int. J. Bus. Intell. Data Min. **4**, 375–390 (2009)

Coelho-Filho, O.R., et al.: Stress myocardial perfusion imaging by CMR provides strong prognostic value to cardiac events regardless of patient's sex. JACC: Cardiovascular Imaging **4**, 850–861 (2011)

Eckel, R.H., et al.: 2013 AHA/ACC guideline on lifestyle management to reduce cardiovascular risk: a report of the American College of Cardiology/American Heart Association Task Force on Practice Guidelines. Circulation **129**, S76–S99 (2014)

Falk, E., Nakano, M., Bentzon, J.F., Finn, A.V., Virmani, R.: Update on acute coronary syndromes: the pathologists' view. Eur. Heart J. **34**, 719–728 (2013)

Frawley, W.J., Piatetsky-Shapiro, G., Matheus, C.J.: Knowledge discovery in databases: An overview. AI Mag. **13**, 57 (1992)

Haixiang, G., Yijing, L., Shang, J., Mingyun, G., Yuanyue, H., Bing, G.: Learning from class-imbalanced data: review of methods and applications. Expert Syst. Appl. **73**, 220–239 (2017)

Hansson, G.K.: Inflammation, atherosclerosis, and coronary artery disease. N. Engl. J. Med. **352**, 1685–1695 (2005)

Hastie, T., Tibshirani, R., Friedman, J.H., Friedman, J.H.: The Elements of Statistical Learning: Data Mining, Inference, and Prediction. Springer, New York (2009)

Huang, J., Cai, Y., Xu, X.: A hybrid genetic algorithm for feature selection wrapper based on mutual information. Pattern Recogn. Lett. **28**, 1825–1844 (2007)

Johnson, K.W., et al.: Artificial intelligence in cardiology. J. Am. Coll. Cardiol. **71**, 2668–2679 (2018)

Khera, A.V., Kathiresan, S.: Genetics of coronary artery disease: discovery, biology and clinical translation. Nat. Rev. Genet. **18**, 331–344 (2017)

Kolossváry, M., et al.: Radiomic features are superior to conventional quantitative computed tomographic metrics to identify coronary plaques with napkin-ring sign. Circul. Cardiovascular Imaging **10**, e006843 (2017)

Krittanawong, C., Zhang, H., Wang, Z., Aydar, M., Kitai, T.: Artificial intelligence in precision cardiovascular medicine. J. Am. Coll. Cardiol. **69**, 2657–2664 (2017)

Kwon, J.-M., et al.: Artificial intelligence algorithm for predicting mortality of patients with acute heart failure. PLoS ONE **14**, e0219302 (2019)

Lecun, Y., Bengio, Y., Hinton, G.: Deep learning. Nature **521**, 436–444 (2015)

Lee, S.-E., et al.: Rationale and design of the Progression of AtheRosclerotic PlAque DetermIned by Computed TomoGraphic Angiography IMaging (PARADIGM) registry: a comprehensive exploration of plaque progression and its impact on clinical outcomes from a multicenter serial coronary computed tomographic angiography study. Am. Heart J. **182**, 72–79 (2016)

Libby, P.: The biology of atherosclerosis comes full circle: lessons for conquering cardiovascular disease. Nat. Rev. Cardiol. **18**, 683–684 (2021)

Motwani, M., et al.: Machine learning for prediction of all-cause mortality in patients with suspected coronary artery disease: a 5-year multicentre prospective registry analysis. Eur. Heart J. **38**, 500–507 (2017)

Narula, J., et al.: Histopathologic characteristics of atherosclerotic coronary disease and implications of the findings for the invasive and noninvasive detection of vulnerable plaques. J. Am. Coll. Cardiol. **61**, 1041–1051 (2013)

Obermeyer, Z., Emanuel, E.J.: Predicting the future—big data, machine learning, and clinical medicine. N. Engl. J. Med. **375**, 1216 (2016)

Ong, P., et al.: International standardization of diagnostic criteria for microvascular angina. Int. J. Cardiol. **250**, 16–20 (2018)

Parmar, D., Wu, T., Blackhurst, J.: MMR: an algorithm for clustering categorical data using rough set theory. Data Knowl. Eng. **63**, 879–893 (2007)

Shameer, K., Johnson, K.W., Glicksberg, B.S., Dudley, J.T., Sengupta, P.P.: Machine learning in cardiovascular medicine: are we there yet? Heart **104**, 1156–1164 (2018)

Taqeti, V.R., Di Carli, M.F.: Coronary microvascular disease pathogenic mechanisms and therapeutic options: JACC state-of-the-art review. J. Am. Coll. Cardiol. **72**, 2625–2641 (2018)

Tran, J.S., Schiavazzi, D.E., Ramachandra, A.B., Kahn, A.M., Marsden, A.L.: Automated tuning for parameter identification and uncertainty quantification in multi-scale coronary simulations. Comput. Fluids **142**, 128–138 (2017)

Trialists, C.T.: The effects of lowering LDL cholesterol with statin therapy in people at low risk of vascular disease: meta-analysis of individual data from 27 randomised trials. Lancet **380**, 581–590 (2012)

Yusuf, S., et al.: Modifiable risk factors, cardiovascular disease, and mortality in 155 722 individuals from 21 high-income, middle-income, and low-income countries (PURE): a prospective cohort study. The Lancet **395**, 795–808 (2020)

Rule-Based System for Pregnancy Monitoring

Dragana Filipović$^{(\boxtimes)}$ ⓘ, Anđela Trajković ⓘ, and Siniša Nikolić ⓘ

Faculty of Technical Sciences, The University of Novi Sad, Novi Sad, Serbia
`{dragana.filipovic,trajkovic.andjela,sinisa_nikolic}@uns.ac.rs`

Abstract. During the pregnancy period, it is important to monitor the baby's and the mother's condition. Many tests are necessary during pregnancy, from screen tests for chromosomal abnormalities detection in a fetus to disease diagnosis tests for the mother. The results of those tests should be taken into consideration when defining prenatal care policies. Inadequate prenatal care is associated with an increased risk of preterm delivery and low birth weight newborns. This paper proposes a software solution for improving prenatal care by allowing cooperation between doctors and expert systems. Integrating expert system knowledge and experience within the rule-based system to guide doctor decisions can produce more adequate and safer prenatal care and reduce the infant mortality rate. The proposed system detects diseases based on symptoms and chromosomal abnormalities based on different parameters. Also, it monitors cardiotocography and calculates its results. This system is an expert system based on rules. During runtime, rules can be modified and changed without affecting the application workings. The proposed system is not a replacement for doctors, it can be seen as a support system for medical personalities.

Keywords: Expert system · knowledge-based systems · Drools · prenatal care · chromosomal abnormalities · cardiotocography · an indication of the disease of a pregnant woman

1 Introduction

During pregnancy, it is vital to monitor the mother's and baby's condition and the baby's development. This is done through regular visits to the doctor, where various tests and analyses are performed on the mother and baby. These visits represent prenatal care, which is essential for the health and well-being of both the mother and the baby. Inadequate prenatal care is associated with an increased risk of preterm delivery and low birth weight of newborns, which results in a higher infant mortality rate. If the mother is getting adequate prenatal care, it can help prevent and address health problems in both mother and baby, and in that way, it can help save and improve the lives of both mother and baby [1]

In the first and the second trimester, double, triple, and quadruple screens, as well as amniocentesis, are used to check whether the baby has anomalies. Fetal anomalies are physical or genetic defects of the fetus and these screens can identify major chromosomal abnormalities such as Down, Patau, and Edwards syndrome.

M. Trajanovic et al. (Eds.): ICIST 2023, LNNS 872, pp. 14–22, 2024.
https://doi.org/10.1007/978-3-031-50755-7_2

Cardiotocography (CTG) is a machine that monitors fetal heart rate and uterine contractions [2]. It can detect problems with fetal heart rate and can report them to the doctor. In the third trimester, CTG is done.

Besides tests that indicate problems with the baby, tests are also run on the mother to monitor her condition. Her weight, blood pressure, and different symptoms that are manifested are being tracked. These symptoms indicate different illnesses. Some of the illnesses that a mother can have during pregnancy are preeclampsia, eclampsia, gestational diabetes, hypertension (high blood pressure), kidney disease, and others.

In the knowledge-based system [3], we can group and store key information about babies' and mothers' conditions in one place. The system is primarily intended for doctors and pregnant women as a web application. Rules in the system are written in Drools [4]. By relying on the application, the mother can be diagnosed with illness depending on manifested and reported symptoms, chromosomal anomalies can be detected in the fetus by calculating risk from the results of the double, triple, and quadruple test, as well as from results of amniocentesis procedure. The doctor can be alarmed when a critical condition for the baby is detected during CTG, where the results of CTG are calculated by the system. The doctor only needs to act as a supervisor of the whole process.

The following section defines the research questions and methodology. The third section provides an overview of related works. The analysis of existing solutions is described in the fourth chapter. The software solution design is in the fifth chapter. The sixth chapter states the implementation specifics. The last chapter discusses systems improvements and possible future work.

2 Research Questions and Methodology

The research in this paper will try to answer to 2 questions:

- Is it possible to make a support system for pregnancy monitoring by relying on the patients' and fetuses' symptoms and calculating CTG results based on values detected with the CTG machine?
- If so, can pregnancy monitoring be done by using a rule-based system where domain knowledge is stored in the form of some formally defined rules?

The methodology consists of the following:

- Review of related work concerning expert systems that are designed to solve problems mimicking the human experts in the field of prenatal care, tests, CTG, and medicine. There are many references to how formally defined rules have been used to represent expert knowledge.
- Analysis of some of the existing popular applications that are used for pregnancy monitoring.
- Collection of the domain knowledge from the field of prenatal care, tests, diagnosis, and CTG that will include consultation with field experts and consulting books.
- Choosing an appropriate rule-based system

3   Related Work

Pregnancy often requires doctors to conduct a large number of tests and track the patient's symptoms, which can be a demanding and exhausting process. The results of these tests must also be calculated, adding to the time and effort required.

In these situations, expert systems [4] and systems based on rules can become relevant and helpful because they can store large amounts of knowledge, which may be accessed quickly and manipulated easily.

Paper [5] describes an expert system for the diagnosis of disorders during pregnancy using the forward chaining method. It can diagnose different disorders such as preeclampsia, eclampsia, abortus, and others depending on symptoms such as excess headaches, proteinuria, blurred vision, vaginal bleeding, and others.

Article [6] presents a web-based forward-chaining expert system for maternal care. In Ethiopia maternal mortality and morbidity rate is high because patient to doctor ratio is 1 doctor to 1000 patients. That is why this system is made, to help diagnose diseases depending on symptoms and to reduce maternal mortality and morbidity rate.

A rule-based algorithm for intrapartum cardiotocography pattern features extraction and classification using MATLAB is described in the paper [7].

Most of the listed works are focused on one specific domain. The first and the second paper are focused on the diagnosis of diseases depending on the symptoms, while the third one is focused on calculating CTG results.

At the moment of writing this paper, not one paper was found that focuses on detecting chromosomal abnormalities and calculating screen test results.

The software solution proposed in this paper follows on the previously mentioned research and will focus on: chromosomal abnormalities detection via double, triple, and quadruple tests and amniocentesis which calculate results based on multiple different parameter values, disease diagnosis based on symptoms and monitoring CTG, calculating its results in real-time and alarm system for the doctor when critical fetus condition is determined.

4   Analysis of Existing Solutions

A large number of applications for pregnancy monitoring have been developed in recent years, they enable patients to track their appointments and tests on the calendar. They give users information on their baby's size and tips for healthy foods.

A few advanced solutions will be described in this chapter. These solutions are CTG Home monitoring [8] and MedTel Remote Pregnancy Monitoring [9].

1. CTG Home monitoring

CTG Home monitoring enables pregnant women and midwives to make a CTG at home or in the clinic. These data are sent directly to a portal in the cloud. The hospital can view these data in real-time or later.

2. MedTel Remote Pregnancy Monitoring

MedTel Remote Pregnancy Monitoring is an application that enables mothers to schedule online consultations with the doctor, book lab tests, order medicine, and track symptoms. Also, it supports document uploading, such as test results, appointments, and scan paperwork. The mother can contact or alert the doctor if she's not feeling well. It offers the mother information needed for her gestational week, it filters for her relevant information so she does not need to search for it herself.

Explained solutions allow the patient to communicate with the doctor directly, but yet again, most of these applications do not offer some deeper insights and do not give calculated results based on patients' and doctors' inputs.

In the next chapter, a software design that gives solutions to these issues will be introduced.

5   Software Solution Design

This system is developed as a web application due to the popularity of this approach in recent years. All the client needs is the internet and a web browser.

As explained in the first chapter it is of vital significance to monitor the fetus and the mother. In the third trimester, it is good to monitor the baby via CTG, this is why the proposed solution has three major subsystems. The first one is responsible for disease diagnosis based on the symptoms and test results, such as the OGTT test for diabetes diagnosis. The second subsystem is responsible for calculating screen test results for double, triple, and quadruple tests and amniocentesis based on various parameters. The third subsystem is used for CTG monitoring, calculating CTG results, and alarming the doctor of critical fetus condition during CTG monitoring.

Identified roles that this software solution supports are the doctor, the nurse, and the patient.

The nurse has patient, pregnancy and, symptoms review, various test results input, and birth information input.

The patient can register, review her pregnancies, test results, diagnosed disease and prescribed medication, CTG results, input manifested symptoms, daily glucose levels, blood pressure, and weight.

During CTG monitoring external applications should be supplying real-time data to the system. If those parameters are above or below the normal range, alarms are activated, and the doctor is notified.

Domain knowledge should be implemented by storing knowledge within the rules, using a rule-based system. The knowledge-based system will store domain knowledge about symptoms, and diseases, calculating screen test results, and CTG results.

6   System Implementation

The system is implemented as a client-side web application that contains the front-end application (Vue.js), server application (Spring boot), rules-based application (Drools), and database (PostgreSQL). The implementation of the system is accessible at link,[1] and a short video demonstrating its' usage is available at link.[2]

---

[1] https://github.com/draganaF/trud.io.

[2] https://drive.google.com/file/d/1-CAxy__HG7UIrDniQ0kEKYmJSn_x9za-/view.

The knowledge base for the rules-based application is collected by consulting with the domain expert and by the review of relevant literature. Consultation with the expert was via Face-to-face interview and paper questionnaire.

Domain knowledge is stored in the shape of formal rules which are machine-readable. Rules-based application is created apart from the server application, thus allowing domain experts to work on domain-based knowledge without knowing the rest of the system, and it makes it possible to modify domain knowledge without interrupting the server-side application.

This system has three different subsystems, so the rules are divided into three different groups of rules. The first group is responsible for calculating screen results by relying on a template mechanism [10], the second group is responsible for disease diagnosis based on symptoms utilizing the backward-chaining mechanism [11], and the third one is responsible for CTG monitoring rules utilizing template and Complex Event Processing (CEP) [12] mechanisms.

6.1   Rules for Calculating Screen Test Results

The double test can be done between the 12th and 14th week of pregnancy. Parameters that are observed are nasal bone presence and shape, nuchal translucency, free beta HCG, and PAPP-A. The nasal bone is formed later in fetuses with chromosomal abnormalities, PAPP-A value lower than normal indicates chromosomal abnormalities, a higher free beta HCG indicates a higher risk for Downs syndrome, while a lower value can indicate Edward or Patau syndromes. Figure 1 shows rule for calculating risk of trisomy with double test. It takes into consideration all the parameters listed above and compares them to the referenced values for that specific week of pregnancy. Risk of trisomy is calculated with results of the comparison being multiplied with coeficients shown in Fig. 1.

Triple test can be done between the 15th and 18th week of pregnancy on the condition that double test is positive, or the mother is older than 35 years. Parameters that are observed are AFP, hCG and Ue3. AFP value lower than normal, and higher HCG and Ue3 values indicate chromosomal abnormalities.

Quadruple test can be done between 15th and 22nd week of pregnancy. If the mother has not done a double test, quadruple test is recommended. Observed parameters are AFP, hCG, Ue3, and inhibinA. Same indicators as in triple test are used here as well as higher inhibinA value.

Amniocentesis is done between the 16th and 22nd week of pregnancy. It can detect birth defects, genetic problems, and infections. Parameters that are observed are AFP, presence of diagnosed gestational diabetes or diabetes, BMI (body mass index) if it is greater than 30, family history of genetic anomalies for parents, or chromosomal abnormalities in previously born children. If AFP is lower than the reference value it can indicate a higher risk of chromosomal abnormalities, while a higher AFP value can indicate a higher probability of neurological anomalies. Because amniocentesis represents an invasive method and it can influence the baby, it is not recommended for high-risk pregnancies.

Each of these parameters is compared to the values specified for that week of pregnancy and based on this comparison risk is calculated.

```
rule "Double test - trisomy risks" salience 10000
    no-loop
    agenda-group "doubleTest"
    when
        $pregnancy: Pregnancy(birth == null, doubleTest != null, Math.floor(Math.abs(Duration.between(LocalDate.now().atStartOfDay(),
        startDate.atStartOfDay()).toDays())/7) >= 12 &&
        Math.floor(Math.abs(Duration.between(LocalDate.now().atStartOfDay(), startDate.atStartOfDay()).toDays())/7) <= 14)
        $week: Number() from $pregnancy.getWeek()
        $d: DoubleTest(result == 'Not yet processed') from $pregnancy.getDoubleTest()
        $age: Number() from $pregnancy.getPatient().checkAge()
        $riskAgeT21: AgeRisk(age == $age, trisomy == 'Trisomy21')
        $riskAgeT18: AgeRisk(age == $age, trisomy == 'Trisomy18')
        $riskAgeT13: AgeRisk(age == $age, trisomy == 'Trisomy13')

        $weeklyParameters: WeeklyParameters(week == $week)

        $coefNt: Double() from $d.getCoefNt()
        $coefPappa: Double() from $d.getCoefPappa($weeklyParameters.getMedianPappa())
        $coefHCGT21: Double() from $d.getCoefHCG($weeklyParameters.getMedianHcg(), 'Trisomy21')
        $coefHCGT1813: Double() from $d.getCoefHCG($weeklyParameters.getMedianHcg(), 'Trisomy18')
        $coefSmoker: Double() from $d.getCoefSmoker($pregnancy.getPatient().isSmoker())

        $riskT21: Double() from $riskAgeT21.getRisk() * 1.27 * 1.24 * 6.33 * $coefNt * $coefPappa *  $coefHCGT21 * $coefSmoker
        $riskT18: Double() from $riskAgeT18.getRisk() * 1.27 * 1.24 * 6.33 * $coefNt * $coefPappa * $coefHCGT1813 * $coefSmoker
        $riskT13: Double() from $riskAgeT13.getRisk() * 1.27 * 1.24 * 6.33 * $coefNt * $coefPappa * $coefHCGT1813 * $coefSmoker

    then

        modify($d) {
            setTrisomy21($riskT21),
            setTrisomy18($riskT18),
            setTrisomy13($riskT13),
            setResult('Calculate risks');
        }
end
```

Fig. 1. Calculating double test risks for trisomies

6.2 Rules for Disease Diagnosis

Rules for disease diagnosis are separated into two parts.

The first part is responsible for rules for the OGTT test which is used to determine if the mother suffers from gestational diabetes. Mothers do this test between the 24th and 28th week of pregnancy, but sometimes this test is done earlier if the mother previously gave birth to a baby whose weight was more than 4 kg, if the mother is obese, or if she is older than 35 years. The results of the OGTT test are interpreted in the following way:

- If results are below 14 mg/dl, the mother is healthy.
- If results are between 140 mg/dl and 199 mg/dl, then a prolonged OGTT test is necessary to be done.
- If results are above 199 mg/dl, the mother suffers from gestational diabetes.

The prolonged OGTT test lasts for 3 h and the blood sample is taken 4 times. Normal values are:

- First sample – value below 95 mg/dl
- Second sample – value below 180 mg/dl
- Third sample – value below 155 mg/dl
- Fourth sample – value below 140 mg/dl

Gestational diabetes is diagnosed if:

- First sample – values above 95 mg/dl
- Values are above normal in two of three samples.

Diet and exercise are recommended first, but the mother's condition is monitored by her daily glucose levels. Every day mother should track her glucose level: before a meal, known as preprandial, one hour after a meal, known as first postprandial, and two hours after a meal, known as second postprandial. If these values are above normal for seven days, the mother should be prescribed Metmorfin.

The second part of this subsystem diagnoses eclampsia, preeclampsia, and kidney disease based on the symptoms. Preeclampsia is a disease that manifests in most cases after the 20th week of pregnancy, it includes symptoms such as swelling, weight gain, headaches, difficulty breathing, nausea, and vomiting. If preeclampsia is not diagnosed on the time it can evolve to eclampsia, which has the same symptoms as preeclampsia but includes loss of consciousness and seizures. If preeclampsia and eclampsia are not kept under control, after 37th week of pregnancy cesarean section or induced pregnancy is done. Kidney disease has similar symptoms as preeclampsia, but it usually manifests before preeclampsia. To diagnose disease in this group all symptoms are put in different groups first. The first group's symptoms are headaches, gaining weight, nausea, and vomiting. The second group's symptoms are low trombone levels, high blood pressure, and high protein level. The third group's symptoms are seizures and loss of consciousness.

Based on these groups and pregnancy week, a decision tree structure is made. Using a backward chaining mechanism a disease can be diagnosed going through this tree based on manifested symptoms and the week of pregnancy.

6.3   Rules for Monitoring CTG

Rules are used to count contractions and monitor fetal heart rate.

The first stage is transforming input values sent from an external application into 1-min aggregation. In this stage, fetal heart rate and variability (the difference between the highest and the lowest point in a beat) are calculated.

The second stage calculates variability and fetal heart rate status. Status can be normal, non-reassuring, or abnormal.

Fetal heart rate rules:

- If the fetal heart rate is below 100 bpm bradycardia is set and fetal heart rate status is set to abnormal.
- If the fetal heart rate is above 160 bpm, tachycardia is set.
- If the fetal heart rate is between 160 and 180 bpm, so fetal heart rate status is set to non-reassuring.
- the fetal heart rate is above 180 bpm, so the fetal heart rate status is set to abnormal.

Variability rules:

- If the variability is below 5 bpm between 30 and 50 min, the variability status is set to non-reassuring.
- If the variability is below 5 bpm for more than 50 min, the variability status is set to abnormal.
- If the variability is above 25 bpm between 15 and 25 min, the variability status is set to non-reassuring.
- If the variability is above 25 bpm for more than 25 min, the variability status is set to abnormal.

Figure 2 shows a template for setting variability status based on the two last rules listed above and sending notification of the current condition of the fetus to the doctor. Via excel sheet values are provided and the two new rules are created that detect variability being above 25 bpm in specific time range.

The third stage calculates CTG results, based on fetal heart rate and variability results. The result can be normal, non-reassuring, or abnormal:

- If both statuses are normal, the CTG result is also normal.
- If one of the statuses is non-reassuring and the other is reassuring, the CTG result is non-reassuring
- If both statuses are non-reassuring or one of them is abnormal, the CTG result is abnormal.

The following rules are used to determine if the doctor needs to be alarmed:

- If fetal bradycardia is present for 3 min doctor is alarmed.
- If fetal bradycardia is prolonged for 9 min baby needs to be delivered.
- If the number of contractions in 10 min is between 3 and 5, the first stage of birth is begun.

```
rule "variability-event-greater-then-25_@{row.rowNumber}"
    agenda-group "ctgMonitoring"
    when
        $curVariabilty: VariabilityEvent( value >= 25)
        VariabilityEvent(
            this != $curVariabilty,
            value >= 25,
            this before[ @{timeFrom}, @{timeTo} ] $curVariabilty
        )
        $ctg: CTG( variabiltyStatus == CTGStatus.@{currentVariabilityStatus} )
    then
        notificationService.sendNotification(new Notification("doctorSocket", "@{title}", "@{message}"));

        modify($ctg) {
            setVariabiltyStatus(CTGStatus.@{newVariabilityStatus})
        };
end

end template
```

Fig. 2. Calculating variability status when variability is greater than 25 bpm

7 Conclusion

This paper describes an expert system based on rules, which detects chromosomal abnormalities, diagnoses diseases and monitors CTG, calculates its results, and alarm critical fetal conditions to the doctor. It monitors all relevant parameters and takes them into account in calculations that are done via implemented rules.

In comparison to the research mentioned in related work, this paper adds new value because it combines the research done in mentioned papers and adds chromosomal abnormalities detection from screen tests, giving unique and all-around insight into prenatal care.

The software solution is currently in the knowledge-gathering phase, where all the knowledge has been collected from a specialist gynecologist utilizing appropriate survey methods, as well as reading appropriate literature. Its usage is intended for doctors and patients. The application could be improved by monitoring accelerations and decelerations of the fetal heart during CTG, rules for sex identification could be added, and rules for other disease diagnoses, such as anemia, could be added. Also, because of the template mechanism it is possible to create new rules.

This system does not validate user inputs, because it is in user's best interest to provide valid data in order to get accurate results. In the future work, validation can be improved for glucose blood level input by providing picture of a measurement.

References

1. Krueger, P.M., Scholl, T.O.: Adequacy of prenatal care and pregnancy outcome. J. Osteopath. Med. **100**(8), 485–492 (2000)
2. Chandraharan, E. (ed.). Handbook of CTG Interpretation: From Patterns to Physiology. Cambridge University Press (2017)
3. Jacob, R.J.K., Froscher, J.N.: A software engineering methodology for rule-based systems. IEEE Trans. Knowl. Data Eng. **2**(2), 173–189 (1990)
4. Li, M., Lu, W., Xiang, D., Wen, Z.: Design and realization of transformer fault diagnostic expert system based on drools. In: 2015 International Conference on Computational Intelligence and Communication Networks (CICN), pp. 1583–1588. IEEE (2015)
5. Basiroh, B., Priyatno, P., Kareem, S.W., Nurdiyanto, H.: Analysis of expert system for early diagnosis of disorders during pregnancy using the forward chaining method. Int. J. Artif. Intell. Res. **5**(1), 44–52 (2021)
6. Misgna, H., Ahmed, M., Kumar, A.: MatES: web-based forward chaining expert system for maternal care. arXiv preprint arXiv:2106.09281 (2021)
7. Alyousif, S., Mohd, M.A., Bilal, B., Sheikh, M., Algunaidi, M.: Rule-based algorithm for intrapartum cardiotocograph pattern features extraction and classification. Health Sci. J. **10**(6), 1 (2016)
8. CTG home monitoring. https://www.ict-healthcare.eu/en/solutions/ctg-home-monitoring. Accessed 21 Dec 2022
9. MedTel. https://medtel.io/pregnancy/. Accessed 21 Dec 2022
10. Templating. https://docs.drools.org/6.5.0.Final/drools-docs/html/ch06.html#d0e6464. Accessed 23 Dec 2022
11. Backward-chaining. https://docs.drools.org/6.5.0.Final/drools-docs/html/ch23.html#d0e 29201. Accessed 23 Dec 2022
12. Complex Event Processing (CEP). https://docs.drools.org/6.5.0.Final/drools-docs/html/ch09. html. Accessed 23 Dec 2022

Finite Element Analysis of Myocardial Work in Cardiomyopathy

Smiljana Tomasevic[1,2(✉)] ⓘ, Miljan Milosevic[2,3] ⓘ, Bogdan Milicevic[1,2] ⓘ,
Vladimir Simic[2,3] ⓘ, Momcilo Prodanovic[2,3,4] ⓘ, Srboljub M. Mijailovic[4,5] ⓘ,
and Nenad Filipovic[1,2] ⓘ

[1] Faculty of Engineering, University of Kragujevac, Kragujevac, Serbia
smiljana@kg.ac.rs
[2] Bioengineering Research and Development Center, Kragujevac, Serbia
[3] Institute for Information Technologies, Kragujevac, Serbia
[4] FilamenTech, Inc., Newton, USA
[5] BioCAT, Department of Biology, Illinois Institute of Technology, Chicago, USA

Abstract. Analysis of myocardial work is essential in determination of left ventricle ejection fraction (LVEF) and non-invasive assessment of different types of cardiomyopathies. Two major classifications of cardiomyopathy are: hypertrophic (HCM) and dilated (DCM) cardiomyopathy. Although there are clinical improvements in cardiomyopathy risk assessment, patients are still under high risk of severe events. Computational modelling and computer-aided drug design can significantly advance the understanding of cardiac muscle activity in HCM and DCM, speed up the drug discovery and reduce the risk of severe events, aiming to improve the treatment of cardiomyopathy. This study is devoted to modelling of HCM using coupled macro and micro simulation through finite element (FE) modelling of fluid-structure interaction (FSI) and molecular drug interaction with the cardiac cells. FSI is used for modelling the HCM with nonlinear material model for heart wall. Analysis of myocardial work and changes of pressure and volume within the LV parametric model of HCM are presented at basic condition and after simulated effects of administered drugs. The obtained results provide better insight into the myocardial work of HCM patients as well as in estimated effects of drug therapy, leading to improved patient monitoring and treatment.

Keywords: Hypertrophic Cardiomyopathy · Parametric Model of Left Ventricle · Computational Modelling · Finite Element Method · Myocardial Work

1 Introduction

1.1 Motivation

Myocardial work provides incremental information of left ventricle ejection fraction (LVEF) and strain which are sensitive to left ventricle (LV) afterload and related to non-invasive assessment of different types of cardiomyopathies. According to the guidelines of the European Society of Cardiology [1], cardiomyopathies are defined as structural and functional abnormalities of the ventricular myocardium.

M. Trajanovic et al. (Eds.): ICIST 2023, LNNS 872, pp. 23–29, 2024.
https://doi.org/10.1007/978-3-031-50755-7_3

Two major classifications of cardiomyopathy are: hypertrophic (HCM) and dilated (DCM) cardiomyopathy. HCM is characterized by enlargement of the heart with increased LV wall thickness, often with asymmetrical hypertrophy of the septum that separates the LV from the right ventricle (RV) and can lead to left ventricular outflow tract obstruction (LVOTO). Figure 1 represents comparison between healthy heart and HCM, with marked thickening of the left ventricular myocardium [2].

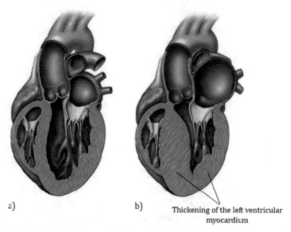

Fig. 1. a) Illustration of a healthy heart; b) Illustration of a heart with hypertrophic cardiomyopathy, with marked thickening of the walls of the left ventricle; cut views [2].

Although there are clinical improvements in cardiomyopathy risk assessment, patients are still exposed to high risk of severe events. Computational modeling and computer-aided drug design can significantly advance the understanding of cardiac muscle activity in cardiomyopathy, speed up the drug discovery and reduce the risk of severe events. The main motivation for such approach relies on application of computational modeling in testing the effects of pharmacological treatment, aiming to contribute to clinical practice, reduce animal experiments and human clinical trials.

1.2   Research Questions

In silico clinical trials are the future of medicine, whereas the virtual testing and simulations are the future of medical engineering. Currently, there is a lack in using the computational platforms and solutions which can assist in a daily clinical practice. In analysis of cardiomyopathy, understanding of disease progression is still limited, as well as the effects of drugs interactions on cardiac tissue.

The main advantage and novelty of presented study are coupled macro and micro simulations into the integrated Fluid Solid Interaction (FSI) system and its application for examination of heart behavior and drug interactions. In contrary to detailed and patient-specific models where FSI analyses are very time-consuming, our models are parametric and based on dimensions of specific LV components. In this way, the simulations can be performed for a large number of patients reducing time needed for developing new models.

1.3 Related Work

Recently established computational platform (SILICOFCM) [3] for *in silico* clinical trials integrates patient-specific data (genetic, biological, pharmacological, clinical, imaging and patient specific cellular data) and allows the testing and optimization of medical treatment to maximize positive therapeutic outcomes. Also, it has been used for risk prediction of cardiac hypertrophic disease [4, 5]. In addition, recently developed computational models, such as the MUSICO platform [6], have significantly advanced our understanding of cardiac muscle activity in HCM and DCM cardiomyopathies.

The idea of an integrative multiscale modelling [7] might help in identifying the symptoms and outcomes of patients with multiple genetic disorders. For the simulation of total heart health or pathology, molecular, cellular, tissue, and organ levels have to be integrated. Simulation of FSI in whole heart during the total heart cycle demands usage of a large number of finite elements (FEs) due to complex heart geometry. Instead of patient-specific geometry, we are employing parametric 3D models of the LV, which are suitable for running large number of simulations by varying selected parameters.

The paper is organized as follows: the applied multiscale modelling is briefly descried in Sect. 2. The results (Pressure-Volume diagrams) with administered two drugs for HCM and comparison with the initial condition (no drug) is presented in Sect. 3. Section 4 discusses relevant work and concludes the paper.

2 Materials and Methods

2.1 Myocardial Work

Myocardial work is a novel technique used in the advanced assessment of LVEF. It includes LV pressure and provides incremental information to LVEF and strain which are sensitive to LV afterload. Figure 2 schematically represents the pressure-volume relationship, i.e. P-V (Pressure-Volume) diagram in the LV during a normal cardiac cycle [8].

The PV diagram includes the diastolic and systolic phases, which are marked by points from A to D. The phase of diastole, or protodiastole, begins at point A, which also represents the end of systole. At that moment, the leaflets of the arterial valves are still open, and they begin to separate from the arterial walls. The phase of isovolumetric relaxation (A-B, Fig. 2.) is the second phase of diastole, where the arterial valves close, while the atrioventricular valves have not opened yet. During this phase, the blood volume and pressure in the ventricles are at their lowest values. Then follows the phase of filling the chambers (B-C, Fig. 2).

The opening of the atrioventricular valves and the sudden transfer of blood from the atria to the relaxed ventricles represents the early phase of rapid filling. The rapid transfer of blood from the atria to the ventricles leads to the pressure drop in the atria and a sudden increase of the ventricles volume. The slow filling phase (diastasis) represents the longest period in the normal duration of the cardiac cycle. The presystolic phase (late rapid filling phase) occurs as a result of atrial systole. Since the contraction of the atria in this phase led to rapid filling of the ventricles with blood, the volume and pressure in them suddenly increase and the atrioventricular valves close. This phase is called isovolumetric contraction (C-D, Fig. 2) and it is the first phase when begins ventricular systole (the atrioventricular valves have closed and the arterial valves have not yet opened).

The ejection phase (D-A, Fig, 2), that is, the blood pumping phase, occurs when the pressure in the LV rises above the pressure in the aorta, which leads to the opening of the aortic valve and the pumping of blood from the LV into the aorta. At the same time, blood is pumped from the right heart (pulmonary blood flow). The ejection phase includes the fast and slow pumping of blood. Slow pumping represents late systole, when relaxation of the ventricles begins and pressure drops in them (returning to point A again, and starting a new cycle).

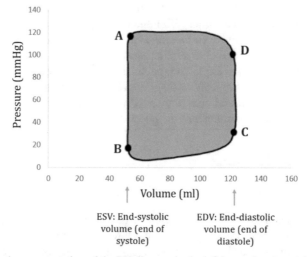

Fig. 2. Schematic representation of the PV diagram in the left heart chamber. A-B isovolumetric relaxation, B-C filling phase, C-D isovolumetric contraction, D-A ejection phase [8].

The green area within the loop (Fig. 2) is equal to the stroke volume, which refers to the amount of blood pumped out of the LV in one cardiac cycle. The effects of isolated changes in preload are best demonstrated in the PV diagram, which relates ventricular volume to the pressure inside the ventricle throughout the cardiac cycle. The maximum right point in the diagram is denoted as the end-diastolic volume (EDV), while the minimum left point represents the end-systolic volume (ESV). Also, as EDV increases, the proportion of blood ejected from the heart slightly increases; this is the ejection fraction (EF) calculated by the following equation [9]:

$$EF = \frac{EDV - ESV}{EDV} \tag{1}$$

2.2 PAK FE Solver and FSI Analysis

The presented study gives the insight into the model-based simulation of myocardial work under HCM, employing the PAK FE software coupled with multi-scale model of muscle contraction. FSI algorithm within the PAK software is used for modelling the LV with nonlinear material model, together with stretches integration along muscle fibers. The methods are integrated within the SILICOFCM platform [3], with aim to propose an advanced approach for the assessment of work indices and biomechanical characteristics of HCM and drugs effects, based on computational modelling.

The parametric LV model for HCM, with specific parametric parts: base part, valves (aortic and mitral) and connecting part (connection between base and valves) has been implemented (Fig. 3). The FE simulations of LV model of HCM using PAK FE solver [10] enable quantitative assessment of the effect of administered drugs on cardiac output including increase in both systolic and diastolic pressures, and the LVEF. The applied boundary conditions are related to adopted nominal inlet and outlet velocities for mitral and aortic valves of LV, as well as applied calcium concentrations for Mavacamten and Disopyramide.

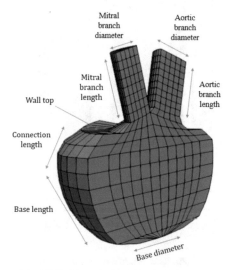

Fig. 3. Parametric 3D model of HCM heart LV with specific parametric parts: base part, valves (aortic and mitral) and connecting part (connection between base and valves).

2.3 Drugs Simulations

There are two major groups of specific drugs by their principal mechanisms of acting. The first group of drugs modulates calcium transients, while the second group changes kinetics of contractile proteins. We have considered the major representative drugs in these two groups: Disopyramide which modulates calcium transients and Mavacamten which changes kinetics of contractile proteins. Both types of selected drugs are used in the treatment of HCM.

3   Results and Discussion

Simulations of the effect of drugs on improving performance of HCM LV parametric model include the drugs that affect calcium transients (Disopyramide) and changes in kinetic parameters (Mavacamten). All simulations are performed using PAK FSI, FE solver and coupled with multi-scale model of muscle contraction. Myocardial work is presented through changes of pressures and volumes (P-V diagrams) for HCM LV model at basic condition (without administered drug) and with using Disopyramide and Mavacamten (Fig. 4).

The predicted P-V diagram for HCM at basic condition shows lower volumes and higher ventricular pressures than normal, with LVEF (LVEF = 59.33%). The principal effects of drugs on HCM after simulations are decrease in peak pressures and shift of P-V loops toward higher volumes, which is in accordance with previous studies and clinical observations [11, 12].

The results provide a quantitative assessment of the effects of different on the cardiac output, including both systolic and diastolic LV pressures and volumes, as well as the LVEF. This approach can give better insight in estimated effects of drug therapy, leading to improved patient monitoring and treatment.

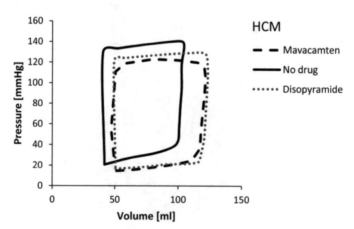

Fig. 4. P-V diagrams for HCM at basic condition (without administered drug) and with using drugs Disopyramide and Mavacamten.

4   Conclusions

In order to achieve more personalized medical treatments and better estimate the patient's condition, this study is devoted to modelling of HCM using coupled macro and micro simulation through FE, FSI and molecular drug interaction with the cardiac cells. FSI is used for modelling the HCM with nonlinear material model for heart wall.

Analysis of myocardial work and changes of pressure and volume within the LV parametric model of HCM are presented at basic condition and after simulated effects of administered drugs. The obtained results provide better insight into the myocardial work of HCM patients as well as in estimated effects of drug therapy, leading to improved patient monitoring and treatment.

Acknowledgment. This work is supported by the European Union's Horizon 2020 research and innovation pro-grammes SILICOFCM (Grant agreement 777204) and SGABU (Grant agreement 952603). The Commission is not responsible for any use that may be made of the information it contains. The research was also funded by Serbian Ministry of Education, Science, and Technological Development, grants [451–03-47/2023–01/200378 (Institute for Information Technologies, University of Kragujevac)] and [451–03-47/2023–01/200107 (Faculty of Engineering, University of Kragujevac)].

References

1. Elliott, P., et al.: 2014 ESC Guidelines on diagnosis and management of hypertrophic cardiomyopathy. Eur. Heart J. **35**(39), 2733–2779 (2014). https://doi.org/10.1093/eurheartj/ehu284
2. Nishimura, R.A., Ommen, S.R., Tajik, A.J.: Hypertrophic cardiomyopathy, a patient perspective. Circulation **108**(19), 133-e135 (2003). https://doi.org/10.1161/01.CIR.0000097621.97566.96
3. SILICOFCM H2020 project: In Silico trials for drug tracing the effects of sarcomeric protein mutations leading to familial cardiomyopathy. 777204, 2018–2022 (2022). www.silicofcm.eu
4. Filipovic, N., et al.: In silico clinical trials for cardiovascular disease. J. Visual. Exp. **183** (2022). https://doi.org/10.3791/63573
5. Filipovic, N., et al.: SILICOFCM platform, multiscale modeling of left ventricle from echocardiographic images and drug influence for cardiomyopathy disease. Comput. Methods Programs Biomed. **227**, 107194 (2022). https://doi.org/10.1016/j.cmpb.2022.107194
6. Mijailovich, S.M., Prodanovic, M., Poggesi, C., Geeves, M.A., Regnier, M.: Multiscale modeling of twitch contractions in cardiac trabeculae. J. General Physiol. **153**(3), e202012604 (2021). https://doi.org/10.1085/jgp.202012604
7. Prodanovic, M., Stojanovic, B., Prodanovic, D., Filipovic, N., Mijailovich, S.M.: Computational modeling of sarcomere protein mutations and drug effects on cardiac muscle behavior. In: 2021 IEEE 21st International Conference on Bioinformatics and Bioengineering (BIBE), 1–6 (2021). https://doi.org/10.1109/BIBE52308.2021.9635428
8. Cheng, Y., Oertel, H., Schenkel, T.: Fluid-structure coupled CFD simulation of the left ventricular flow during filling phase. Ann. Biomed. Eng. **33**(5), 567–576 (2005). https://doi.org/10.1007/s10439-005-4388-9
9. Djorovic, S.: Myocardial work and aorta stenosis simulation. In: Cardiovascular and Respiratory Bioengineering, pp. 135–147. Elsevier (2022). https://doi.org/10.1016/B978-0-12-823956-8.00010-9
10. BIOIRC Kragujevac Serbia 2022, PAK Finite Element Software
11. Tomasevic, S., et al.: Computational Modeling on Drugs Effects for Left Ventricle in Cardiomyopathy Disease. Pharmaceutics **15**(3), 793 (2023). https://doi.org/10.3390/pharmaceutics15030793
12. Warriner, D.R., et al.: Closing the loop: modelling of heart failure progression from health to end-stage using a meta-analysis of left ventricular pressure-volume loops. PLoS ONE **9**(12), e114153 (2014). https://doi.org/10.1371/journal.pone.0114153

Saccade Identification During Driving Simulation from Eye Tracker Data with Low-Sampling Frequency

Ilija Tanasković[1,2] [iD], Nadica Miljković[1,3(✉)] [iD], Kristina Stojmenova Pečečnik[3] [iD], and Jaka Sodnik[3] [iD]

[1] School of Electrical Engineering, University of Belgrade, Bulevar Kralja Aleksandra 73, 11120 Belgrade, Serbia
nadica.miljkovic@etf.bg.ac.rs
[2] The Institute for Artificial Intelligence of Serbia, Fruškogorska 1, 21000 Novi Sad, Serbia
[3] Faculty of Electrical Engineering, University of Ljubljana, Tržaška C. 25, 1000 Ljubljana, Slovenia

Abstract. This paper introduces an improved velocity-based method with optimal velocity threshold parameter for saccade detection to calculate saccade number per second, saccade duration, and velocity peak from the eye tracker data during driving simulation. The algorithm is tested on data recorded with low-sampling frequency of 50 Hz and on the interpolated with sampling frequency of 200 Hz. The obtained results for saccade-related features are in the expected ranges reported in literature. However, results showed statistically significant changes in all parameters (except in standard deviation of saccade duration) as a consequence of interpolation to 200 Hz. Moreover, slight change in threshold parameter (from 5.75 to 6.0) for interpolated data led to statistically significant changes in all velocity-related saccade features. Although results showed that selection of processing technique plays a major role in the analysis of eye tracker data and influences saccade parameters, a more thorough algorithm evaluation is required for guided threshold estimation for saccade identification from eye tracker data recorded with low sampling frequency.

Keywords: driving simulation · eye movements · saccade duration · sampling frequency · velocity threshold

1 Introduction

Saccades are the fastest voluntary movements in the human body as saccades main function is to shift the eyes to the target of interest [1, 2]. Although a great potential for measurement and analysis of saccadic eye movements by eye tracker lies in the assessment of neurodegenerative disorders [1], saccades can be used to evaluate for example a task-related fatigue in drivers [3] and drivers' cognitive load [4] in driving simulators. However, the main bottleneck in saccade identification lies in poor reproducibility and validity of available algorithms [5, 6]. The simplest, the fastest, and straightforward

© The Author(s), under exclusive license to Springer Nature Switzerland AG 2024
M. Trajanovic et al. (Eds.): ICIST 2023, LNNS 872, pp. 30–39, 2024.
https://doi.org/10.1007/978-3-031-50755-7_4

procedure to identify saccades is based on the Velocity-Threshold Identification (I-VT) algorithms that are suitable to identify as saccades present rapid eye movements corresponding to [6–8]. Currently, there is no consensus on how to determine threshold for the I-VT procedure, especially as threshold may depend on the data acquisition parameters [6]. Sampling frequency (fs) has been identified as one of the main factors for accurate saccadic eye movement identification [1, 8]. This is especially challenging in low-cost eye trackers with low fs leading researchers to trade-off between accuracy and cost. Although internationally standardized protocol recommends $fs > 100$ Hz [9], measurements showed that accurate peak velocities and saccade durations can be preserved with data sampled at 50/60 Hz either by interpolation from 50 Hz to 1 kHz [2] and by comparing results with higher fs [8]. This study is motivated by the promising results presented by Wierts et al. and Leubeet al. [2, 8] and by the robust procedure for saccade detection for data sampled at 1250 Hz [5]. We propose an adaptation of the algorithm introduced by Nyström and Holmqvist [5] for low fs eye tracking data (50 Hz) with relatively high noise levels for detection of saccadic eye movements.

The main novelty of the proposed algorithm is that it introduces appropriate threshold adaptations to low fs (50 Hz) on signal obtained from the Tobii eye tracker for saccade identification. We report comparison of obtained results with published findings for feature ranges calculated from both low (50 Hz) and high (200 Hz) fs data (obtained by interpolation) for validity. Finally, the algorithm is tested on data from 10 healthy subjects older than 65 years recorded during driving simulation.

1.1 Research Questions

The goal of this paper is to develop a data-driven algorithm inspired by the robust iterative subject-specific I-VT procedure to automatically determine threshold introduced by Nyström and Holmqvist [5] for saccade identification for high fs and relatively clean data (only 10 out of 300 recruited participants were selected for analysis in [5]). The present research addresses the following research questions:

1. Can an improved I-VT algorithm function as an assessment tool for saccadic eye movements recorded with eye tracker with low fs (50 Hz) during driving simulation in 10 elderly subjects?
2. What effect does interpolation technique to increase fs from 50 Hz to 200 Hz (by assumption that the shortest saccade duration is 10 ms [5–7]) have on the saccade-related features extracted during the driving simulation? This question is motivated by successful interpolation of eye tracker data for saccade peak velocity estimation introduced in [2].

2   Methodology

The I-VT algorithm for saccade detection is developed in MATLAB ver. R2022b (Math-Works Inc., Natick, USA). Data are obtained from 10 healthy individuals participating in long driving simulation study lasting approximately between 4 and 23 min. Subjects' eye movements were recorded with Tobii Pro Glasses 2 eye tracker [10] and stored for the analysis. All participants signed Informed Consents in accordance with the Code of

Ethics of the World Medical Association (Declaration of Helsinki). The experiment was designed and completed following the Code of ethics for researchers and Guidelines for ethical conduct in research involving people issued by the University of Ljubljana. Before the study, the participants were informed about the study goals and asked to sign the Informed consent provided by the Ethical Committee of University of Ljubljana.

Subjects performed driving simulation in a motion-based Nervtech driving simulator [11] (Nervtech Ltd., Trzin, Slovenia), shown in Fig. 1. The simulator consists of three 49-inch TV screens, which enable a 145 degree visual field of view, real car parts (steering wheel, pedals and driver's seat), and a motion platform with 4 degrees of freedom (DOF) that simulates the car movements in relation to the different road conditions. The simulation software resembles real-life environments and enables the creation of variety of traffic situations and weather conditions [11]. Altogether used car driving simulator provides a controllable and dynamic platform for the driving performance evaluation.

Fig. 1. The photography of the driving simulator (Nervtech Ltd., Trzin, Slovenia) set-up used for recording eye movements in dynamical driving environment.

In this experiment, a 13 km long driving scenario during daytime with clear weather was used. The route started off on a two-lane road in a sub-urban area with moderate to heavy traffic. The speed limit was 50 km/h. Three kilometers into the drive, the route encountered three pedestrian crossings without traffic lights. After that, the intensity of the traffic decreased and was low to moderate until the 10th kilometer of the route. At the beginning of the 6th kilometer, there was a complicated intersection with multiple exits. The route continued onto a new four-lane road. At the 10th kilometer, the vehicle approached the city center full of pedestrians, schools, shops, and business commercial buildings and entered a 30 km/h speed limited zone. After 1 km (at the 11th kilometer in the route) the traffic again decreased and remained like this until the final destination, which was a parking lot parallel to the street the vehicle was on.

2.1   Pre-processing and Noise Cancellation

The processing pipeline of the proposed algorithm is presented in Fig. 2. Eye tracker data are stored together with a list of timestamps with x and y coordinates of participants' gaze locations. Coordinates are provided in normalized form, so prior to the processing we preformed transformation into pixels-based resolution (1920 × 1080). To compensate the varying fs, data are re-sampled into equidistant points for further processing. Initial varying sampling frequency is ether re-sampled to the constant fs of 50 Hz or increased by linear interpolation to 200 Hz. All further processing steps (Fig. 2) are performed on both data with fs of 50 Hz and 200 Hz. To smooth the signal and to reduce the noise, we filter x and y positions using one-dimensional median filter (medfilt1 function from MATLAB) with 60 ms window width. Although Nyström and Holmqvist [5] proposed filtering using the second order Savitzky-Golay filter with 20 ms window, our sampling period is 20 ms, so we choose a larger window of 60 ms to include at least three points in window. Then, angular velocity is calculated from Cartesian coordinates using (1).

$$\dot{\theta} = \sqrt{(\phi_1 \dot{x})^2 + (\phi_2 \dot{y})^2} \tag{1}$$

In (1), $\dot{x}$ and $\dot{y}$ stand for Cartesian velocities (the first derivative of x and y) and $\Phi 1$ and $\Phi 2$ stand for pixel-to-angle transformation factors, which are calculated based on the Tobii User Manual [10] as 82°/1920 and 52°/1080, respectively. Velocities higher than 1000°/s are considered to be physiologically impossible, so they are omitted from the analysis as proposed in [5]. Also, to avoid noisy periods effect on the detection of saccades, the samples lower then median of the whole signal are set to zero [5].

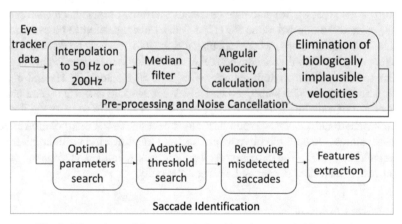

Fig. 2. Flow chart of the proposed algorithm for saccade identification.

2.2   Saccade Identification

Setting a threshold that would detect saccades and ignore small movements of eyes during gaze is rather difficult task as it is influenced by the acquisition parameters and distribution of samples [6]. The proposed algorithm initializes with handpicked threshold of 200°/s (Nyström and Holmqvist proposed the initial threshold to be set from 100°/s to 300°/s [5]) and calculate mean and standard deviation (SD) of samples below threshold as shown in (2). New threshold is then calculated as a sum of calculated mean and SD multiplied by factor n. The operation continues until two consecutive thresholds do not differ by more than 1°/s in absolute value [5]. The calculation of adaptive threshold is shown in (2).

$$AT_{i+1} = mean(signal < AT_i) + nsd(signal < AT_i) \qquad (2)$$

where *sd* stands for standard deviation, *ATi* stands for the adaptive threshold value in i^{th} iteration of the algorithm, and *n* stands for multiplier parameter. If calculated threshold is under 100°/s, we consider that algorithm did not converge, because 100°/s is minimal peak velocity of saccades [5]. After the threshold is determined, we use *findpeaks* function with parameters *MinDistance* of 50 ms (being minimal distance between saccades [5]) and *MinPeakHeight* with converged threshold value. The *findpeak* function returns peaks locations, peaks velocity values, and saccade widths. The visualized processing steps are presented in Fig. 3.

    Parameter *n* resembles angular velocity distribution, and it is affected by the noise as it presents multiplication factor of standard deviation. In [5] *n* was set to 6 and in [7] to 2.5. We hypothesized that during driving simulation subjects are instructed to perform driving task, so the noise contamination is inevitable and the search for optimal *n* may compensate movement artifacts. Therefore, the search for optimal *n* is performed across whole dataset between values 2.5 and 6 with step of 0.5. The criterion is based on the minimization of the misdetection of saccades based on their physiological duration from 10 ms to 100 ms [1, 5, 8]. We calculate the relative number of misdetected saccades for each subject and the optimal value for entire dataset is considered to be average of all individual values. The steps, where calculated threshold reached 100°/s, are excluded because we consider they do not converge. Averaged value is then rounded to the nearest step. The search is done on both datasets with *fs* of 50 Hz and 200 Hz and obtained optimal parameters are 5.75 and 6, respectively. We leave 5.75 as optimal parameter for 50 Hz (instead of rounding it to 5.5 or 6.0) because the value is exactly on the limit, and we did not want to induce bias via rounding it to either one. To test interpolation effect on extracted features, we additionally perform saccade identification on interpolated data (200 Hz) with n = 5.75. The extracted features for each identified saccade are saccade duration, saccade number per second, and peak velocity.

2.3 Statistical Analysis

Statistical analysis is performed on mean, median, and SD values of saccade duration and peak velocity obtained from each subject. Also, we preformed statistical analysis for saccade frequency. All statistical tests are performed using R programming language v4.1.2 [12] and R Studio environment (R Studio, Inc., Boston, MA, USA) with library *effsize* [13]. For normally distributed data, independent two sample t-test is used with Cohen's d (Cd) to estimate the effect size; otherwise, Wilcoxon's Rank-Sign test with Cliff's delta (*Cdelta*) is deployed to compare mean, median, and SD of velocities

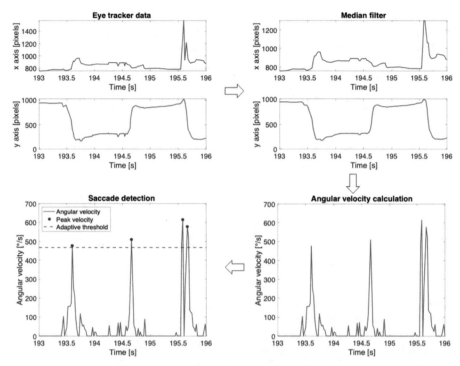

Fig. 3. Saccade detection algorithm processing steps: Cartesian coordinates *i.e.*, eye tracker data; Data after filtration using median filter; Angular velocity calculation; and Saccade detection. Features extracted from detected saccades are peak velocities (476.32°/s, 509.22°/s, 614.02°/s, and 577.05°/s) and saccade widths (42.70 ms, 47.62 ms, 37.54 ms, and 68.75 ms).

and saccade durations, as well as saccade occurrence per second between two groups. The normality is tested by Shapiro–Wilk's normality test. If not stated otherwise, p is set to 0.05. Moreover, the matched statistical tests for dependent samples are used to statistically compare parameters obtained from two different sampling frequencies.

3 Results

The example of the search for optimal multiplier n for both 50 Hz and 200 Hz are shown in Table 1 for subject ID201. Table 1 for values of multiplier $n = 2.5$ and $n = 3.0$ reveals that saccades were not detected as algorithm did not converge (such inputs are marked with NA – Not Available).

Two types of features are extracted each detected saccade: (1) velocity based-features (mean peak velocity, median peak velocity, and standard deviation of peak velocity) and (2) duration-based features (saccade mean duration, saccade median duration, and standard deviation of saccade duration). Additionally, for each saccade we calculated saccade frequency parameter. All features are averaged across 10 participants and presented in Table 2 in form of mean ± standard deviation.

Table 1. Example of adaptive threshold and percentage of missed saccades for subject ID201. Not Available (NA) values are used to emphasize misdetected saccades in cases when adaptive threshold did not converge (please, see text for more details).

Multiplier n	Adaptive threshold [°/s]		Misdetected saccades [%]	
	50 Hz	200 Hz	50 Hz	200 Hz
2.5	100	100	NA	NA
3.0	100	100	NA	NA
3.5	165	147	3.07	13.65
4.0	241	228	3.12	11.14
4.5	325	319	1.96	8.61
5.0	398	401	1.52	7.65
5.5	466	471	0.73	6.71
6.0	536	545	0.81	5.76

Table 2. Saccade-related features calculated for driving simulation. SD stands for standard deviation; n is multiplier parameter for adaptive threshold and fs is sampling frequency.

Features / fs [Hz]	50 ($n = 5.75$)	200 ($n = 5.75$)	200 ($n = 6.0$)
Saccade frequency [saccades/s]	0.72 ± 0.30	1.07 ± 0.28	0.94 ± 0.27
Mean Peak velocity [°/s]	442.41 ± 134.16	468.13 ± 133.46	496.50 ± 136.44
Median Peak velocity [°/s]	389.81 ± 144.61	411.01 ± 151.76	441.74 ± 157.42
SD Peak velocity [°/s]	168.65 ± 25.98	193.06 ± 30.17	188.33 ± 33.74
Saccade mean duration [ms]	37.70 ± 1.85	25.09 ± 2.33	24.96 ± 2.48
Saccade median duration [ms]	33.78 ± 2.59	19.70 ± 2.50	19.66 ± 2.43
Saccade SD duration [ms]	16.38 ± 1.70	15.75 ± 1.55	15.54 ± 1.92

All features showed normal distribution according to Shapiro-Wilk's normality test ($p > 0.05$). The resulting p values of paired two-sample t-tests and Cohen's d parameters are shown in Table 3. When comparing results obtained from original data (50 Hz) with interpolated data (200 Hz) all parameters changed significantly ($p < 0.05$), except for the SD of saccade duration for both multipliers (5.75 and 6.0) applied on interpolated data. Moreover, statistical comparison of parameters obtained from interpolated data with fs of 200 Hz for $n = 5.75$ and for $n = 6.0$ revealed statistically significant differences in all velocity-based parameters, while duration-related parameters remained unchanged.

4 Discussion and Conclusion

In Table 1 we can notice that the increase of the multiplier n directly affects the increase the threshold, which is expected based on the formula (2) for threshold. Also, the increase in n leads to reduction of relative number of misdetected saccades, which can be explained by the fact that as the threshold increases the algorithm does not detect small spikes during gaze periods that correspond to saccades with higher noise contamination. Also, it can be noticed that for small values of parameter n (2.5 and 3.0) as suggested in [7] our algorithm does not converge and that the optimal value is more inclined towards value suggested in [5].

Table 3. Statistical comparison of parameters (between sampling frequency of 50 Hz and 200 Hz; and between $n = 5.75$ and $n = 6.0$ for data with sampling frequency 200 Hz). SD stands for Standard Deviation; Cd stands for Cohen's d and n is multiplier parameter for adaptive threshold. Non statistically significant changes are highlighted in Bold.

Comparison	Saccade frequency [saccades per s]	Mean peak velocity [°/s]	Median peak velocity [°/s]	SD peak velocity [°/s]	Saccade mean duration [ms]	Saccade median duration [ms]	Saccade SD duration [ms]
50 Hz vs. 200 Hz ($n = 5.75$)	$p =$ 6.41e-08 $Cd =$ -1.191	$p =$ 2e-04 $Cd =$ -0.192	$p =$ 0.003 $Cd =$ -0.129	$p =$ 1e-04 $Cd =$ -0.816	$p =$ 1.719e-10 $Cd =$ 5.636	$p =$ 7.243e-11 $Cd =$ 5.514	**$p =$ 0.085 $Cd =$ 0.382**
50 Hz vs. 200 Hz ($n = 6.0$)	$p =$ 1.914e-06 $Cd =$ -0.716	$p =$ 8.988e-07 $Cd =$ -0.395	$p =$ 7.128e-06 $Cd =$ -0.24	$p =$ 0.002 $Cd =$ -0.562	$p =$ 2.083e-10 $Cd =$ 5.206	$p =$ 1.893e-11 $Cd =$ 5.559	**$p =$ 0.07167 $Cd =$ 0.457**
200 Hz ($n = 5.75$ vs $n = 6.0$)	$p =$ 5.448e-08 $Cd =$ 0.323	$p =$ 2.512e-08 $Cd =$ 0.169	$p =$ 1.43e-06 $Cd -$ -0.151	$p =$ 0.0127 $Cd -$ -0.1	**$p =$ 0.432 $Cd =$ 0.05**	**$p =$ 0.7418 Cd ■ -0.014**	**$p =$ 0.261 $Cd =$ -0.092**

Based on results presented in Table 2 we can see that with the increase of sampling frequency results in the increase of saccade frequency and velocity-based features and decrease in duration-based features. When 50 Hz and interpolated 200 Hz data are compared (Table 2), the obtained trends are in line with previous research as larger number of saccades is detected for eye tracker with 120 Hz than for eye tracker with 60 Hz [8]. Moreover, the results for lower fs (60 Hz) showed longer saccade durations than for higher rate (120 Hz) in [8] which is also in line with our results presented in Table 2. We should stress that data in [8] are obtained from actual recordings (not interpolated) and that the expected trend shows that successful interpolation may be possible. Our results show somewhat shorter saccade durations and higher peak velocities than in [5], but this is expected as the driving task is more demanding and complex than scene perception. A similar trend is observed in [3] but driving simulation in [3] was mostly monotonous in comparison to our dynamic drive in the populated area.

Table 3 shows statistically significant changes in all features except for SD of saccade duration which suggests that this feature may be the most robust parameter in relation to the *fs*. Also, when comparing features between parameter $n = 5.75$ and $n = 6.0$ obtained from 200 Hz data, we can notice non-significant changes in duration-based features. This is relatively intuitive as the parameter n directly determine value of adaptive threshold which then discerns peaks that would be detected as saccades. However, this also suggests that majority of saccades during driving simulation are of lower amplitudes, thus being more prone to noise [1, 8].

Our modification of I-VT algorithm originally proposed by Nyström and Holmqvist [5] showed promising results in combination with optimization of parameter *n* for both low sampling frequency, as well as for interpolated data. Although interpolation signif-icantly affected results, it showed an expected trend with real-life high *fs* data and such resolution boost may be a viable option for low-cost devices [8].

An appropriate evaluation of the algorithm on larger sample and in controlled con-ditions is the future step in our research. Even with presented preliminary findings, we contribute to the current state-of-the art in the field of saccade detection by proposing an adaption of the I-VT algorithm for eye tracking data with lower *fs*.

Acknowledgment. This research was funded by HADRIAN (Holistic Approach for Driver Role Integration and Automation Allocation for European Mobility Needs) EU Horizon 2020 project, grant number 875597.

N. M. acknowledges the support from Grant No. 451-03-47/2023-01/200103 funded by the Ministry of Science, Technological Development and Innovation of the Republic of Serbia.

References

1. Imaoka, Y., Flury, A., de Bruin, E.D.: Assessing saccadic eye movements with head-mounted display virtual reality technology. Front. Psychiatry. **11**, 572938 (2020). https://doi.org/10.3389/fpsyt.2020.572938
2. Wierts, R., Janssen, M.J.A., Kingma, H.: Measuring saccade peak velocity using a low-frequency sampling rate of 50 Hz. IEEE Trans. Biomed. Eng. **55**, 2840–2842 (2008). https://doi.org/10.1109/TBME.2008.925290
3. Hu, X., Lodewijks, G.: Exploration of the effects of task-related fatigue on eye-motion features and its value in improving driver fatigue-related technology. Transport. Res. F: Traffic Psychol. Behav. **80**, 150–171 (2021). https://doi.org/10.1016/j.trf.2021.03.014
4. Reyes, M.L., Lee, J.D.: Effects of cognitive load presence and duration on driver eye move-ments and event detection performance. Transport. Res. F: Traffic Psychol. Behav. **11**, 391–402 (2008). https://doi.org/10.1016/j.trf.2008.03.004
5. Nyström, M., Holmqvist, K.: An adaptive algorithm for fixation, saccade, and glissade detec-tion in eyetracking data. Behav. Res. Methods **42**, 188–204 (2010). https://doi.org/10.3758/BRM.42.1.188
6. Salvucci, D.D., Goldberg, J.H.: Identifying fixations and saccades in eye-tracking protocols. In: Proceedings of the Symposium on Eye Tracking Research & Applications - ETRA '00, pp. 71–78. ACM Press, Palm Beach Gardens, Florida, United States (2000). https://doi.org/10.1145/355017.355028

7. Coe, B.C., Huang, J., Brien, D.C., White, B.J., Yep, R., Munoz, D.P.: Automated analysis pipeline for extracting saccade, pupil, and blink parameters using video-based eye tracking. Neuroscience (2022). https://doi.org/10.1101/2022.02.22.481518
8. Leube, A., Rifai, K., Wahl, S.: Sampling rate influences saccade detection in mobile eye tracking of a reading task. JEMR. **10** (2017). https://doi.org/10.16910/jemr.10.3.3
9. Antoniades, C., et al.: An internationally standardised antisaccade protocol. Vis. Res. **84**, 1–5 (2013). https://doi.org/10.1016/j.visres.2013.02.007
10. Tobii eye tracker. https://www.tobii.com/. Accessed 9 Feb 2023
11. Vengust, M., et al.: NERVteh 4DOF motion car driving simulator. In: Adjunct Proceedings of the 6th International Conference on Automotive User Interfaces and Interactive Vehicular Applications, pp. 1–6. ACM, Seattle WA USA (2014). https://doi.org/10.1145/2667239.266 7272
12. Team, R.D.C.: A language and environment for statistical computing (2009). http://www.R-project.org
13. Torchiano, M.: effsize: Efficient effect size computation. R package version 0.8. 1 (2020)

Digital Technologies to Support the Decision-Making Process for Dental Implants Treatment Planning

Sabrina Tinfer[1], Anderson Luis Szejka[1]([⊠]) [iD], Osiris Canciglieri Junior[1] [iD], and Miroslav Trajanovic[2] [iD]

[1] Industrial and Systems Engineering Graduate Program (PPGEPS), Pontifical Catholic University of Parana (PUCPR), Curitiba, Brazil
anderson.szejka@pucpr.br
[2] Faculty of Mechanical Engineering, University of Nis, Nis, Serbia

Abstract. The use of digital technologies by dentistry has primarily grown over the past years, allowing an improvement in dental implant procedures for dental failure correction. The dental implant is complex and non-trivial since multiple variables are involved. Therefore, the integration between dentistry and IT has allowed the creation of expert systems that assist the dentist during the planning and execution stages of the dental implant process. Thus, this research presents the use of digital technologies to support decision-making for dental implant treatment planning. Therefore, based on the concepts and techniques of the implant dentistry area, image processing and intelligent systems systematize the determination of the best implants for multiple failures and the planning of the dental implant process. The preliminary results demonstrate the solution's potential in finding the best solution for planning the dental implant process.

Keywords: Dental Implant · Support Decision · Product Model · Image Processing

1 Introduction

The continuous evolution of the processing capacity of computerized systems allied to artificial intelligence, image processing techniques and collaborative engineering has allowed the conception of expert systems capable of developing activities automatically, supporting the decision-making process [1, 2]. In parallel to this evolution, Computed Tomography (CT) has revolutionized the diagnosis by images [3]. Using CT images allied with image processing can apply algorithms for digital reconstruction, 3D visualization and data extraction from the image supporting the medical in the patient diagnostic [1, 4, 5].

Although some areas are consolidated with the use of the CT for the analysis and diagnoses, others areas, such as dental implants, use the experience of dentistry to define the most suitable dental implant for the patient based on visual analysis of a CT or a

M. Trajanovic et al. (Eds.): ICIST 2023, LNNS 872, pp. 40–52, 2024.
https://doi.org/10.1007/978-3-031-50755-7_5

Magnetic Resonance Imaging (MRI) linked with the physical information analysis [6]. However, for a dental implant definition is necessary to consider multiple information such as tooth failure and analyze the region to be implanted, checking the bone volume, location of nerves, bone, and tooth limits [7]. This process is complex and inaccurate because, with conventional methods, it is impossible to precisely define the bone geometry, bone density and nerve locations. Therefore, the dentist must make the entire decision based on personal information to make the right decision [8–10].

Based on this context, this article presents the study of the conceptual proposal of an expert system of design oriented to the dental implant process, whose objective is to seek from the concepts and techniques of image processing and dental implant to formulate a conceptual model that provides support for decision making through an expert system. This model has inference mechanisms that can capture the information contained in a representation, convert, translate, and/or share them with other terms, offering subsidies to the dentist in choosing the most appropriate dental implant.

As the main contributions of this research, we can highlight: (a) improvement of the dental implant process based on the analysis of precise information of the patient's dental arch extracted from tomographic images; (b) reduction of the surgical procedure time, due to a previous planning of the dental implant process as well as the reduction of the dental implant absorption time due to trauma reductions; (c) reduction of the dental implant rejection risks.

2 Digital Technologies to Support the 3D Reconstruction in Dentistry

Computed Tomography (CT) is a radiographic technique which consists of the acquisition of images allowing the three-dimensional interpretation of the region of interest through sets of slices [3]. It has enabled advances in imaging diagnosis, revolutionizing not only the radiology practice but also the field of medicine and dentistry, combining image processing techniques in the development of tools which provide medical data to assist in the decision-making processes [11].

The standard DICOM allowed the evolution of image processing algorithms because the information obtained from the hardware, regardless of manufacturer, is equal, allowing efforts to be concentrated on developing systems to subsidize surgeons, dentists, and nurses. Additionally, DICOM image files can be converted into different formats, enabling their exhibition on computers without dedicated applications and allowing them to be sent by the network to remote computers [12]. However, depending on the choice of format, there may be a considerable loss of important information for analyzing this image [13].

In the case of dentistry, a branch of the odontology is a field that has increased the use of CT in the treatment of edentulous patients with rehabilitation by a dental implant. In this case, CT is used for 3D image reconstruction, providing a better visualization to the dental surgeon of the patient's bone structure [14]. This visualization allows them to overcome some limitations found in the conventional planning treatments of dental implants, mainly in the pre-implantation stages, which are based on 2D data obtained by MRI. Thus, this environment of graphic multi-visualization, provided by image reconstruction, increases the interactivity of the dental surgeon with the surgical planning [15].

In the Dental Implant field, the advantage of using fixed prostheses, according to [7], is the longevity that they present compared to partially fixed prostheses. Fixed prostheses reduce the risk of caries, improve hygiene, reduce the risk of sensitivity and contact with the roots of existing teeth, improve the aesthetics of the prosthetic pillars, the hygiene of the bone in the edentulous space, reduce the risk of tooth loss of the prosthesis in addition to the psychological aspect [16, 17]. The disadvantages are the high cost, prolonged treatment time and the possibility of implant insertion failure due to poor planning or execution [18]. In [7], the authors report the advantage that there is no resorption process of the structures surrounding the missing tooth, i.e., there is no absorption of the soft bone in this region. The fixed implanted prosthesis can be divided into segmented and non-segmented prostheses. The segmented prosthesis comprises three distinct parts: implant, pillar, and crown, whereas the non-segmented prosthesis consists of only two elements: implant and crown (built from a pillar connected to the prosthesis), facilitating the aesthetic result [19].

The use of computed tomography in the process of dental implantation has made the procedure increasingly safe, as is the case of other areas that already use these images obtained, for three-dimensional modelling (3D), as in the case of cranial reconstruction [4], where the entire rebuilding of the bone is performed and virtually corrects the flaws that may exist in the bone, exporting this data in CAD file, enabling the manufacture and then the grafting of this piece [1]. These virtual reality technologies have been widely disseminated, improving the interpretation of the patient's tomographic image, improving performance in the planned treatment, and reducing recovery time. However, to develop this software that presents reasoning to link the information extracted from the CT images with the dentist's experience, it is necessary to model the information and knowledge from the medical into system rules. The following section demonstrates the approach to extract the data from the CT images and to determine the most suitable dental implant based on the knowledge and experience modelled by the dentist in making decision rules.

3   Rule-Based System Concept

The core of a rule-based system is usually a set of IF-THEN rules that have the knowledge provided by domain experts or extracted from historical data [20]. An inference engine allows the system to produce an output using IF-THEN rules supporting a decision-making process [21]. A rule-based system requires an important Knowledge Representation (KR) about the process; otherwise, the rules will be modelled incorrectly. Therefore, KR is related to connected distinctive techniques, including logic, ontology, and computation [22]. This research's logic is more accurate since it uses equations, tables and other information extracted from historical data or modelled by an expert to apply in the decision-making process.

4 Digital Technologies and Rule-Based System for the Decision-Making Process in Dental Implant

Inaccurate and reduced information makes it problematic and imprecise to define the dental implant, which may lead to premature failure, bone loss, implant rejection and infections, as observed in the work of [7, 19, 23]. The existing computer systems only provide the dentist with a three-dimensional dental arch reconstruction. However, these systems do not offer subsidies or interactivity with the dentist in his decision-making process for determining the dental implant, as occurs in computer-aided diagnosis systems [24], which are essential tools used and accepted in medicine, because they provide support based on a tested knowledge to physicians in their diagnoses. Therefore, the decision-making system to support the dental implant definition must be able to analyze the CT images and select, based on the obtained characteristics, the set of implants and pillars that best fit the patient, supporting the dentist's decision. In addition, it must advise the surgeon dentist in planning the surgical procedure.

Based on this context, this research proposes integrating digital technologies and a rule-based system to define the most suitable dental implant for the patient and the implantation process. With digital technologies is possible to extract information from the DICOM about bone density, patient dental arc geometry, positioning of nerves, and geometry of the mouth arc. With a rule-based system is possible to model the procedures and methods from the dentist surgeon to identify the dental implant and implantation procedure automatically. Figure 2 represents the conceptual architecture of the Dental Implant Definition and Description of the Implantation Process Expert System (DI2DP-ExS).

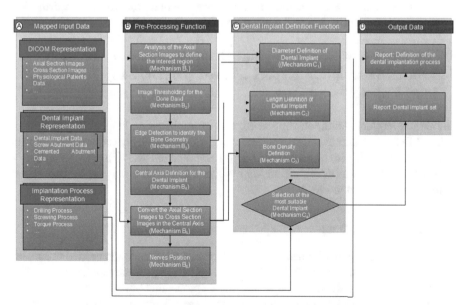

Fig. 1. Architecture of the Dental Implant Definition and Implantation Process Description Expert System (DI2DP-ExS).

Figure 1 is organized into four main structures or functions: (i) Mapped Input Data; (ii) Pre-processing Function; (iii) Data Implant Definition Function; and (iv) Output data.

- **Mapped Input Data** (Detail "A" of Fig. 1) - This structure has information data to support the whole functions and methods for the dental implant and implantation process definitions. Each of these representations in the Mapped Input Data contains information related to the patient, the geometry of the dental arc, and the dental implant procedures. For example, the DICOM representation includes the CT files and patient information. The Dental Implant representation defines the dental implants, and the abutments and Implantation Process Representation contains the definitions to proceed with the dental implant procedure.
- **Pre-Processing Function** (Detail "B" of Fig. 1) – This function gathers different methods to pre-process the mapped input data to prepare and extract information to support the Dental Implant Definition Function. The pre-processing function is composed of six mechanisms (B1 to B6).
- **Dental Implant Definition Function** (Detail "C" of Fig. 1) – This function is the core of the DI2DP-ExS. Based on the information extracted in pre-processing function, the Diameter and Length of the Dental Implant and the Bone Density are defined. With these three parameters and Dental Implant Representation is possible to determine the most suitable dental implant set using the rule-based system approach. The dental Implant Definition Function comprises four mechanisms to proceed with this definition.
- **Output Data** (Detail "D" of Fig. 1) – This structure gathers the reports of the results from the Dental Implant Definition Function. In addition, based on the dental implant set, it is possible to define the implantation process with the tools and the logical steps supporting the dentist during the insertion of the dental implant.

5   Application of Dental Implant Definition and Implantation Process Description Expert System (DI2DP-ExS) in an Experimental Case

The Dental Implant Definition and Description of Implantation Process Expert System (DI2DP-ExS) were applied in an experimental case of a partially edentulous patient with a gap in the mandible in the canine region, as illustrated in Fig. 2. For this evaluation, a prototype software was developed in MathWorks' MATLAB following the conceptual architecture proposed in Fig. 1.

5.1   Mapped Input Data

This structure has the requirements and specifications of information needed to support the functions that make up the DI2DP-ExS. Each of these representations contains information related to the patient and the dental implant procedures or techniques, such as, for example, the DICOM representation that contains the CT files and patient information.

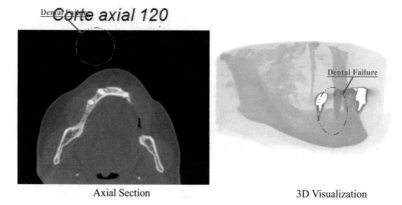

Axial Section   3D Visualization

Fig. 2. The experimental case of a partially edentulous patient.

- **DICOM Representation** gathers all images of the patient in DICOM 3.0 format. The control parameters used in the research are contained in the header of the CT file with the physiological data of the patient, as well as data from the tomographic who performed the acquisition and the parameters of how the images are conditioned in the DICOM standard. We used *Width, Height, PixelSpacing, SliceThickness, Colortype, and BitDepth* in this research because they guide the system during processing. The axial sections guide the delineation of bone geometry, showing the limits between teeth and between the border of the external and internal bone. In addition to this information, with the delineation obtained by the axial section and the region of implant insertion, it is possible to generate the cross-section through the inference conversion mechanism, extracting new parameters from this new image.
- **Dental Implant Representation** has the information inherent to the implant models (types, diameters, length, etc.), which will be used to create the database. This study is focused on the application of the body of the implant and the abutment. The diameter, length and bone density data obtained by the dental implant definition function, added with information on the interocclusal region provided by the dental surgeon, are the key parameters for the search in the database. In this context, Fig. 3 presents the structure for the database of the dental implant representation to the expert system.

5.2 Pre-Processing Function

The pre-processing function comprises six mechanisms, as presented in Detail B of Fig. 1. The mechanisms are responsible for treating the DICOM images and preparing for the Dental Implant Definition Function, which is responsible for extracting the diameter, length, and bone density.

- **Mechanism B_1** analyses the axial section images to define the interest region. The surgeon dentist defines the region of interest through the observation/analysis of the axial sections contained in the DICOM representation. The region must present details of the tooth failure, as well as the bone geometry and, in case of partial edentulousness, the teeth that limit the insertion space of the implant.

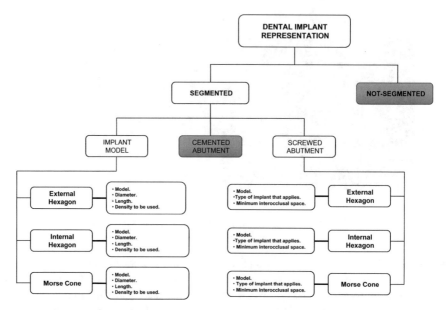

Fig. 3. Information of Dental Implant Representation.

- **Mechanism B₂** is responsible for the image thresholding for the bone band. The information of the CT images follows the Hounsfield scale, which contains information on tissues, bones, nerves, and their referred intensities of grey colour in the scale. A typical bone varies from 100uH to 500uH, and a dense bone from 501uH to 1600uH. By thresholding, the original image f0(x,y) for this range, according to Eq. 1, a new image f1(x,y) is obtained from the original image (Detail "A" of Fig. 4) with only the bone and teeth information (Detail "B" of Fig. 4).

$$f1(x, y) = \begin{cases} 0 \rightarrow sef0(x, y) < 100 \\ 255 \rightarrow se100 \leq f0(x, y) \leq 1600 \\ 0 \rightarrow sef0(x, y) \geq 1600 \end{cases} \quad (1)$$

- **Mechanism B₃** carries out the edge detection to identify the bone geometry. Sobel's edge detection technique is an adequate technique, which will show only the contour of the bone and the roots of teeth, allowing a punctual analysis to determine the insertion centre of the implant. Detail "C" of Fig. 4 presents the results of Sobel's edge detection application.

- **Mechanism B₄** is dedicated to defining the central axis for the dental implant position. The central axis is based on two other reference lines that contour the curve of existing teeth or the edge of the bone. From the reference lines ($RL_1(x)$ and $RL_2(x)$), it is possible to obtain the mean points ($MP_1(x,y)$ and $MP_2(x,y)$), as illustrated in Fig. 7, which allow the construction of the straight line of the symmetry axis through the

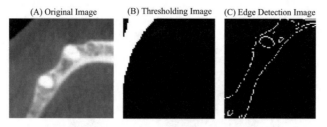

(A) Original Image   (B) Thresholding Image   (C) Edge Detection Image

Fig. 4. Images results obtained with the application of the different techniques for pre-processing.

solution of a linear system, according to Eq. (2), obtaining the parameters a and b which are the coefficients of the straight line of the symmetry axis to be substituted in Eq. (3). Figure 5 represents the reference line.

$$
\begin{cases}
a + b(x_{i1} + \frac{x_{f1}-x_{i1}}{2}) = RL_{i1} \\
a + b\left(x_{i2} + \frac{x_{f2}-x_{i2}}{2}\right) = RL_{i2}
\end{cases}
\tag{2}
$$

$$CA(x) = a + bx \tag{3}$$

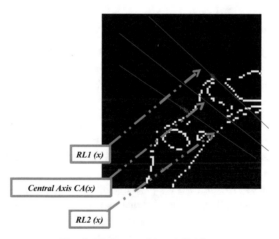

RL1 (x)

Central Axis CA(x)

RL2 (x)

Fig. 5. Reference Line definition.

- **Mechanism B₅** is dedicated to converting the axial section images into cross-section images. The cross-section is determined using the central axis (CA(x)) to generate a plane called WZ, where the equation of the central axis forms W. Z will be created by the axial slices (Ranging from 1 to Z), as shown in Fig. 6. The limits of the axial slice image are obtained from the control information (header) in the variable *Width* (X) and *Height* (Y) and the number of existing Axial Sections (Z) is obtained from the filename.

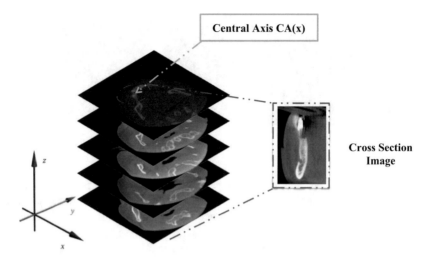

Fig. 6. Cross Section Image Calculation.

- **Mechanism B₆** is dedicated to analysing the cross-section images to identify the position of nerves. This information will be the limit of the implant in its length.

5.3 Dental Implant Definition Function

This function has four mechanisms (Detail "C" of Fig. 1) to determine the most suitable dental implant: (a) Mechanism C_1: Diameter of Dental Implant; (b) Mechanism C_2: Length of the Dental Implant; (c) Mechanism C_3: Bone Density and (d) Mechanism C_4: Dental Implant Selection.

- **Mechanism C_1** is dedicated to determining the diameter of the dental implant. The implant diameter is calculated from the geometric information of the central axis CA(x), the reference lines $RL_1(x)$ and $RL_2(x)$ and the bone and teeth edges. For total edentulous patients, only the thickness of the bone in the implant insertion region is considered. In contrast, for partially edentulous patients, it is necessary to evaluate the thickness of the bone and the distance between the teeth using the smallest measurement obtained between these two parameters. Bone thickness is determined using two points obtained between the intersection of the central axis and the external and internal edges of the bone $PI_1(x,y)$ and $PI_2(x,y)$. In the case of partial edentulous instances in which there is a need to investigate the distance between neighbouring teeth, the reference lines ($RL_1(x)$ and $RL_2(x)$) are used to determine the length that the centre of the insert concerning the neighbouring teeth, i.e., how far the centre of the insert is from the auxiliary lines. From the thickness of the bone and the distance between teeth, it is possible to determine the diameter of the implant to be used. Figure 7 summarizes the rules to define de dental implant diameter.

- **Mechanism C_2** is dedicated to determining the dental implant length. Length calculation is the inference mechanism that analyses and translates the information contained

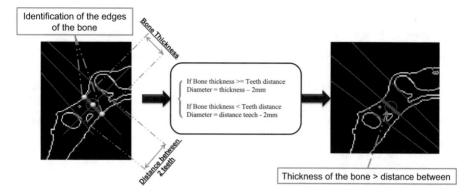

Fig. 7. Diameter of Dental Implant Definition.

in the cross-section by analysing the depth of the bone. This calculation determines the length of the dental implant body based on the bone's geometry and the nerves' location. Localizing the nerves is crucial, as any sizing error can lead to irreversible damage. The total length is obtained through the cross-section analysis until defining the height. Figure 8 summarizes the rule to determine the dental implant length.

- **Mechanism C_3** is allocated to define the bone density that identifies the type of dental implant appropriate for the bone where the implant body will be inserted. The density is obtained by calculating the histogram of the bone H(x), which provides the frequency that each intensity value appears in the image. Table 1 presents the Misch bone density classification to define the type of bone.

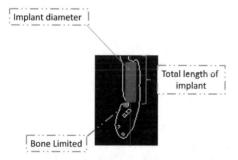

Fig. 8. Length of Dental Implant Definition.

- **Mechanism C_4** is responsible for gathering all information from the other mechanism and defining the most suitable dental implant. Figure 9 shows the decision process to determine the most suitable dental implant for the patient.

Table 1. Misch bone density classification.

Bone Type	Bone Density in UH	Description
D1	> 1250 UH	Dense cortical bone
D2	= 850 to 1250 UH	Thick, dense to porous cortical bone at the ridge crest and thin trabecular bone inside
D3	= 350 to 850 UH	The thin, porous cortical bone at the rim surrounds the thin trabecular bone
D4	= 150 to 350 UH	Thin trabecular bone
D5	< 150	Immature non-mineralised bone

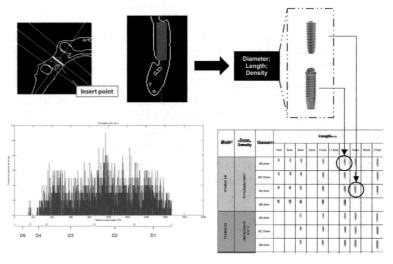

Fig. 9. Dental Implant Definition rules.

6   Conclusion

This research presented a proposal for a Dental Implant Definition and Description of the Implantation Process Expert System that helps the decisions of surgeons to define the implant most suitable for the patient. Traditional dental implant processes do not provide sufficient subsidies for correctly determining the implant. This failure in the decision leads to premature fatigue of the implants, prolonged surgical procedures with high trauma and, in some situations, incorrect sizing of the implant body, providing nerve disruption that can cause partial paralysis of the mouth for a specific time to total paralysis for an indefinite period.

In this way, we required the concepts and techniques in Digital Technologies and Rule-Based Systems to define the implant body and abutment. To formulate a conceptual model of dental implant process-oriented design, with DICOM representation, Dental Implants Representation, and Implantation Process, where the former provides information inputs to the latter, which performs information conversion, translation and sharing, to determine the implant that fits the patient's requirements. This conceptual model was implemented in a computational system resulting in an expert system that presents interactivity with the user and provides at the end of the inference mechanisms procedures the indication of a group of sets of dental implants (implant body and abutment) and information subsidy (diameter, length, density, geometry, nerve location) for the oral and facial surgeon to choose among these implants the most suitable. However, this solution must be experimented with multiple test cases to validate and correct the presented parameters.

Acknowledgments. The authors would like to thank the Coordination for the Improvement of Higher Education Personnel (CAPES), National Council for Scientific and Technological Development (CNPq) and Pontifical Catholic University of Parana for the financial support to this research.

References

1. Canciglieri, M.B., de Mour Leite Staben, A.F.C., Szejka, A.L., Canciglieri Júnior, O.: An approach for dental prosthesis design and manufacturing through rapid manufacturing technologies. Int. J. Comput. Integr. Manuf. **32**(9), 832–847 (2019). https://doi.org/10.1080/095 1192X.2019.1636410

2. de Andrade, J.M.M., et al.: A multi-criteria decision tool for FMEA in the context of product development and industry 4.0. Int. J. Comput. Integr. Manuf. **35**(1), 36–49 (2022). https://doi.org/10.1080/0951192X.2021.1992664

3. Mehta, V., Ahmad, N.: Cone beamed computed tomography in pediatric dentistry: concepts revisited. J. Oral Biol. Craniofac. Res. **10**, 210–211 (2020). https://doi.org/10.1016/j.jobcr.2020.03.013

4. Szejka, A.L., Rudek, M., Canciglieri, O., Jr.: Engineering inference mechanisms reasoning system in design for dental implant. WIT Trans. Built Environ. **145**, 549–557 (2014). https://doi.org/10.2495/ICBEEE20130701

5. Dantas, T., Rodrigues, F., Araújo, J., Vaz, P., Silva, F.: Customized root analogue dental implants - Procedure and errors associated with image acquisition, treatment, and manufacturing technology in an experimental study on a cadaver dog mandible. J. Mech. Behav. Biomed. Mater. **133**, 105350 (2022). https://doi.org/10.1016/j.jmbbm.2022.105350

6. Neto, J.D.P., et al.: Diagnostic performance of periapical and panoramic radiography and cone beam computed tomography for detection of circumferential gaps simulating osseointegration failure around dental implants: a systematic review. Oral Surg Oral Med Oral Pathol Oral Radiol **132**(6), e208–e222 (2021). https://doi.org/10.1016/j.oooo.2021.08.012

7. Li, J., Jansen, J.A., Frank Walboomers, X., van den Jeroen, J.J.P., Beucken,: Mechanical aspects of dental implants and osseointegration: A narrative review. J. Mech. Behav. Biomed. Mater. **103**, 103574 (2020). https://doi.org/10.1016/j.jmbbm.2019.103574

8. Baqain, Z.H., Moqbel, W.Y., Sawair, F.A.: Early dental implant failure: risk factors. Br. J. Oral Maxillofac. Surg. **50**, 239–243 (2012). https://doi.org/10.1016/j.bjoms.2011.04.074

9. Jia, S., Wang, G., Zhao, Y., Wang, X.: Accuracy of an autonomous dental implant robotic system versus static guide-assisted implant surgery: a retrospective clinical study. J. Prosthet. Dent.. **S0022-3913**, 00284–00286 (2023). https://doi.org/10.1016/j.prosdent.2023.04.027

10. Hossain, N., et al.: Recent development of dental implant materials, synthesis process, and failure – a review. Results Chem **6**, 101136 (2023). https://doi.org/10.1016/j.rechem.2023.101136

11. Watanabe, H., Fellows, C., An, H.: Digital technologies for restorative dentistry. Dent. Clin. North Am. **66**, 567–590 (2022). https://doi.org/10.1016/j.cden.2022.05.006

12. Santos, M., Rocha, N.P.: A big data approach to explore medical imaging repositories based on DICOM. Procedia Comput. Sci. **219**, 1224–1231 (2023). https://doi.org/10.1016/j.procs.2023.01.405

13. Alhossaini, S.J., Neena, A.F., Issa, N.O., Abouelkheir, H.M., Gaweesh, Y.Y.: Accuracy of markerless registration methods of DICOM and STL files used for computerized surgical guides in mandibles with metal restorations: an in vitro study. J. Prosthet. Dent. **S0022-3913**, 00636–00639 (2022). https://doi.org/10.1016/j.prosdent.2022.09.017

14. Szejka, A.L., Rudek, M., Canciglieri Jr., O.: Methodological proposal to determine a suitable implant for a single dental failure through CAD geometric modelling. In: 20TH ISPE International Conference on Concurrent Engineering, pp. 303–313. IOS Press, Amsterdam (2013)

15. Fajar, A., Sarno, R., Fatichah, C., Fahmi, A.: Reconstructing and resizing 3D images from DICOM files. J. King Saud Univ. Comput. Inf. Sci. **34**, 3517–3526 (2022). https://doi.org/10.1016/j.jksuci.2020.12.004

16. Surapaneni, H., Yalamanchili, P., Basha, M., Potluri, S., Elisetti, N., Kiran Kumar, M.: Antibiotics in dental implants: a review of literature. J Pharm Bioall Sci. **8**, 28 (2016). https://doi.org/10.4103/0975-7406.191961

17. Wittneben, J.-G., Joda, T., Weber, H.-P., Brägger, U.: Screw retained vs. cement retained implant-supported fixed dental prosthesis. Periodontol. 2000 **73**(1), 141–151 (2017). https://doi.org/10.1111/prd.12168

18. Silveira, M., Campaner, L., Bottino, M., Nishioka, R., Borges, A., Tribst, J.P.: Influence of the dental implant number and load direction on stress distribution in a 3-unit implant-supported fixed dental prosthesis. Dent. Med. Probl. **58**(1), 69–74 (2021). https://doi.org/10.17219/dmp/130847

19. Ochiai, K.T., Ozawa, S., Caputo, A.A., Nishimura, R.D.: Photoelastic stress analysis of implant-tooth connected prostheses with segmented and nonsegmented abutments. J. Prosthet. Dent. **89**, 495–502 (2003). https://doi.org/10.1016/S0022-3913(03)00167-7

20. Yang, L.-H., et al.: Highly explainable cumulative belief rule-based system with effective rule-base modeling and inference scheme. Knowl.-Based Syst. **240**, 107805 (2022). https://doi.org/10.1016/j.knosys.2021.107805

21. Shoaip, N., El-Sappagh, S., Barakat, S., Elmogy, M.: Reasoning methodologies in clinical decision support systems: a literature review. In: U-Healthcare Monitoring Systems, pp. 61–87. Elsevier (2019). https://doi.org/10.1016/B978-0-12-815370-3.00004-9

22. Silva, B., Hak, F., Guimarães, T., Manuel, M., Santos, M.F.: Rule-based system for effective clinical decision support. Procedia Comput. Sci. **220**, 880–885 (2023). https://doi.org/10.1016/j.procs.2023.03.119

23. Pye, A.D., Lockhart, D.E.A., Dawson, M.P., Murray, C.A., Smith, A.J.: A review of dental implants and infection. J. Hosp. Infect. **72**, 104–110 (2009). https://doi.org/10.1016/j.jhin.2009.02.010

24. Deng, F., Wang, D.: Computer aided diagnosis. In: Radiopaedia.org. Radiopaedia.org (2018). https://doi.org/10.53347/rID-61706

SGABU Computational Platform as a Tool for Improved Education and Research in Multiscale Modelling

Tijana Geroski[1,2(✉)] (ID), Jelena Živković[1,2], Christian Hellmich[3] (ID),
Themis Exarchos[4,5] (ID), Hans Van Oosterwyck[6] (ID), Djordje Jakovljević[7] (ID),
Miloš Ivanović[2,8] (ID), and Nenad Filipović[1,2] (ID)

[1] Faculty of Engineering, University of Kragujevac, Kragujevac, Serbia
tijanas@kg.ac.rs
[2] Bioengineering Research and Development Center (BioIRC), Kragujevac, Serbia
[3] Vienna University of Technology, Vienna, Austria
[4] University of Ioannina, Ioannina, Greece
[5] Ionian University, Corfu, Greece
[6] Katholieke Universiteit Leuven, Leuven, Belgium
[7] Coventry University, Coventry, UK
[8] Faculty of Science, University of Kragujevac, Kragujevac, Serbia

Abstract. There is a need to develop an integrated computational platform that will contain both datasets and multiscale models related to bone (modelling), cancer, cardiovascular diseases, and tissue engineering. The SGABU platform is a robust information system capable of data integration, information extraction, and knowledge exchange, with the goal of designing and developing suitable computing pipelines to give accurate and adequate biological information from the patient's molecular to organ level. Datasets integrated into the platform are directly obtained from experimental and/or clinical studies and are mostly in tabular or image file format. Multiscale models range from models that can be described using partial or ordinary differential equations, to complex models that use finite element modelling. The majority of the SGABU platform's simulation modules are built as Common Workflow Language workflows. This implies creating a CWL implementation on the Functional Engine Service backend and creating an acceptable User Interface. The key advantage of SGABU platform is the utilization of new, contemporary, modular, and unique technology for various levels of architecture.

Keywords: computational platform · multiscale modelling · user-friendly interface · open science

1 Introduction

The goal of multiscale computational modeling is to link complicated networks of effects at multiple spatial and/or temporal scales. These networks, for example, frequently include intracellular molecular signaling, crosstalk, and other interactions between

M. Trajanovic et al. (Eds.): ICIST 2023, LNNS 872, pp. 53–63, 2024.
https://doi.org/10.1007/978-3-031-50755-7_6

neighboring cell populations, as well as higher levels of emergent phenomena across different tissues and collections of tissues or organs interacting with each other throughout the body [1]. Powerful computers and effective numerical tools make it possible to solve complex biological issues in actual geometries and in remarkable detail within short time. Globally, education and research in this field are of considerable interest. All of this motivates the development of a rapid and resilient system that integrates multiscale models from different fields to aid medical professionals in decision making, while maintaining accuracy and precision.

1.1   Literature Review

Certain useful biomedical research platforms have arisen. For example, PANBioRA is a modular platform that standardizes biomaterial evaluation and enables pre-implantation, personalized diagnostics for biomaterial-based applications [2]. The SILICOFCM platform is a revolutionary cloud-based in silico clinical trial solution for the design and functional optimization of total cardiac performance, as well as monitoring the efficacy of pharmaceutical therapy [3]. The Bioengineering and Technology (BET) platform fosters innovation, advances research, and connects the greater interdisciplinary community participating in translational bioengineering [4]. It is primarily concerned with the needs of the cancer research community.

However, to the best of our knowledge, there is no platform that incorporates multiple domains and mixes models from various modeling fields. Furthermore, there are no platforms that are dedicated to both research and education at the same time. The goal of the project "Increasing Serbia's Scientific, Technological, and Innovation Capacity in the Domain of Multiscale Modeling and Medical Informatics in Biomedical Engineering (SGABU)" [5] is to create a platform that will include examples from four different fields: cardiovascular, cancer, tissue, and bone modeling. Motivated by this latest idea and aforementioned demand for having a user friendly and adaptable platform that suits the needs of the students, researchers, but also doctors and specialists, SGABU platform is proposed to integrate multiscale models and datasets from four different fields (cardiovascular, bone, tissue and cancer modelling).

2   Materials and Methods

The integration of multiscale models is performed according to standardized procedures, which enables secure delivery of all applications to anyone with an Internet-connected device and a browser. All data, analytical tools, and techniques generated under the SGABU project framework should be findable, accessible, interoperable, and reusable (FAIR principles), not only for practical reasons but also to encourage an open-science attitude. The FAIR principles serve as a guideline for data producers and researchers to ensure that their data is as interoperable as feasible. Individual tools should be standardized in their organization and interconnection. This entails packaging software using Linux container technologies like Docker or Singularity, and then coordinating workflows and pipelines with domain-specific workflow languages like WDL (Workflow Description Language) and CWL (Common Workflow Language).

We used Common Work-flow Language[1] as a specification pathway for all of our processes when developing the SGABU platform backend. It makes use of Docker containers as fundamental building pieces to give a concise explanation of any scientific procedure. In the following sections, we will demonstrate how to construct a typical CWL process using pre-existing software components.

In general, creating a CWL process consists of three separate steps:

1. Containerization,
2. Creating a CWL tool out of Docker image, and
3. Merging the tools into a CWL workflow.

Existing computer models for bone modelling, cancer modelling, cardiovascular diseases modelling and tissue engineering with patient specific databases are integrated. The main aim is to: (i) develop robust information system capable for data integration, information extraction and knowledge sharing, (ii) design and develop proper computation pipelines to provide valid and sufficient biomedical information from the molecular to organ level of the patient, (iii) produce sophisticated mathematical models to predict the progression of the diseases, their relation with the collected biological markers and, desirably, tolerance/resistance to various drug families and the existing risks to the patient. The integration process can be divided into two main subtasks:

1. Integration of the datasets,
2. Integration of the multiscale models.

Integration of the Datasets. Depending on the requirements, datasets can be given in a simple form, represented as tabular data, while most of the datasets required further tuning carried out by front-end developers employing technologies such as Angular, Plotly.js, Paraview Glance, etc. Angular [6] is one of the most popular forntend development tools for creating single page dynamic apps. The SGABU platform may be utilized on any device thanks to automatic screen adjustment (responsiveness). Plotly.js [7], which is based on JavaScript, may be used to visualize certain data. The advantage over other visualization tools is the variety of graphs available, such as maps, pie charts, bar charts, bubble charts, and so on. Plotly.js has several features such as plot download as a PNG, zoom, pan etc. Datasets can also be downloaded in their original form, which is in accordance with Open Science policy [8].

Integration of the Multiscale Models. The majority of the SGABU platform's simulation modules are built as Common Workflow Language (CWL) workflows. This solution is an obvious choice since it makes use of Docker containerization and a standardized manner of describing inputs, outputs, and intermediate outcomes, resulting in intrinsic findability, accessibility, inter-operability, and reusability (FAIR principles). The effort involved in offering CWL-type processes is divided into two main actions: creating a CWL implementation on the FES (Functional Engine Service) backend and creating an acceptable user interface (UI). The second activity consists of developing the UI elements as proper workflow input forms with validation of numeric values, filetypes, etc. as well as output visualization tabs with tabular views, interactive diagrams, 3D views, and animations.

---

[1] https://www.commonwl.org/user_guide/

3   Results and Discussion

Homepage of SGABU platform is depicted in Fig. 1. Each investigated field section is divided into subsections - datasets and multiscale modes. After accessing any of the modules, the help page opens automatically to point the user towards the guidelines of how to use a specific model/dataset, together with the theoretical background and references for further reading. The access to the Biomaterial corrosion module in the Bone Modelling section on the SGABU platform is provided through the main dashboard. This module will serve as an example of integration of one model on the platform.

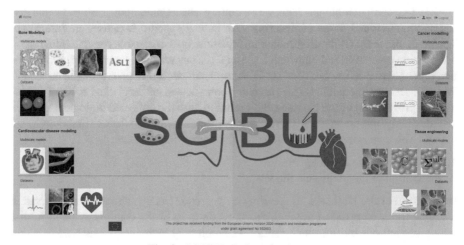

Fig. 1. SGABU platform homepage

3.1   Use Case of a Model – Biomaterial Corrosion

The model of corrosion has been developed based on Cellular Automata (CA) theory and the methodology of the model and main research results are published in [9]. In this paper we focus on the methodology for integration of one such model within SGABU platform.

The computational modelling of multi-pit corrosion in medical implants based on cellular automata (CA) is divided into two sub-models - pit initiation and growth models, where the evolution of cell CA each occurs through a series of synchronous updates of all cells, governed by a set of rules. Since the visualization of corrosion is presented in image, the state of each cell will be in an interval $0 - 255$, meaning an 8-bit image (uncorroded cell has the value of 0 and totally corroded cell the value of 255). This means that we look at the surface and look at where the corrosion pit will appear (corrosion pit initiation) and how it will develop (corrosion growth model). User Interface for CorrosionPit example is shown in Fig. 2.

The window is divided into 2 sections. The basic execution unit in SGABU is a workflow. In the left section of the window, users are able to see names and status of the workflows. Possible statuses of the workflow are:

Fig. 2. User Interface for CorrosionPit module

- Not yet executed
- Terminated
- Running
- Finished OK
- Finished Error

In the right section of the window, users can create new workflows for this submodule. Each of the forms needs to be filled out in order for simulation to run. Users are expected to fill out the following forms:

- Workflow name
- pH value
- Potential of the metal
- Potential of the solution
- Absolute temperature
- Concentration of the reaction species
- Diffusivity of the reaction species
- Charge of the reaction species
- Time of simulation

All forms except for Workflow name are numerical and value ranges are provided for the users. The exception handling is integrated into the user interface (UI) of the SGABU platform. Specific examples can be seen in Fig. 3. (Empty forms, Non-numerical forms, Out of range values).

Once everything is correctly filled, the workflow can be started. The user can monitor the current status of the workflow in the left section of the window (Fig. 4).

At the beginning we assume no corrosion, meaning the first step is the black image (all zeros). After that, in pit initiation model, each cell x is associated with an initial potential state $I(u, t) = I(u, t - 1) + \alpha$ at time t. For each un-corroded cell u, initiation potential state values in its Von Neumann neighbourhood together with its own is considered

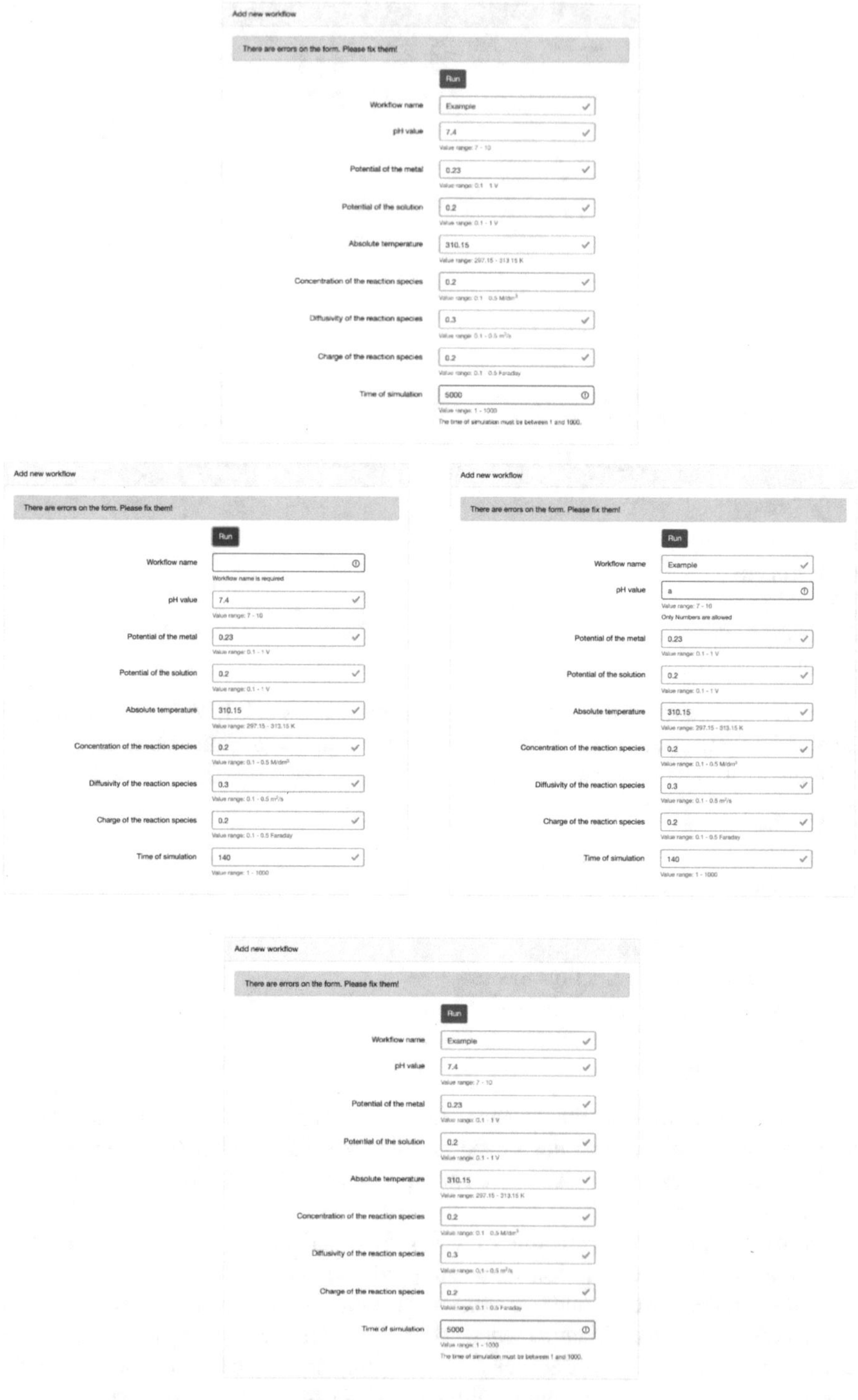

Fig. 3. Exception handling

Fig. 4. Running the workflow

$I(u, t) + \sum I(u + \delta_i, t)$ and if that sum divided by 3 overcomes the threshold value, then a corrosion pit is initiated at cell u and its corrosion state $S(u, t)$ is set to a small positive number between 3 and 5. For the next time steps, pit initiation model is applied on the uncorroded cells again, and at the same time pit growth model is applied on the cells where corrosion has been initiated with formula:

$$S(t + 1, x) = S(t, x) + k_1 f[S(t, x)]$$
$$+ k_2 \sum_i f[S(t, x + c_i)] + k_3 \sum_j f[S(t, x + d_j)] + k_4 \Delta \tag{1}$$

where $c_i = (0, -1), (1, 0), (0, 1), (-1, 0), d_j = (1, 1), (1, -1), (-1, -1), (-1, 1)$ for $i, j = 1, 2, 3, 4$. This means that we use Moore neighborhood to describe how the surrounding cells influence the cell of interest. Coefficient k_1 is described as:

$$k_1 = \lambda \times (pH - 7)^2 \times step(4, 8.5) \times e^{\varphi_M - \varphi_S} \times (1/T) \times C \times D \times z \tag{2}$$

In these eq. λ is a discount factor ranges from 1 to 3; pH is the pH value of the solution; step (4, 8.5) is a function with value 0 between 4 and 8.5, and 1 otherwise; φ_M and φ_S are the potentials of the metal and solution, respectively; T is the absolute temperature; C is the concentration of the reaction species; D is the diffusivity of the reaction species; z is the charge of the reaction species. The parameters k_2, k_3 and k_4 are in similar forms as k_1, but with different discount factors.

Different environmental factors are included in the model to describe their effect on the corrosion. Inputs to the corrosion pit model with allowed ranges and default values, as well as adequate unites are given in Table 1.

Table 1. Inputs for the corrosion pit model

Name of the parameter	Label	Description	Unit	Range	Default value
pH value	pH	pH value of the solution	/	7–10	7.4
Potential of the metal	φ_M	change in a corrosion system of the metal	V	0.1 - 1	0.23
Potential of the solution	φ_S	change in a corrosion system of the solution	V	0.1 - 1	0.2
Absolute temperature	T	Absolute temperature of the environment	K	297.15 -313.15 (24 °C – 40 °C)	310.15
Concentration of the reaction species	C	Concentration of one of the species participating in a corrosion reaction	M/dm^3	0.1–0.5	0.2
Diffusivity of the reaction species	D	The rate of diffusion-controlled corrosion of reaction species	m^2/sec	0.1–0.5	0.3
Charge of the reaction species	z	Charge resulting from the reaction of species	Faraday	0.1–0.5	0.2
Time of simulation	t	Number of time steps to run the simulation	/	1000	100

Results are displayed in the form of images and analysis of the resulted images that represent the corrosion in time. The model describes adequately the multi-pit corrosion pit initiation and growth. Figure 5 shows the corrosion states at four different time steps, where black indicates uncorroded cells and white indicates fully corroded cells.

Fig. 5. Corrosion states (every 20 iterations are plotted)

After the workflow has been successfully executed, a user can visualize the results on the platform itself. The results window, includes the following sections:

- Overview of the input parameters (Fig. 6)
- Results in the table form (Fig. 7),
- Results in the form of figures, following corrosion steps in time (Fig. 8).

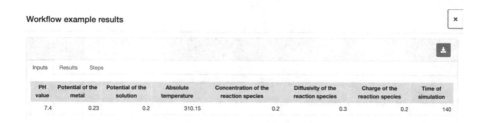

Workflow example results   ×

Inputs   Results   Steps

PH value	Potential of the metal	Potential of the solution	Absolute temperature	Concentration of the reaction species	Diffusivity of the reaction species	Charge of the reaction species	Time of simulation
7.4	0.23	0.2	310.15	0.2	0.3	0.2	140

Fig. 6. Inputs section – table form

Workflow example results   ×

Inputs   Results   Steps

results.txt

	Step 0	Step 1	Step 2	Step 3	Step 4	Step 5	Step 6	Step 7
Mean	0	0	3.76215	10.9426	32.7257	77.4634	127.693	176.535
Standard deviation	0	0	3.77803	9.81423	25.8477	59.895	90.6756	88.7539
Skew	NaN	NaN	0.132853	0.0342759	-0.140565	-0.302727	-0.446	-0.666329
Corroded area (%)	0	0	51.55	59.305	65.29	65.2975	81.7875	97.955
Kurtosis	NaN	NaN	1.25303	1.54321	1.78475	1.51692	1.45557	1.67242
Energy	1	1	0.292049	0.194695	0.132258	0.126106	0.0390663	0.0367367
Entropy	-0	-0	2.28213	3.39195	4.54945	5.20052	6.50129	6.48973
Power	0	0	4.54831e+10	3.45691e+11	2.78249e+12	1.53406e+13	3.92438e+13	6.24668e+13
Contrast	0	0	0	0.0101508	0.854422	5.65814	16.181	18.0485
Wavelet features (S1)	0	0	2.98982	8.70658	26.0659	61.7169	101.587	140.095
Wavelet features (S2)	0	0	0.583552	1.50283	3.64247	7.8081	12.0291	12.4918

Fig. 7. Results section – table form

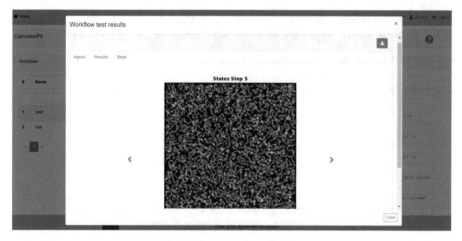

Fig. 8. Steps section - figures

The results can also be downloaded in the form of figures and txt files. In order to estimate quantitatively how the corrosion has progressed in time, we calculate statistical measures presented in textual file:

- Mean corrosion (calculated based on histogram probability as the average grey value
- Standard deviation (describes the spread in the data, which is related to the contrast)
- Skew (measures the asymmetry about the mean value in the distribution)
- Percentage of corroded material (sum of the corroded pixels (all non-zero values) divided by number of pixels and multiplied by 100)
- Kurtosis (shows whether the data are peaked or flat relative to a normal distribution)
- Energy (in the case of corrosion image, this feature indicates the degree of corrosion at the pit level)
- Entropy (entropy of each sub-band provides a measure of the image characteristics in that sub-band)
- Power (indicates the texture property in an image and in this case, the level of corrosion itself)
- Contrast (difference between maximum and minimum pixel intensity in an image)
- Wavelet features (calculated through the processes of singular values decomposition (SVD) and first two eigenvalues are reported)

4 Conclusions

There is a need to create an integrated computational platform that would include datasets as well as multiscale models relating to bone (modeling), cancer, cardiovascular disorders, and tissue engineering. SGABU platform represents the first platform of its kind that integrates both multiscale datasets and models. The main impact of the proposed methodology is the established e-infrastructure capacity and designed protocols for integrating novel multiscale solutions into a platform that will be scalable and capable of adopting new ideas and solutions, and capable of accepting research challenges. Future development would be directed towards faster calculations of the models on the platform, as well as even better user-platform interactivity.

Acknowledgement. This research is funded by Serbian Ministry of Education, Science, and Technological Development [451–03-68/2022–14/200107 (Faculty of Engineering, University of Kragujevac)]. This research is also supported by the project that has received funding from the European Union's Horizon 2020 research and innovation programmes under grant agreement No 952603 (SGABU project). This article reflects only the author's view. The Commission is not responsible for any use that may be made of the information it contains.

References

1. Versypt, A.N.F.: Multiscale modeling in disease. Curr. Opin. Syst. Biol. **27**, 100340 (2021)
2. Personalised and generalized integrated biomaterial risk assessment (PANBioRa). https://www.panbiora.eu/. Accessed 10 Aug 2022
3. In Silico trials for drug tracing the effects of sarcomeric protein mutations leading to familial cardiomyopathy (SILICOFCM). https://silicofcm.eu/. Accessed 15 Nov 2022
4. Bioengineering and Technology platform (BET). https://www.epfl.ch/research/facilities/ptbet/. Accessed 07 Dec 2022
5. Increasing scientific, technological and innovation capacity of Serbia as a Widening country in the domain of multiscale modelling and medical informatics in biomedical engineering (SGABU). http://sgabu.eu/. Accessed 05 Jan 2023
6. Seshadri, S: Angular: Up and running: Learning angular, step by step: O'Reilly Media (2018)
7. Plotly homepage. https://plotly.com/. Accessed 15 Dec 2022
8. European Commission, The EU's open science policy. https://research-and-innovation.ec.europa.eu/strategy/strategy-2020-2024/our-digital-future/open-science_en. Accessed 23 Jan 2023
9. Šušteršič, T., et al.: An in-silico corrosion model for biomedical applications for coupling with in-vitro biocompatibility tests for estimation of long-term effects. Front. Bioeng. Biotechnol. **9**, 718026 (2021)

The Importance of Genetic and Clinical Data Features in Risk Stratification of Patients with Hypertrophic Cardiomyopathy

Ognjen Pavić[1,2(✉)] ⓘ, Lazar Dašić[1,2] ⓘ, Tijana Geroski[2,3] ⓘ, Anđela Blagojević[3] ⓘ, and Nenad Filipović[2,3] ⓘ

[1] Institute of Information Technologies Kragujevac, University of Kragujevac, Kragujevac, Serbia
opavic@kg.ac.rs
[2] Bioengineering Research and Development Center (BioIRC), Kragujevac, Serbia
[3] Faculty of Engineering, University of Kragujevac, Kragujevac, Serbia

Abstract. Hypertrophic cardiomyopathy is a genetic cardiovascular disease which affects the heart's left ventricle. This paper presents the results obtained from examining the importance of clinical and genetic data points in risk stratification of patients with hypertrophic cardiomyopathy. The significance of features was gathered in consultations with cardiologists as well as from the evaluation of created classification models built for the purposes of risk assessment. The main goal of the study was to find hidden knowledge within the dataset that could be used to further improve classification results and to compare the aforementioned knowledge with the information gathered from doctors with the goal of potentially improving the manual diagnostics approach. The study was conducted on genetic and clinical data separately as well as on a combined dataset. The importance of parameters was calculated with two different classification models, and was also calculated using two different methods of manual data annotation. All of the acquired results show both similarities and differences from one another. The acquired results were evaluated based on the predictive abilities of classification models.

Keywords: cardiomyopathy · machine learning · data importance · diagnostics · biomarker analysis · classification · risk assessment

1 Introduction

Hypertrophic cardiomyopathy (HCM) is a genetic cardiovascular disorder that is characterized by the hypertrophy of the heart muscle walls of the heart's left ventricle [1]. The hypertrophy, or thickening of the left ventricle walls affects the stiffness and the rigidity of the heart muscle tissue as a consequence of the fact that thickened walls cannot relax properly during the cardiac cycle. Hypertrophy of the left ventricle walls directly impacts the blood flow through the heart and causes obstructions. In the majority of cases, HCM has a stable course over the years without any major complications; however it is essential for HCM to be properly diagnosed on time because it can lead to the development of arrhythmias, heart failure, stroke and death [1].

M. Trajanovic et al. (Eds.): ICIST 2023, LNNS 872, pp. 64–72, 2024.
https://doi.org/10.1007/978-3-031-50755-7_7

When calculating the severity of HCM, the main classification target is the risk of suffering a sudden cardiac death. Sudden cardiac death is a death attributed to a cardiovascular cause which happens within one hour after the onset of symptoms [2].

HCM can be diagnosed through genetic testing as well as through echocardiography; therefore this paper aims to calculate the importance of features for classifying patients into high-risk and low-risk classes based on the risk of suffering a sudden cardiac death using genetic test findings, clinical echocardiography findings, and both methods in tandem with one another.

Most of the available research [3–5] focuses mainly on classifying patients into two risk classes with regards to the risk of suffering a sudden cardiac death and glosses over feature importance, viewing it only as a tool for minor prediction accuracy improvement. In their paper, Smole et al. [3] used a very similar methodology for patient classification to the methodology that was employed during this study, with differences arising in the methods of data sample utilization. Namely, in their research, each patient was viewed as a single entity, while our study views each visit to the doctor as a possible state that any patient can find themselves in, at some point in time and thus utilizes each visit as a separate entity, thereby creating a larger number of training and testing samples for resulting classification models. In contrast to their approach, Kochav et al. [4] used random forest and extreme gradient boosted trees algorithms for patient stratification, however their models were trained using mainly data on events that have happened to patients in the past. A study was conducted by João et al. [5] in which left ventricular maximum wall thickness (MWT) was used as the primary feature for risk stratification. Aurore et al. [6] used mathematical models combined with clustering methods to divide patients into four distinct risk classes. They used data gathered from healthy volunteers as well as HCM patients for comparison. This data was comprised of a genetic dataset and a clinical dataset which contained ECG along with CMR images and extracted T and QRS biomarkers. In their study, Tse et al. [7] utilized a multilayer perceptron approach to solving a risk stratification problem regarding incidents induced by atrial fibrillation and stroke. While their study was aimed at cardiovascular diseases in general, the stratification of risk of death amidst atrial fibrillation includes hypertrophic cardiomyopathy into the disease interest group. Even though each of the aforementioned studies achieved satisfactory classification results in their own rights, they all lack a certain degree of result explanation as well as decision making process explainability. With a deeper assessment of feature importance for different feature sets, our approach to risk stratification provides additional insight into both the achieved results and the way classification models make decisions. The assessment of the significance of certain features contained in the available dataset also provides a baseline for future improvement in terms of providing stable grounds for the construction of explanation modules aimed at explaining results in a more detailed way to patients and medical professionals alike.

On the other hand, there exists medical research [8] specifically aimed at discovering new ways of diagnosing heart diseases like HCM. This paper aims to utilize the AI centered approach to risk stratification while also paying close attention to explanations of machine learning models regarding the way they learn and classify given data. This approach opens the possibilities of discovering hidden knowledge in the data that can improve the manual diagnostics process while also providing an automated machine learning based model that has the possibility to serve as a decision support tool.

2 Methodology

The dataset is comprised of demographic, genetic and clinical data as well as clinical investigations and disease related events. The dataset contains a total of 13386 samples, collected from 2302 distinct patients, gathered during multiple attended checkups. The retrospective data that was used for training the machine learning algorithms was provided by the Careggi University Hospital, University of Florence, Italy. Retrospective data was gathered over a 13 month period, during which clinical test were performed in regular intervals. ECG and Doppler tests were performed during months 1 and 12, Holter test were performed during months 2 and 13, while CMRI findings were gathered during month 6. The inclusion criteria for patients were a primary diagnosis of HCM or the existence of a HCM diagnosed relative.

The available dataset first needed to be processed and brought into a state usable for classification model training. The set contained instances of missing data that was filled in using transcription of past or future values. In cases in which transcription of data was not possible, missing data was filled in using other common data imputation techniques. Namely missing numeric data was filled in using the mean value of the observed variable; missing categorical data was filled in using the category which is most numerous. When it comes to binary data imputation, a system was devised to input values of 0 or 1, while paying attention not to assign different values of binary variables to the same patient when filling in data missing in follow-ups and while also paying close attention to the distribution of new values so that the distribution stays the same after imputation as it was before imputation. The dataset also contained extreme, physically impossible values that were eliminated from the set before model training. Finally, none of the patient data in the available dataset was labeled with risk classes, so labeling had to be conducted as a way of creating the classification target.

The first approach to data labeling was using doctors' instructions. Cardiologists named the following 9 criteria:

1. Past diagnosis of syncope
2. New York heart association (NYHA) class value greater than 3
3. Family history of sudden cardiac death while the patient is younger than 40 years of age
4. Interventricular septum (IVS) thickness or posterior wall (PW) thickness less than 30mm
5. Left atrium diameter greater than 40mm
6. Ejection fraction lower than 50%
7. Left ventricular outflow tract pressure gradient (LVOT PG) in resting state higher than 30mmHg
8. N-terminal-pro hormone BNP (NT-proBNP) value greater than 900pg/ml
9. The existence of atrial fibrillation in any form

Of these 9 criteria if 4 or more were true, the patient is classified as having a high risk of suffering a sudden cardiac death.

The second approach to data labeling was using the information on disease related events. If an event corresponding to high risk of sudden cardiac death occurred in the event dataset, the patient would be labeled as high risk from that point in time onwards. Events that were taken into account as being closely related to having a high risk of suffering a sudden cardiac death were the following:

1. Arrhythmia – non-sustained ventricular tachycardia (NSVT)
2. Arrhythmia – sustained ventricular tachycardia (SVT)
3. Abnormal Holter
4. Abnormal exercise tolerance test (ETT)
5. Heart failure

Additionally the high-risk class was attributed to patients who were marked as dead from suffering a sudden cardiac death, and also patients who had an implantable car-dioverter defibrillator, readings gathered from an implantable cardioverter defibrillator, had a heart transplant or were marked to receive a heart transplant.

Using these two approaches, two datasets were created for later comparison, for both feature importance and model prediction accuracy.

For the purposes of classification, two different ensemble classification groups of models were created. The first group was built using the random forest algorithm, while the second group was created using the extreme gradient boosted trees method. All

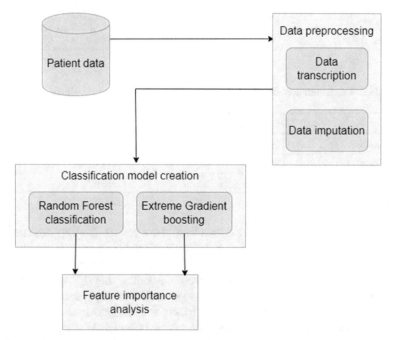

Fig. 1. Patient classification and feature importance analysis methodology diagram

of these models were trained and evaluated using 6 different datasets. Namely, each model was trained using only genetic data, using only clinical data or using both genetic and clinical data together. Training was conducted with both labels created from cardiologists' instructions and those created from disease related event data. In the end, there were 12 results gathered from distinct combinations of inputs, outputs and model creation algorithms. The entire methodology diagram is shown in Fig. 1.

Data importance was calculated for each of the resulting 12 classification models. It is important to note that not all of the models achieved good classification accuracy, therefore when assessing the importance of certain data features, prediction metrics were also taken into account.

3 Results and Discussion

After conducting classification using all 12 combinations of inputs, labeling methods and classification algorithms, the models were evaluated based on prediction accuracy. The achieved results are shown in Table 1 for random forest-based models, as well as in Table 2 for extreme gradient boosted trees-based models.

Table 1. Prediction accuracy for random forest classification models

	Gold standard (doctor)	Event labeling
Genetic data	90.56%	60.98%
Clinical data	92.66%	92.55%
Combined data	94.37%	94.21%

Table 2. Prediction accuracy for extreme gradient boosted trees classification models

	Gold standard (doctor)	Event labeling
Genetic data	90.52%	60.92%
Clinical data	92.18%	91.64%
Combined data	96.69%	96.53%

The following conclusions were drawn from these results. Both approaches to creating classification models are very close in terms of prediction accuracy, with extreme gradient boosting exceling in full dataset classification, while the random forest is better at classification using individual parts of the dataset. Training the models using data labeled through disease related events is worse no matter the input data used, especially when training models using genetic data. With these results, data importance for different sets of inputs was valued closely between models and data labeling approaches, except the importance calculated for genetic data with event labeling which was completely ignored due to the extremely poor results.

Table 3 shows the most important clinical data features. There exist some variations in most important clinical features between random forest (RF) and extreme gradient boosted trees (XGB) algorithms, so only features which were of high importance in every training case were chosen. The most prominent clinical features include left atrium (LA) volume, LA diameter, interventricular septum thickness (IVS), left ventricular ejection fraction (LVEF), left ventricular internal diameter end systole (LVIDs), left ventricular internal diameter end diastole (LVIDd), left ventricular end systole volume (LVESV), left ventricular end diastole volume (LVEDV), ECG rhythm shape and n-terminal pro hormone BNP (NTBNP) concentration.

It is important to note that, even though it does not play the most important role in decision making while training random forest classification models, ECG rhythm plays by far the most important role when training extreme gradient boosted trees classification models, with it being responsible for slightly more than 15% of the decision making process.

Table 3. Feature importance of clinical data

| Feature name | Feature importance | | | |
| | Gold standard labeling | | Event labeling | |
	RF	XGB	RF	XGB
LA volume	6.2%	3.9%	5.2%	3.9%
LA diameter	5.4%	5.6%	5.6%	5.6%
IVS	3.9%	3.2%	3.8%	3.2%
LVEF	3.7%	3.3%	3.4%	3.3%
LVIDs	2.3%	1.3%	2.4%	1.3%
LVIDd	2.1%	1.3%	2.1%	1.3%
NTBNP	2.3%	2.7%	2.3%	2.7%
ECG Rhythm	2%	15.1%	1.8%	15.1%
LVEDV	2%	1.6%	2.1%	1.6%
LVESV	1.9%	1.6%	2.1%	1.6%

Table 4 shows the most important genetic data features which were evaluated as having a high degree of importance for both classification model creation algorithms. From the following results it can be seen that the importance of genes varies greatly between classification algorithms, especially in the case of MYL3 and TPM1 genes. However in both cases MYBPC3 and ACTC1 genes are responsible for making the greater part of the decision during classification, making up nearly 52% of the decision making process for extreme gradient boosted trees classification and 77% of the decision making process for random forest classification.

Table 5 shows the importance of features within in the grand scheme when classification is conducted using all of the available data. The importance of most prominent features is more stable across the board when training classification algorithms with the

Table 4. Feature importance of genetic data

Feature name	Feature importance	
	RF	XGB
MYBPC3	66.9%	29.6%
ACTC1	10.3%	22%
TNNI3	6%	5.8%
TNNT2	1.7%	3%
MYL3	1%	12%
MYL2	1.6%	1.9%
TPM1	1.9%	11.3%

exception of ECG rhythm which still plays a disproportionately more important role for the decision making process of extreme gradient boosted tree classifiers. It is important to note that the most important feature importance values are lower when using the entire available dataset because of the increase in the number of features used for training as well as the more balanced role each feature plays in the decision making process.

Table 5. Feature importance of the full dataset

Feature name	Feature importance			
	Gold standard labeling		Event labeling	
	RF	XGB	RF	XGB
LA volume	3.7%	2.6%	4.3%	2.6%
LA diameter	5.1%	4%	4.4%	4%
IVS	3.2%	2.5%	3%	2.5%
LVEF	2.5%	2.7%	2.6%	2.7%
ECG rhythm	2.3%	7.8%	2.2%	7.8%
MYBPC3	2.8%	4.6%	2.8%	4.6%
NYHA	2.1%	4.6%	2.1%	4.6%
Age	2.6%	2.3%	2.7%	2.3%

The classification models which were trained using the entire available dataset achieved higher prediction accuracy than models which were trained using only clinical or only genetic data. The most notable conclusion that can be drawn from the final results are that clinical data plays a much bigger role in the decision making process than genetic data and that demographic data which was included only in the decision making process of models trained using the entire dataset also plays a big role in achieving accurate classification, most prominently patient age and New York heart association class.

4 Conclusion

Although there were multiple studies conducted on the subject of patient risk stratification for risk of suffering a sudden cardiac death caused by hypertrophic cardiomyopathy, none of those studies focus on uncovering the importance of features used for said stratification. In many of those cases satisfactory results are achieved but are not elaborated upon. When tackling problems of this nature it is important to have a degree of explainability to both further the knowledge on the subject matter and increase the likelihood of the created technology to be adopted by medical professionals and patients alike.

Our classification models for risk stratification achieved great results especially in the case when the entire feature set was used during model training. When it comes to the explainability of the models, feature importance calculation, although not the only approach, provides a deeper insight into the inner working of the developed models thereby making the utilized black box approaches more see through.

In order to further improve the presented models in the future, we plan to train new classification models, which will be trained using best combinations of patient attributes based on the discovered significance of said attributes. Additionally, an explanation module is planned to accompany the improved classification models which will make the decision making process even more understandable to both patients and medical professionals potentially increasing the degree of trust in the system.

Acknowledgements. The research was funded by the project that has received funding from the European Union's Horizon 2020 research and innovation programmes under grant agreement No 952603 (SGABU project). This paper is supported by the SILICOFCM project that has received funding from the European Union's Horizon 2020 research and innovation programme under grant agreement No 777204. This research is also supported by the Ministry of Science, Technological Development and Innovation of the Republic of Serbia, [Contract No. 451–03-47/2023–01/200378, (Institute for Information Technologies Kragujevac, University of Kragujevac)]. This article reflects only the author's view. The Commission is not responsible for any use that may be made of the information it contains.

References

1. Maron, B.J., Maron, M.S.: Hypertrophic cardiomyopathy. Lancet **381**(9862), 242–255 (2013)
2. Virmani, R., Burke, A.P., Farb, A.: Sudden Cardiac Death. Cardiovasc Pathol **10**(5), 211–218 (2001)
3. Smole, T., et al.: A machine learning-based risk stratification model for ventricular tachycardia and heart failure in hypertrophic cardiomyopathy. Comput. Biol. Med. **135**, 104648 (2021)
4. Kochav, S.M., et al.: Predicting the development of adverse cardiac events in patients with hypertrophic cardiomyopathy using machine learning. Int. J. Cardiol. **327**, 117–124 (2021)
5. João, B.A., et al.: Diagnosis and risk stratification in hypertrophic cardiomyopathy using machine learning wall thickness measurement: a comparison with human test-retest performance. Lancet Digit. Health **3**(1), 20–28 (2021)
6. Aurore, L., et al.: Distinct ECG phenotypes identified in hypertrophic cardiomyopathy using machine learning associate with arrhythmic risk markers. Front. Physiol. **9**, 213 (2018). https://doi.org/10.3389/fphys.2018.00213

7. Tse, G., et al.: Multi-modality machine learning approach for risk stratification in heart failure with left ventricular ejection fraction ≤ 45. ESC Heart Fail **7**, 3716–3725 (2020)
8. Matthia, E.L., et al.: Circulating biomarkers in hypertrophic cardiomyopathy. J. Am. Heart Assoc. **11**, e027618 (2022)

Can Haptic Actuator Be Used for Biofeedback Applications in Swimming?

Matevž Hribernik[1], Milivoj Dopsaj[2], Anton Umek[1], and Anton Kos[1]([⊠])

[1] Faculty of Electrical Engineering, University of Ljubljana, Ljubljana, Slovenia
anton.kos@fe.uni-lj.si
[2] Faculty of Sport and Physical Education, University of Belgrade, Beograd, Serbia

Abstract. Wearable devices have become indispensable tools in everyday life and sports, providing users with information and feedback using various modalities, built-in sensors and algorithms. Biomechanical systems consist of four components: users, sensors, actuators, and processing devices. In this study, the focus is on the actuators. Biomechanical feedback systems and applications can provide information to the user through different modalities, such as visual, auditory, or haptic. We have developed a feedback device that uses haptic actuators and can be used in underwater environments. Our exploratory study has shown that athletes can perceive haptic actuators and understand simple commands even when they are underwater. In this continuation of the study, we focus on the swimmers' ability to not only sense the actuators, but also to understand the information and translate it into a change in movement. 51 young swimmers tested this device in our experiments. Results show that the information from the haptic actuators can be perceived during swimming and that users can follow the commands from the haptic feedback by changing their motion. In this study, this ability was demonstrated by changing the swimming technique.

Keywords: haptic actuator · user interface · feedback · modality · vibrotactile · water sport · wearables · usability study

1 Introduction

Wearable devices are important tools for everyday life. This is especially true for recreational, but also for professional sports [1, 2]. Wearable devices can track and measure the user through built-in sensors and algorithms while connected to other, more powerful devices such as smartphones or computers. Some wearable devices can provide information to the user through built-in screens, sounds, or vibrations. These are the key building blocks of user interface: the ability to interact with the device, in this case via sensors, and the ability of the device to present information to the user using a particular modality. This basic understanding of device interaction also applies to biomechanical feedback applications [3, 4]. Biomechanical feedback systems and applications consist of four basic components: Users, Sensors, Actuators, and Processing Devices, all of which are assumed to be interconnected. The user has sensors attached to the body or interacts

© The Author(s), under exclusive license to Springer Nature Switzerland AG 2024
M. Trajanovic et al. (Eds.): ICIST 2023, LNNS 872, pp. 73–80, 2024.
https://doi.org/10.1007/978-3-031-50755-7_8

with external sensors or sensory systems. The sensors measure activity and generally output digital data. This data is processed and converted into useful information that can be rendered to the user. The information can be presented through different modalities, either visual, auditory or haptic. Visual means some sort of screen or display, auditory means sound from speakers or headphones, and haptic means vibrotactile actuators. The actuators can either be worn on the user's body or placed outside. The user receives this feedback and acts accordingly by changing their movements, completing the feedback loop.

Our previous work shows that human movement in many sports, including swimming, has been sufficiently explored [3, 5–7], and some technological challenges have also been addressed. We have learned that there is a lack of sufficient studies on real-time feedback for the athlete in aquatic sports. In this paper, we address the use of haptic feedback in swimming. In general, the use of haptic interfaces is limited in the field of biomechanical feedback, but especially in swimming. We have developed a feedback device that can be used in biomechanical feedback systems, implement haptic actuators, and be used in underwater environments. Our main motivation for developing this idea and device is the novelty of this field, especially in aquatic environments. To our knowledge, there are no similar studies that test the usability of haptic interfaces in water. This environment presents a particular challenge because electronic devices are not designed to operate underwater. Water can act as a filter for actuators and is perceived differently in water than out of it. There is also a particular challenge in that high-speed wireless communications are compromised underwater. We were also motivated by the fact that our exploratory study [8, 9] showed conclusive results that athletes can perceive haptic actuators in water or outdoors and distinguish them to some degree. Our results also show that simple commands using only one actuator at a time are much more likely to be perceived correctly by the wearer. We have also demonstrated that the haptic actuators with vibration motors can be placed either at the waist or at the head when presented to the swimmer. Here we present a continuation of this study, focusing on the swimmers' ability to not only feel the sensors, but also to understand what they feel as a command and translate that information into a change in motion.

2   Methods

The proposed device was designed using several ideas from other authors, but we could not find a similar wearable device for an aquatic environment that uses haptic actuators [5]. To the best of our knowledge, we are the first to investigate haptic interfaces in swimming. Some authors have studied auditory interfaces as feedback during swimming and other authors have used haptic actuators as feedback during walking [10, 11]. Our device has been improved over previous versions [8, 9] and is now more compact and comfortable to wear, and designed it to be worn on the head, as shown in Fig. 1. For optimal comfort, the actuators are placed inside the elastic band, and the device is positioned on top of the head, as shown in Fig. 2, and later covered with a swim cap, as shown in Fig. 4.

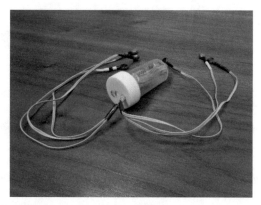

Fig. 1. Haptic interface device, water tight compartment in the center with control electronics and connected wires to the haptic actuators.

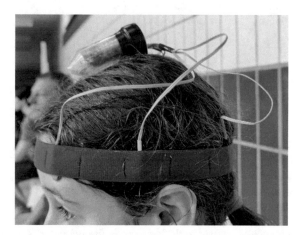

Fig. 2. Haptic device mounted on the head with haptic actuators arranged around the head.

As our previous tests indicate [8, 9], users in the water can feel this type of haptic stimuli during physical activity. We tested the idea in an exploratory study to see if the device could be used in different environments, placement and movements. We found that the haptic vibrations can be used while swimming. In this study, we focus on the usability aspects of the proposed haptic user interface. The research question for this study is whether the swimmer can understand the information presented by this device during swimming exercises. In the previous study, we only tested which actuators the users can feel and how they can distinguish them. With this study, we are testing whether users can understand these vibrations as commands and change their movements accordingly.

This usability study shows how a particular device can benefit users and tests whether that device can be used by users in a particular environment. To test the proposed device, we conducted a test with 51 young swimmers (12–23 years old). In this test, the haptic

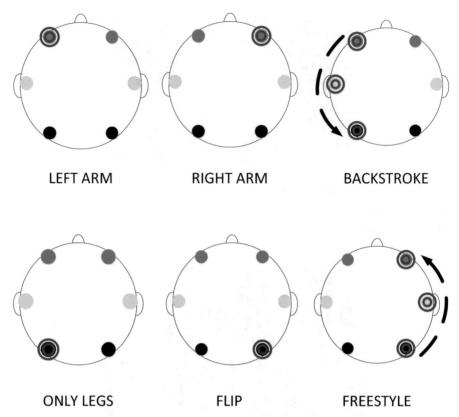

Fig. 3. All 6 symbols used in this usability study. Each one corresponds to a swimming technique, swimmers should adopt while swimming or should show while outside.

device and actuators were attached to the head under the swim cap. As shown in another study [8]. Six actuators were attached around the head, as shown in Fig. 2. We performed two tests with each participant, first outside and then in water, both during physical activity: walking outside the water and swimming in the water. During each test, participants were presented with 18 random commands from a set of 6 different symbols (actuator combinations) that had shown the greatest potential in a previous study. The symbols used are shown in Fig. 3 and can be considered logical. When one of the front actuators was triggered, it meant swimming with only one arm, when a sequence of actuators was triggered from front to back, it meant backstroke, when the sequence was from back to front, and it meant freestyle swimming. For the back triggers, they had to remember well, because the left back trigger meant swimming with legs only, and the right back trigger meant doing a flip.

Before the start of the test, all symbols were presented to the users and the swimming technique to be used when a particular symbol was active was explained. This part is called get to know session and is shown in Fig. 4. After this session, a learning session was conducted. It included 12 random symbols (each symbol two times) presented to the participants, and they performed this learning session as if they were conducting

Fig. 4. Get to know session above is designed to provide participants with an in-depth knowledge of the symbols that will be presented to them during the study. The figure below shows the test being conducted in water with two participants swimming independently and two groups of examiners.

the test outside the water while walking; the learning session was repeated as many times as necessary for the user to learn all symbols and required actions. As for the tests, first one was performed outside while walking and then another was performed in the water while swimming. Each time a symbol was triggered, the swimmer was required to change swimming technique. The symbols were not allowed to be repeated. The examiner observed the participants during the two tests and noted their technique.

3   Results

Our study produced quality results with a high success rate. Participants had to follow commands given to them by haptic actuators. The commands could be either a single haptic trigger or a sequence of triggers, which we refer to as symbols. The results are better than expected, as all symbols collectively achieved a success rate of 96.96% in all tests, with 2.34% errors and 0.71% misses. Errors are symbols that were misinterpreted as the wrong symbol, and misses are considered to be when participants did not change their movement even though they were expected to do so, i.e., they ignored the command. The success rate outside the water was 99.35%, and 94.55% inside the water.

Since all participants performed similar tests, the only difference was in the order of the symbol sequences. Table 1 shows how successful participants were in both tests. Almost half of the participants made no errors. Others made either one error (13 participants) or more, with only a single participant scoring 5 or 8 errors. On average, this translates to 1.10 ± 1.62 errors per user, with those who scored 3 or more errors being above the norm.

Table 1. Number of errors caused by participants

Number of errors	Number of participants
0	25
1	13
2	6
3 or more	7

The error rate of each symbol can also be evaluated. Figure 5 shows how all 6 symbols performed in outside and inside tests. Each symbol was triggered three times in each test, either in water or outside, for a total of 6 times for each participant. It can be seen that some symbols performed better than others. For all symbols, the error rate was very low when performed outside. The two front symbols (left arm, right arm) had no errors when the test was performed outside. The maximum number of 2 errors was recorded for the flip and backstroke symbols when triggered outside. More interesting are the results for the tests in water, where the results ranged from 0.65% (e.g., 1 error, freestyle) to 10.46% (e.g., 16 errors, flip), with freestyle performing better than all others and flip performing worst, those two also had statistically significant results below and above the normal distribution.

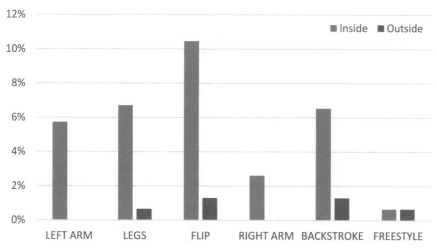

Fig. 5. Error rate according to each symbol both swimming (blue) and walking (red).

4 Discussion

The presented device and usability study presented a unique challenge to the participants. This usability study is an important step in the development of biomechanical feedback with haptic actuators, especially in challenging environments such as water and during physical activity.

The results are promising and demonstrate the success of our previous exploratory study and this usability study. Users were able to understand commands and change their movements with a very high success rate of nearly 95% in water while swimming and nearly 100% outside. This success rate is much higher than in the previous exploratory study. This was to be expected since the users were able to learn the symbols at the beginning and also the number of symbols presented was reduced, using only the most successful ones from the exploratory study here. Somewhat surprisingly, almost half of the participants made no errors. This gives hope that the success rate could be even higher if participants had more time to learn the symbols and take the test again. Four out of six symbols could be considered logical. Only flip and legs were more difficult to learn and therefore had higher symbol errors. Interestingly, backstroke and freestyle show a significant difference in errors even though they are similar symbols. There is no logical explanation for these results, except that backstroke requires significantly more movement changes (body rotation) than freestyle.

Our study shows that simple commands can be understood during swimming and swimmers can follow commands from the haptic wearable device. These important findings will enable the future development of wearable biomechanical feedback devices used in real-time applications for aquatic sports. This wearable device is currently already capable of providing haptic information to the user and communicating with the control and processing device. In the future, kinematic sensors will need to be implemented to complete the feedback loop. These future studies will combine this work and previous work investigating swimming motion [12].

References

1. Lightman, K.: Silicon gets sporty. IEEE Spectr. **53**, 48–53 (2016). https://doi.org/10.1109/MSPEC.2016.7420400
2. Aroganam, G., Manivannan, N., Harrison, D.: Review on wearable technology sensors used in consumer sport applications. Sensors **19**, 1983 (2019). https://doi.org/10.3390/s19091983
3. Kos, A., Umek, A.: Biomechanical Biofeedback Systems and Applications. Springer, Cham. https://doi.org/10.1007/978-3-319-91349-0
4. Giggins, O.M., Persson, U.M., Caulfield, B.: Biofeedback in rehabilitation. J. Neuroeng. Rehabil. **10**, 60 (2013). https://doi.org/10.1186/1743-0003-10-60
5. Hribernik, M., Umek, A., Tomažič, S., Kos, A.: Review of real-time biomechanical feedback systems in sport and rehabilitation. Sensors **22**, 3006 (2022). https://doi.org/10.3390/s22083006
6. Kos, A., Umek, A.: Wearable sensor devices for prevention and rehabilitation in healthcare: swimming exercise with real-time therapist feedback. IEEE Internet Things J. **6**, 1331–1341 (2019). https://doi.org/10.1109/JIOT.2018.2850664
7. Kos, A., Umek, A.: Reliable communication protocol for coach based augmented biofeedback applications in swimming. Procedia Comput. Sci. **174**, 351–357 (2020). https://doi.org/10.1016/j.procs.2020.06.098
8. Hribernik, M., Umek, A., Kos, A.: Zasnova haptičnega uporabniškega vmesnika za biomehansko povratno vezavo v vodnih športih. In: 31st International Electrotechnical and Computer Science Conference, ERK 2022, Portorož, pp 167–170 (2022)
9. Hribernik, M, Kos, A., Umek, A., Sodnik, J.: Haptic user interface for biofeedback in aquatic sports: a design concept. In: Proceedings of the Information Society of Serbia – ISOS, ICIST 2022, pp 193–197 (2022)
10. Chen, D.K.Y., Haller, M., Besier, T.F.: Wearable lower limb haptic feedback device for retraining foot progression angle and step width. Gait Posture **55**, 177–183 (2017)
11. Ashapkina, M.S., Alpatov, A.V., Sablina, V.A., Melnik, O.V.: Vibro-tactile portable device for home-base physical rehabilitation. In: 2021 10th Mediterranean Conference on Embedded Computing, pp. 1–4 (MECO) (2021)
12. Umek, A., Kos, A.: Wearable sensors and smart equipment for feedback in watersports. Procedia Comput. Sci. **2018**, 496–502 (2018)

Overview of Deep Learning Methods for Retinal Vessel Segmentation

Gorana Gojić[1]([⊠]) [iD], Ognjen Kundačina[1] [iD], Dragiša Mišković[1] [iD],
and Dinu Dragan[2] [iD]

[1] The Institute for Artificial Intelligence Research and Development of Serbia, Novi Sad, Serbia
{gorana.gojic,ognjen.kundacina,dragisa.miskovic}@ivi.ac.rs
[2] Faculty of Technical Sciences, University of Novi Sad, Novi Sad, Serbia
dinud@uns.ac.rs

Abstract. Methods for automated retinal vessel segmentation play an important role in the treatment and diagnosis of many eye and systemic diseases. With the fast development of deep learning methods, more and more retinal vessel segmentation methods are implemented as deep neural networks. In this paper, we provide a brief review of recent deep learning methods from highly influential journals and conferences. The review objectives are: (1) to assess the design characteristics of the latest methods, (2) to report and analyze quantitative values of performance evaluation metrics, and (3) to analyze the advantages and disadvantages of the recent solutions.

Keywords: Medical imaging · Retinal vessels · Segmentation · Deep learning · Machine learning · Overview

1 Introduction

Retinal vessel segmentation is a key step in the screening and early diagnosis of many eye and systemic diseases. Abnormal changes in the retinal vascular network may be indicators of retinopathy of prematurity [1, 2], glaucoma [3], age-macular degradation [3], and diabetic retinopathy [4]. Therefore, accurate segmentation of retinal vascularization is of great importance in screening and treatment procedures. Many imaging modalities can be used to capture retinal images, including fundus. A fundus image is a 2D projection of a 3D inner eye surface, with the retina being the top surface layer. Due to retinal semi-transparence property, the fundus image also shows non-informative anatomical structures belonging to layers below the retina often alleviating the segmentation process.

Some of the challenges in retinal blood vessel segmentation from fundus images are (1) low contrast, (2) image artifacts such as noise, blur, and uneven illumination, (3) variable blood vessel width and shape, and (4) blood vessel bifurcations and crossover points [5–7]. The aging of the world population and the increasing trend of vision impairment has resulted in an increased workload for ophthalmologists [8]. Increased workload leads to a higher probability of human error and increased risk to patient health. This

M. Trajanovic et al. (Eds.): ICIST 2023, LNNS 872, pp. 81–92, 2024.
https://doi.org/10.1007/978-3-031-50755-7_9

has motivated higher research interest in automated retinal vessel segmentation methods to facilitate ophthalmologists in decision-making through computer-aided diagnosis (CAD) systems [9]. With the appearance of machine and deep learning methods novel approaches to retinal vessel segmentation capable of learning latent features have emerged. Recent years have demonstrated a growing development trend of these deep learning methods for retinal vessel segmentation [5]. Combined with natural robustness improvement techniques [10], these methods could enter broader clinical practice.

Retinal vessel segmentation methods are constantly evolving with new advancements being proposed frequently, especially in the deep learning domain. While review papers in the field already exist [5, 6, 11–13] they provide exhaustive reviews involving many studies from a variety of journals and conferences with varying degrees of reliability and reproducibility of reported results. A new review paper includes just the most recent publications from highly-influential conferences and journals providing insights into the latest development and trends relying solely on highly relevant studies in the field. This can help identify research gaps more efficiently and direct future research. Additionally, the review can serve as a quick introduction to the latest approaches to retinal vessel segmentation for researchers entering the field, but it can also serve as an extension to some of the existing review studies.

The rest of the paper is organized as follows. In Sect. 2 we give a definition of the retinal vessel segmentation task. Section 3 gives an overview of DNN types that are commonly used in segmentation for this specific task. Follows the Sect. 4 with numerical results for commonly reported metrics in 39 papers. We provide a brief discussion in Sect. 5, and conclude in Sect. 6.

2   Retinal Vessel Segmentation

Retinal vessel segmentation involves classifying pixels in a fundus image as either blood vessel or background. A CNN produces a grayscale probability map indicating the likelihood of each pixel in a corresponding fundus image being a blood vessel. The probability map is binarized to obtain a segmentation mask for the fundus image. Binarization is performed by threshold a probability map for a predetermined pixel intensity threshold value. Pixels with intensities above the threshold are labeled as blood vessels, while those below are labeled as background.

3   Overview of Segmentation Architectures

In this section, we discuss different architectural choices for retinal vessel segmentation. We first introduce the UNet network which is a commonly used base architecture. Usually, it is modified to improve performance on the retinal vessel segmentation task by substituting existing layers with more efficient alternatives (e.g., dilated convolution) [14, 15] or by adding additional blocks (e.g., attention blocks) [16–19]. We are focusing on a specific subset of CNNs for retinal vessel segmentation. These CNNs use attention mechanisms, context information, and domain knowledge to improve segmentation. We chose these techniques because they are among the most commonly used modifications to improve UNet segmentation results.

3.1   UNet

UNet architecture has been predominantly used in medical image segmentation. It is designed to accommodate small-size medical image datasets and strongly relies on data augmentation to compensate for the limited size of medical image datasets. UNet is a fully convolutional neural network with the encoder-decoder architecture. Here, the encoder learns to transform the input image into an embedding that is then expanded by the decoder into a probability map. The encoder of the originally proposed architecture in [20] consists of five blocks, each performing convolutions and max-pooling operations to learn features of different complexity and reduce feature maps dimensionality. Each of the encoder blocks is directly connected to a corresponding decoder block by skip connections. Skip connections facilitate learning by providing higher-dimension encoder feature maps alongside lower-dimension decoder feature maps to generate decoder block output. In the originally proposed decoder blocks, max-pooling from the encoder block is replaced with up-convolutions to expand learn features into a probability map.

The majority of architectures for retinal vessel segmentation modify UNet to improve its performance. Some of the improvements include adding attention mechanism [16−19] to weight the importance of learned features, exploiting knowledge about blood vessel appearance and geometry [21−23], and using richer context information [14, 15, 24].

3.2   Domain-Knowledge-Based Approaches

The idea of incorporating domain knowledge in retinal vessel segmentation is motivated by the fact that learning CNNs in an unconstrained manner may result in learning features that are irrelevant to the segmentation task, e.g. the network might learn to extract different types of noise like morphological structures such as hemorrhages and microaneurysms. Domain-knowledge-based approaches offer a degree of performance robustness compared to approaches based solely on local image information. The performance of the latter is easily affected by the quality and quantity of training data that are critical in medical image applications. An additional challenge in retinal vessel segmentation is obtaining groundtruth masks for supervised learning, which remains the predominant approach in the field. Incorporating domain knowledge into segmentation architectures is an alternative to the attention mechanism since both approaches rely on learning a subset of features useful for segmentation. However, the former relies on information on geometry or pixel intensity, while the latter is completely domain-knowledge oblivious.

One approach to exploiting domain knowledge is by introducing retinal vessel geometry structure in segmentation architectures [21, 22]. A work in [21] proposes a geometric representation layer that incorporates domain-specific knowledge to learn curvilinear features from fundus images. Knowledge is incorporated by modifying the loss function to include two regularizers to enhance learned kernel response on diverse orientations in the range 0–180°, and to penalize the generation of false positives. Additionally, the authors propose a policy to train the network to be robust to domain noise, such as hemorrhages and micro-aneurysms, by training on image patches that have a high chance of being interpreted as false positives. In [22] graph convolutional network (GCN) [23] is used to produce an additional set of features that are used jointly with local CNN features to generate final predictions. Here, a pretrained CNN is used to obtain a probability map that is used to generate an input graph for GCN by thresholding a probability map, applying skeletonization by morphological thinning, generating graph vertices by sampling over blood vessel pixels and generating edges between vertices. While having the advantage of providing vessel structural information, this approach is highly dependent on the quality of CNN-generated probability maps. It also has limited practical applicability, due to higher time and memory demands to run the segmentation pipeline.

Historically, the first to appear domain-knowledge approaches were hand-crafted filters [25–28]. Here, we review just a single study on image matting that has been published recently. Image matting aims to extract a foreground given an image trimap, which is usually a manually annotated image having all pixels sorted out in the foreground, background, and unknown classes. Matting is formulated as a linear aggregation of a foreground and background image with coefficients k and $1-k$ respectively, and the output of the algorithm are coefficient values [29]. Since it is demanding to create trimaps, for fundus image vascularization, work in [29] proposes generating coarse trimaps automatically by preprocessing the input image and using coarse segmentation as an input to the image matting algorithm.

3.3 Context-Based Approaches

Context-based approaches rely on capturing more context information that would help generate more accurate segmentations. Here we discuss two implementations identified from the literature. The first includes strategies to adapt CNN's receptive field, as it has been shown previously that a fixed-size receptive field is not an optimal choice for multi-scale curvilinear structures as blood vessels [14, 24]. The second uses multi-scale feature extraction architectures to extract vessels of different diameters. Additionally, two implementations are often used jointly for optimal performance [14, 15, 24].

To address adaptive receptive field size, [14] dynamically selects the most appropriate receptive field based on the characteristic of the target feature map. Paper contribution is a receptive field size selection criteria based on a correlation between learned weights for every two adjacent scales of the receptive field. However, the sole idea of choosing between multiple-sized kernels has been proposed earlier in [30], showing that CNNs incorporating selective kernels perform better in capturing multi-scale objects from natural images. Dilated, or atrous convolutions [31], are used to capture more context by using a larger receptive field while keeping the same number of parameters compared to a regular convolution kernel. Dilated convolution, unlike regular convolution, introduces gaps in the receptive field. The extent of these gaps is determined by the dilation rate. When the dilation rate is set to one, dilated and regular convolution are equivalent. It has been used in multi-scale feature extraction in multiple papers on retinal vessel segmentation [14,15]. In [24] optimal choice of context is learned through reinforcement learning [32]. The proposed approach iteratively improves a given coarse segmentation mask by learning a policy to optimally sample image patches from the input image that are then segmented to assemble a segmentation mask. The policy determines patch extraction parameters, such as patch position and scale. By learning patch extraction properties instead using fixed patch extraction strategy, the approach shares the idea with [14] that learns receptive field size, but on a higher, image patch level.

The first layers of CNN learn high-resolution, detailed features based on the local context covered by receptive field size. Successive layers learn coarser, semantically meaningful features by aggregating the features learn from previous layers. In retinal vessel segmentation, CNN design results in reduced performance in thin blood vessel segmentation. To address this design choice, multi-scale implementations are proposed to handle blood vessel shape and diameter variability. Additionally, it is shown that some multi-scale architectures exhibit improved natural robustness on certain image corruptions and perturbations in natural images [33, 34], which proves usable in segmentation from fundus images having different kinds of morphological noise structures [14]. Implementations of multi-scale architectures in retinal vessel segmentation vary. In [14] each encoder layer produces three output feature maps for a given input feature map, using dilated convolution with different atrous rates to cover different context areas. The resulting maps are concatenated into a single feature map passed to the next CNN layer. The multi-level semantic supervision module reconstructs probability maps from each decoder layer, using multi-scale feature maps to generate a more accurate segmentation mask. The authors also argue that using vanilla skip connections passes many non-informative, or even harmful features that the encoder learns, like noise. Thus, they propose an additional block to process features transferred by skip connections prior to decoding. Work in [15] proposes a feature pyramid cascade module to process encoder feature maps. Multi-scale average pooling is performed to obtain three feature maps, that are further processed by successive dilated convolutions and concatenated into an output feature map. A disadvantage of the multi-scale approach is increased computational and memory complexity, which is often alleviated through dilated or octave convolutions reducing parameter number compared to regular convolutions. Another approach is to limit the use of multi-scale features, such as in [15] where just the encoder output features are processed in a multi-scale manner.

3.4 Attention-Based Approaches

The attention mechanism in deep learning is based on the assumption that not all parts of the input data are equally important to solve the target task. The concept is originally introduced in language translation to eliminate a bottleneck problem, leading to performance loss for long input sequences. To alleviate the issue, parts of the input sequence are provided directly to the decoder when generating the output. Corresponding attention weights are learned for each part of the input, telling the decoder how important it is. Attention can be global if the entire input sequence is used to calculate the attention, or local if an input subsequence is used in the calculation. While specific implementations of the attention mechanism can vary, one of the most influential works on attention in sequence-to-sequence processing is introduced in Transformer architecture that uses attention solely to generate predictions [35]. The proposed attention mechanism has been successfully adapted to the computer vision field through vision transformers (ViT) [36]. However, as ViTs have emerged recently, the majority of works still propose CNN-based solutions for retinal vessel segmentation with incorporated attention mechanism. Attention is often introduced through attention blocks that vary in design and position in a base architecture [16–19].

According to [37] there are four main attention categories in computer vision: spatial, channel, temporal, and branch attention, with the first two being used in segmentation from fundus images [16]. While spatial attention weights the importance of learned spatial features, channel attention weights the importance of in-channel image features. In [16,17] the authors propose using a single spatial attention block to selectively focus on important features learned by the encoder. To calculate attention maps, average and max pooling are used to derive intermediate maps from feature maps. Intermediate maps are then concatenated and convolved to obtain final attention maps that are used to weight encoder output before passing it to the decoder. In [17] spatial attention is used to build a dual-direction attention block that looks for inter-feature dependences in horizontal and vertical directions. The proposed attention block is implemented using average pooling and inserted in each decoder block of UNet architecture. Similarly, in [18] attention blocks precede each upsampling layer in the decoder. The attention is incorporated in a guided filter [38] that uses high-level encoder feature maps to help filter out low-level decoder feature maps. Work in [19] modifies UNet by adding two encoders, one to encode spatial, and the other to encode context information. The network uses channel attention implemented as self-attention to learn the importance of feature dependencies over channel dimension.

4   Performance Analysis

Here, we describe a baseline dataset for retinal vessel segmentation for which we collect and compare the performance of different models. We discuss metrics considered in the paper and analyze collected results.

4.1   Dataset

The DRIVE dataset is a collection of 40 color fundus images used as a baseline in retinal vessel segmentation. The images were taken with a non-mydriatic 3CCD camera and belong to adults aged 25 to 90 with developed retinal vascularization. Each image has manual binary segmentation masks for blood vessels and a field of view (FoV) mask. By providing predefined test-train split, the dataset provides a consistent benchmark for evaluating different methods in this field.

4.2   Metrics

The performance of retinal vessel segmentation is commonly evaluated using standard, pixel-wise metrics for semantic segmentation. Metrics are based on a confusion matrix containing information on truly and falsely classified pixels relative to segmentation ground truth. These include true positives (TP), true negatives (TN), false positives (FP), and false negatives (FN). Evaluation metrics such as classification accuracy, sensitivity, specificity, and F1 score are calculated based on the confusion matrix. The area under the ROC curve (AUC) is another commonly reported metric calculated by plotting the ROC curve for the confusion matrix and different thresholds and measuring the area beneath the curve.

Retinal vessel segmentation datasets, including the DRIVE dataset, exhibit a high class imbalance. This is due to a significantly larger number of pixels labeled as background, compared to those pixels labeled as blood vessels. For the DRIVE dataset, the average percentage of blood vessel pixels is 8.7% without the FoV mask and with a standard deviation of 1.07. The value is produced by calculating the average percentage of blood vessel pixels for each of the 40 images in a dataset and then averaging those values. Class imbalance affects the choice of performance metrics [39]. When calculating accuracy, a trivial segmentation model classifying all pixels as a background would on average have 91.1% accuracy, leaving a range of approximately 8.9% for inter-model comparison. The most suitable metrics should account for the presence of disbalance to avoid the strong influence of the major class classification performance in the result. Some class imbalance-aware metrics include the F1 score and Matthews Correlation Coefficient (MCC). While F1 is calculated based on TP, FN, and FP, MCC takes into account all four confusion matrix categories proportionally to the number of samples in both classes.

4.3 Numerical Results

In Fig. 1 we show distributions of results reported in 39 papers on retinal vessel segmentation grouped by metric. All papers are published in the last three years in leading conferences and journals in the computer vision and machine learning fields. Some examples include IEEE Transactions on Medical Imaging (TMI), Conference on Computer Vision and Pattern Recognition (CVPR), and Conference on Medical Image Computing and Computer Assisted Intervention (MICCAI).

All distributions are presented as boxplots with characteristic points known as lower whisker, lower quartile, median, upper quartile, and upper whisker. These points partition the whole distribution in the approximately 25%, 50%, and 75% samples respectively. Since the lower and upper whiskers in Fig. 1 do not correspond to the distribution minimum and maximum, outliers exist and are marked with dots. In Table 1 we provide additional distribution properties such as distribution sample size, mean, and standard deviation.

All metric values are in [0, 1] range. According to the observed sample, the most reported metrics are classification accuracy (28/39), sensitivity (27/39), AUC (26/39), and specificity (25/39). All metrics except sensitivity have a comparatively low inter-quartile range of 1.5% or below, meaning that 50% of reported results differ by 1.5% in metric value in case of accuracy or even less ($<1\%$) for other metrics. For accuracy, AUC, and F1 score, 99% of distribution samples are clustered in the 5% metric value range. While having a low inter-quartile range specificity has a more dispersed, right-skewed distribution with less than 25% of methods achieving less than 90% in background pixel classification. According to the comparison sample, sensitivity is the only metric having an almost symmetric and relatively dispersed distribution, not as tightly clustered around the median as is the case for accuracy, AUC, and F1. Dispersion of sensitivity distribution demonstrates differences in compared model abilities to correctly segment

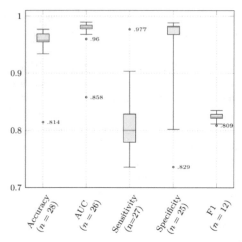

Fig. 1. Reported performance value distributions per metric with n being the number of papers in which the particular metric is reported.

Table 1. Numerical properties of metric distributions in Fig. 1 with n being a sample size, $\bar{x}$, and σ mean value and standard deviation, respectively, rounded to three digits.

	Accuracy	AUC	Sensitivity	Specificity	F1
n	28	26	27	25	12
$\bar{x}$	0.955	0.971	0.804	0.957	0.824
σ	0.029	0.034	0.050	0.064	0.007

blood vessel pixels. Alongside here discussed metrics, other evaluation metrics reported in the papers include MCC ($n = 2$), precision ($n = 2$), and intersection over the union ($n = 3$). However, due to the low number of samples, they are not considered in this paper.

In Table 2 we show results reported in the papers categorized into domain-knowledge, context-based, and attention-based approaches to gain preliminary insights into the efficiency of the approaches in retinal vessel segmentation. For each metric, we bolded out the best-reported result and underlined the second-best. Attention-based approaches yield the highest values in accuracy, AUC, and F1 score. The first three metrics take into account the classification efficiency of both background and blood vessel pixels and are not designed to account for class imbalance problems. As for the F1 score, attention-based approaches indicate superior results. However, it is not possible to derive reliable conclusions due to the small sample size. The second-best results in accuracy, AUC, and sensitivity are reported for domain-knowledge methods. Superior sensitivity results have been achieved in [22] using a combination of purely convolutional and graph convolutional networks. However, the same method reports the second-lowest result for specificity, introducing noise in the classification of background pixels. As sensitivity

Table 2. Performance metric values reported in representative papers for each covered approach. Bolded values are highest reported, the underlined ones are the second highest.

Method	Approach	Accuracy	AUC	Sensitivity	Specificity	F1
Cherukuri et al. [21]	Domain-knowledge	0.9723	0.987	0.8425	0.9849	0.822
Shin et al. [22]		0.9271	0.9802	**0.9382**	0.9255	-
Fan et al. [29]		0.96	0.858	0.736	0.981	-
Wu et al. [14]	Context-based	0.9697	0.9837	0.8289	0.9838	-
Wang et al. [24]		0.8353	-	-	0.8419	-
Wang et al. [15]		0.9681	0.9817	0.8107	0.9845	-
Guo et al. [16]	Attention-based	0.9698	0.9864	0.8212	0.984	0.8263
Li et al. [17]		**0.9769**	**0.9895**	0.8145	**0.9883**	-
Zhang et al. [18]		0.9692	0.9856	0.81	0.9848	-
Wang et al. [19]		0.827	0.9567	0.794	0.9772	**0.9816**

directly implicates the model's performance in blood vessel pixel segmentation, the analyzed sample indicates that incorporating domain knowledge into the network facilitates true positive classification.

5 Discussion

The commonly used metrics in retinal vessel segmentation include accuracy, AUC, sensitivity, and specificity. Accuracy and AUC are suboptimal choices due to insensitivity to class imbalance, while sensitivity and specificity specifically evaluate the classification performance of blood vessels and background pixels. While sensitivity can be a significant indicator in classification, specificity is less informative as a result of class imbalance, and the tendency of the methods to classify the background with low-noise levels due to the larger training sample size for the negative class. While the F1 score is designed to address the class imbalance, the observed results on a limited sample number show low dispersion around the distribution mean, resulting in very similar results between compared solutions, indicating that class imbalance is too high to properly reflect on the F1 score. Alternative general-purpose metrics, such as MCC, or specialized metrics might be more informative choices of model performance.

Considering the limited sample size for each of the retinal vessel approaches presented in this paper, domain-knowledge-based and attention-based approaches tend to yield better results compared to pure context-based approaches. While representative of domain-knowledge results utilizing graph convolutional neural networks achieves significantly higher sensitivity compared to other proposed solutions, it is computationally demanding to train and report lower specificity, generating background noise. With sufficient training resources, it could be possible to combine ideas from domain knowledge and attention-based approaches to maximize sensitivity while preserving high specificity.

6 Conclusions

In this study, we reviewed recent literature on deep learning methods for retinal vessel segmentation and identified the main conceptual CNN improvements for efficient retinal vessel segmentation. We discussed domain knowledge integration, a more comprehensive segmentation context, and attention mechanism in CNNs can enhance the segmentation performance. To support the theoretical review, we provided a numerical analysis of the results and discussed approaches in the context of reported results. Preliminary results showed that among the isolated approaches pure context-based approaches are the least efficient, while domain knowledge and attention-based approaches have similar performance in general. However, certain ideas from those groups exhibit extraordinary numerical performance and thus could be combined in the future into a more efficient solution.

While this study identifies a subset of dominant conceptual approaches to retinal vessel segmentation, more approaches might be isolated from the literature, that were not discussed in this paper due to paper length constraints. Our assumption is that combining the advantages of different approaches, more efficient and robust solutions can be designed.

Acknowledgments. This paper has received funding from the European Union's Horizon 2020 research and innovation programme under Grant Agreement number 85696. The paper has also been supported by the Ministry of Science, Technological Development, and Innovation of Republic of Serbia through project no. 451-03-47/2023-01/200156 "Innovative scientific and artistic research from the FTS domain".

References

1. Cheung, C.S., Butty, Z., Tehrani, N.N., Lam, W.C.: Computer-assisted image analysis of temporal retinal vessel width and tortuosity in retinopathy of prematurity for the assessment of disease severity and treatment outcome. JAAPOS **15**(4), 374–380 (2011)
2. Gojić, G., et al.: Deep learning methods for retinal blood vessel segmentation: evaluation on images with retinopathy of prematurity. In: Proceedings of the SISY 2020, pp. 131–136 (2020)
3. Fraz, M., et al.: Blood vessel segmentation methodologies in retinal images – a survey. Comput. Meth. Prog. Biomed. **108**(1), 407–433 (2012)

4. Li, Q., You, J., Zhang, D.: Vessel segmentation and width estimation in retinal images using multiscale production of matched filter responses. Exp. Syst. Appl. **39**(9), 7600–7610 (2012)
5. Mookiah, M.R.K., et al.: A review of machine learning methods for retinal blood vessel segmentation and artery/vein classification. Med. Image Anal. **68**, 101905 (2021)
6. Srinidhi, C.L., Aparna, P., Rajan, J.: Recent advancements in retinal vessel segmentation. J. Med. Syst. **41**(4), 1–22 (2017)
7. Petrović, V.B., et al.: Robustness of deep learning methods for ocular fundus segmentation: evaluation of blur sensitivity. Concurr. Comput. **34**(14), e6809 (2022)
8. Taylor, H.R., Keeffe, J.E.: World blindness: a 21st century perspective. Br. J. Ophthalmol. **85**(3), 261–266 (2001)
9. Hallak, J.A., Azar, D.T.: The AI revolution and how to prepare for it. Transl. Vis. Sci. Technol. **9**(2), 16 (2020)
10. Gojić, G., Vincan, V., Kundačina, O., Mišković, D., Dragan, D.: Non-adversarial robustness of deep learning methods for computer vision. In: IcETRAN, 2023
11. Khandouzi, A., Ariafar, A., Mashayekhpour, Z., Pazira, M., Baleghi, Y.: Retinal vessel segmentation, a review of classic and deep methods. Ann. Biomed. Eng. **50**(10), 1292–1314 (2022)
12. Chen, C., Chuah, J.H., Ali, R., Wang, Y.: Retinal vessel segmentation using deep learning: a review. IEEE Access **9**, 111985–112004 (2021)
13. Soomro, T.A., et al.: Deep learning models for retinal blood vessels segmentation: a review. IEEE Access **7**, 71696–71717 (2019)
14. Wu, H., Wang, W., Zhong, J., Lei, B., Wen, Z., Qin, J.: SCS-Net: a scale and context sensitive network for retinal vessel segmentation. Med. Image Anal. **70**, 102025 (2021)
15. Wang, W., Zhong, J., Huisi, Wu., Wen, Z., Qin, J.: RVSeg-Net: an efficient feature pyramid cascade network for retinal vessel segmentation. In: Martel, A.L., Abolmaesumi, P., Stoyanov, D., Mateus, D., Zuluaga, M.A., Kevin Zhou, S., Racoceanu, D., Joskowicz, L. (eds.) MICCAI 2020. LNCS, vol. 12265, pp. 796–805. Springer, Cham (2020). https://doi.org/10.1007/978-3-030-59722-1_77
16. Guo, C., Szemenyei, M., Yi, Y., Wang, W., Chen, B., Fan, C.: SA-UNet: spatial attention U-Net for retinal vessel segmentation. In: Proceedings of the ICPR 2021, pp. 1236–1242 (2021)
17. Li, K., Qi, X., Luo, Y., Yao, Z., Zhou, X., Sun, M.: Accurate retinal vessel segmentation in color fundus images via fully attention-based networks. IEEE J. Biomed. Health Inform. **25**(6), 2071–2081 (2021)
18. Zhang, S., et al.: Attention guided network for retinal image segmentation. In: Shen, D., Liu, T., Peters, T.M., Staib, L.H., Essert, C., Zhou, S., Yap, P.-T., Khan, A. (eds.) MICCAI 2019. LNCS, vol. 11764, pp. 797–805. Springer, Cham (2019). https://doi.org/10.1007/978-3-030-32239-7_88
19. Wang, Bo., Qiu, S., He, H.: Dual encoding u-net for retinal vessel segmentation. In: Shen, D., Liu, T., Peters, T.M., Staib, L.H., Essert, C., Zhou, S., Yap, P.-T., Khan, A. (eds.) MICCAI 2019. LNCS, vol. 11764, pp. 84–92. Springer, Cham (2019). https://doi.org/10.1007/978-3-030-32239-7_10
20. Ronneberger, O., Fischer, P., Brox, T.: U-Net: convolutional networks for biomedical image segmentation. In: Navab, N., Hornegger, J., Wells, W.M., Frangi, A.F. (eds.) Medical Image Computing and Computer-Assisted Intervention – MICCAI 2015: 18th International Conference, Munich, Germany, October 5–9, 2015, Proceedings, Part III, pp. 234–241. Springer, Cham (2015). https://doi.org/10.1007/978-3-319-24574-4_28
21. Cherukuri, V., Kumar B.G., V., Bala, R., Monga, V.: Deep retinal image segmentation with regularization under geometric priors. IEEE Trans. Image Process. **29**, 2552–2567 (2020)
22. Shin, S.Y., Lee, S., Yun, I.D., Lee, K.M.: Deep vessel segmentation by learning graphical connectivity. Med. Image Anal. **58**, 101556 (2019)

23. Kipf, T.N., Welling, M.: Semi-supervised classification with graph convolutional networks. In: Proceedings of the ICLR (2017)
24. Wang, F., Gu, Y., Liu, W., Yu, Y., He, S., Pan, J.: Context-aware spatio-recurrent curvilinear structure segmentation. In: Proceedings of the CVPR 2019, pp. 12 640–12 649 (2019)
25. Sheng, B., et al.: Retinal vessel segmentation using minimum spanning superpixel tree detector. IEEE Trans. Cybern. **49**(7), 2707–2719 (2019)
26. Wang, X., Jiang, X.: Enhancing retinal vessel segmentation by color fusion. In: Proceedings of the ICASSP 2017, pp. 891–895 (2017)
27. Zhang, J., Dashtbozorg, B., Bekkers, E., Pluim, J.P.W., Duits, R., ter Haar Romeny, B.M.: Robust retinal vessel segmentation via locally adaptive derivative frames in orientation scores. IEEE Trans. Med. Imaging **35**(12), 2631–2644 (2016)
28. Yin, B., et al.: Vessel extraction from non-fluorescein fundus images using orientation-aware detector. Med. Image Anal. **26**(1), 232–242 (2015)
29. Fan, Z., Lu, J., Wei, C., Huang, H., Cai, X., Chen, X.: A hierarchical image matting model for blood vessel segmentation in fundus images. IEEE Trans. Image Process. **28**(5), 2367–2377 (2019)
30. Li, X., Wang, W., Hu, X., Yang, J.: Selective kernel networks. In: Proceedings of the CVPR 2019, pp. 510–519 (2019)
31. Yu, F., Koltun, V.: Multi-scale context aggregation by dilated convolutions. In: Proceedings of the ICLR 2016 (2016)
32. Goodfellow, I., Bengio, Y., Courville, A.: Deep Learning. MIT Press (2016)
33. Ke, T.-W., Maire, M., Yu, S.X.: Multigrid neural architectures. In: Proceedings of the CVPR 2017, pp. 6665–6673 (2017)
34. Huang, G., Chen, D., Li, T., Wu, F., van der Maaten, L., Weinberger, K.: Multi-scale dense networks for resource efficient image classification. In: Proceedings of the ICLR 2018 (2018)
35. Vaswani, A., et al.: Attention is all you need. In: Proceedings of the NeurIPS, vol. 30 (2017)
36. Dosovitskiy, A., et al.: An image is worth 16×16 words: transformers for image recognition at scale. In: Proceedings of the ICLR 2021 (2021)
37. Guo, M.-H., et al.: Attention mechanisms in computer vision: a survey. Comput. Vis. Media **8**(3), 331–368 (2022)
38. He, K., Sun, J., Tang, X.: Guided image filtering. IEEE Trans. Pattern Anal. Mach. Intell. **35**(6), 1397–1409 (2012)
39. Chicco, D., Jurman, G.: The advantages of the Matthews Correlation Coefficient (MCC) over f1 score and accuracy in binary classification evaluation. BMC Genomics **21**, 1–13 (2020)

Convolutional Neural Network for Atherosclerotic Plaque Multiclass Semantic Image Segmentation in Transverse Ultrasound Images of Carotid Artery

Lazar Dašić[1,2(✉)] ⬤, Ognjen Pavić[1,2] ⬤, Andjela Blagojević[2,3] ⬤,
Tijana Geroski[2,3] ⬤, and Nenad Filipović[2,3] ⬤

[1] Institute for Information Technologies Kragujevac, University of Kragujevac,
Jovana Cvijića Bb, 34000 Kragujevac, Serbia
`lazar.dasic@kg.ac.rs`
[2] Bioengineering Research and Development Center (BioIRC), Prvoslava Stojanovića 6,
34000 Kragujevac, Serbia
[3] Faculty of Engineering, University of Kragujevac, Sestre Janjić 6, 34000 Kragujevac, Serbia

Abstract. Arterial stenosis is one of the most common diseases and if it is not discovered in time and adequately treated, it may have critical consequences, such as a debilitating stroke and even death. This is the reason why early detection is a number one priority. This disease occurs as a result of plaque deposition within the coronary vessel. The process of manually annotating plaque components is both resource and time consuming, therefore, an automatic and accurate segmentation tool is necessary. The goal of this research is to create a model that sufficiently identifies and segments atherosclerotic plaque components such as fibrous and calcified tissue and lipid core, by using Convolutional Neural Network (CNN) on transverse ultrasound imaging data of carotid artery. U-net model was trained with dataset of 60 ultrasound samples, collected and annotated by medical experts during TAXINOMISIS project, and achieved 96.94% and 57.38% Jaccard similarity coefficient (JSC) for segmentation of background and fibrous classes, respectively. On the contrary, model had difficulties with segmentation of lipid and calcified plaque components due to dataset being imbalanced and small, which is shown with respective JSC values of 19.05% and 32.68%. Future research will focus on expanding current dataset with additional annotated ultrasound samples, with the goal of improving segmentation of lipid and calcified plaque components.

Keywords: carotid atherosclerotic plaque · Convolutional neural network · Ultrasound imaging data

1 Introduction

Atherosclerotic vascular disease (AVS), or atherosclerosis, is a serious condition that affects millions of patients each year [1]. If not treated in a timely manner, it may have fatal consequences, such as an ischemic stroke. This is the reason why early detection

M. Trajanovic et al. (Eds.): ICIST 2023, LNNS 872, pp. 93–101, 2024.
https://doi.org/10.1007/978-3-031-50755-7_10

of AVS is very important. Due to the obesity, poor life choices and other factors, there is an increasing number of young individuals that are at the risk of atherosclerosis [2].

AVS is caused by significant blood flow reduction due to accumulation of atherosclerotic plaque within the coronary vessel. Over time, this plaque deposition, could lead to Transitional Ischemic Attack and stroke. Globally 1 in 4 adults over the age of 25 will have a stroke in their lifetime [3]. For a long time, it was thought that estimation of the Intima-Media Thickness (IMT) of the Common Carotid Artery (CCA) was best approach for the preclinical atherosclerosis diagnosis. In the recent studies it was shown that composition of carotid plaque has stronger association with vascular disease events [4], because stroke occurs when atherosclerotic plaques in the arteries suddenly rupture, leading to the obstruction of the blood flow to the heart or to the brain [5]. These vulnerable plaques that are prone to rapturing are usually characterized by thin fibrous cap and a large lipid area, with some calcified tissue present as well [6]. Correct identification of lipid, fibrous and calcified atherosclerotic plaque components is essential to pre-estimate the risk of cardiovascular disease, which would allow patients to be treated in a preventive and adequate manner.

Vast majority of techniques for identification of plaque components have been developed for multi-contrast magnetic resonance imaging (MRI) data. In some research, instead of MRI, nuclear imaging and multi-detector computed tomography were used for detection. Even though aforementioned imaging data give quality and high-resolution view of atherosclerotic plaque, all of them have high cost and scanning limitations that hinder their usage in daily clinical practice. In contrast, B-mode ultrasound (US) is generally used for detection of plaque in vessels due to its availability, remarkably lower cost and ease of use. These advantages are the reason why the goal of this research is to develop dependable method for identification and segmentation of the atherosclerotic plaque components, by using Deep Learning methods on US imaging data of carotid artery.

Numerous research, that differ in both methodology and used imaging data, have been conducted on the topic of plaque segmentation. On MRI images, Clarke et al. [7] achieved satisfactory results using minimum distance classifier algorithm. On the other hand, Hofman et al. [8] tested multiple supervised learning algorithms on MRI images, but every model had problem correctly classifying calcification component. Rezaei et al. [9] proposed a set of algorithms for segmentation, feature extraction, and plaque type classification. A hybrid model using k-nearest neighbor (KNN) and the fuzzy c-means (FCM) algorithm was used to accurately segment the plaque area of intravascular ultrasound images. Perhaps, the best results were achieved by Athanasiou et al. [10] who used random forest algorithm for classification of features that were extracted from Optical coherence tomography (OCT) imaging data. They attained 0.71 and 0.81 Jaccard similarity coefficient for lipid and calcification components, respectively.

Despite aforementioned advantages, US imaging data are not extensively used in research on the topic of plaque segmentation, due to their low image quality with incorporation of significant noise [11]. This is the reason why most research that use US mainly focus on Intima-Media thickness (IMT) [12, 13] and not on vulnerable plaque assessment. Zhou et al. [14] were able to successfully localize plaque segments, but were not able to classify plaque composition. Lekadir et al. [15] presented convolutional neural network (CNN) model for automatic classification of plaque composition that showed promising results. Problem with their approach is inability of CNN model to work with US image of the whole carotid wall, but with image of each plaque segment individually. This is inconvenient, because it requires manual extraction of plaque segments. Also due to the fact that their model only performs classification and not segmentation, there is a lack of clear visual representation of carotid wall plaque constitution. Nevertheless, this paper and numerous other research showed that convolutional neural networks are state-of-the-art in image segmentation.

In this paper, we describe use of a deep learning CNN (U-net) for segmentation of fibrous, lipid and calcification atherosclerotic plaque components on ultrasound images of carotid artery wall. The resulting segmented images show clear constitution of the atherosclerotic plaque, which makes process of patient risk assessment much more straightforward.

2 Materials and Methods

2.1 Dataset

First step in developing method for segmentation of the plaque components on US images is acquisition of a dataset containing original and annotated US images. Dataset used in this research was collected during TAXINOMISIS project, from Faculty of Medicine, University of Belgrade and Ethniko kai Kapodistriako Panepistimio Athinon, Greece [16]. Original dataset includes captured common carotid artery, carotid bifurcation and branches in transversal and longitudinal projections of 108 patients who went through US examination. Ultrasound examination was done in both B mode and Color doppler mode, so dataset consists of both types of images. To protect patient rights and safety, all imaging data were anonymized. From this dataset, only CCA images in transversal projections were used in the research, because they gave the clearest view of atherosclerotic plaque. Transverse projection allows visualization of the entire vessel wall as well as localization of plaques. This resulted in 67 images in total.

To ensure regularity and correctness in imaging data, data annotation was done by two independent clinical professionals. Process of data annotation has included labeling of different plaque components such as fibrous, calcified and lipid tissues. Labeling was done using "LabelMe" annotation program, which gave clear separation and distinction between different plaque components, as shown in Fig. 3c. In the case of differences in the image interpretation, medical professional would come to a mutual agreement during an in-person meeting. It should be noted that fibrous component was prevailing component among all patients included in the dataset.

2.2 Image Preprocessing

Image preprocessing is composed of steps shown in Fig. 1.

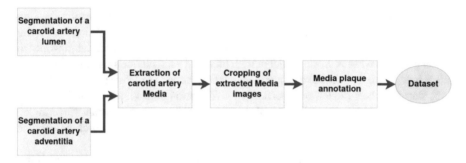

Fig. 1. Image preprocessing workflow

Since deposition of plaque takes place in carotid artery wall, it is necessary to make the wall focal point by removing tissue that is surrounding carotid artery in the imaging data. Extraction of carotid artery media was done by using two different CNN models that segment lumen and adventitia of the artery. Figure 2 depicts an ultrasound image with annotated adventitia and lumen.

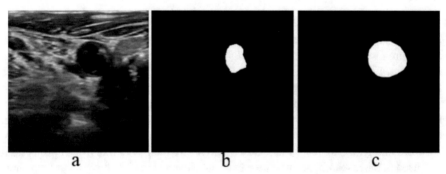

Fig. 2. Example of annotated images. (a) original ultrasound image, (b) annotated carotid lumen area, (c) annotated carotid adventitia area

Outputs of two CNN models, after successful training on aforementioned data, are segmented lumen and adventitia of carotid artery. These outputs are then combined in a way that forms carotid artery media "ring" while removing unnecessary background tissue from the image. With this removal, images were left with large amount of background pixels which causes dataset to be highly imbalanced. This is the reason why unnecessary background was cropped out, leaving only Media in the images as shown in Fig. 3b.

Process of plaque components segmentation is quite complex and time-consuming which is one of the reasons why U-net with data augmentation was used as a convolutional neural network model of choice. The dataset is split into two subsets: 90% of samples (60 images) were used as training dataset, while remaining 10% (7 images) represent the test dataset. Validation dataset was not used in this research, since its creation would

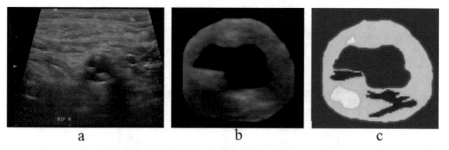

a   b   c

Fig. 3. Example of US image used in training process. (a) original ultrasound image, (b) extracted carotid artery media, (c) annotated plaque image

result in additional reduction in number of samples in the, already small, training dataset. Optimal hyperparameters for the CNN model were found in the process of trial and error.

2.3 Methods

Multiclass semantic segmentation aims to assign semantic label to every pixel in an image. The topic of this research, atherosclerotic plaque components segmentation is defined as multiclass segmentation problem, where CNN is used to detect following four classes as shown in Fig. 3c: background (part of the image outside the carotid artery media), calcified, lipid and fibrous plaque components. Input images have dimension of 256×256 pixels. Also, it was found that using images in RGB format, rather than grayscale, gave better results for segmentation problem. Pixel map for the model was labeled as follows:

- background (0) is annotated with dark blue color,
- fibrous plaque (1) is annotated with light blue color,
- lipid plaque (3) is annotated with yellow color,
- calcified plaque (4) is annotated with red color.

U-net architecture is used as plaque segmentation model as it was shown in numerous previous reports that U-net achieves great results for the segmentation problem on biomedical imaging data [17]. The U-net model consists of a contracting path to extract image features and an expanding path to upsample feature maps to their original size. U-net architecture that gave the best results using our dataset is shown in Fig. 4.

To mitigate the effects of datasets small size, different image augmentations were applied to the input imaging data at the very beginning of the model. Combination of random horizontal flip and random rescaling by 10% was used in data augmentation layer. Left branch of the model is encoder with seven blocks, where each block contains two convolutional layers with kernel of size 3×3 pixels followed by 2×2 max pooling layer. Encoder blocks use convolutional layers with 24, 64, 128, 256, 512, 768 and 768 filters respectively. On the right side is a decoder branch that is symmetric to the contracting path. In each decoder block, 2×2 upconvolution and skip connection are followed by two more convolutional layers with 3×3 filters, and the last decoder block produces the segmentation mask with 1×1 convolution and sigmoid activation

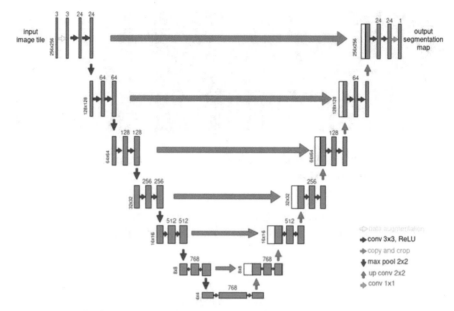

Fig. 4. U-net model used for segmentation of plaque components

function. Activation function of choice is rectified linear unit (ReLU) for each convolutional layer. Segmentation map preserves the same height and width due to the fact that all convolutional layers are padded. This results in segmentation maps of the same dimensions as the input image. Even though data augmentation prevents overfitting to some degree, every other convolutional layer was followed by dropout layer as an extra measure for overfitting problem.

Training process lasted for 100 epochs. Numerous optimizers were tested, but it was shown that results did not differ by much, so Adam was used in the final model. On the other hand, results were drastically different depending on the choice of a loss function. Some of the loss functions that are often used in segmentation tasks, like categorical cross-entropy, had some issues with this dataset due to high class imbalance. In Fig. 5, this imbalance is clearly showed by number of pixels that are part of the background or fibrous tissue class.

To try to lessen the effects of this imbalanced dataset, custom weighted loss function that combines dice loss and categorical focal loss was used. Weights were calculated according to number of pixels for each class, resulting in classes 2 and 3 (lipid and calcified plaque components) having larger weights values due to lower number of pixels belonging to these two classes. These calculated weights for each class are: [0.38 0.95 5.14 7.24].

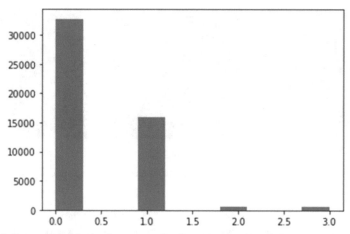

Fig. 5. Imbalance present in the dataset. y-axis shows the number of pixels present in one image sample and x-axis shows four classes present in imaging data

3   Results and Discussion

In segmentation tasks Jaccard similarity coefficient (JSC) is often used, so it was decided to use it as evaluation metric in this research as well. JSC is computed as Intersection over Union between annotated image (ground truth) and segmentation mask predicted by U-net model. Mean JSC value, as well as JSC values for each class, are shown in Table 1.

Table 1.   Class-wise and mean JSC scores

Classes	JSC score values [%]
Background class	96.94
Fibrous plaque class	57.38
Lipid plaque class	19.05
Calcified plaque class	32.68
Mean	51.52

Results in Table 1 evidently show that U-net model generally segments fibrous component correctly, but has difficulties correctly segmenting lipid and calcified plaque components. This problem occurs due to small dataset size, dataset imbalance and low quality of ultrasound imaging data. Looking at predicted segmentation mask of one ultrasound image example presented in Fig. 6c, couple of problems could be spotted. Generally, model correctly classifies type and location of every plaque component, but have trouble correctly segmenting shape and size of lipid and calcified plaque classes. Some of lipid plaque components are too large, due to the fact that surrounding fibrous tissue was wrongly classified as lipid plaque.

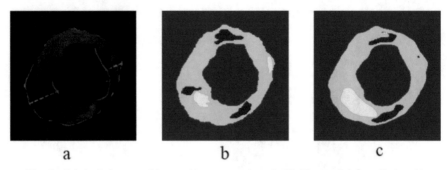

Fig. 6. Original ultrasound image (a), annotated mask (b), U-net model prediction (c)

These results are hard to compare to other methods since there isn't other research done on the topic of plaque components segmentation in ultrasound imaging data, but looking at segmented masks and JSC scores, it can be deduced that some classes, trained with current dataset, could not be segmented with the accuracy that could be used in clinical practice. However, current research set the basis for further investigation with the focus on improving segmentation of lipid and calcified components, as well as overall segmentation ability of the model. One of the steps that would undoubtably enhance segmentation capabilities of the model would be training on larger dataset.

4   Conclusions

In this research deep learning method for segmentation of different plaque components of carotid artery in ultrasound imaging data was developed. Created model, based on modified U-net architecture, was trained on 60 ultrasound samples annotated by medical professionals. The training dataset showed to be heavily imbalanced which prompted the usage of a custom weighted loss function that combines dice loss and categorical focal loss. While achieving good results for segmentation of background and fibrous classes (96.94% and 57.38%, respectively), developed model had problem correctly segmenting lipid and calcified plaque components (19.05% and 32.68%, respectively), mainly due to the heavy class imbalance and small dataset size. There is a strong indication that larger dataset would provide better segmentation results.

Future research will focus on expanding current dataset with additional annotated ultrasound samples, with the goal of improving segmentation of lipid and calcified plaque components.

Funding Statement. This paper is supported by the project that has received funding from the European Union's Horizon 2020 research and innovation programme under grant agreement No 755320 (TAXINOMISIS project). This article reflects only the author's view. The Commission is not responsible for any use that may be made of the information it contains. The research was funded by the Ministry of Science, Technological Development and Innovation of the Republic of Serbia, contract number [Agreement No. 451-03-47/2023-01/200378 (Institute for Information Technologies Kragujevac, University of Kragujevac)]. T. Geroski also acknowledges the support from L'OREAL-UNESCO awards "For Women in Science" in Serbia.

References

1. Virani, S.S., et al.: Heart disease and stroke statistics—2021 update: a report from the American Heart Association. Circulation **143**(8), e254–e743 (2021)
2. Yach, D., Hawkes, C., Gould, L., Hofman, K.J.: The global burden of chronic diseases: overcoming impediments to prevention and control. JAMA **291**(21), 2616–2622 (2004)
3. GBD 2016 Lifetime Risk of Stroke Collaborators: Global, regional, and country-specific lifetime risks of stroke, 1990 and 2016. N. Engl. J. Med. **379**(25), 2429–2437 (2018)
4. Vancraeynest, D., Pasquet, A., Roelants, V., Gerber, B.L., Vanoverschelde, J.-L.J.: Imaging the vulnerable plaque. J. Am. Coll. Cardiol. **57**(20), 1961–1979 (2011)
5. Petty, G.W., Brown, R.D., Jr., Whisnant, J.P., Sicks, J.D., O'Fallon, W.M., Wiebers, D.O.: Ischemic stroke subtypes: a population-based study of incidence and risk factors. Stroke **30**(12), 2513–2516 (1999)
6. Kwee, R.M.: Systematic review on the association between calcification in carotid plaques and clinical ischemic symptoms. J. Vasc. Surg. **51**(4), 1015–1025 (2010)
7. Clarke, S.E., Hammond, R.R., Mitchell, J.R., Rutt, B.K.: Quantitative assessment of carotid plaque composition using multicontrast MRI and registered histology. Mag. Reson. Med. Off. J. Int. Soc. Magn. Reson. Med. **50**(6), 1199–1208 (2003)
8. Hofman, J.M.A., et al.: Quantification of atherosclerotic plaque components using in vivo MRI and supervised classifiers. Magn. Reson. Med. **55**(4), 790–799 (2006)
9. Rezaei, Z., Selamat, A., Taki, A., Rahim, M.M.S., Kadir, M.R.A.: Automatic plaque segmentation based on hybrid fuzzy clustering and k nearest neighborhood using virtual histology intravascular ultrasound images. Appl. Soft Comput. **53**, 380–395 (2017)
10. Athanasiou, L.S., Bourantas, C.V., Rigas, G., Sakellarios, A.I., Exarchos, T.P., Siogkas, P.K.: Methodology for fully automated segmentation and plaque characterization in intracoronary optical coherence tomography images. J. Biomed. Opt. **19**(2), 026009 (2014)
11. Hashimoto, B.E.: Pitfalls in carotid ultrasound diagnosis. Ultrasound Clin. **6**(4), 463–476 (2011)
12. Loizou, C.P., Pattichis, C.S., Pantziaris, M., Tyllis, T., Nicolaides, A.: Snakes based segmentation of the common carotid artery intima media. Med. Biol. Eng. Comput. **45**(1), 35–49 (2007)
13. Golemati, S., Stoitsis, J., Sifrakis, E.G., Balkizas, T., Nikita, K.S.: Using the Hough transform to segment ultrasound images of longitudinal and transverse sections of the carotid artery. Ultrasound Med. Biol. **33**(12), 1918–1932 (2007)
14. Zhou, R., et al.: Deep learning-based carotid plaque segmentation from B-mode ultrasound images. Ultrasound Med. Biol. **47**(9), 2723–2733 (2021)
15. Lekadir, K., et al.: A convolutional neural network for automatic characterization of plaque composition in carotid ultrasound. IEEE J. Biomed. Health Inf. **21**(1), 48–55 (2017)
16. TAXINOMISIS project. https://taxinomisis-project.eu/. Accessed 23 May 2023
17. Ronneberger, O., Fischer, P., Brox, T.: U-Net: convolutional networks for biomedical image segmentation. In: International Conference on Medical Image Computing and Computer-Assisted Intervention, pp. 234–241 (2015)

Applied Artificial Intelligence

Using AutoML for AI Service Deployment

Vladimir Mitrović$^{(\boxtimes)}$, Milan Zdravković , and Dragan Mišić

Faculty of Mechanical Engineering, University of Niš, Niš, Serbia
`mitrovic.vladimir.mv@gmail.com`

Abstract. AutoML frameworks are making the life easier for users in demand for machine learning models, but without prior knowledge how to choose and train one. There are many available open-source frameworks with different working principles and performances. From the point of view of an average user, selecting and using the right framework might not be a trivial task. We elaborate on this issue and propose our solution for it. We present an idea of AutoML service that offers a selection of many open-source AutoML frameworks and provides two specific features, (1) Framework Recommendation System, which recommends suitable framework based on user's dataset, machine learning task and time budget, and (2) Feedback Data Base, which stores data about use cases of the service. We explore how the data from Feedback Data Base can be used to further improve Framework Recommendation System, as well as the potential benefits that AutoML researchers and developers can have from this data. The features of the service are outlined in a list of requirements that will be followed in the development process.

Keywords: AutoML · web service · AutoML frameworks · benchmark

1 Introduction

Automated Machine Learning (AutoML) is a concept that gained much attention over the years. Typical AutoML tool is aimed at automating tasks such as data preprocessing, feature engineering, selection of Machine Learning (ML) algorithm(s) and tuning of its hyperparameters. The capabilities of AutoML tools come with several commodities. The common one is the lack of need for expertise in ML or data science in order to obtain a quality trained model. On the other hand, even the ML experts can benefit. Example would be a fast acquiring of a working model to see the potential of the data for the problem at hand, without investing too much development time.

To this day, a variety of AutoML tools emerged differing in implementation, features and easiness to use. The most popular tools can be roughly categorized into two groups. The first group is consisted of open-source frameworks including, among others, AutoGluon [1], AutoSklearn (versions 1.0 [2] and 2.0 [3]), FLAML [4], H2O AutoML [5], LightAutoML [6] and TPOT [7]. These frameworks come as Python packages and are built around other popular machine learning packages (e.g., scikit-learn, XGBoost, LightGBM). The selection and tuning of ML models for given task and dataset are automatically performed within framework. Still, depending on the framework capabilities,

M. Trajanovic et al. (Eds.): ICIST 2023, LNNS 872, pp. 105–116, 2024.
https://doi.org/10.1007/978-3-031-50755-7_11

some skills in data preprocesing and/or coding are needed, for example, in converting all data to numerical form or dealing with missing values. The other group of AutoML tools includes commercialized cloud-based services that come from industry leaders like Microsoft's "Azure ML" [8], Google's "Vertex AI" [9], and "Sage-Maker Autopilot" [10] from Amazon Web Services. These services eliminate the need for programming, offer graphic user interface and even further simplify execution of other tasks such as data exploration or model deployment.

With so many options at hand, the question one might ask is how to choose the right AutoML tool for particular task. In fact, for the reasons we will discuss, this can be quite a challenge. The aim of this paper is to provide users a solution to this question. We propose the idea of AutoML service that provides possibility to use any of the available open-source frameworks. Aside from the usual features that are expected from the service of this type (e.g., graphic user interface, data exploration, model deployment etc.), our solution comes with specific features that complement the idea of helping the users to choose the right framework. Namely, presented features are Framework Recommendation System and Feedback Data Base. In addition, benefits for the AutoML researchers and developers are also explored.

The remainder of the paper is organized as follows. Section 2 covers the challenges of choosing the right AutoML framework and gives a brief overview of the proposed service. In Sect. 3 we present specific features that provide solution to the prior question and describe possible benefits to users, researchers and developers. Section 4 discusses other specifics of the service such as model deployment and implementation of frameworks. We give conclusions, overview of current progress and remarks about future work in Sect. 5.

2   The Challenge of Choosing AutoML Framework

The User REQuirement (UREQ) of an average user for any AutoML tool may be expressed in an input/output form as:

UREQ1. For the given dataset, ML task and time budget, provide the ML model with the best possible performance, in terms of a) selected accuracy metrics and b) computing cost.

While the aspect of computing cost is not considered in this paper, it is a crucial issue (especially with the upcoming interest for re-training Large Language Models) that will be taken into account in the future research.

Off course, the performance can be quantified with appropriate metric depending on the type of ML task (e.g., classification or regression) and time budget is not the only possible constraint type of input, but it is probably the one that is addressed the most. AutoML frameworks are state-of-the-art software tools in terms of implemented design and robustness, and for certain they are all up to the task to answer this requirement. Nevertheless, it is expected that they showcase performance differences in certain use cases when compared to each other. The user requirement given in UREQ1 is the basis for AutoML benchmark studies that perform head-to-head comparison of these frameworks under the same combination of inputs (datasets, tasks, time budgets). While it makes sense for these studies to serve as a guidance to AutoML users, in this section we will discuss why choosing framework is still not an easy decision and what can be done to improve the situation for the users.

2.1 Benchmark of AutoML Frameworks

If an average user on a quest to select AutoML framework seeks guidance from benchmark studies, a requirement in a similar input/output form can be conceived as:

UREQ2. For the given dataset, ML task and time budget, provide the best AutoML framework for fulfilling UREQ1.

To illustrate the struggle users might experience with finding answer to UREQ2 from benchmark studies, the papers given in [11–13] and [14] are discussed in this context. The first thing that can be noticed is the presentation form of the benchmark results. Common ways to present the results are, among others, box plots of frameworks' performances over all used datasets [11–14], averaged head-to-head comparison [13] or scatter plots of head-to-head comparison for each used dataset [11, 13]. From the point of UREQ2, these methods don't provide information on individual performances of frameworks per dataset and time budget. To address this issue, authors in [12] subdivided the results into pairs of groups based on single threshold values of five dataset characteristics (categorical proportion, feature dimensionality, class imbalance, missing proportion and sample size). More sophisticated approach was taken in [14] where the authors implemented Bradley-Terry trees [15]. The tree structure is built by splitting benchmark results into groups by selecting a dataset characteristic and its threshold value that introduces the statistically significant differences in frameworks' performances. From all the papers presented here, this approach is the closest one to fulfilling the UREQ2. In addition, authors in [13] and [14] provided extensive tabulated results of frameworks' performances per dataset, task and time budget combination. Although this data is very relevant to UREQ2, the downsides are large table sizes and separate table for datasets' characteristics. This arguably makes it more difficult and time consuming for users to manually analyse and decide which framework might be a good option for them.

The other issue found in the benchmark studies comes from the fact that AutoML frameworks have configurable parameters that have effect on performance of the final trained model. The approaches are not consistent in this aspect, although discussed papers mostly base the benchmark around default configurations. If time budget and performance metrics are ruled out, most of the changed parameters include hardware resources such as memory and number of CPUs [11–14]. Some framework specific options are tuned in [11], while authors in [14] also consulted framework developers to provide configuration modes for their respected framework, which better suit obtained performance or lower complexity of the model. So, if the user deducted from the benchmark which framework to use, additional attention has to be paid that the framework decision is tied to its particular configuration. Furthermore, benchmark studies are performed on different hardware, and frameworks' configurations and performances are also tied to this resource constraints. Since it is not expected for user to use the exact same hardware, the output of UREQ2 provided by the benchmark might not be relevant to the environment in which the user wants to use the framework.

Issues covered in this section can be summarized as standardization problems. This includes the methods for presenting results, framework configurations, time budgets and used hardware, but also the fact that the framework set used for benchmark is not consistent. The lack of standard, as explained, makes it more difficult for users to make the right choice. Standardization in AutoML benchmark is covered in great detail by authors in [14], which set them out to develop their own benchmark framework. The proposed solution in this paper builds on top of that idea, with even more user-oriented approach. With the UREQ2 constantly on our minds, we introduce our AutoML service, what problems it addresses and how do we see its benefits.

2.2 Our AutoML Service

To better explain the intention, in Table 1 we provide a list of requirements on which the service should be built. The list is not limited only by the stated requirements, but it does contain the ones that are the most important for our solution.

Table 1. List of Service REQuirements (SREQ) for the proposed AutoML service.

Req. No	Description
SREQ1	The service has to offer a choice between a number of different open-source AutoML frameworks
SREQ2	The service has to be expandable with additional frameworks
SREQ3	The service has to have the recommendation system able to suggest suitable framework for the given dataset, ML task and time budget
SREQ4	The recommendation system has to take into account only previously determined configurations of frameworks
SREQ5	Model training for recommendation system and users' tasks have to be performed on the same hardware
SREQ6	The user has to have an option to accept recommendation by the service or to freely choose any other framework
SREQ7	The user has to have an option to manually configure selected framework, to extent allowed by possible limitations of the service
SREQ8	The service has to have options for model deployment
SREQ9	The service has to have additional features common to similar services, such as graphic user interface, data exploration, data preprocessing etc

SREQ1 and SREQ2 make the ground base for the proposed service. SREQ3 is directly tied to UREQ2. This requirement is covered with Framework Recommendation System (FRS), which is one of the two special features of the service we will discuss. SREQ4 and SREQ5 ensure that the user doesn't fall into configurations and hardware pitfalls we pointed out while discussing AutoML benchmark. Requirements SREQ3-5 are covered in Sect. 3. SREQ6 and SREQ7 provide the user with options to use frameworks as freely as possible, according to individual preferences. SREQ8 and SREQ9 are requirements expected from this type of service. This is more discussed in Sect. 4, especially from the perspective of deployment options. Block diagram of the service showing user interactions and internal features is given in Fig. 1.

The second special feature of the service is Feedback Data Base (FDB) that stores information about user tasks performed on the service. As it will be discussed, its main purpose is to provide data for continuous improvement of FRS. We will also explore the benchmarking options of the service and the potential benefit that AutoML developers might find from FDB. These two features, FRS and FDB, are intended to be essential to our service, distinguishing it from the others. They do answer to UREQ2 by providing users the help in selecting AutoML framework. Therefore, FRS and FDB are covered in detail in the following section. We also make a note that as a service of this kind, other common commodities are offered such as easy-to-use interface with no need to code anything. This allows users with various technical background and broad range of skills to easily exploit the service.

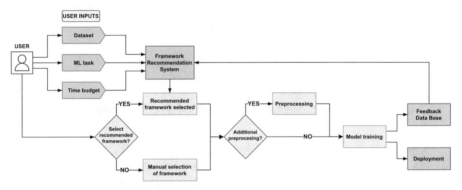

Fig. 1. Functional block diagram of the proposed AutoML service.

The flow of tasks in using the AutoML service is described as follows. The user uploads the dataset and provides other inputs such as ML task (with target features) and time budget. These inputs are fed to FRS, which outputs the recommendation. The user has the option to accept recommendation or to manually select preferred framework. Additional data preprocessing takes place, if needed. Finally, the trained model is ready for deployment and the information about service usage is stored in FDB.

3   Service Design Essentials

In this section the focus is on presenting novel features of the proposed service that introduce the added value to the user. Namely, the Framework Recommendation System (FRS) and Feedback Data Base (FDB) are covered.

3.1   Framework Recommendation System

The task of FRS is to recommend suitable framework for new data set based on its characteristics, machine learning task and time budget (SREQ3, UREQ2). At its core,

FRS can be consisted of several machine learning models, each one dedicated to the single type of machine learning task supported by the service. For example, classification and regression tasks may have their own independent machine learning models within FRS. We dub the model within FRS as the Model for Recommending Framework (MRF). To illustrate the working principle of MRF, we describe its building process. The steps of MRF building process include obtaining the benchmark training data followed with training of machine learning model. The functional block diagram of FRS is given in Fig. 2.

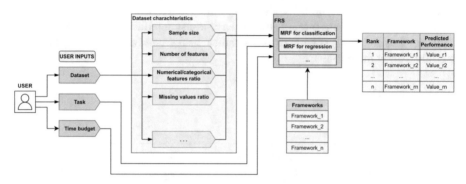

Fig. 2. Functional block diagram of Framework Recommendation System (FRS).

Characteristics of user's dataset are determined and provided to FRS, along with ML task and time budget. Within FRS, appropriate Model for Recommending Framework (MRF) is selected depending on ML task. For each implemented framework on the service, MRF predicts its performance based on dataset characteristics and time budget. The output of FRS is a ranked list of frameworks according to predicted performance.

Benchmark Training Data. The training data represents the benchmark results of frameworks on the variety of datasets and time budgets. The datasets can be selected from the OpenML platform, as this is the widely adopted approach in other benchmark studies [11–14]. For each dataset we keep a record of characteristics such as sample size, number of features, numerical/categorical ratio and ratio of missing values. The chosen characteristics are the ones that are also commonly addressed in other benchmark studies, for example in [12–14], but additional characteristic may be included. Next, we select values for time budgets to be used in benchmark. Values used for time budget in other studies range from 15 min to 4 h [12–14]. As noted in [12], the upper bound of this range provides enough time for the frameworks' performances to converge, so we follow this guideline, but we also note that bigger time budgets can be included in case of a need (e.g., for very large datasets or frameworks failing to find solutions). Finally, for all implemented frameworks on the service, we perform benchmark for each combination of dataset and time budget, in order to determine chosen performance metric. Single combination of framework, dataset and time budget can be run multiple times to obtain aggregated result. We make a note that entire benchmark should be performed on the same service hardware (SREQ5). If we denote with N, M and T the number of datasets, frameworks and time budgets respectively, then the final size of benchmark training data is N*M*T samples.

When it comes to the performance metrics, we are in favour of using metrics that give normalized values (e.g., in (0, 1) range), as they provide advantage in terms of comparing and ranking frameworks. In case of classification, a number of metrics can be used like accuracy, precision, recall, area under the ROC curve etc. In case of regression, some of the options are coefficient of determination (R2) or normalized MAE, MSE or RMSE.

An important step before performing benchmark on actual datasets is to select default configurations for frameworks (SREQ4). We discuss this more in Sect. 4 when we talk about choosing frameworks for the service. For the moment, we make a note that for the benchmark, single framework can be used in more than one configuration. This may cover the possibility that frameworks' performances might be boosted on certain datasets by having some parameters specifically configured.

Training a Machine Learning Model. For this step, first we consider whether the model should solve classification or regression task. In case of classification, one way to do this is to label a framework that performed the best for each dataset used in training. While this may work fine for training data, the problem may arise in case of incremental training using data from FDB. The FDB stores information about users' individual model training on the service. For obvious reasons, it is not expected that users perform training on their dataset with all frameworks. Labelling the best framework for some dataset without having the results for other frameworks doesn't seem like a fair approach. This imbalance of FDB along with its other capabilities will be discussed later in more details.

On the other hand, the MRFs can be made as regression models. With inputs being the dataset characteristics, time budget and particular framework, while the output being the predicted value of performance metric for that framework. This is depicted on diagram in Fig. 2. By varying framework input to the MRF model, while keeping the dataset characteristics and time budget the same, the MRF returns performance predictions for all implemented frameworks. Therefore, a framework with the best performance prediction or a ranked list of frameworks can be presented to the user as a final recommendation output of the FRS. The positive side of MRF as regression model is that we do not need to explicitly give labels, instead, individual framework performance is predicted. We believe that with this approach the mentioned imbalance issue of FDB is handled in more acceptable manner. For this reason, we are in favour of regression MRFs.

The model itself, regardless of the ML task it solves, can be any that supports incremental training to avoid unnecessary and potentially time-consuming retraining on combined data from benchmark and FDB. Short inference times are also desirable, in order for users to have the best possible experience on the service. Several different algorithms will be cross-compared and the best would be selected. We make an interesting note here that the MRF can also be obtained by using some AutoML framework, which would be in a spirit of the service itself.

3.2 Feedback Data Base

As the service is being exploited by the users, the idea is to store any relevant information about individual usages in a dedicated publicly available data base. For example, when the user starts the training with the uploaded dataset, selected framework and time budget, upon completion, following information is stored as an entry in the FDB: dataset characteristics, type of machine learning task, used framework, allocated time budget, achieved performance metrics and the final trained model(s) or pipeline(s). We believe that the data from FDB can be beneficial to the users, AutoML developers and researchers, and here we present several use cases, but also give remarks about possible limitations. The FDB will be accessible through an API call. Therefore, it is important to mention that no sensible information is going to be stored, including whole datasets, specific dataset entries and any users' private information.

Further Improvement of FRS. Most of the features of data in FDB are also features of benchmark training data used to build MRFs within FRS. This means that as the service is being used, the additional data stored in FDB can be used for incremental training of MRFs. The benefit from this feature is directly tied to the users, as they can be provided with better framework recommendations and consequentially, a model with better quality could be trained. As previously stated, the MRF itself would have to support incremental training, to avoid retraining on combined data from benchmark and FDB. It is important to note that the data from FDB can become quite imbalanced over time. If we make assumptions that users' datasets can be quite different and that only one framework is used per dataset, it is clear that the data from FDB might not be as balanced as the training data acquired from benchmark, which has entries for each combination of dataset, framework and time budget. Therefore, the gained quality from incremental training with such data can be questionable. Imbalance in FDB can come from other reasons. For example, users might favour one or several frameworks instead of others. This issue can be dealt with in a few ways. For example, entries from FDB could be filtered or users could be asked to participate in FRS improvement by using all frameworks with their datasets. Regardless of imbalance level, only use cases with default framework configurations can be used for improvement of FRS, as it was initially built on benchmark training data obtained with default framework configurations (SREQ4).

Conducting Studies. In the previous section, we briefly covered papers dealing with benchmarking AutoML frameworks. Benchmarking, as well as variety of other analytics can be conducted using the FDB and the service itself. This is something that researchers of the AutoML or other areas could find helpful. When it specifically comes to benchmarking, the problem with potential imbalance of the FDB also applies here. On the other hand, if the researcher wants to perform benchmark on their own datasets, then it is easily selectable which frameworks (with configurations, SREQ4) and time budgets to use, thus eliminating potential imbalance. Once again, we make a note that all training is done on the same hardware (SREQ5).

Feedback to AutoML Developers. AutoML frameworks are sophisticated software tools that implement complex algorithms, and there is no doubt that a lot of thought, development and testing time was invested in making them a quality product as they are. On the other hand, it is often a case when developing a product that the use cases in production can quite differ from the ones tested in the development phase. This might also be the case with AutoML frameworks. The FDB can provide feedback information on how are the frameworks used by users, what are the characteristics of datasets, what performance is achieved and which models or pipelines are trained. We hope that this possibility can find a way to be of use to AutoML developers, potentially leading to even further improvement of the frameworks.

4 Other Service Specifics

In this section, we cover additional features mentioned in the context of the service requirements in Table 1. Specifically, we cover model deployment (SREQ8), selection of frameworks (SREQ1) and data preprocessing (SREQ9).

4.1 Model Deployment

When it comes to deployment or using the trained model in general, we plan to provide several options to the users (SREQ8). The most versatile option is to make an API for each trained model, which can be used to access the model outside of the service. We make a note here that allowable data size and throughput have to be tested, as this makes room for some possible limitations. Other option is to use model directly on the service. In that case, the user selects the trained model and enters or uploads new data in order to make predictions. The third option is to allow users to download a binary file of the model, which can be imported into users' own applications. The trained models from frameworks surely can be stored this way on the service, so it is convenient to provide a download link to the users. Unfortunately, for users to be able to properly load the model outside of the service, the same environment has to be recreated as the one in which the model was created on the service. Specifically, this addresses the version of Python and used packages (e.g., framework package, scikit-learn, Numpy etc.). Therefore, this information has to be clearly provided to the users.

4.2 Implementing AutoML Frameworks on the Service

In the first section of this paper, we mentioned a number of currently existing AutoML frameworks. In general, all of them can be implemented on the service, as they produce a trained model which can be saved and used later. Still, some considerations need to be taken into account. Final step in any machine learning task is to deploy a model in production. Ideally, a production model, aside from being able to make reliable predictions, has some other characteristics, such as light size and short inference times. These characteristics can be an issue. For example, some of the frameworks are designed to make an ensemble of machine learning models, with some of the models possibly being ensembles themselves (e.g., ensemble of tree algorithms). Executing such framework with bigger time budget can produce a model of large size, which implies the need for bigger storage space, the slower loading of model and the longer inference times. One way to deal with this is to find a configuration of the framework in which the model is more optimized for deployment, but not crippled in prediction capabilities.

When it comes to configurations of frameworks, care needs to be taken in choosing the right ones to use as default on the service (SREQ4). The default configurations will also be used for benchmarking datasets in order to build FRS. A good practice in finding a suitable default configuration is to consult the AutoML developers themselves. This is in accordance with approach taken by the authors in [14] which was implemented in creating their benchmark framework. In addition, AutoML frameworks will be analysed in a manner described above and tested with different options.

4.3 Data Preprocessing

Unfortunately, not all framework implementations are capable of using the raw data without any preprocessing. Some common preprocessing steps include labelling categorical features, converting categorical values to numerical and removing/imputing missing values. From the perspective of the service, this is something that has to be automatically addressed, not only during the model training, but also at deployment time. If some framework requires preprocessing raw data, this step has to be included next to its configuration used for benchmark and building FRS (SREQ4). Users need to be aware of the used preprocessing, especially if they are to download the trained model and use it outside of the service (SREQ8).

5   Conclusions and Future Work

To the best of our knowledge and research, there exists no AutoML service similar to the one presented in this paper. The novel idea behind our service is to provide users easy and fast way to select the most appropriate framework based on the given dataset, machine learning task and time budget. In addition, the users can use the selected framework to train the model without the specific need for machine learning or programming skills. We discussed why choosing among the many framework options is still not an easy task for an average user, even with the help of existing benchmark studies. For that reason, the service requirements given in Table 1 are oriented toward user requirement that arise when choosing which AutoML framework to use for their particular task. The specific features of the service that correspond to these requirements are Framework Recommendation System and Feedback Data Base. We provide discussion about their purpose, implementation and limitations. Other features of the service are also covered, such as model deployment and data preprocessing. These features are commonly found on other AutoML services, but in this situation they come with certain special remarks.

There is no doubt that many limitations will be faced during the development of the service. In this paper we addressed only some of the specific limitations that are introduced by service concept, framework implementations and server capabilities. It is expected that more limitations will emerge as the progress evolves.

The required steps that have to be taken in service development include selecting frameworks with configurations, selecting benchmark datasets, performing benchmark, training Models for Recommending Frameworks, setting up a Feedback Data Base and, in general, programming the service itself and its features according to discussed service requirements. To demonstrate the working principle of the presented service, we provide a prototype of the service, which is currently in a proof-of-concept phase, showing that the frameworks can be used in intended manner. The project's source code can be found at https://github.com/mitvlada/automl_service. Here, we give a brief description of the progress. Several frameworks have been implemented (SREQ1). The user can upload the dataset and select task (binary classification or regression), time budget and target feature. Recommendations from FRS are displayed (SREQ3). The MRFs within FRS are implemented with simulated data, actual benchmark still needs to be performed. The user can select recommended framework or choose any other at will (SREQ6). The trained model is saved and can be either loaded on the service for making predictions on new data or downloaded to local disk (SREQ8). In general, frameworks follow the similar implementation pattern on the service, implying possibility for future expansions (SREQ2). The user interactions are carried out within graphic interface, and some pre-processing is done automatically (SREQ9). In the future work, we plan to build on this concept and implement all of the mentioned steps and required features of the service.

References

1. Erickson, N., et al.: AutoGluon-Tabular: robust and accurate AutoML for structured data. arXiv preprint arXiv:2003.06505 (2020)
2. Feurer, M., Klein, A., Eggensperger, K., Springenberg, J., Blum, M., Hutter, F.: Efficient and robust automated machine learning. In: 28th Advances in Neural Information Processing Systems, pp. 2962–2970 (2015)
3. Feurer, M., Eggensperger, K., Falkner, S., Lindauer, M., Hutter, F.: Auto-Sklearn 2.0: hands-free AutoML via meta-learning. J. Mach. Learn. Res. **23**(1), 11936–11996 (2022)
4. Wang, C., Wu, Q., Weimer, M., Zhu, E.: FLAML: a fast and lightweight AutoML library. In: 3rd Proceedings of Machine Learning and Systems, pp. 434–447 (2021)
5. LeDell, E., Poirier, S.: H2O AutoML: scalable automatic machine learning. In: 7th Proceedings on ICML Workshop on Automated Machine learning (2020)
6. Vakhrushev, A., Ryzhkov, A., Savchenko, M., Simakov, D., Damdinov, R., Tuzhilin, A.: LightAutoML: AutoML solution for a large financial services ecosystem. arXiv preprint arXiv:2109.01528 (2021)
7. Olson, R.S., Moore, J.H.: TPOT: a tree-based pipeline optimization tool for automating machine learning. In: Hutter, F., Kotthoff, L., Vanschoren, J. (eds.) Automated Machine Learning. TSSCML, pp. 151–160. Springer, Cham (2019). https://doi.org/10.1007/978-3-030-05318-5_8
8. https://azure.microsoft.com/en-us/products/machine-learning/automatedml. Accessed 20 May 2023
9. https://cloud.google.com/vertex-ai. Accessed 20 May 2023
10. https://aws.amazon.com/sagemaker/autopilot/. Accessed 20 May 2023
11. Balaji, A., Allen, A.: Benchmarking automatic machine learning frameworks. arXiv preprint arXiv:1808.06492 (2018)
12. Truong, A., Walters, A., Goodsitt, J., Hines, K., Bruss, C.B., Farivar, R.: Towards automated machine learning: evaluation and comparison of AutoML approaches and tools. In: IEEE 31st International Conference on Tools with Artificial Intelligence (ICTAI), pp. 1471–1479 (2019)

13. Zöller, M.A., Huber, M.F.: Benchmark and survey of automated machine learning frameworks. J. Artif. Intell. Res. **70**, 409–472 (2021)
14. Gijsbers, P., et al.: AMLB: an AutoML benchmark. arXiv preprint arXiv:2207.12560 (2022)
15. Eugster, M.J., Leisch, F., Strobl, C.: (Psycho-) Analysis of benchmark experiments: a formal framework for investigating the relationship between data sets and learning Algorithms. Comput. Stat. Data Anal. **71**, 986–1000 (2014)

Comparison of Deep Learning Algorithms for Facial Keypoints Detection

Matija Dodović[(✉)] [ID], Maja Vukasović [ID], and Dražen Drašković [ID]

School of Electrical Engineering, University of Belgrade, Belgrade, Serbia
{matija.dodovic,maja.vukasovic,drazen.draskovic}@etf.bg.ac.rs

Abstract. A study on facial keypoints recognition using five deep learning models is presented in this paper, focusing on emphasizing the significance of striking the right balance between model complexity and training data volume. Three convolutional neural networks - LeNet-5, AlexNet, and VGG-16 - are also included in the models that are being presented, along with two custom architectures. The first architecture has a very simple design, while the second one has a regular depth. The study utilizes a small publicly available dataset and focuses on the relationship between model complexity and accuracy in facial keypoints detection. While the more complex custom Neural Network model achieved high accuracy in facial keypoints identification, the simpler custom Neural Network model could not achieve high performance.

Keywords: Deep Learning · CNN · Facial Keypoints

1 Introduction

Facial keypoints detection is a crucial aspect of computer vision and has various applications, including facial recognition, emotion recognition, and face tracking. In facial recognition, keypoints are used to identify and match faces, making it possible to verify a person's identity or to find similarities between faces. In emotion recognition, keypoints are used to analyze facial expressions and emotions, providing insights into human behavior and communication. In face tracking, keypoints are used to track facial movements, enabling the development of interactive applications. The accurate detection of facial keypoints is also important for studying emotions and social behavior in psychology, creating natural and intuitive interfaces in human-computer interaction, and providing secure biometric authentication in security.

Facial keypoints detection involves detecting specific points on a person's face, such as the mouth's corners, the nose's tip, and the eyes' center. The number of keypoints can vary depending on the specific use case and the level of detail required. Commonly, 15 keypoints are detected, including both the left and right eye, eyebrow, nose, and mouth. Other keypoints that can be detected include the jawline, cheekbones, and eyelids. The precise locations of these keypoints can be used to determine facial expressions, emotions, and movements, which have various applications in computer vision, human-computer interaction, psychology, and security.

© The Author(s), under exclusive license to Springer Nature Switzerland AG 2024
M. Trajanovic et al. (Eds.): ICIST 2023, LNNS 872, pp. 117–125, 2024.
https://doi.org/10.1007/978-3-031-50755-7_12

One of the biggest challenges in facial keypoints detection is the high variability in facial expressions and poses, which can result in different keypoints configurations and positions. This variability can make it difficult for a facial keypoints detection model to accurately detect keypoints, especially in real-world scenarios where expressions can change dynamically and rapidly. Additionally, factors such as lighting, occlusion, and background can also have a significant impact on the accuracy of keypoints detection. For example, poor lighting conditions can result in shadows and glare that obscure facial features, making it challenging to detect keypoints. Occlusion, such as glasses or hair, can also cover up keypoints, reducing the visibility of crucial elements.

This paper presents a comprehensive study of facial keypoints detection using five neural network models. The main contributions of this research are:

- General insights into the relationship between model complexity and training data size in facial keypoint detection.
- A solution for facial keypoint detection trained on a small dataset.
- A comparison of the performance of classical neural networks and convolutional neural networks with the same dataset size.

In Sect. 2 an overview of the existing solutions is described. Section 3 presents the dataset and used deep learning models. The results of each model are presented in Sect. 4. Section 5 concludes the paper.

2   Related Work

Active Shape Models (ASMs) presented by Cootes et al. [1] are one of the first solutions in this area. It presents a method for a face representation using the Gaussian mixture model. This representation can be used for rough keypoint estimation. Cootes et al. [2] presented Active Appearance Models (AAMs), an extended face model including both shape and texture, which improve the accuracy of keypoint estimation.

The first machine learning solution is Constrained Local Models (CLMs) described in Cristinacce et al. [3]. It incorporates linear regression, decision trees, and support vector regression (SVR). The choice of algorithm depends on the characteristics of the data. Boosted Regression on Features of Gradients (BoFG) [4] solution uses the ensemble learning method that combines multiple weak classifiers and strong classifiers. Weak classifiers represent gradient-based features to describe local appearance of facial regions, while strong classifier is a regression model for predicting the keypoints.

The pioneer deep learning solution in facial keypoints detection is the classical convolutional neural network architecture (convolutional layers at the beginning followed by fully-connected layers) created by Saragih et al. [5]. The improvement in CNN architecture, in terms of its deepness and parameters of each layer, is presented by Wu et al. [6]. They achieved high performance on the large dataset AFW [7]. Gao et al. [8] presented further improvements including exceptionally high performance in real-time applications and avoidance diffusion of lightening problem. They utilized different WFLW [9] and AFLW [10] datasets.

The heatmap regression approach, presented by Gupta et al. [11], involves training a deep neural network on two datasets BIWI [12] and AFLW [10], in order to predict the heatmap representation of each keypoints. The keypoints are afterward obtained by finding the peaks in the heatmap. Gupta et al. [11] successfully localize an area of each keypoints, nevertheless the presence of noise reduce the quality of precise location estimation.

Zhang et al. [13] proposed a deep multi-task learning approach for facial keypoint detection and face alignment, while the Multi-task Cascaded Convolutional Networks (MTCNN) [14] model combines this approach with CNNs. The same deep neural network performs both tasks with high performance because the shared features can provide more information and context for both functions. This idea of shared features was an inspiration for the modern solutions called PifPaf [15], HigherHRNet [16], and Hourglass [17], which are capable of the whole human pose estimation and detecting facial keypoints with high performance. This paper presented another solution for facial keypoints detection trained on a small dataset and gives general insights on how to achieve high performance on a relatively small dataset.

3   Dataset and Models

The publicly available dataset used in this paper is a part of the Kaggle competition [18]. In this section, the dataset and models used will be described.

3.1   Dataset

The dataset contains 7049 grayscaled images in resolution 96×96 pixels, labeled with 15 keypoints. Each keypoint is represented with 2 coordinates, assuming that the coordination start is at the top-left corner of an image. Keypoints represent two eyes, nose and mouth. Figure 1 presents an example of labeled data with all significant facial keypoints.

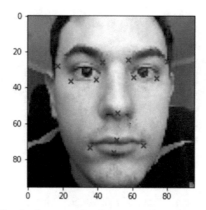

Fig. 1. An example of data used in evaluation

In 68.2% of labeled images there is at least one coordinate missing, therefore any kind of data removal will not leave enough data for training. In order to preserve the original amount of labeled data when some coordinates are missing, the *forward-fill* strategy was used. The entire dataset is divided into three subsets for training (90%), validation (5%), and testing (5%).

3.2 Models

In order to create a Neural Network model which operates on a small dataset, two custom models were created, one with very simple architecture (hereinafter referred to as SmallNet) and one more complex (hereinafter referred to as DeepNet). In the research, not only the performance of these networks was compared, but three well-known convolutional neural networks (CNN) were also included in the evaluation. LeNet-5 [19], AlexNet [20] and VGG-16 [21] CNNs were selected. LeNet-5 has a very simple architecture with two convolutional layers and two fully-connected layers. The other two CNNs have 8 and 16 hidden layers, respectively, which makes the architecture of these networks exceptionally complex. The hyperparameters and the number of parameters for each layer were set to the values presented in the original papers.

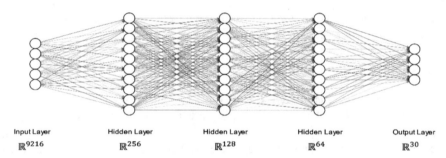

Input Layer	Hidden Layer	Hidden Layer	Hidden Layer	Output Layer
$\mathbb{R}^{9216}$	$\mathbb{R}^{256}$	$\mathbb{R}^{128}$	$\mathbb{R}^{64}$	$\mathbb{R}^{30}$

Fig. 2. SmallNet model architecture

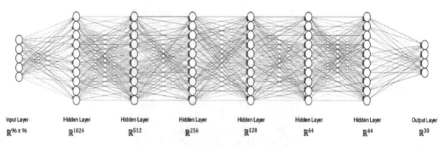

Input Layer	Hidden Layer	Hidden Layer	Hidden Layer	Hidden Layer	Hidden Layer	Hidden Layer	Output Layer
$\mathbb{R}^{96 \times 96}$	$\mathbb{R}^{1024}$	$\mathbb{R}^{512}$	$\mathbb{R}^{256}$	$\mathbb{R}^{128}$	$\mathbb{R}^{64}$	$\mathbb{R}^{64}$	$\mathbb{R}^{30}$

Fig. 3. DeepNet model architecture

Figure 2 presents the architecture of SmallNet. This network has 3 fully-connected hidden layers with 256, 128, and 64 units, respectively. The activation function for each layer is ReLu. The total amount of parameters for training is approximately 2.5 M.

Figure 3 shows the architecture of DeepNet model. The number of parameters in this model is roughly 10 M and it contains 6 fully-connected hidden layers. The activation function for each layer is ReLu. The number of parameters in each layer respectively are 1024, 512, 256, 128, 64, and 64.

4   Evaluation

We trained all models using one CPU with 8 GB RAM, 6 cores, running on 2.1 GHz frequency, and with integrated GPU. Each model is trained across 100 epochs and the batch size is set to 32. The optimizer for each training is Adam with the default values of all hyperparameters [22]. The standard loss function for problems in this category can be either MAE (Mean Absolute Error) or MSE (Mean Squared Error). MSE is more sensitive to larger deviations between predicted and actual values, while MAE is equally sensitive to large and small deviations, which is why MAE was chosen for the loss function [23].

Table 1 contains the relevant metrics about every model and the value of the loss function on the test set. The loss value represents the absolute distance (in pixels) between predicted and the corresponding value from the dataset. This value is treated as the performance of an algorithm. The execution time for training SmallNet is around 80 s, DeepNet and LeNet-5 takes around 110 s each. AlexNet takes 700 s, while VGG-16 consumes 4000 s for training. This is tightly coupled with the fact that the first three models have significantly fewer parameters for training.

Table 1. Algorithm performance

sAlgorithm	Number of hidden layers	Number of training parameters [x 10^6]	Loss function value [px]
SmallNet	3	2,5	5,88
DeepNet	6	10,1	1,41
LeNet-5	4	0,9	1,18
AlexNet	7	24,8	1,69
VGG-16	15	50,5	1,73

The loss function value trend through epochs for the VGG-16, SmallNet, DeepNet, LeNet-5 models are presented in Figs. 4, 5, 6, and 7. Loss function value is in the logarithmic scale due to the huge variations in the beginning and the end of the training. Change of the loss function value for AlexNet is similar to VGG-16, so this redundant graphic will not be presented.

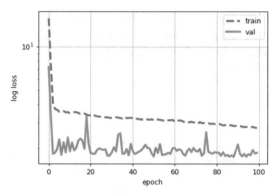

Fig. 4. Loss function for VGG-16 model

In the graphics showing the loss function value trend for VGG-16, it can be observed that the validation loss value is lower than the training loss value during the entire training process. This indicates the model underfit, which occurs because the model is too complex for the small amount of data in our dataset. On the other hand, loss function value for SmallNet indicates that model tends to overfit. This can be seen in the validation loss graph which is flattening through epochs, while training loss is dropping all the time.

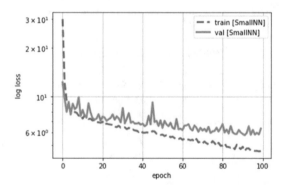

Fig. 5. Loss function for SmallNet model

Loss function value changes through the epochs for LeNet-5 model indicate that the model is trained properly. Both validation and training values are dropping during the entire training process. On the other hand, DeepNet loss value behaves in a different way. In the first part of the training process, this model acts as SmallNet. Both training and validation loss values drop sharply around the 50th epoch. It indicates that the DeepNet model has learned some new features, that the SmallNet model was not capable of.

According to the results in Table 1, the loss value calculated on the test set shows that the model makes an error in prediction for around one pixel. SmallNet performance is worse than LeNet-5, because the calculated error in the test set is around five pixels. This model has a very simple architecture and cannot achieve high performance. Error in prediction of DeepNet model is very close to LeNet-5, but the number of parameters in this NN model is 10 times greater than the CNN showing similar performance.

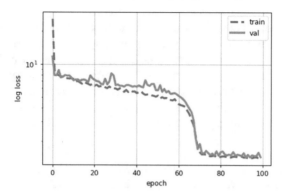

Fig. 6. Loss function for DeepNet model

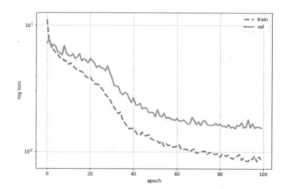

Fig. 7. Loss function for LeNet-5 model

5   Conclusion and Future Work

This research aims to present a solution for facial keypoints detection. The authors realized two custom deep learning models and used three well-known convolutional neural networks, all five with different complexity, to provide general insights on achieving high performance on a relatively small dataset. Classical neural networks and convolutional neural networks were also compared. Although the SmallNet model was too simple, while AlexNet and VGG-16 were too complex, high performance was achieved with DeepNet and LeNet-5. Based on this, we can conclude that models specialized for facial keypoints detection such as PifPaf, HigherHRNet, and Hourglass cannot achieve high performance results, as they are too complex, and they are trained on huge datasets. As a result, it can be concluded that if a dataset is small, very complex architectures cannot achieve precise results, which indicates that optimal architecture should be between the simpler and very complex. The final values of the loss function show that DeepNet and LeNet-5 have very similar performances. This means that a very complex classical neural network can be as powerful as convolutional, but with the cost of a prohibitively large number of trainable parameters.

The future improvements of this research will be in models' performance by adding the Inception modules technique [24] or Skipping connections between the layers [25], similarly to what has been implemented in DenseNet [26]. Augmentation techniques, such as mirroring, stretching, and rotating, can effectively increase the dataset size, which would improve the performance of the neural network models. Additionally, due to the very nature of the problem, the use of ensemble methods, which involve training multiple models and averaging the results, could be the subject of an extension of this research.

Acknowledgement. This research was supported by Belgrade Data Innovation Hub, School of Electrical Engineering - University of Belgrade, within the project EUROPEAN FEDERATION OF DATA DRIVEN INNOVATION HUBS, EUHubs4Data, Horizon 2020 Program.

References

1. Cootes, T.F., Taylor, T., Cooper, D., Graham, J.: Active shape models - their training and application. Comput. Vis. Image Underst. **61**, 38–59 (1995)
2. Cootes, T.F., Edwards, G.J., Taylor, C.J.: Active appearance models. IEEE Trans. Pattern Anal. Mach. Intell. **23**(6), 681–685 (2001)
3. Cristinacce, D., Cootes, T.F.: Feature detection and tracking with constrained local models. In: British Machine Vision Conference, p. 3 (2006)
4. Xiong, X., De la Torre, F.: Supervised descent method and its applications to face alignment. In: Proceedings of the IEEE Conference on Computer Vision and Pattern Recognition, pp. 532–539 (2013)
5. Saragih, J.M., Simon, L., Cohn, J.F.: Face alignment through subspace constrained mean-shifts. In: IEEE International Conference on Computer Vision, pp. 1034–1041 (2009)
6. Wu, Y., Hassner, T., Kim, K., Medioni, G., Natarajan, P.: Facial landmark detection with tweaked convolutional neural networks. IEEE Trans. Pattern Anal. Mach. Intell. **40**(12), 3067–3074 (2017)
7. Zhu, X., Ramanan, D.: Face detection, pose estimation, and landmark localization in the wild. In: Proceedings of the IEEE Conference on Computer Vision and Pattern Recognition, pp. 2879–2886 (2012)
8. Gao, J., Yang, T.: Research on real-time face key point detection algorithm based on attention mechanism. Comput. Intell. Neurosci. **2022**, 1–11 (2022)
9. Wu, W., Qian, C., Yang, S., Wang, Q., Cai, Y., Zhou Q.: Look at boundary: a boundary-aware face alignment algorithm. In: Proceedings of the IEEE Conference on Computer Vision and Pattern Recognition, pp. 2129–2138 (2018)
10. Kostinger, M., Wohlhart, P., Roth, P.M., Bischof, H.: Annotated facial landmarks in the wild: a large-scale, real-world database for facial landmark localization. In: Proceedings of the International Conference on Computer Vision Workshops, pp. 397–403 (2011)
11. Gupta, A., Thakkar, K., Gandhi V., Narayanan, P.J.: Nose, eyes and ears: head pose estimation by locating facial keypoints. In: IEEE International Conference on Acoustics, Speech and Signal Processing (ICASSP), pp. 1977–1981 (2019)

12. Fanelli, G., Weise, T., Gall, J., Van Gool, L.: Real time head pose estimation from consumer depth cameras. In: Mester, R., Felsberg, M. (eds.) Pattern Recognition, pp. 101–110. Springer, Heidelberg (2011). https://doi.org/10.1007/978-3-642-23123-0_11

13. Zhang, Z., Luo, P., Loy, C.C., Tang, X.: Facial landmark detection by deep multi-task learning. In: Fleet, D., Pajdla, T., Schiele, B., Tuytelaars, T. (eds.) ECCV 2014. LNCS, vol. 8694, pp. 94–108. Springer, Cham (2014). https://doi.org/10.1007/978-3-319-10599-4_7

14. Li, X., Yang, Z., Wu, H.: Face detection based on receptive field enhanced multi-task cascaded convolutional neural networks. IEEE Access **8**, 174922–174930 (2020)

15. Kreiss, S., Bertoni, L., Alahi, A.: PifPaf: Composite fields for human pose estimation. In: Proceedings of the IEEE Conference on Computer Vision and Pattern Recognition, pp. 11977–11986 (2019)

16. Cheng, S., Xiao, B., Wang, J., Shi, H., Huang, T.S., Zhang, L.: HigherHRNet: Scale-aware representation learning for bottom-up human pose estimation. In: Proceedings of the IEEE Conference on Computer Vision and Pattern Recognition, pp. 5386–5395 (2020)

17. Newell, A., Yang, K., Deng, J.: Stacked hourglass networks for human pose estimation. In: Leibe, B., Matas, J., Sebe, N., Welling, M. (eds.) ECCV 2016. LNCS, vol. 9912, pp. 483–499. Springer, Cham (2016). https://doi.org/10.1007/978-3-319-46484-8_29

18. Kaggle dataset. https://www.kaggle.com/competitions/facial-keypoints-detection/data. Accessed 19 May 2023

19. LeCun, Y., Bottou, L., Bengio, Y., Haffner, P.: Gradient-based learning applied to document recognition. In: Proceedings of the IEEE, pp. 2278–2324 (1998)

20. Krizhevsky, A., Sutskever, I., Hinton, G.E.: Imagenet classification with deep convolutional neural networks. Commun. ACM **60**(6), 84–90 (2017)

21. Simonyan K., Zisserman, A.: Very deep convolutional networks for large-scale image recognition. In: 3rd International Conference on Learning Representations (ICLR) 2015

22. Kingma, D.P., Ba, J.: Adam: a method for stochastic optimization. In: 3rd International Conference on Learning Representations (ICLR) 2015

23. Willmott, C.J., Matsuura, K.: Advantages of the mean absolute error (MAE) over the root mean square error (RMSE) in assessing average model performance. Climate Res. **30**(1), 79–82 (2005)

24. Szegedy, C., et al.: Going deeper with convolutions. In: Proceedings of the IEEE Conference on Computer Vision and Pattern Recognition, pp. 1–9 (2015)

25. He, K., Zhang, X., Ren, S., Sun, J.: Deep residual learning for image recognition. In: Proceedings of the IEEE Conference on Computer Vision and Pattern Recognition, pp. 770–778 (2016)

26. Huang, G., Liu, Z., Van Der Maaten, L., Weinberger, K.Q.: Densely connected convolutional networks. In: Proceedings of the IEEE Conference on Computer Vision and Pattern Recognition, pp. 4700–4708 (2017)

Artificial Intelligence in Public Administration: A Bibliometric Review in Comparative Perspective

Aleksander Aristovnik$^{(\boxtimes)}$ ⓘ, Lan Umek ⓘ, and Dejan Ravšelj ⓘ

Faculty of Public Administration, University of Ljubljana, Ljubljana, Slovenia
{aleksander.aristovnik,lan.umek,dejan.ravselj}@fu.uni-lj.si

Abstract. The paper provides an in-depth examination of artificial intelligence (AI) research in public administration by comparison with the business sector. The results of bibliometric analysis on 1758 documents for public administration and 2163 documents for the private sector from the Scopus database published until 2022 reveal the growth of the research after 2000, whereby the growth was much faster in public administration than in the business sector. Most research has been conducted by several prominent authors from the United States, followed by the United Kingdom and China and published in several prominent journals. Further insights reveal the development of different AI (related) technologies over time, main research hotspots and most characteristic keywords for public administration and the business sector. The findings have the potential to benefit not only the scientific community by serving as a valuable resource for identifying associated research gaps but also evidence-based policymaking aimed at effectively addressing the challenges associated with AI in public administration in the future.

Keywords: Artificial Intelligence · Public Administration · Bibliometric Review · Comparative Perspective

1 Introduction

The recent trends and challenges emphasize the need to exploit the potential of disruptive technologies in public administration. Artificial intelligence (AI) encompasses various technologies that can facilitate policymaking [1–3]. Therefore, AI can improve the quality of public services, foster trust among citizens, boost efficiency, and influence competitiveness and the generation of public value [4]. Additionally, AI can enhance communication and collaboration among various stakeholders, minimize time and expenses, facilitate resource allocation, and handle complex tasks [5, 6].

Nevertheless, the implementation of AI in public administration has not progressed as rapidly as in the business sector, resulting in a more recent exploration of AI in public administration [7]. The emphasis of public administration on prioritizing public value means that adopting AI practices and digital transformation strategies from the business sector is not straightforward [8]. The distinct challenges posed by AI in public administration are not yet well understood, in contrast to the business sector [9–11]. Most

M. Trajanovic et al. (Eds.): ICIST 2023, LNNS 872, pp. 126–140, 2024.
https://doi.org/10.1007/978-3-031-50755-7_13

existing AI research primarily focuses on technical aspects and computer science-based solutions [9]. Accordingly, the existing knowledge on implementing AI in public administration is relatively scarce compared to research on implementing AI in the business sector. Accordingly, the paper aims to address this issue and provide a comprehensive and in-depth examination of AI research in public administration by comparison with the business sector. More specifically, besides the overview of the existing literature, the paper aims to examine the gap between public administration and the business sector regarding the research and the use of AI.

The paper is structured as follows. After the introduction section, the next section explains the materials used and the methods applied. The following section presents the main results of the bibliometric analysis. The paper ends with a conclusion in which the main findings and implications are summarized.

2 Materials and Methods

Comprehensive bibliometric data on AI research in public administration and the business sector were retrieved from Scopus, a world-leading bibliographic database of peer-reviewed literature. The Scopus database was chosen due to its coverage of scientific research, which is broader than other databases such as Web of Science [12]. The initial search in both databases confirmed this since Scopus retrieved more documents for the intended search conditions than Web of Science. Moreover, compared to the Scopus database, the Web of Science has also been described as a database that significantly underrepresents scientific disciplines of the Social Sciences [13]. Therefore, the Scopus database seems to be more relevant and tailored to the needs of the bibliometric analysis of fiscal research.

Bibliometric data were retrieved in January 2023 using the advanced online search engine. The search strategy was based on title, abstract, and keywords search with consideration of the search keyword artificial intelligence (or its abbreviation AI), followed by selected relevant keywords related to public administration (e.g., public administration, public sector, public service, etc.) or business sector (e.g., business, company, private sector, etc.). Although the search was initially limited to the subject area of Social Sciences, some relevant documents from other relevant subject areas (i.e., Computer Sciences and Arts and Humanities) were added manually. As a result, two different datasets of documents were formed, i.e. wide and narrow. While the wide dataset encompasses Social Sciences, as well as Computer Sciences and Arts and Humanities, the narrow dataset covers Social Sciences only, limiting the influence of documents with pronounced technical content (especially in the context of the business sector) on the results of the bibliometric analysis.

The presented search strategy resulted in 3921 documents included in the wide dataset (with 1758 and 2163 documents related to public administration and business sector, respectively) and 2235 documents included in the narrow dataset (with 1196 and 1039 documents related to public administration and business sector, respectively). Various bibliometric approaches and software tools were applied to analyze relevant documents on AI in public administration and the business sector. The bibliometric analysis was performed using the Python data analysis libraries Pandas and Numpy [14] and visualized using the Python visualization library Matplotlib [15] and VOSviewer [16].

3   Results

3.1   Scientific Production

The involvement in AI research in public administration and the business sector has garnered significant interest across multiple countries worldwide. Notably, the United States, the United Kingdom, and China stand out as the countries with the highest level of engagement in AI research. According to the average year of publication, it becomes evident that the United States and China have been actively involved in AI research longer than the United Kingdom. However, when considering public administration, the Russian Federation emerges as the country with the most recent AI research contributions. On the other hand, South Korea takes the lead in terms of the most recent AI research conducted in the business sector. Moreover, the most productive sources in AI research related to public administration are Government Information Quarterly, Sustainability (also highly relevant for the context of the business sector) and ACM International Conference Proceeding Series.

The distribution of documents and citations in AI research by sector, presented in Fig. 1, reveals the rise of AI research in the public administration and business sector over time, particularly after 2015. According to the cumulative number of documents, it is evident that despite the prominent role of the business sector in AI research, the share of documents related to public administration significantly increased over time. In 2000, the proportion of documents related to public administration was 30%, which increased to 35% by 2010. Subsequently, in 2017, the percentage reached 40%. However, following the Covid-19 pandemic, the share of documents related to public administration further rose to 45%. Similar upward trends can also be observed in the cumulative number of citations. While only 7% of all citations were related to public administration in 2000, by 2010, this percentage had more than doubled, reaching 16%. In the subsequent years, the significance of AI in public administration became even more pronounced, particularly with the outbreak of the Covid-19 pandemic. Notably, the share of citations related to public administration in the cumulative number has surpassed 35% in the past three years. Compared with the business sector, the overview of the cumulative number of documents and citations reveals higher growth of AI research in public administration, especially in recent years. The presented trends are comparable with some existing bibliometric findings [17].

AI generally encompasses a wide range of technologies, from fundamental to highly advanced [1], resulting in several AI (machine learning, deep learning, artificial neural networks, chatbot, natural language processing) and AI-related (big data analytics, decision support system, cloud computing) technologies. The distribution of publications in AI research by AI (related) technologies is presented in Fig. 2.

Between 2000 and 2010, the decision support system emerged as a prominent AI (related) technology, particularly within the business sector. However, there were also some notable attempts to apply decision support systems in public administration during that period. Concurrently, big data analytics and machine learning started gaining

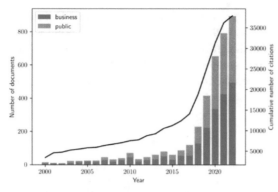

Fig. 1. Distribution of documents and citations in AI research by sector (2000–2022).

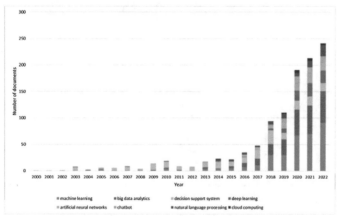

Note: Documents focusing only on general AI are excluded from the presentation; therefore, the number of documents may differ from other presentations.

Fig. 2. Distribution of documents in AI research by AI (related) technologies (2000–2022).

traction within the business sector. Moving into the period from 2011 to 2017, decision support systems continued to grow in importance in public administration, with big data analytics and machine learning also playing significant roles. Additionally, this period witnessed the emergence of natural language processing and artificial neural networks within the business sector. Between 2018 and 2022, all AI (related) technologies became relevant for both public administration and the business sector. This period experienced a notable surge in deep learning, which had minimal presence in the literature before 2018. Additionally, big data analytics and machine learning experienced significant advancements and adoption during this time.

A more detailed comparison between AI research in public administration and the business sector for the last five years is presented in Fig. 3. Among AI (related) technologies, big data analytics, machine learning, deep learning, and decision support systems were predominantly utilized in the business sector, while natural language processing and chatbots seem to be more prominent in public administration, especially after 2020 as a response to the Covid-19 pandemic. Moreover, the prominence of artificial neural networks seems to be similar for public administration and the business sector, while there is almost no cloud computing in the context of public administration. Finally, the growth of machine learning and big data analytics in public administration is much higher than in the business sector.

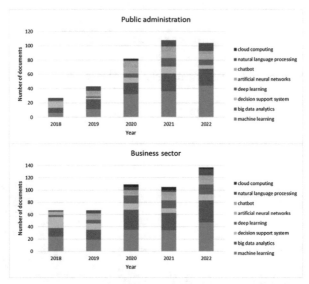

Note: Documents focusing only on general AI are excluded from the presentation; therefore, the number of documents may differ from other presentations.

Fig. 3. Distribution of documents in AI research by AI (related) technology in public administration and business sector (2018–2022).

The most relevant documents in AI research are presented in Table 1. These are the top 5 cited documents for public administration and the business sector. When it comes to public administration, the most relevant documents tend to concentrate on general AI issues without a specific sectoral focus [18–20], with one document examining the opportunities of the AI system IBM Watson for the health sector [21] and one document addressing machine learning in the context of housing and community amenities [22]. Moreover, in the context of the business sector, all of the most relevant documents focus on general AI issues [23, 24], with some implications for agriculture [25], marketing [26] and transportation [27].

3.2 Network Analysis

The network analysis, i.e., keyword co-occurrence analysis, is conducted separately for the public administration and business sector context. Note that the nodes indicate keywords and the links the co-occurrence relations between them. The node size is

Table 1. The most relevant documents in AI research (2000-2022).

	Authors	Title	Year	Source title	Cited by	Technology	Sector
Public	Floridi et al.	AI4People—An ethical framework for a good AI society: Opportunities, risks, principles, and recommendations	2018	Minds Mach	552	General AI	/
	Dwivedi et al.	Artificial Intelligence (AI): Multidisciplinary Perspectives on emerging challenges, opportunities, and agenda for research, practice and policy	2021	Int J Inf Manage	550	General AI	/
	Sun and Medaglia	Mapping the challenges of Artificial Intelligence in the public sector: Evidence from public healthcare	2019	Gov. Inf. Q	227	AI system IBM Watson	Health
	Jiang et al.	Mining point-of-interest data from social networks for urban land use classification and disaggregation	2015	Comput. Environ. Urban Syst	219	Machine learning	Housing and community amenities
	Cath et al.	Artificial Intelligence and the 'Good Society': the US, EU, and UK approach	2018	Sci. Eng. Ethics	216	General AI	/

(continued)

Table 1. (*continued*)

	Authors	Title	Year	Source title	Cited by	Technology	Sector
Private	Makridakis	The forthcoming Artificial Intelligence (AI) revolution: Its impact on society and firms	2017	Futures	505	General AI	/
	Warner and Wäger	Building dynamic capabilities for digital transformation: An ongoing process of strategic renewal	2019	Long Range Plann	482	General AI	/
	Klerkx et al.	A review of social science on digital agriculture, smart farming and agriculture 4.0: New contributions and a future research agenda	2019	NJAS Wageningen J. Life Sci	350	General AI	Agriculture
	Roberts	Behind the screen: Content moderation in the shadows of social media	2019	NA (Book)	244	General AI	Marketing
	Tang and Veelenturf	The strategic role of logistics in the industry 4.0 era	2019	Transp. Res. Part E Logist. Transp. Rev	201	General AI	Transportation

Note: Most relevant documents in the narrow database.

proportional to the number of keyword occurrences, showing research intensity (node degree), while the link width is proportional to the co-occurrences between keywords (edge weight). Additionally, the node colour indicates the cluster to which a particular keyword belongs [28, 29].

Five distinct research hotspots have been identified in the context of public administration research, each with its unique focus and subject matter (see Fig. 4). The first cluster, technology and data management (red), encompasses various technologies like machine learning, deep learning, artificial neural networks, and the Internet of Things. It also encompasses topics related to data management, such as open data, digital government, and forecasting. The second cluster, digital innovation and transformation (green), covers topics related to big data analytics, blockchain, robotics, digital economy, and sustainability, all related to digital innovation and transformation. The third cluster, governance and policy (blue), includes topics related to governance, public policy, regulation, public administration, human rights, education, ethics, and industry 4.0. The fourth cluster, AI and automation (yellow), explores topics related to artificial intelligence, automation, and their applications in government services and the public sector. Finally, the fifth cluster, transparent and accountable decision support system (purple), concentrates on topics related to algorithms, transparency, accountability, and decision support systems.

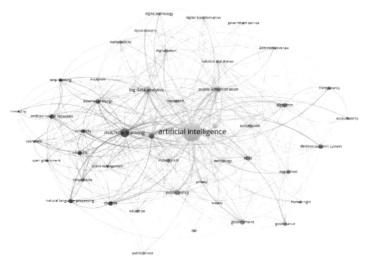

Fig. 4. Keyword co-occurrence analysis of AI research in public administration. (Color Figure Online)

Moreover, five distinct research hotspots have also been identified in the business sector research context (see Fig. 5). The first cluster, digital business and sustainability (red), encompasses topics related to digital business models, strategies, and technologies. This includes areas such as digital transformation, e-commerce, digitalization, logistics, as well as sustainability and circular economy. The second cluster, data ethics and governance (green), includes topics related to data ethics, privacy, governance, and regulation, exploring areas like big data analytics, blockchain, fintech, human rights, and the impact on small and medium-sized enterprises (SMEs). The third cluster, AI and knowledge management (blue), covers AI-related topics, including artificial neural networks, decision support systems, and machine learning. Additionally, it explores knowledge management, corporate governance, and simulation. The fourth cluster, automation and disruption (purple), explores topics related to automation and disruptive technologies like natural language processing, social media, and machine learning. Finally, the fifth cluster, innovation and robotics cluster (yellow), concentrates on topics related to the Internet of Things, innovation, robotics, drones, intellectual property, automation, and their effects on employment.

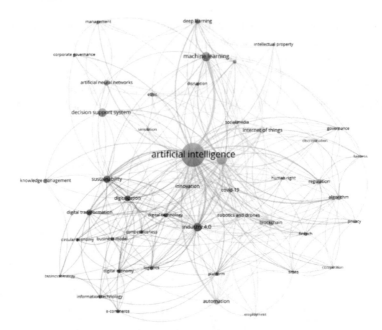

Fig. 5. Keyword co-occurrence analysis of AI research in the business sector. (Color Figure Online)

3.3 Yule's Q Analysis

Since selected aspects of the research hotspots may overlap between public administration and the business sector, further analysis of AI-related and general keywords facilitated by Yule's Q analysis helps identify the characteristics of the public administration and business sector context. In this analysis, a positive Yule Q coefficient indicates that a keyword is highly representative of the field of public administration, while a negative coefficient suggests that the keyword is more closely associated with the business sector. Based on the statistical (in)significance, it is further possible to identify relevant keywords for both public administration and the business sector.

First, Yule's Q analysis is performed on a set of 22 general keywords (see Table 2). Characteristic general keywords for public administration are related to democratic governance (democracy, transparency, accountability, trust, ethic), data privacy and regulation (data protection, regulation) and surveillance and Covid-19 (surveillance, covid-19). Moreover, characteristic general keywords for the business sector are related to operations and efficiency (innovation, automation, knowledge management) and future planning and sustainability (sustainability, prediction, simulation). Finally, characteristic general keywords that are relevant for both public administration and the business sector are related to data security (cybersecurity, security, privacy), rights and fairness (human right, fairness) and risk and optimization (risk, optimization).

Table 2. Results of Yule's Q analysis for general keywords.

General keywords	Yule Q	P value
transparency	**0.502**	0.001
democracy	**0.471**	0.029
accountability	**0.321**	0.095
trust	**0.268**	0.060
ethic	**0.221**	0.009
data protection	**0.367**	0.066
regulation	**0.221**	0.030
surveillance	**0.321**	0.095
covid-19	**0.275**	0.009
innovation	**−0.169**	0.041
automation	**−0.214**	0.065
knowledge management	**−0.447**	0.009
sustainability	**−0.193**	0.078
prediction	**−0.290**	0.029
simulation	**−0.402**	0.055
cybersecurity	0.250	0.156
security	0.164	0.130
privacy	0.147	0.329
human right	0.093	0.708
risk	−0.045	0.685
optimization	−0.203	0.222
fairness	−0.434	0.139

Note: The bold number indicates a statistically significant coefficient ($p < 0.1$).

Second, Yule's Q analysis is also performed on a set of 8 keywords, representing main AI (related) technologies (see Table 3), by considering selected subperiods to extract additional time-dimensional insights. The results reveal the following. From 2014 to 2016, machine learning emerged as a defining characteristic of the business sector. Between 2017 and 2019, both decision support systems and machine learning remained prominent in the business sector. From 2020 to 2022, cloud computing became increasingly associated with the business sector, while artificial neural networks emerged as a distinguishing feature of public administration. However, some AI (related) technologies (i.e., big data analytics, chatbot, deep learning and natural language processing) can be considered as technologies that are relevant for public administration and the business sector to a similar extent.

Table 3. Results of Yule's Q analysis for main AI (related) technologies.

AI and AI-related technologies	2014–2016		2017–2019		2020–2022	
	Yule Q	P value	Yule Q	P value	Yule Q	P value
artificial neural networks	−0.248	0.705	−0.328	0.178	**0.256**	0.096
big data analytics	0.389	0.223	−0.055	0.691	−0.115	0.189
chatbot	/	/	0.060	1.000	0.100	0.628
cloud computing	1.000	0.154	0.095	1.000	**−0.665**	0.028
decision support system	0.096	0.685	**−0.284**	0.064	−0.004	1.000
deep learning	1.000	0.394	−0.584	0.123	−0.145	0.266
machine learning	**−0.539**	0.068	**−0.267**	0.042	0.092	0.203
natural language processing	−0.513	0.205	0.222	0.449	0.178	0.330

Note: The bold number indicates a statistically significant coefficient (p < 0.1).

3.4 Regression Analysis

Binary logistic regression analysis is used to empirically predict AI (related) technologies based on the 20 most relevant keywords related to public administration (trust, surveillance, smart city, regulation, public service, ethic, digital government, accountability), the business sector (sustainability, supply chain management, knowledge management, innovation, industry 4.0, e-commerce, business model), general issues (security, risk, privacy, human right, fairness). Therefore, every AI (related) technology has a corresponding indicator variable (as indicator variables $Y_1, Y_2, \ldots, Y_6$), which takes values 1 (a document is related to an AI (related) technology) and 0 (otherwise). Thus, the indicator variables are considered separate dependent variables for logistic regression models. Moreover, $p = 55$ keywords (as indicator variables $X_1, X_2, \ldots, X_{20}$) were identified as important for the AI context and used as predictor variables in the models. Accordingly, 7 different binary logistic models with 20 predictors were tested (see Table 4). This approach provides information on the most characteristic keywords for a particular AI (related) technology [30]. The formula of binary logistic regressions corresponds to (Eq. 1):

$$P(Y_i = 1|X_1, X_2, \ldots, X_p) = \frac{\exp\left(\beta_{0i} + \beta_{1i} \cdot X_1 + \beta_{2i} \cdot X_2 + \ldots + \beta_{pi} \cdot X_p\right)}{1 + \exp\left(\beta_{0i} + \beta_{1i} \cdot X_1 + \beta_{2i} \cdot X_2 + \ldots + \beta_{pi} \cdot X_p\right)} \text{ for } i = 1, 2, 3, 4, 5, 6, 7$$

(1)

The results reveal the following. AI is commonly associated with documents that discuss topics such as trust, regulation, public service, ethics, digital government, innovation, business models, and human rights. Machine learning is often linked to documents that address concepts like smart cities and fairness. Big data analytics tends to be connected to documents discussing smart cities, sustainability, and privacy. Decision support systems are frequently associated with documents focusing on digital government, knowledge management, and risk. Deep learning is more likely to be related to

Table 4. Results of binary logistic regression analysis.

		artificial intelligence	machine learning	big data analytics	decision support system	deep learning	chatbot	cloud computing
Public	trust	0.733						
	surveillance					1.610		
	smart city		0.700	1.288		2.453		
	regulation	1.116						
	public service	0.764					2.266	
	ethic	0.754						
	digital government	0.618			0.803		1.963	
Private	sustainability			0.719				
	knowledge management				1.710			
	innovation	0.464	−1.195		−2.599			
	industry 4.0				−2.562			
	e-commerce							2.139
	business model	0.640						
General	risk				0.942			
	privacy			0.716				
	human right	0.691						
	fairness		1.633					

Note: The binary logistic regression analysis is performed on 20 relevant keywords predicting AI (related) technologies. Only statistically significant coefficients ($p < 0.1$) are presented. A positive (negative) coefficient indicates that a document with a selected keyword is likely (unlikely) to be related to a certain AI (related) technology.

documents addressing surveillance and smart cities. Chatbots are often linked to documents discussing public service and digital government. Lastly, cloud computing is commonly associated with documents that address e-commerce.

4 Conclusion

Recent globalization and digitalization trends, which the Covid-19 pandemic has further emphasized, have revealed the existing digital deficiencies and emphasized the importance of leveraging the potential of AI in public administration to improve public value. Namely, AI in public administration may help address socioeconomic challenges more efficiently, contribute to better lives for society, and sustain healthy economic growth [31]. Nevertheless, public administration stayed a certain distance from technological change. While technological developments in the field of AI rapidly improve, a large gap remains between the implementation in public administration and that in the private sector.

Since challenges posed by AI in public administration are not yet well understood, in contrast to the business sector [9–11], the paper provides a comprehensive and in-depth examination of AI research in public administration by comparison with the business sector. The results reveal that, in some aspects, the public administration is already catching up with the private sector as regards the implication of different AI (related) technologies (e.g., chatbots, big data analytics, machine learning), where chatbots are emphasized in the government-to-consumer relationships and machine learning in financial and fiscal affairs and health.

This bibliometric study is limited only to documents indexed in the Scopus database. While Scopus is widely recognized as a top database for peer-reviewed literature, it may not encompass the entirety of AI research in public administration. Therefore, incorporating other databases such as Google Scholar or Web of Science could potentially uncover additional insights not captured in this bibliometric study. Regardless of this limitation, the findings have the potential to benefit not only the scientific community by serving as a valuable resource for identifying associated research gaps but also evidence-based policymaking aimed at effectively addressing the challenges associated with AI in public administration in the future.

Acknowledgment. The authors acknowledge financial support from the Slovenian Research and Innovation Agency (research core funding No. P5-0093 and project No. J5-2560).

References

1. Agrawal, A., Gans, J., Goldfarb, A.: What to expect from artificial intelligence. MIT Sloan Manag. Rev. **58**(3), 22–27 (2017)
2. Dwivedi, Y.K., Hughes, L., Ismagilova, E., et al.: Artificial Intelligence (AI): Multidisciplinary perspectives on emerging challenges, opportunities, and agenda for research, practice and policy. Int. J. Inf. Manage. **57**, 101994 (2019)
3. Misuraca, G., van Noordt, C.: Overview of the use and impact of AI in public services in the EU. EUR 30255 EN. Publications Office of the European Union, Luxembourg (2020)
4. Zuiderwijk, A., Chen, Y.C., Salem, F.: Implications of the use of artificial intelligence in public governance: a systematic literature review and a research agenda. Gov. Inf. Q. **38**(3), 101577 (2021)
5. Criado, J.I., Gil-Garcia, J.R.: Creating public value through smart technologies and strategies: from digital services to artificial intelligence and beyond. Int. J. Public Sect. Manag. **32**(5), 438–450 (2019)
6. Kankanhalli, A., Charalabidis, Y., Mellouli, S.: IoT and AI for smart government: a research agenda. Gov. Inf. Q. **36**(2), 304–309 (2019)
7. Desouza, K.C., Dawson, G.S., Chenok, D.: Designing, developing, and deploying artificial intelligence systems: lessons from and for the public sector. Bus. Horiz. **63**(2), 205–213 (2020)
8. Fatima, S., Desouza, K.C., Dawson, G.S.: National strategic artificial intelligence plans: a multi-dimensional analysis. Econ. Anal. Policy **67**, 178–194 (2020)
9. Aoki, N.: An experimental study of public trust in AI chatbots in the public sector. Gov. Inf. Q. **37**(4), 101490 (2020)

10. Wang, W., Siau, K.: Artificial intelligence: a study on governance, policies, and regulations. In: paper presented at the 13th Annual Conference of the Midwest Association for Information Systems, Saint Louis, Missouri (2018)

11. Wirtz, B.W., Weyerer, J.C., Sturm, B.J.: The dark sides of artificial intelligence: an integrated AI governance framework for public administration. Int. J. Public Adm. **43**(9), 818–829 (2020)

12. Falagas, M.E., Pitsouni, E.I., Malietzis, G.A., Pappas, G.: Comparison of PubMed, scopus, web of science, and Google scholar: strengths and weaknesses. FASEB J. **22**(2), 338–342 (2008)

13. Mongeon, P., Paul-Hus, A.: The journal coverage of web of science and scopus: a comparative analysis. Scientometrics **106**(1), 213–228 (2016)

14. McKinney, W.: Python for data analysis: Data wrangling with Pandas, NumPy, and IPy-thon. O'Reilly Media Inc., Sebastopol (2012)

15. Hunter, J.D.: Matplotlib: a 2D graphics environment. Comput. Sci. Eng. **9**(3), 90–95 (2007)

16. Jan, N., van Eck, L., Waltman: Software survey: VOSviewer a computer program for bibliometric mapping. Scientometrics **84**(2), 523–538 (2010). https://doi.org/10.1007/s11192-009-0146-3

17. De la Vega Hernández, I.M., Urdaneta, A.S., Carayannis, E.: Global bibliometric mapping of the frontier of knowledge in the field of artificial intelligence for the period 1990–2019. Artif. Intell. Rev. **56**(2), 1699–1729 (2023)

18. Floridi, L., Cowls, J., Beltrametti, M., et al.: AI4People—an ethical framework for a good AI society: opportunities, risks, principles, and recommendations. Mind. Mach. **28**, 689–707 (2018)

19. Dwivedi, Y.K., Hughes, L., Ismagilova, E., et al.: Artificial Intelligence (AI): multidisciplinary perspectives on emerging challenges, opportunities, and agenda for research, practice and policy. Int. J. Inf. Manage. **57**, 101994 (2021)

20. Cath, C., Wachter, S., Mittelstadt, B., Taddeo, M., Floridi, L.: Artificial intelligence and the 'good society': the US, EU, and UK approach. Sci. Eng. Ethics **24**(2), 505–528 (2018). https://doi.org/10.1007/s11948-017-9901-7

21. Sun, T.Q., Medaglia, R.: Mapping the challenges of artificial intelligence in the public sector: evidence from public healthcare. Gov. Inf. Q. **36**(2), 368–383 (2019)

22. Jiang, S., Alves, A., Rodrigues, F., Ferreira, J., Jr., Pereira, F.C.: Mining point-of-interest data from social networks for urban land use classification and disaggregation. Comput. Environ. Urban Syst. **53**, 36–46 (2015)

23. Makridakis, S.: The forthcoming artificial intelligence (AI) revolution: Its impact on society and firms. Futures **90**, 46–60 (2017)

24. Warner, K.S., Wäger, M.: Building dynamic capabilities for digital transformation: an ongoing process of strategic renewal. Long Range Plan. **52**(3), 326–349 (2019)

25. Klerkx, L., Jakku, E., Labarthe, P.: A review of social science on digital agriculture, smart farming and agriculture 4.0: new contributions and a future research agenda. NJAS-Wageningen J. Life Sci. **90**, 100315 (2019)

26. Roberts, S.T.: Behind the Screen: Content Moderation in the Shadows of Social Media. Yale University Press, New Haven (2019)

27. Tang, C.S., Veelenturf, L.P.: The strategic role of logistics in the industry 4.0 era. Transp. Res. Part E Logistics Transp. Rev. **129**, 1–11 (2019). https://doi.org/10.1016/j.tre.2019.06.004

28. Wang, C., Lim, M.K., Zhao, L., Tseng, M.L., Chien, C.F., Lev, B.: The evolution of Omega the international journal of management science over the past 40 years: a bibliometric overview. Omega **93**, 102098 (2020)

29. Ravšelj, D., Umek, L., Todorovski, L., Aristovnik, A.: A review of digital era governance research in the first two decades: a bibliometric study. Future Internet **14**(5), 126 (2022)

30. Aristovnik, A., Ravšelj, D., Umek, L.: A bibliometric analysis of COVID-19 across science and social science research landscape. Sustainability **12**(21), 9132 (2020)
31. Palomares, I., et al.: A panoramic view and swot analysis of artificial intelligence for achieving the sustainable development goals by 2030: progress and prospects. Appl. Intell. **51**(9), 6497–6527 (2021). https://doi.org/10.1007/s10489-021-02264-y

Machine Learning Model as a Service in Smart Agriculture Systems

Aleksandra Stojnev Ilić[1][(✉)] ⓘ, Dragan Stojanović[1] ⓘ, Natalija Stojanović[1] ⓘ,
and Miloš Ilić[2] ⓘ

[1] Faculty of Electronic Engineering, University of Niš, Niš, Serbia
aleksandra.stojnev@elfak.ni.ac.rs
[2] Faculty of Technical Sciences Kosovska Mitrovica, University of Pristina, Kosovska
Mitrovica, Serbia

Abstract. Machine learning applications that rely on trained machine learning
models are a major area of interest within the field of Smart Agriculture Systems.
One of the greatest challenges in this area is to ensure that the machine learning
model is up-to-date and easily and effectively deployed to all smart systems that
rely on it. However, the rapid changes, or drifts in the data can have a serious effect
on the accuracy of the model, leading to unforeseeable consequences in system
behavior. Whilst some research has been carried out on ML model updates, this
is still an open challenge in modern IoT-based Smart Systems. This paper reviews
different ways in which ML models can be served and updated, and proposes a
detailed architecture for Edge part of Edge-Cloud-based Smart Agriculture System. To support this architecture, a prototype system that provides a machine
learning model as a service in the agricultural domain is developed and tested.

Keywords: Smart Agriculture · Machine learning · Model as a Service

1 Introduction

Recent advances in machine learning have led to proliferation of studies targeting
machine learning applications in Smart Agriculture Systems. There is a growing body
of research that recognizes the importance of machine learning applications in the agricultural domain and of having up-to-date information in the decision-making process.
Machine learning models are created by processing the information collected and used
for training. However, the data itself is not fixed, but it changes in time. In the literature,
this is often referred to as data drift, or concept drift, and it refers to a change in the
statistical properties of data over time in contrast to data being used to train a machine
learning model. This can occur when the data distribution changes or when the underlying relationship between the input features and the target variable changes. Data drift can
cause a trained machine learning model to perform poorly when applied to new, unseen
data. It can also cause a model to make incorrect predictions or lead to a decrease in the
model's overall accuracy, which can have serious effects in Smart Agriculture Systems.

M. Trajanovic et al. (Eds.): ICIST 2023, LNNS 872, pp. 141–148, 2024.
https://doi.org/10.1007/978-3-031-50755-7_14

Once a data drift has been detected, there are several ways to address it, such as retraining the model on new data, fine-tuning the model, using an ensemble of models or via continual learning. The most common method for updating a machine learning model is to retrain it on new data. This can be done by obtaining more labeled examples for the task the model is being used for, and then using these examples to update the model's parameters. This process can be automated and scheduled to update the model regularly. Another way to update a machine learning model is to fine-tune it on a new set of problems or data. This approach involves using a pre-trained model as a starting point and then training it further on a smaller dataset. This can be done using transfer learning, which allows a model to learn new tasks by leveraging knowledge learned from previous tasks. Another method is to combine multiple models because ensemble methods can often improve the performance of the models. The ensemble can be updated by adding new models to it or by adjusting the weights of the existing models. A more recent approach is to use Continual Learning, which allows the model to learn from new data over time, in an incremental fashion, without forgetting the previously learned information. Furthermore, these methods are not mutually exclusive, and they can be stacked to achieve better results. However, in systems that operate in real-time, such as the ones in the agriculture field, it is of utmost importance to provide the up-to-date model seamlessly, no matter how it is generated. System should be resilient to the drifts in the data, meaning the machine model that is used can be updated online. Therefore, the primary challenge in Smart Agriculture Systems is to timely detect the drifts in the incoming data. The next challenge is to retrain the model and provide it seamlessly to the system. One of the viable solutions that might address this problem is to create a service that will be used as a model provider, a method that in literature addressed as machine learning as a service or MLaaS.

This paper gives an overview of the challenges that are documented in literature regarding Smart Agriculture System, and describes a crucial part of the designing this type of systems – part that is responsible for machine learning model usage and lifecycle. The main contributions of this paper are:

- A brief overview of related studies targeting either agricultural domain or MLaaS as a concept.
- Specification for a service that should be used as machine learning model provider and a medium for seamless updates of the model used by the rest of the system.
- Light prototype of Smart Agriculture System that gives a proof of concept for the designed service.

The rest of this paper is structured as follows. The second section presents related work. Third section sets forth the idea of Smart Agriculture System and Machine Learning as a Service as a part of that system. The fourth section gives an overview of the created prototype. Fifth section gives main conclusions and ideas for future research.

2   Related Work

There has been a growing body of literature on the application of machine learning as a service (MLaaS) that defines the state-of-the-art techniques related to this problem.

Philipp et al. [1] give a structured literature review in this field and groups the studies into four key concepts: Platform, Applications, Performance Enhancements and Challenges. Bacciu et al. [2] propose a conceptual architecture of a machine learning service, integrated within the IoT reference model. Both studies are not specific to the agricultural domain, but all outlines are fully applicable.

Liakos et al. [3] and Benos et al. [4] provide a comprehensive reviews of research dedicated to machine learning applications in agricultural production systems. Both studies are a beneficial starting point for exploring the advantages of using machine learning in agriculture. Pathmudi et al. [5] give a comprehensive literature review of the key technologies involved in smart and precision agriculture, including various sensors, controllers, communication standards, and IoT based intelligent machinery, as well as the architecture and importance of data analytics in agriculture IoT, case studies of current agricultural automation utilizing IoT, key challenges and open issues in agriculture IoT technology. The architecture and importance of data analytics in agriculture IoT, case studies of current agricultural automation utilizing IoT, key challenges and open issues in agriculture IoT technology were discussed. The findings provide support for the selection of IoT technologies for specific applications.

Another important topic for building MLaaS is model selection. Assem et al. [6] introduce the TCDC (train, compare, decide, and change) approach, which can be thought as a 'Machine Learning as a Service' approach, as well as the results of testing and evaluating the recommenders based on the TCDC approach (in comparison with the traditional default approach) applied to 12 open-source datasets drawn from diverse domains including health care, agriculture, aerodynamics, and others. Authors presented results that indicate that the proposed approach selects the best model in terms of predictive accuracy in 62.5% for regression tests performed and 75% for classification tests. Raschka [7] reviews general, domain – independent techniques that can be used for model evaluation, model selection, and algorithm selection techniques and discusses the main advantages and disadvantages of each technique with references to theoretical and empirical studies.

When building large-scale system, different characteristics of horizontal and vertical scaling must be taken into account. Chaterji et al. [8] list and discuss various aspects of digital agriculture at scale, starting with strategic importance of digital agricultural solutions, necessity for data sharing among individuals, and all aspects of big data processing in agriculture. It underlines the importance of data privacy and ownership, especially in cases where the training algorithm must be fed to MLaaS computing platforms, and lists various questions related to goals and nexus-sensitive challenges of digital agriculture. MLaaS can be a challenge for scaling, both horizontally and vertically, which is discussed by Erran et al. [9]. A solution is proposed to enable scalable MLaaS, which first scales feature computation for different use cases, and aims to build accurate models using global data and account for individual regions. Finally, different models of deployment and real-time model serving are discussed. Although it is not related to agriculture, various conclusions can be used as first steps when designing a model for smart agriculture systems. For real time prediction, authors proposed high performance RPC service using TChannel as the networking framing protocol.

Paleyes et al. [10] survey published reports of deploying different machine learning solutions across different industries and applications, and list identified challenges in the machine learning workflow. Different model learning phases are identified, including model selection, training, hyper-parameter selection, model verification, and model deployment. Authors discussed challenges that arise during the data management, model learning, model verification, and model deployment stages, as well as considerations that affect the whole deployment pipeline including ethics, end-users' trust, law, and security. Each stage is illustrated with examples across different fields and industries by reviewing case studies, experience reports, and academic literature. Authors conclude that ML as a field would benefit from a cross-disciplinary dialog with such fields as software engineering, human-computer interaction, systems, policy making.

These studies confirm that there is a growing body of literature for listed challenges, but the model for a system that uses machine learning models as a service while can autonomously detect drifts in the data and therefore trigger model change is still listed as an open question. Furthermore, these studies demonstrate the potential of MLaaS in agriculture, but still fail to address MLaaS particularly in precision agriculture, crop growth predictions, crop disease diagnoses, and predictive maintenance of agricultural equipment. However, it is important to note that this is still a relatively new area of research, and more studies are needed to fully explore the capabilities of MLaaS in agriculture. Furthermore, agriculture-oriented studies mainly propose a monolith system, without consideration of performance, scalability, flexibility, security, privacy, etc. In order to build scalable smart agriculture system, it is necessary to combine domain related knowledge with already developed best practices in ML area.

3 Smart Agriculture Systems

Smart Agriculture Systems, encompassing precision agriculture, data analytics, and data ownership, have the potential to transform agricultural throughput by applying data science for mapping input factors to crop throughput and that too in a region-specific and crop-specific manner, while bounding the available resources. Furthermore, as the data volumes and varieties escalate with the development and availability of sensors in agricultural fields, data engineering techniques must be able to stand the low latency requirements of the end users and applications.

There are different challenges in the data workflow that must be accessed when designing Smart Agriculture System. Firstly, these systems operate with vast amounts of data that is gathered from various inputs, very often from large numbers of inexpensive and unreliable sensors placed in farms. Some of the commonly used measurements from agricultural sensors include meteorological data, soil sampling, different plantation, fertilization and yield maps, imagery scouting, etc. Obtained data should be sanitized, loaded, processed, stored, summarized and/or analyzed. That means that Smart Agriculture System should be able to provide data warehousing, preprocessing, analysis, and post processing. Very often, for this purpose different machine learning models are used. Secondly, different challenges related to the architectural design of the system must be addressed. Some of them might use the advantages of edge-fog-cloud paradigm, or architectures based on microservices, others might be designed as monoliths, if that is applicable. General architectures should allow the system to be scalable and adapt to various needs.

In the era of connected devices where applications rely on a multitude of devices aggregating and processing data sets across highly heterogeneous networks, distributed deployment and containerization of the different information channels should improve the resiliency and stability of the systems, while the partitioning of the data stream for computing at different degrees of latency can provide resources for time-sensitive actions. Since most of the data obtained in digital agriculture is real time, stream processing is preferred over batch processing. That means that all predictions should be in real-time. If they are based on machine learning models, it will be of utmost importance for the system to have access to the model. Furthermore, since the domain itself imposes a high possibility of data drifts, it is necessary to provide adequate way to update the model, but without impacting the rest of the system. In this paper, we rely on microservice architecture to pave the way for designing Smart Agriculture system that can fulfill most of the requirements. To do so, it is important to define modules that are responsible for prediction and machine learning model. Due to all specifics that agricultural domain imposes on the system, such as data drifts, personification, anonymization, etc., it is essential to encapsulate machine learning model and detach it from the rest of the system, while enabling its usage. For that manner, we create a service that is responsible for model updates and usage.

3.1 General System Overview

The general idea is to create a scalable Smart Agriculture System that can address different agricultural data challenges. In this paper we focus primarily on the machine learning model lifecycle. Considering the amount of data generated on the edge, and limitations brought in by storage, bandwidth and privacy concerns, the system that can support adaptive analysis of big streaming data can fully exploit all the benefits that are offered by edge-fog-cloud architecture. The general idea for this architecture is to split the usage and generation of the model between edge/fog and cloud, and details about this part of the system are presented in [11].

Firstly, input data collected from various input devices is sent to a message broker, where preprocessing or filtering if needed are performed. Processing component has a main goal of near real-time analysis of the incoming data. For all predictions, this component uses MLaaS node. Cloud component will ingest data to evaluate current machine learning model, and if necessary, create new one, and replace the old one.

Service that provides machine learning model as well as methods for its usage should be decoupled from the rest of the system, but for its design, it is important to take into the consideration the actual use cases, and other components that might interact with it.

3.2 Machine Learning Model as a Service

One important part in the Smart Agriculture system is a microservice capable of provisioning the pretrained ML models that can be accessed and used via an API. Such general-purpose service can be used for a variety of tasks that are applicable in agriculture, such as predictive analytics, detection, and recognition, and can ingest diverse types of data such as sensor data, imagery, different feeds, crowdsourced data, etc. When a drift in data is detected, the service can be adapted to use the new model. Exposing machine learning model as a service has numerous benefits regarding resilience of the entire system. It can allow developers to monitor and shift its execution from edge to cloud, or to scale accordingly. Its future implementations can be implemented as a hierarchy of services that can allow personalization of the models regarding user preferences.

The presented service has three key API endpoints: one for setting of the new model, one for retrieving the model, and one for providing the data the model should be used on and obtaining the results of a prediction. This API allows other parts of the Smart Agriculture System to access the service. It is a REST API that uses HTTP requests. It is built in a lightweight form, so it can be run both on the edge and in the cloud, and therefore scaled properly, as needed. The advantages of running this service on the edge of a network, near the devices that collect the data are reduced latency, improved security, and privacy, as well as reliability. If run on the cloud, resource constraints can be released, more voluminous data can be processed, and it can be scaled to address the needs of the system. Figure 1 shows an overview of this service and its interaction with the rest of the Smart Agriculture System.

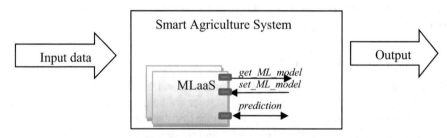

Fig. 1. MLaaS in Smart Agriculture System

There can be one or more machine learning models exposed either as an independent service or the part of a system, and such system should provide an option for model update. That update can be autonomous, continual, on specific timespan or completely manual, depending on the system itself.

4   MLaaS in Smart Agriculture Systems – Implementation and Evaluation

To create a valid proof of concept for the designed service, a mockup system that mimics Smart Agriculture System is implemented. Apache Kafka was chosen as a message broker, Apache Spark is chosen for data processing and machine learning model generation. The mockup has a component that behaves like a data source that is generating the data, a component that accepts the generated stream and for each received value uses MLaaS for prediction. The third component stores all the data, detects the drifts, and retrains the model when necessary. In this way, we can test all the crucial methods of the service. As for the use case, the simplest form of data is used.

Data generator is a simple python application that generates random numeric tuples and publishes them to a Kafka topic. This generator is implemented to be an interactive console tool, so it allows users to create data drifts by specifying the chances for the same. This option is important as we want to be able to track how the service behaves when we update the model. It also has a parameter that defines how often a new tuple is generated. The numeric data tuples are selected because we expect the actual data in the agricultural domain to have similar format, for example information about temperature, humidity, wind speed, etc. The data processor ingests generated data from the topic, and for each value communicates with the MLaaS. After receiving the prediction, it sends it along with the original data to the third component. This might not be the actual case, but MLaaS is not targeted by these assumptions, and it can make an implementation much easier.

To create a prototype of the defined MLaaS, the service is implemented using Python and Flask. Python is a popular and versatile programming language that has a large ecosystem of libraries and frameworks for machine learning. All the major platforms for large-scale machine learning have their python interface. Flask is a lightweight web framework that is easy to use and has a small footprint, making it well suited for building web APIs for machine learning services. Both Python and Flask have a large and active developer community, which can make it easier to find help and resources. Flask's built-in development server and support for WSGI make it easy to test and debug the service during development. Flask can be easily deployed to various platforms, such as AWS, GCP or Azure. The model itself is built using Spark platform, and it is saved in its format. This does not affect the interface itself, as we send serialized version of the model, no matter how it is generated. Data storage and batch analysis component receives the input stream and performs data drift detection, in our case based on percentage of wrong predictions for parameters. If the drift is detected, a new Spark task is launched that should perform training on all available historical data and send the obtained model to the service. After the update, the percentage of errors should drop.

To simulate the system, all the components are containerized. The entire system is run on machine equipped with Intel Core i7-4710HQ CPU and 16GB physical memory. With the data generator set to create a new record every 10ms, and probability of data drifts set to 5%, 17 records were processed before the model was successfully changed. This change includes the retraining process and actual update.

5 Conclusion

This paper continues the previous efforts for designing an architecture of Smart Agriculture System capable to support modern technologies, requirements and urge for unlocking the full potential of agricultural resources while providing agronomists possibility to react fast to unplanned situations in the fields.

In this paper we give a general idea of different challenges related to design of Smart Agriculture System, different advantages to using machine learning model as a service, as well as a prototype that support the microservice approach in the design of this type of applications. Future work will include further development of the rest of the system, and its implementation beyond the prototype stage.

References

1. Philipp, R., Mladenow, A., Strauss, C. Völz, A.: Machine learning as a service: challenges in research and applications. In: Proceedings of the 22nd International Conference on Information Integration and Web-based Applications & Services, pp. 396–406 (2020)
2. Bacciu, D., Chessa, S., Gallicchio, C., Micheli, A.: On the need of machine learning as a service for the internet of things. In Proceedings of the 1st International Conference on Internet of Things and Machine Learning, pp. 1–8 (2017)
3. Liakos, K.G., Busato, P., Moshou, D., Pearson, S., Bochtis, D.: Machine learning in agriculture: a review. Sensors **18**(8), 2674 (2018)
4. Benos, L., Tagarakis, A.C., Dolias, G., Berruto, R., Kateris, D.: Bochtis, B: Machine learning in agriculture: a comprehensive updated review. Sensors **21**(11), 3758 (2021)
5. Pathmudi, V.R., Khatri, N., Kumar, S., Abdul-Qawy, A.S.H., Vyas, A.K.: A systematic review of IoT technologies and their constituents for smart and sustainable agriculture applications. Scientific African **19**, e01577 (2023)
6. Assem, H., Xu, L., Buda, T.S., O'Sullivan, D.: Machine learning as a service for enabling Internet of Things and People. Pers. Ubiquit. Comput.Ubiquit. Comput. **20**, 899–914 (2016)
7. Raschka, S.: Model Evaluation, Model Selection, and Algorithm Selection in Machine Learning. Mach. Learn. arXiv:1811.12808, (2020)
8. Chaterji, S., et al.: Artificial intelligence for digital agriculture at scale: techniques, policies, and challenges. Comput. Soc. 1–15 (2020)
9. Erran, L., Chen, E., Hermann, J., Zhang, P., Wang, L.: Scaling machine learning as a service. In: Proceedings of the 3rd International Conference on Predictive Applications and APIs, PMLR 67, pp. 14–29 (2017)
10. Paleyes, A., Raoul-Gabriel, U., Neil, D.L.: Challenges in deploying machine learning: a survey of case studies. ACM Comput. Surv. **55**(6), 15–29 (2016)
11. Stojnev Ilic, A., Stojanovic, D., Stojanovic, N., Ilic, M.: A big data system architecture for adaptive streaming data analytics. In: Proceedings of the XVI International SAUM Conference on Systems, Automatic Control and Measurements, pp. 15–18 (2022)

A Semi-supervised Framework for Anomaly Detection and Data Labeling for Industrial Control Systems

Jiyan Salim Mahmud$^{(\boxtimes)}$ (iD), Ermiyas Birihanu, and Imre Lendak (iD)

Data Science and Engineering Department, Faculty of Informatics, Eötvös Loránd University, Pázmány Péter Street 1/A, Budapest 1117, Hungary

`{jiyan,ermiyasbirihanu,lendak}@inf.elte.hu`

Abstract. To ensure uninterrupted service delivery in critical sectors like electricity, water, and oil, safeguarding information systems against anomalies is imperative. Detecting anomalies within Industrial Control Systems (ICSs) is vital, but it's challenging without a comprehensive understanding of their causes. This necessitates a well-annotated dataset encompassing diverse anomaly types, often dependent on domain experts. Unfortunately, such datasets are scarce. To address this challenge, this study introduces a specialized framework for unsupervised anomaly detection and anomaly categorization within data collected from monitoring ICSs. The framework was validated using data from a Secure Water Treatment (SWaT) testbed, where multiple cyberattacks were intentionally introduced. An Isolation Forest model was utilized, achieving 77% accuracy in anomaly identification. These anomalies were then isolated from normal samples, and a K-means clustering model categorized similar attacks and labeled anomaly clusters. The most suitable supervised model for the data was determined through experimentation with various classifiers, including SVM, Random Forest, Decision Tree, KNearest Neighbor, and AdaBoost. Remarkably, K-Nearest Neighbor (KNN) outperformed all, achieving 98% accuracy. This framework automates anomaly detection, categorization and data labeling, elevating data quality and accuracy in ICS anomaly detection while reducing the need for manual expert intervention and addressing the challenge of limited well-annotated datasets and improving the overall security of vital infrastructure sectors.

Keywords: Anomaly detection · Data Labeling · Industrial Control Systems · Semi- supervised · Clustering · Classification

1 Introduction

Industrial control system (ICS) usually consists of a large number of components such as sensors, actuators and distributed control systems [1]. Having faults in any of these units might lead to physical damage in the system, cause injury or death or negative impact on the natural environment. With the increasing rate of new cyber-security threats in recent years, industry and researchers started to show more interest in protecting these systems from potential cyber adversaries.

© The Author(s), under exclusive license to Springer Nature Switzerland AG 2024
M. Trajanovic et al. (Eds.): ICIST 2023, LNNS 872, pp. 149–160, 2024.
https://doi.org/10.1007/978-3-031-50755-7_15

There are several different types of cyber-security threats, which can be detected using one of the two main techniques, namely knowledge-based and anomaly-based. Knowledge-based systems are able to detect only known anomalies such as known attack types and they are not suitable for new and zero-day attacks. Anomaly-based detection techniques on the other hand can detect novel attacks. Compared with traditional statistical models, using data mining techniques is very effective to analyze big data and investigate different patterns with a low dependency on domain knowledge [2]. This can be done with respect to one of the following approaches of machine learning: (1) supervised such as classification which needs a labeled dataset, (2) unsupervised with clustering and outlier detection techniques which do not require to have a labeled dataset, and (3) semi-supervised approaches involve training a model using a combination of labeled and unlabeled data [3]. To train any machine learning model a suitable dataset is needed. In the case of supervised learning the dataset must be labeled. However, most of the time, labeling a dataset requires significant time and/or computing resources, and we need to have domain experts. Additionally, most available data labeling techniques are not tailored for security monitoring data in ICS settings. As most real-life ICS monitoring datasets we encounter are unlabeled, any approach which helps system owners and/or operators in their data labeling approaches is valuable.

The contributions of this study are: i) Identifying and categorizing types of anomalies to identify the root cause of each anomaly, ii) Addressing accurate automated anomaly labeling for supervised anomaly detection for ICSs with multi-class classifiers, and iii) Unlike prior research primarily centered on SWaT- 2015 and SWaT-2017 datasets, our study focuses specifically on SWaT-2019, which has received less attention in the literature.

2 Related Works

Hawkins [4] defines an anomaly 'as an observation which deviates so much from other observations as to arouse suspicions that it was generated by a different mechanism'. Anomaly detection is the process of finding any abnormal activities in the given data in the shortest time possible. This topic is widely studied in statistics and data mining [5]. Anomaly detection algorithms can be classified as i) distribution-based methods such as auto encoders and adversarial feature learning, ii) distance-based methods (self-organizing maps, CMeans and adaptive resonance theory) and iii) density-based methods (local outlier factor/probability, DBSCAN and iv) tree-based with hyper methods such as isolation forest (IForest) [6].

A considerable amount of research focused on the problem of anomaly detection in industrial control systems. A dual Isolation Forest (DIF) framework was proposed in [7] for industrial control systems, particularly the SWaT dataset, aiming to improve anomaly detection with reduced computational complexity compared to existing methods for high-dimensional data. Ripan et al. introduced an Isolation Forest-based model for classifying cyber anomalies in a US Air Force network intrusion dataset [8]. They initially removed outliers using Isolation Forest and cleaned the dataset then assessed classification model performance with different classifiers. Our approach goes further, using Isolation Forest to identify and categorize anomalies within an unlabeled dataset.

A hybrid unsupervised clustering-based anomaly detection method was presented in [9], they combined multiple clustering algorithms to identify anomalous patterns in unlabeled data effectively.

The strengths of different clustering techniques were leveraged by the hybrid approach to enhance the accuracy and robustness of anomaly detection.

Labeling data is an important phase in supervised anomaly detection. It is a time consuming and costly procedure; many labeling techniques were proposed in the literature for different use cases. A clustering-based approach is presented in [10] to address the challenge of limited labeled data in network anomaly detection. By grouping similar network traffic instances and estimating labels for clusters, pseudo-labeled data can be generated for training supervised models. A novel approach was introduced in [11] for anomaly detection with partially observed anomaly types. The methodology combined clustering and filtering approaches to address the challenges posed by partially labeled data. By clustering observed anomalies and filtering unlabeled data, potential anomalies are identified. A weighted multi-class model is then constructed, leveraging instance weights and the similarity between labeled and unlabeled data to achieve accurate anomaly classification. Baek et al. [12] introduced an approach for enhancing supervised anomaly detection in enterprise and cloud networks. They proposed a method that addresses the labeling challenge by leveraging unsupervised learning techniques. By analyzing network traffic patterns and applying clustering algorithms, potential anomalies are automatically labeled. These labeled instances are then utilized to train supervised anomaly detection models.

There is a significant body of research dedicated to data labeling; however, many existing approaches are not specifically tailored for cybersecurity or anomaly detection in industrial control systems, highlighting the need for specialized labeling methods in these domains.

3   Data Exploration

In this study the Secure Water Treatment (SWAT) dataset was used, which was generated and collected at the iTrust Lab, Singapore University of Technology and Design [13]. The dataset collected in 2019 contains 77 features and it was collected during one day between 4:35 AM to 8:35 AM, with 6 different attacks injected between 7:06 AM and 8:30 AM. The dataset was collected from 39 actuators and 38 sensors distributed in different stages of the test bed. The attacks targeted 6 points of the system. Two of the attacks targeted sensors by spoofing the values while the remaining attacks changed actuators states. A brief overview of each attack is presented in Table 1. Figure 1 provides a visual representation of the affected sensors and actuators, the vertical line marks the start of the attack's injections. By looking at the diagram there are no changes in the data before 7:06 AM, while the immediate changes are clearly visible after this time for each attack. However, in FIT 401 and LIT 301 continuous changes are visible in the data even after stopping the attacks, which is due to after-attack outcomes and the resulting domain-specific anomalies. As all stages of the test-bed were connected, any unexpected change in one stage could affect the values of different sensors in the other stages of the system. The data is a combination of categorical (for actuators) which

All Attacks

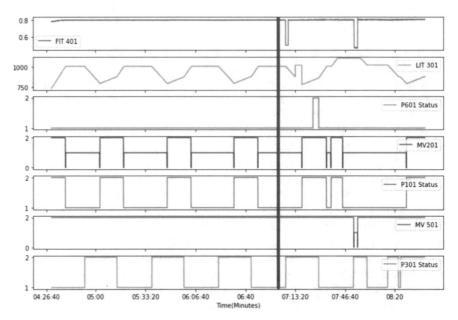

Fig. 1. A visualization of the affected sensors and actuators.

Table 1. Details of the targeted points

Attacks	Features	Type	Description	Time	Intent
1	FIT401	Sensor	UV de-chlorinator	7:06:46–7:10:36 Am	Switching off UV401[1]
2	LIT301	Sensor	Level Transmitter	7:17–7:19:32 AM	Underflow in T301[2]
3	P601	Actuator	Water Pump	7:26:57–7:30:48 AM	Increasing water in the tank
4	MV201 P101	Actuator Actuator	Water flow Water pump	7:39:50–7:46:20 AM 7:39:50–7:46:20 AM	Overflow tank T301
5	MV501	Actuator	Motorized Valve	7:54–7:56 AM	Drain water from RO[3]
6	P301	Actuator	UF feed Pump	8:02:56–8:16:18AM	Halt stage 3 (UF process)

---

[1] De-chlorinator; Removes chlorine from water.

[2] Ultra-Filtration Tank 301.

[3] Reverse Osmosis Feet Tank.

shows different states of each actuator and ordinal data (for sensor readings). In order to detect anomalies, we created a separated dataset from 7:00 AM and then using the time stamp of each attack we were able to manually label the data to evaluate our results at the end. Further details about the dataset are given in Sect. 5.1.

4 Proposed Anomaly Detection and Labeling Framework

Detecting anomalies in ICS is vital for ensuring the security, stability, and consistency of these systems. Having limited prior information about threats and not having labeled data are key challenges. In this study a framework is proposed to label ICS monitoring data. Although the SWaT-2019 dataset provides limited information about the timing of each attack, it is not fully labeled. Therefore, our objective was to detect anomalies and fully label the dataset.

A flow diagram of our framework is demonstrated in Fig. 2. It consists of four main stages, in the first stage the data is analysed and preprocessed using preprocessing techniques such as feature selection, one-hot encoding and normalization. In the second stage the anomaly samples are detected and separated from the normal data points. Two data sets are created, the "Attack" and'Normal' data sets. In the third stage only the Attack dataset was clustered and then labeled each cluster based on the time stamp records which is provided with the dataset, the labeled anomalies then combined with normal data to create a fully labeled dataset. In the fourth stage the dataset was fed to different classification algorithms and the best-performing model was chosen. A detailed explanation of each stage is provided in sub Sects. 4.1 to 4.4.

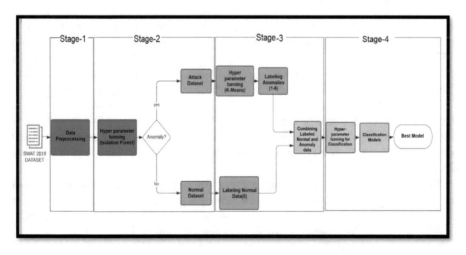

Fig. 2. The proposed semi-supervised anomaly detection framework.

4.1 Data Preprocessing

After the dataset was analyzed, it was found that 33 of the 77 features only had a single value. Therefore, they were removed from the dataset, categorical attributes were then label encoded. In order to find the highly correlated features Pearson correlation coefficient was used and features with higher than 0.99 correlation were removed. Since the dataset included both analog and discrete data, one hot encoding technique was used on the discrete features, and the data was normalized and scaled using min-max scaling [14, 15], thereby scaling the data to the [0,1] range.

4.2   Anomaly Detection

After conducting experiments with several unsupervised anomaly detection algorithms such as local outlier factor, DBSCAN and IF we observed that the IF algorithm consistently outperformed both in terms of detection accuracy and overall performance. This led us to select the IF algorithm for our anomaly detection task. it has low computational complexity, superior performance, and reliable results with both small samples and high-dimensional data [16]. These factors are crucial, particularly in real- time analyses where efficiency is essential.

IF leverages the rarity nature of anomalies by constructing an ensemble of random decision trees known as isolation trees, which resemble binary search trees. Furthermore, the isolation forest algorithm exhibits versatility by effectively handling both categorical and numerical data. Given that our dataset contains both types of data, therefore, isolation forest algorithm emerged as a suitable and effective choice for anomaly detection in our case.

4.3   Labeling the Anomalies

In order to group and label the anomalies K-means clustering was used which is a distance-based algorithm that groups data points based on their distances to centroid [17]. This algorithm was chosen because of its simplicity and solid performance for unsupervised learning [18]. The detected anomalies from isolation forest were fed to K-means. Grid search [19] was used to select the suitable number of clusters.

The attacks targeted specific points and in specific time slots. We theorized the data points for each attack would be very similar and K-means would be able to identify these similarities and cluster each attack. The time stamps of each cluster were com- pared with the list of attacks from the description and each attack was identified in different clusters.

In order to evaluate the results of k means clustering we used purity measure which can be used to evaluate the quality of each cluster, A grouping is considered pure when all data objects belonging to the same class are contained within the same cluster [20]. To calculate the purity of each cluster the following formula is used:

$$Purity(j) = 1/Nj * (max(nij)) \tag{1}$$

where: Purity (j) = purity value for the j cluster, Nj = the total number of data points that belongs to the j cluster, i, j = index of the cluster A higher purity value signifies the existence of a high-quality cluster.

4.4 Anomaly Classification Models

In this stage we created a dataset by combining both normal and labeled anomaly samples. This dataset was used to train several different machine learning classifiers to determine the most suitable model for the dataset. We experimented with different types of classifiers, i) distance-based used (Support Vector Machine (SVM) [21] and K-nearest Neighbours (KNN)) [22], ii) Tree-based (Decision Tree(DT) [23], iii) boosting ensemble classifiers (AdaBoost [24]). It must be mentioned that all these models are generally used for binary classifications while in our case there were more than two classes, there fore multi-class classification was used [25, 26].

4.5 Performance Metrics

Accuracy, precision, True Positive Rate (TPR), False Positive Rate (FPR), and F1-Score were applied to assess the effectiveness of our framework across various stages. Machine learning models demonstrating higher levels of accuracy, precision, recall, and F1-score were considered more favorable.

5 Experiments

5.1 Anomaly Detection with Isolation Forest

From the SWAT-2019 dataset we selected the time period after 7:00 am when the attacks were observed. The resulting dataset comprised a total of 6000 samples. To ensure data quality we preprocessed the data before anomaly detection, Isolation Forest was used to detect the anomalies in the dataset. We performed grid search hyper-parameter tuning for the model, the range and selected hyper-parameters are shown in Table 2.

Table 2. Range of Isolation Forest hyper-parameters in grid search

N-estimator	Max_ Samples	Contamination
[50, 30, 5]	[50, 30, 5]	[0,1,2, 0.3, 0.4, 0.5]

The Isolation Forest (IF) model identified a total of 2099 data points as anomalies out of the 6000 records in the dataset. After analyzing the anomalies, post-attack anomalies were successfully identified within the results. For example, in attack-2 which targeted the water level in a raw tank the anomalies are not visible immediately (Fig. 1). Therefore, it can be concluded that there are two types of anomalies in the dataset, the first type is the immediate changes in the system in the exact time of each attack injection and the second group is due to post-attack changes which caused domain specific anomalies such as increasing the water level or gradual changes in the chlorine level in the water. IF successfully detected anomalies of both types. In order to confirm the accuracy of our results, we require labels, even though the dataset lacks them. To address this limitation, we cross-referenced the attack timestamps within the dataset and conducted a visual inspection of each attack occurrence to manually designate them as anomalies the total numbers of anomalies in the dataset based on the time stamps was 2099. After evaluating model performance, the results show that IF was able to correctly detect 1735 anomalies with an accuracy of 77% and F-1 Score of 66%.The goal was to group each attack and label them separately therefore, we created a new dataset with 1735 anomalous samples and fed it to the K-means model in the next step. We used the elbow method [27] and cross validation with grid search and chose 10 for the number of clusters which had the highest silhouette coefficient [29] and were selected in the elbow curve as can be seen in Fig. 3b. A diagram of the silhouette score of clusters based on the anomaly dataset can be seen in Fig. 3a. We analysed the samples in each cluster, some attacks were grouped in different clusters, and in two clusters (3 and 5) the time period of the samples were outside any attack's time slots.

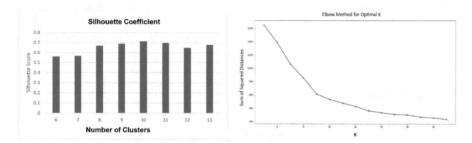

Fig. 3. Elbow method and Silhouette score of the clusters results.

After analyzing these samples, we were able to identify the samples in cluster 3 as domain-specific anomalies of attack 1, 2 and 6, and in cluster 5 the data points belonged to attack-3.

The details of each identified anomaly cluster are shown in Table 3. Upon comparing our findings with the timestamps of each attack, we determined that our K-means model successfully grouped similar anomalies into different clusters, achieving a purity value of 0.89. As a result, a labeled anomaly dataset was created, the number of samples for each class of anomalies are shown in Table 4 which suggests an imbalance ratio between them especially number of samples for attack 5 is lower than the rest of the classes.

Table 3. Details of each anomaly cluster.

Cluster	0	1	2	3	4	5	6	7	8	9
Number of data points	97	375	395	265	137	142	79	129	49	67
Attacks	6	4	2	1,2,6	5	3	6	3	1	1

Table 4. Number of samples labeled for each attack.

Attacks	1	2	3	4	5	6	Total
Target	FIT401	LIT301	P601	MV201-P101	MV501	P301	
Number of Samples	324	429	270	375	137	200	1735

5.2 Finding the Optimal Classification Model for the Dataset

Supervised models tend to outperform in anomaly detection scenarios, but they require training on a labeled dataset. In order to find the best supervised model for our study we created a new labeled dataset by combining both labeled anomalies and normal data sets obtained from the clustering stage. This dataset consisted of 6000 samples with 1735 anomalous and 4265 normal samples. The data was divided into training and testing sets using a stratified sampling approach, with a ratio of 75% for training and 25% for testing. We used the default hyper-parameters for all the models in the first round and then we applied hyper-parameter tuning in order to improve the performance of each model. After conducting hyper-parameter tuning, the optimal hyper-parameters for each model were determined. Among the classifiers, K-Nearest Neighbors (KNN) demonstrated the highest accuracy of 97%, followed by Support Vector Machine (SVM) with 92% accuracy.

AdaBoost, on the other hand, exhibited the lowest accuracy of 66% with default hyper-parameters in the initial round. However, after hyper-parameter tuning, the performance of all models improved. KNN and SVM surpassed the other classifiers with accuracy rates of 98% and 97% respectively. A detailed evaluation of the classification results can be found in Table 5. The results indicate high precision and recall rates for the normal class across all models, regardless of hyper-parameter tuning. A comparison between SVM and KNN for each attack revealed that KNN achieved high precision and recall for all attacks except for attack 6, where recall dropped to 84%. In contrast, SVM exhibited a recall score of only 38% for attack 6, suggesting its failure in correctly identifying this specific attack. However, after hyper-parameter tuning, the performance of both models improved significantly. SVM achieved an 88% recall for attack 6, while KNN's recall remained high at 98%. In both rounds, AdaBoost performed poorly, failing to detect several attacks (1, 3, 4, and 6 in round 1; 2, 3, and 5 in round 2). Despite the high accuracy of Decision Tree (DT) in the second round, it also misclassified attacks 1 and 6. The fact of having an imbalanced multi-class dataset and the similarity between the anomalies might have contributed to the low performance of AdaBoost and DT, requiring further investigation.

Table 5. The results of the classification stage for each Class

Model	Class	With default hyper-parameters			After hyper-parameter tuning			Support
		Precision (%)	Recall (%)	F1-Score (%)	Precision (%)	Recall (%)	F1-Score (%)	
SVM	0 (Normal)	95	94	94	99	97	98	1067
	1 (Attack 1)	84	100	91	98	100	99	81
	2(Attack 2)	84	93	88	91	99	95	108
	3 (Attack 3)	63	78	69	77	94	85	67
	4 (Attack 4)	100	90	95	100	100	100	94
	5 (Attack 5)	100	100	100	100	100	100	34
	6 (Attack 6)	76	38	51	100	88	94	50
DT	0 (Normal)	89	70	78	89	100	94	1067
	1 (Attack 1)	49	98	65	0	0	0	81
	2(Attack 2)	48	100	65	100	100	100	108
	3 (Attack 3)	1.00	100	100	99	100	99	67
	4 (Attack 4)	100	0	0	100	100	100	94
	5 (Attack 5)	100	100	100	100	100	100	34
	6 (Attack 6)	29	100	45	0	0	0	50
KNN	0 (Normal)	98	98	98	99	99	99	1067
	1 (Attack 1)	96	96	96	100	98	99	81
	2(Attack 2)	89	100	94	99	94	97	108
	3 (Attack 3)	88	88	88	87	97	92	67
	4 (Attack 4)	100	100	100	100	100	100	94
	5 (Attack 5)	100	100	100	100	100	100	34
	6 (Attack 6)	100	84	91	100	98	99	50
Ada Boost	0 (Normal)	83	80	81	82	99	90	1067
	1 (Attack 1)	0	0	00	63	80	71	81
	2(Attack 2)	25	100	39	0	0	0	108
	3 (Attack 3)	0	0	0	0	0	0	67
	4 (Attack 4)	0	0	0	100	85	92	94
	5 (Attack 5)	100	100	100	0	0	0	34
	6 (Attack 6)	0	0	0	100	44	61	50

6   Conclusion and Future Work

Throughout the years, numerous methods were developed for detecting anomalies in ICS. However, these existing approaches encounter practical challenges due to reliance on manual labeling of training data or inadequate performance without knowledge of historical anomalies. In this study we investigated the possibility of detecting anomalies in an unlabeled dataset, labeling those anomalies and training different classifier models. After preprocessing the data, we used Isolation Forest to detect anomalies, the model was able to detect anomalies with 77% accuracy. The detected anomalies were labeled using K-means and then a fully labeled dataset was created combining the results from labeled clusters and normal data. In order to find the optimal classification model, we trained several classifiers such as SVM, Decision Tree, K-Nearest Neighbour and AdaBoost. KNN outperformed the rest with 98% accuracy and AdaBoost performed the worst with 81% accuracy, having an imbalanced multi-class dataset and similarity between anomalies can be the reason for poor performance of other classifiers. Our proposed data labelling and anomaly detection framework can be used to detect and label different types of anomalies in an unlabeled data set specially in ICSs, however in order to improve this approach further analyses with different datasets is required.

Possible future directions for research include enhancing the framework's performance through alternative detection algorithms and hyper-parameter tuning techniques, achieving a balanced dataset via under or oversampling methods, exploring diverse datasets, conducting time series analysis, and evaluating the proposed framework in real-time anomaly detection scenarios. Additionally, considering alternative approaches like active learning and ensemble methods for data labeling can also be beneficial.

References

1. Al-Abassi, A., Karimipour, H., Dehghantanha, A., Parizi, R.M.: An ensemble deep learning-based cyber-attack detection in industrial control system. IEEE Access **8**, 83965–83973 (2020)
2. Liu, X., Ding, Y., Tang, H., Xiao, F.: A data mining-based framework for the identification of daily electricity usage patterns and anomaly detection in building electricity consumption data. Energy Build. **231**, 110601 (2021)
3. Guo, P., Wang, L., Shen, J., Dong, F.: A hybrid unsupervised clustering-based anomaly detection method. Tsinghua Sci. Technol. **26**(2), 146–153 (2020)
4. Hawkins, D.M.: Identification of outliers. Springer Netherlands, Dordrecht (1980). https://doi.org/10.1007/978-94-015-3994-4
5. Pathan, A.K: The state of the art in intrusion prevention and detection, vol. 44. CRC Press, Boca raton (2014)
6. Khaledian, E., Pandey, S., Kundu, P., Srivastava, A.K.: Real-time synchrophasor data anomaly detection and classification using isolation forest, kmeans, and loop. IEEE Trans. Smart Grid **12**(3), 2378–2388 (2020)
7. Elnour, M., Meskin, N., Khan, K , Jain, R.: A dual-isolation-forests based attack detection framework for industrial control systems. IEEE Access **8**, 36639–36651 (2020)
8. Ripan, R.C., Sarker, I.H., Musfique, M.: An isolation forest learning based outlier detection approach for effectively classifying cyber anomalies. In: Abraham, A., Hanne, T., Castillo, O., Gandhi, N., Tatiane Nogueira Rios, T. (ed.) HIS 2020. AISC, vol. 1375, pp. 270–279. Springer, Cham (2021). https://doi.org/10.1007/978-3-030-73050-5_27
9. Guo, P., Wang, L., Shen, J., Dong, F.: A hybrid unsupervised clustering-based anomaly detection method. Tsinghua Sci. Technol. **26**(2), 146–153 (2021)
10. Baek, S., Kwon, D., Suh, S.C., Kim, H., Kim, I., Kim, J.: Clustering-based label estimation for network anomaly detection. Digit. Commun. Netw. **7**, 37–44 (2020)
11. Zhang, Y.-L., Li, L., Zhou, J., Li, X., Zhou, Z.-H.: Anomaly detection with partially observed anomalies. In: Companion Proceedings of the The Web Conference 2018, pp. 639–646 (2018)
12. Baek, S., Kwon, D., Kim, J., Suh, S.C., Kim, H., Kim, I.: Unsupervised labeling for supervised anomaly detection in enterprise and cloud networks. In: 2017 IEEE 4th International Conference on Cyber Security and Cloud Computing (CSCloud), pp. 205–210. IEEE (2017)
13. Mathur, A.P., Tippenhauer, N.O.: Swat: a water treatment testbed for research and training on ics security. In: 2016 International Workshop on Cyber-Physical Systems for Smart Water Networks (CySWater), pp. 31–36. IEEE (2016)

14. Ahsan, M.M., Parvez Mahmud, M.A., Saha, P.K., Gupta, K.D., Siddique, Z.: Effect of data scaling methods on machine learning algorithms and model performance. Technologies **9**(3), 52 (2021)
15. Ganapathi Raju, V.N., Prasanna Lakshmi, K., Jain, V.M., Kalidindi, A., Padma, V.: Study the influence of normalization/transformation process on the accuracy of supervised classification. In: 2020 Third International Conference on Smart Systems and Inventive Technology (ICSSIT), pp. 729–735. IEEE (2020)
16. Liu, F.T., Ting, K.M., Zhou, Z.H.: Isolation forest. In: 2008 Eighth IEEE International Conference on Data Mining, pp. 413–422. IEEE (2008)
17. Grira, N., Crucianu, M., Boujemaa, N.: Unsupervised and semi-supervised clustering: a brief survey. Rev. Mach. Learn. Techn. Proc. Multimedia Content **1**, 9–16 (2004)
18. Rajabi, A., Eskandari, M., Ghadi, M.J., Li, L., Zhang, J., Siano, P.: A comparative study of clustering techniques for electrical load pattern segmentation. Renew. Sustain. Energy Rev. **120**, 109628 (2020)
19. Feurer, M., Hutter, F.: Hyperparameter optimization. In: Hutter, F., Kotthoff, L. (ed.) Automated machine learning. TSSCML, pp. 3–33. Springer, Cham (2019). https://doi.org/10.1007/978-3-030-05318-5_1
20. Sebayang, F.A., Lydia, M.S., Nasution, B.B.: Optimization on purity k-means using variant distance measure. In: 2020 3rd International Conference on Mechanical, Electronics, Computer, and Industrial Technology (MECnIT), pp. 143–147 (2020)
21. Lucas, B.: Proximity forest: an effective and scalable distance-based classifier for time series. Data Mining Knowl. Dis. **33**(3), 607–635 (2019)
22. Zuber, M., Sirdey, R.: Efficient homomorphic evaluation of k-nn classifiers. Proc. Priv. Enhancing Technol. **2021**(2), 111–129 (2021)
23. Rai, K., Syamala Devi, M., Guleria, A.: Decision tree based algorithm for intrusion detection. Inter. J. Adv. Netw. Appli. **7**(4), 2828 (2016)
24. Javed, A.R., Jalil, Z., Moqurrab, S.A., Abbas, S., Liu, X.: Ensemble adaboost classifier for accurate and fast detection of botnet attacks in connected vehicles. Trans. Emerging Telecommun. Technol., e4088 (2020)
25. Maglaras, L.A., Jiang, J.: Intrusion detection in scada systems using machine learning techniques. In: 2014 Science and Information Conference, pp. 626–631. IEEE (2014)
26. Zhang, J., Xia, K., He, Z., Yin, Z., Wang, S.: Semi-supervised ensemble classifier with improved sparrow search algorithm and its application in pulmonary nodule detection. Mathematical Prob. Eng. **2021** (2021)
27. Syakur, M.A., Khotimah, B.K., Rochman, E.M.S., Satoto, B.D.: Integration k-means clustering method and elbow method for identification of the best customer profile cluster. IOP Conf. Ser. Mater. Sci. Eng. **336**, 012017 (2018)
28. Shahapure, K.R., Nicholas, C.: Cluster quality analysis using silhouette score. In: 2020 IEEE 7th International Conference on Data Science and Advanced Analytics (DSAA), pp. 747–748. IEEE (2020)

Multilingual Transformer and BERTopic for Short Text Topic Modeling: The Case of Serbian

Darija Medvecki[(✉)] [iD], Bojana Bašaragin[iD], Adela Ljajić[iD], and Nikola Milošević[iD]

The Institute for Artificial Intelligence Research and Development of Serbia, Fruškogorska 1, 21000 Novi Sad, Serbia

{darija.medvecki,bojana.basaragin,adela.ljajic,
nikola.milosevic}@ivi.ac.rs

Abstract. This paper presents the results of the first application of BERTopic, a state-of-the-art topic modeling technique, to short text written in a morphologically rich language. We applied BERTopic with three multilingual embedding models on two levels of text preprocessing (partial and full) to evaluate its performance on partially preprocessed short text in Serbian. We also compared it to LDA and NMF on fully preprocessed text. The experiments were conducted on a dataset of tweets expressing hesitancy toward COVID-19 vaccination. Our results show that with adequate parameter setting, BERTopic can yield informative topics even when applied to partially preprocessed short text. When the same parameters are applied in both preprocessing scenarios, the performance drop on partially preprocessed text is minimal. Compared to LDA and NMF, judging by the keywords, BERTopic offers more informative topics and gives novel insights when the number of topics is not limited. The findings of this paper can be significant for researchers working with other morphologically rich low-resource languages and short text.

Keywords: BERTopic · Topic Modeling · Serbian Language · Natural Language Processing

1 Introduction

As an unsupervised task, topic modeling is an invaluable tool in many areas, especially where user-generated content (emails, user comments, reviews, complaints, etc.) needs to be analyzed without prior annotation. While there is significant work done in this area for English, especially using Latent Dirichlet Allocation (LDA) [1], a classical topic modeling method, this is an unexplored area for Serbian. The authors in [2] initiated this work by applying LDA and Nonnegative Matrix Factorization (NMF) [3] on tweets in Serbian to find the hidden reasons for COVID-19 vaccine hesitancy.

Recently, BERTopic [4] has been proposed as a new, more flexible model for detecting topics in unannotated text. Unlike more conventional methods that use bag-of-words approaches to describe documents, BERTopic uses state-of-the-art pre-trained language

M. Trajanovic et al. (Eds.): ICIST 2023, LNNS 872, pp. 161–173, 2024.
https://doi.org/10.1007/978-3-031-50755-7_16

models to create document embeddings, which enables capturing semantic relationships between words. As this framework relies on context, by definition, it should not require substantial data preprocessing. In contrast, both LDA and NMF require extensive preprocessing and significant parameter tuning. Some preliminary research showed that BERTopic generalizes better than LDA judging by topic coherence [5, 6] and topic diversity scores [6], and that it creates more clear-cut topics and gives more novel insights compared to NMF and LDA [7].

Since BERTopic has not been applied to Serbian yet, our aim was to explore its usability and performance, particularly on short minimally preprocessed text. As a morphologically rich language, Serbian normally requires the lemmatization step for most NLP tasks. Using pre-trained language models as embedding models for BERTopic could render this step unnecessary. Our results can serve as pointers for other researchers working with short text and morphologically rich languages.

2 Related Work

Several methods can provide insight into structures hidden in large amounts of text by grouping them into topics. Two of the most used traditional topic modeling methods are Latent Dirichlet Allocation (LDA) [1] and Nonnegative Matrix Factorization (NMF) [3]. As language-agnostic models, both have been applied in the context of different languages and texts of various lengths. The downside to these models, often mentioned in research papers, can be summarized into three points. First, they have difficulties modeling short text due to the data sparseness problem [8]. Second, LDA and NMF both require the number of topics as one of their initial parameters. Finding this number requires extensive parameter tuning, making the models difficult to optimize [9]. Third, both models require significant preprocessing, including stemming and/or lemmatization, which can produce unreliable and ambiguous results, depending on the language and the quality of the algorithms used [10].

The rise of self-attention-based models and the concept of pre-training in the late 2010s and early 2020s gave rise to a number of pre-trained language models (PLMs). PLMs made significant advances in many fields of natural language processing by introducing pre-trained contextual embeddings. BERTopic [4] is the most recent topic modeling method that leverages PLMs to create document embeddings. Combining such embeddings with a class-based TF-IDF procedure allows BERTopic to better deal with sparse data.

So far, BERTopic has been used in diverse domains (hospitality, sports management, finance, and medicine) to gain insight into different types of text (customer reviews, students' answers, consumer and general domain data), mostly in English [11–14]. Unlike LDA or NMF, applying BERTopic to other languages requires language-specific sentence embeddings, an issue that can be overcome by leveraging monolingual or multilingual PLMs that fit the specific use case. In a pilot study by [5], BERTopic was tested on news texts using several mono- and multilingual PLMs trained on Arabic data. Authors of [15] applied BERTopic that employs AraBERT PLM to tweets as one of the steps in designing a cognitive distortion classification model. Thanks to its ability to integrate multilingual PLMs, researchers in [16] used BERTopic to make a hoax news classification pipeline for Indonesian.

When applying BERTopic to news and tweets along with NMF and LDA, authors in [4] reported high topic coherence scores across all datasets, with the highest ones on slightly preprocessed text of tweets. Reseachers in [6] compared the performance of LDA and BERTopic with two clustering algorithms (HDBSCAN and k-means) on student comments and news and found that BERTopic using HDBSCAN achieved the highest topic coherence and topic diversity scores. Authors in [5] found that BERTopic with AraVec2.0 as a word embedding model outperformed NMF, LDA and other BERTopic embedding models in terms of NPMI topic coherence scores.

The results achieved by BERTopic so far suggest that it is a powerful model able to overcome the difficulties of more traditional options.

3 Methodology

We started the process of exploring BERTopic by defining the research questions. After applying several architecture variants, we specified the architecture that would best fit our needs and the dataset. Since the first study of using topic modeling on Serbian short text was performed on tweets, we used the same dataset as an opportunity to compare BERTopic against LDA and NMF on the same data.

3.1 Research Questions

There were two research questions we wanted to address:

RQ 1: How does preprocessing affect the quality of BERTopic topic representations in case of a morphologically rich language?
Since BERTopic relies on an embedding approach which takes context into account, in theory, it should provide informative results even without significantly changing the text structure first. In the case of tweets, researchers in [4] and [7] claim that employing minimal preprocessing is sufficient for English. Proving that BERTopic can overcome the need for more thorough preprocessing (lemmatization) in morphologically rich languages would make topic modeling for Serbian a more straightforward task as it would: 1) assume minimum human involvement in text preparation, 2) prevent relying on restoration of diacritics and lemmatization, which can possibly create faulty lemmas and change the structure of topics.

RQ 2: How does BERTopic compare to LDA and NMF on lemmatized text?
Although [2] found some differences in topic quality and stability when using LDA and NMF, their performance was quite similar. We wanted to explore if, under the same conditions, BERTopic would give more informative topics in terms of keywords, and offer some new insights.

3.2 The Model

BERTopic has a modular architecture made up of several core layers that can be built upon. Those layers are sequential and include embedding extraction, dimensionality reduction and clustering, and creation of topic representation. We discuss the details of our architecture (Fig. 1) in the following subsections.

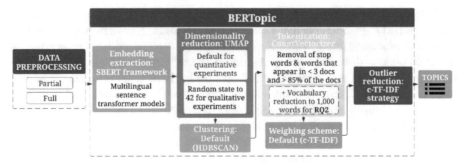

Fig. 1. Workflow and details of our BERTopic architecture.

Embedding Models. Although BERTopic supports the implementation of several different embedding techniques, by default it uses sentence transformers [17]. They are often optimized for semantic similarity, which can significantly help in the clustering task. Since there is no sentence transformer trained solely on Serbian, we could either try to optimize a word embedding model for Serbian or use one of the available multilingual sentence transformers that were trained on multilingual data, including Serbian. In all our experiments, we used a sentence transformer architecture. The three multilingual sentence transformer models that are trained on parallel data for 50+ languages including Serbian are:

- *distiluse-base-multilingual-cased-v2*: knowledge distilled version of multilingual Universal Sentence Encoder that encodes the sentences into 512-dimensional dense vector space. Its size is 480 MB (135 million parameters).
- *paraphrase-multilingual-MiniLM-L12-v2*: multilingual version of paraphrase-MiniLM-L12-v2 that maps sentences to a 384-dimensional dense vector space. The size of this model is 420 MB (117 million parameters).
- *paraphrase-multilingual-mpnet-base-v2*: the largest model with the size of 970 MB (278 million parameters). It is a multilingual version of paraphrase-mpnet-base-v2 that maps sentences to a 768-dimensional dense vector space.

Dimensionality Reduction and Clustering. To reduce the embedding dimensionality, we used UMAP, a default BERTopic dimensionality reduction algorithm. We formed the clusters using HDBSCAN with default parameters.

Creation of Topic Representation. CountVectorizer, as the default BERTopic vectorizer model, and c-TF-IDF, which models the importance of words in clusters instead of individual documents, are together responsible for extracting topic representations from the previously created clusters of documents. We used CountVectorizer to define word filtering options: stop words, filtering of the words that appear in less than 3 tweets and in more than 85% of the tweets.

Outlier Reduction. When used with HDBSCAN, BERTopic creates a bin for topic outliers, which can sometimes contain over 74% of the dataset [6]. To prevent this, the outlier reduction step can optionally be added on top of the BERTopic architecture.

We reduced the outliers to almost 0 using the *reduce_outliers* BERTopic function with c-TF-IDF as the reduction strategy.

3.3 The Data

To fit the model, we used the dataset created for the study presented in [2]. The dataset is composed of 3,286 tweets that express negative attitudes towards COVID-19 vaccination in Serbia. The authors report that the dataset was manually checked for topics, therefore we knew that it contains 15 broad topics. We applied two preprocessing scenarios to the tweets:

1. Partial preprocessing, which consisted of transliterating from Cyrillic to Latin, removal of links, mentions and emojis, conversion of hashtags into words, removal of numbers and punctuation, and lowercasing.
2. Full preprocessing, which consisted of partial preprocessing and lemmatization.

Transliteration was performed using *srtools* [18], lemmatization was performed using the *classla* library for non-standard Serbian [19], and we used custom Python regex for the remaining preprocessing steps.

3.4 Evaluation

We performed quantitative evaluation of our models using two metrics – topic coherence (TC) and topic diversity (TD) – both commonly used to evaluate topic models [4, 6, 20]. According to [21], TC represents average semantic relatedness between topic words. The specific flavor of TC we used was NPMI [22]. NPMI ranges from -1 to 1, where a higher score signifies that words in a topic are more strongly related. TD [20] measures the percentage of unique words in the top-n words across the topics. It ranges from 0 to 1, where 1 signifies more varied and 0 indicates redundant topics.

For RQ1, we evaluated all three BERTopic embedding models in two preprocessing settings – partial preprocessing and full preprocessing, as defined in Sect. 3.3. For RQ2, we compared TC and TD between BERTopic models with LDA and NMF. Since both LDA and NMF require fully preprocessed text, we used that preprocessing scenario for all the models. Researchers in [2] compare LDA and NMF using the vocabulary reduction to 1,000 words, so we set the same parameter in BERtopic models (see Fig. 1) for the sake of comparability. In both experiments, we averaged TC and TD across 3 runs for 10–50 topics with steps of 10 for every model.

Besides using TC and TD as quantitative measures, we also manually checked the topics for keyword diversity, overall interpretability, and novelty. To obtain repeatable results, we set the UMAP random state to 42. For both research questions, we compare the topics and keywords of best performing models after quantitative evaluation. By default, BERTopic does not put any limitations on the number of topics, which can result in hundreds of topics for larger datasets [7]. We predefined this number in both RQ for the sake of comparability. For RQ1, we set the number of topics to 15 for both models to match the number of topics manually identified by [2]. For RQ2, we matched the most optimal number of topics of the best performing traditional topic model. We also reduced the vocabulary to 1,000 words for the same reason.

4   Results & Discussion

4.1   RQ1

Quantitative Evaluation. For partially preprocessed text, the third and the largest model gave the best TC (-.133) and TD (.896) scores (see Table 1). The other two embedding models share the TC score of -.145. The second-best TD score was achieved by *paraphrase-multilingual-MiniLM-L12-v2*. For fully preprocessed text, *distiluse-base-multilingual-cased-v2* had the best results for TC and it shares the best TD score with our second model. All three models achieve high TD values in both preprocessing scenarios, suggesting diverse keywords regardless of the model. TD scores are slightly higher for partially preprocessed text, with *paraphrase-multilingual-mpnet-base-v2* achieving the highest one. On the other hand, TC scores are slightly better for all three models in the case of fully preprocessed text, suggesting more coherent topics.

Table 1. Topic coherence (TC) and topic diversity (TD) for different BERTopic embedding models and two preprocessing scenarios.

BERTopic embedding model	Partial preprocessing		Full preprocessing	
	TC	TD	TC	TD
distiluse-base-multilingual-cased-v2	-.145	.887	**-.042**	**.868**
paraphrase-multilingual-MiniLM-L12-v2	-.145	.895	-.063	**.868**
paraphrase-multilingual-mpnet-base-v2	**-.133**	**.896**	-.058	.860

Qualitative Evaluation. We compared the models with the highest TC and TD scores per preprocessing scenario during the quantitative evaluation: *paraphrase-multilingual-mpnet-base-v2* for partially preprocessed text and *distiluse-base-multilingual-cased-v2* for fully preprocessed text. To start, we paired together the topics yielded by the two models based on keywords. In Table 2 we show five illustrative topics that cover over 60% of documents in each scenario. By looking at 10 representative keywords, we can see that even without lemmatization we obtained informative topics.

As for keyword diversity and interpretability, except for several morphological variations of three nouns and a verb in the keywords of the first model (underlined), the words are varied for both models. The bolded keywords are the ones that clearly point to the interpretation of each topic. For example, the fourth topic shows less keyword diversity for the partially preprocessed text since there are four different morphological forms of the same word (*dete* – child) in the top 10 keywords. Despite this, the keywords are still informative and indicate the concern over the side effects of mandatory vaccination for children. While keywords in this topic for fully processed text may seem more varied, the number of tweets and inspection of representative documents prove that this variety stems from several distinct topics merged into one. The same can be noticed in the first topic, but this time for partially preprocessed text.

Table 2. Overview of ten keywords and the number of tweets per topic for five illustrative topics obtained by *paraphrase-multilingual-mpnet-base-v2* and *distiluse-base-multilingual-cased-v2*.

paraphrase-multilingual-mpnet-base-v2		*distiluse-base-multilingual-cased-v2*	
Keywords	No. of tweets	Keywords	No. of tweets
eksperiment (experiment), **nuspojave** (side effects), zna (knows), sto (hundred), **dnk** (DNA), **mrna** (mRNA), godina (year), dr (dr), ce (will), **bil** (Bill)	1097	**eksperimentalan** (experimental), **ispitivanje** (examining), faza (phase), lek (drug), medicinski (medical), nijedan (none), proći (pass), struka (experts), **nuspojava** (side effect), **proizvođač** (producer)	215
imunitet (immunity), **zaštita** (protection), **simptomi** (symptoms), <u>**koronu**</u> (corona), imam (I have), **štiti** (protects), <u>**korone**</u> (corona), **vakcinaciju** (vaccination), **antitela** (antibodies), **prirodni** (natural)	451	**virus** (virus), **imunitet** (immunity), **soj** (strain), **simptom** (symptom), grip (flu), **štititi** (protect), napraviti (to make), hiv (HIV), **prirodan** (natural), **preležati** (to develop immunity)	606
<u>**smrti**</u> (death), **slučajeva** (cases), <u>**umrli**</u> (died), godine (year), <u>**umrlo**</u> (died), broj (number), **smrtnih** (death, adj), <u>**smrt**</u> (death), miliona (million), **umrlih** (dead)	212	**umreti** (to die), broj (number), **bolnica** (hospital), **umirati** (die), slučaj (case), **ubiti** (to kill), **smrtan** (deadly), **zaraziti** (to infect), **respirator** (respirator), tri (three)	263
<u>**decu**</u> (children), <u>**deca**</u> (children), <u>**dece**</u> (children), **štete** (damage), **pravo** (right), <u>**deci**</u> (children), vakciniše (vaccinates), svoju (one's own), vakcinacije (vaccination), vakcinom (vaccine)	153	**dete** (child), **gejts** (Gates), **bil** (Bill), misliti (think), **dnk** (DNA), **nuspojava** (side effect), srbija (Serbia), **posledica** (consequence), narod (people), **menjati** (change)	1035
<u>**maske**</u> (masks), **štite** (protect), <u>**maska**</u> (mask), nose (wear), **distanca** (distance), **mere** (measures), nosi (wears), **štete** (damage), ratom (war), svjet (world)	94	**maska** (mask), nositi (to wear), **distanca** (distance), **štititi** (to protect), dobro (well), **mera** (measure), **odbiti** (to reject), **pasoš** (passport), naravno (of course), ruka (hand)	144

Judging by the results, it seems that parameters need to be separately defined for different levels of preprocessing. Applying the same parameters to both scenarios affects the keywords, which is reflected in the number of documents as well. However, even under these conditions, the keywords are diverse and informative for both preprocessing scenarios, indicating that BERTopic can be successfully applied to partially preprocessed text in Serbian.

4.2 RQ2

Quantitative Evaluation. The results in Table 3 show that differences in TD and TC scores between BERTopic models are slight, with *distiluse-base-multilingual-cased-v2* performing best, as in the case of RQ1. Compared to LDA and NMF, BERTopic achieved better TC scores. NMF performed slightly worse (-.065 compared to the lowest BERTopic score of -.054), while LDA showed a more significant drop (-.104). Even though our TD scores were high for all the three BERTopic models for Serbian dataset, LDA achieved a slightly better result (.897). The only model with a TD score lower than .8 was NMF. Comparing RQ2 results to the TC and TD scores for the RQ1 (Table 1), it can be concluded that vocabulary reduction only slightly influenced these metrics in case of fully preprocessed text.

Table 3. Comparison of TC and TD values across three BERTopic embedding models, LDA and NMF on fully preprocessed text and with vocabulary reduced to 1,000 words.

Model	TC	TD
distiluse-base-multilingual-cased-v2	**-.050**	.861
paraphrase-multilingual-MiniLM-L12-v2	-.054	.859
paraphrase-multilingual-mpnet-base-v2	<u>-.051</u>	.858
LDA	-.104	**.897**
NMF	-.065	.795

Qualitative Evaluation. We extracted the topics generated by the best performing BERTopic model during RQ2 quantitative evaluation, which is *distiluse-base-multilingual-cased-v2*. We set the number of topics for BERTopic to 14 to match the most optimal number of topics for LDA for this dataset [2], since LDA showed the best TD score during quantitative evaluation.

Authors in [2] gather all the LDA and NMF topics into five groups and 16 subgroups that we used to name and align BERTopic topics (Table 4). When defining the topic names, [2] looked at 20 keywords and representative documents, which we did as well. Topics detected by BERTopic match the ones in [2] in 69% of cases, meaning that 31% of the topics found by LDA and NMF were not detected by BERTopic. BERTopic did not isolate any topics that deal with the number of doses, which were the topics detected by both LDA and NMF. One BERTopic topic contains several topics in one, combining a conspiracy theory of DNA change with concern over not having a choice with vaccinating children and doubt about effectiveness for new strains, similarly as in RQ1. In the case of the topic dealing with mistrust of authorities, BERTopic breaks it into a total of eight topics, each covering a specific aspect of mistrust. One of them is a completely novel topic that groups tweets regarding concerns about different vaccine manufacturers (84 tweets). While some topics seem uninterpretable by looking at the keywords (e.g., third topic under *Mistrust of government and political decision makers*), there is a clear idea of the topic based on the keywords for most topics.

With the same parameter settings, but without limiting the number of topics, BERTopic found 41 topics for this dataset. By closer inspection of these topic representations, BERTopic detected all reasons identified by [2] as separate topics. Although this number of topics is not appropriate for this dataset, this approach could be an important starting point for further analysis and parameter optimization, as it could provide more detailed or new insights into topics hidden in the dataset. In some cases, identifying smaller topics may be very important in some fields.

Another important BERTopic parameter that significantly affects the number of generated topics is the minimum topic size (*min_topic_size*), which is the minimum number of documents to form a topic. The default parameter value in BERTopic, which we used in our experiment, is 10. Increasing this value results in a lower number of clusters/topics when HDBSCAN is used as the clustering algorithm. When we set this parameter to 15, BERTopic generated 23 topics without any outliers.

To answer RQ2, BERTopic's performance is comparable to LDA and NMF on this dataset, but not under the same parameter settings. When the number of topics is not limited and when the minimum topic size parameter value is changed, BERTopic could potentially provide new insights.

Table 4. LDA and NMF topics for the COVID-19 dataset (taken from [2]) and the corresponding BERTopic topics with the number of tweets.

Reasons for vaccine hesitancy identified by LDA and NMF	BERTopic topic representations	No. of tweets
Concern over general side effects	umreti (to die), smrt (death), nuspojava (side effect), umirati (to be dying), bolnica (hospital), slučaj (case), život (life), smrtan (deadly), tri (three), nevakcinisan (unvaccinated)	427
Concern over side effects for children	dete (child), gejts (Gates), bil (Bill), nov (nov), dnk (DNA), soj (strain), bolest (illness), menjati (change), priča (story), sloboda (freedom)	962
Concern over side effects due to many required doses	-	-

(continued)

Table 4. *(continued)*

Reasons for vaccine hesitancy identified by LDA and NMF	BERTopic topic representations	No. of tweets
Concern over vaccine effectiveness: natural immunity is better protection	virus (virus), imunitet (immunity), simptom (symptom), hiv (HIV), otrov (poison), prirodan (natural), napraviti (make), štititi (protect), zaraziti (infect), bolest (illness)	517
Concern over vaccine effectiveness: vaccines are not effective against new COVID-19 strains	dete (child), gejts (Gates), bil (Bill), nov (nov), dnk (DNA), soj (strain), bolest (illness), menjati (change), priča (story), sloboda (freedom)	962
Concern over vaccine effectiveness: vaccines are not effective since so many doses are required	-	-
Concern over side effects of insufficiently tested vaccines	-	-
Concern over effectiveness of insufficiently tested vaccines	eksperiment (experiment), eksperimentalan (experimental), ispitivanje (examining), faza (phase), testiranje (testing), vršiti (perform), medicinski (medical), trajati (last), pravo (right), nijedan (none)	241
Violation of freedom by imposing the use of insufficiently tested vaccines	dete (child), gejts (Gates), bil (Bill), nov (nov), dnk (DNA), soj (strain), bolest (illness), menjati (change), priča (story), sloboda (freedom)	962
Mistrust of medical experts and institutions	nauka (science), bog (god), naučan (scientific), vera (faith), dokazati (prove), laž (lie), dokaz (proof), verovanje (belief), reč (word), govoriti (speak)	209
	dr (dr), doktor (doctor), lekar (doctor), medicina (medicine), medicinski (medical), antivakser (anti-vaxxer), nauka (science), mrn (mRN), efikasnost (effectiveness), crn (black)	240
	maska (mask), nositi (wear), distanca (distance), štititi (protect), odbiti (refuse), mera (measure), značiti (to mean), naravno (of course), virus (virus), daleko (far)	138
	Mesec (month), javan (public), zdrav (healthy), zakon (law), radnik (worker), javno (public), vlada (government), pitati (ask), doneti (bring), zdravstven (health, adj)	44

(continued)

Table 4. (*continued*)

Reasons for vaccine hesitancy identified by LDA and NMF	BERTopic topic representations	No. of tweets
Mistrust of government and political decision makers	kineski (Chinese, adj), ruski (Russian, adj), kinez (Chinese), eksperiment (experiment), sinopharm (Sinopharm), brat (brother), astrazeneca (AstraZeneca), član (member), predsednik (president), testirati (test)	84
	milion (million), epidemija (epidemics), milijarda (billion), svinjski (swine), država (state), režim (regime), suzbiti (supress), kupiti (buy), hiljada (thousand), isplatiti (pay off)	101
	fašist (fascist), otrovan (poisonous), kreten (jerk), globus (globus), tv (TV), fašizam (fascism), rnk (RNA), kolonija (colony), zombirati (to zombie), glup (stupid)	95
	pasoš (passpost), ukinuti (to cancel), smisao (meaning), rio (Rio [Tinto]), obavezan (obligatory), mera (measure), ostati (to stay), glupost (stupidity), nuditi (to offer), zdravlje (health)	54
Vaccines are a money-making scheme	-	-
Vaccines, especially mRNA vaccines, change DNA	dete (child), gejts (Gates), bil (Bill), nov (nov), dnk (DNA), soj (strain), bolest (illness), menjati (change), priča (story), sloboda (freedom)	962
COVID-19 does not exist; thus, vaccines are unnecessary	-	-
Vaccines are a means of population reduction and control	čekati (wait), nadati (to hope), panika (panick), očekivati (expect), nov (new), red (line), zaraditi (earn), mera (measure), depopulacija (depopulation), strah (fear)	64
Vaccines are an instrument of world powers and their agenda	nemački (German), rat (war), rus (Russian), ukrajina (Ukraine), ruski (Russian), rusija (Russia), promoter (promoter), agenda (agenda), idiot (idiot), shvatiti (realize)	106

5   Conclusions and Future Work

In this paper, we tested the performance of BERTopic on short text in Serbian. We were interested in whether BERTopic can yield meaningful topics when applied to morphologically rich slightly processed short text and how well it performs in comparison with LDA and NMF on fully processed text. To answer the first question, we compared the performance of BERTopic with different embedding models on fully and partially preprocessed text and found that BERTopic can produce meaningful and informative topics even with slight preprocessing. In this case, the larger model, the better the performance. As for the second question, we concluded that applying the same parameters as the ones used for LDA is not the optimal scenario for BERTopic. When the number of topics was not limited, BERTopic was able to provide novel insights.

There are several directions we would like to explore in the future. We plan to apply BERTopic to different datasets to check if our conclusions can be generalized. We also plan to explore its prediction capabilities on new, unseen documents. Since BERTopic supports using different transformer models as the embedding models, we also plan to test the applicability and performance of the currently only language model trained on Serbian data – BERTić [23].

References

1. Blei, D.M., Ng, A.Y., Jordan, M.I.: Latent Dirichlet Allocation. J. Mach. Learn. Res. **3**, 993–1022 (2003). https://doi.org/10.1162/jmlr.2003.3.4-5.993
2. Ljajić, A., Prodanović, N., Medvecki, D., Bašaragin, B., Mitrović, J.: Uncovering the reasons behind COVID-19 vaccine hesitancy in Serbia: sentiment-based topic modeling. J. Med. Internet Res. **24**, e42261 (2022). https://doi.org/10.2196/42261
3. Févotte, C., Idier, J.: Algorithms for nonnegative matrix factorization with the beta-divergence. Neural Comput. **23**, 2421–2456 (2011). https://doi.org/10.1162/NECO_a_00168
4. Grootendorst, M.: BERTopic: neural topic modeling with a class-based TF-IDF procedure (2022). https://doi.org/10.48550/arXiv.2203.05794
5. Abuzayed, A., Al-Khalifa, H.: BERT for Arabic topic modeling: an experimental study on BERTopic technique. Proc. Comput. Sci. **189**, 191–194 (2021). https://doi.org/10.1016/j.procs.2021.05.096
6. de Groot, M., Aliannejadi, M., Haas, M.R.: Experiments on generalizability of BERTopic on multi-domain short text (2022). https://arxiv.org/abs/2212.08459
7. Egger, R., Yu, J.: A topic modeling comparison between LDA, NMF, Top2Vec, and BERTopic to demystify Twitter posts. Front. Sociol. **7**, 886498 (2022). https://doi.org/10.3389/fsoc.2022.886498
8. Chen, Y., Zhang, H., Liu, R., Ye, Z., Lin, J.: Experimental explorations on short text topic mining between LDA and NMF based schemes. Knowl.-Based Syst. **163**, 1–13 (2019). https://doi.org/10.1016/j.knosys.2018.08.011
9. Egger, R., Yu, J.: Identifying hidden semantic structures in Instagram data: a topic modelling comparison. TR (2021). https://doi.org/10.1108/TR-05-2021-0244
10. Chauhan, U., Shah, A.: Topic modeling using Latent Dirichlet Allocation: a survey. ACM Comput. Surv. **54**, 1–35 (2022). https://doi.org/10.1145/3462478
11. Sánchez-Franco, M.J., Rey-Moreno, M.: Do travelers' reviews depend on the destination? An analysis in coastal and urban peer-to-peer lodgings. Psychol. Mark. **39**, 441–459 (2022). https://doi.org/10.1002/mar.21608
12. Bulut, O., MacIntosh, A., Walsh, C.: Using Lbl2Vec and BERTopic for semi-supervised detection of professionalism aspects in a constructed-response situational judgment test (2022). https://doi.org/10.31234/osf.io/n5fqe
13. Sangaraju, V.R., Bolla, B.K., Nayak, D.K., Kh, J.: Topic modelling on consumer financial protection bureau data: an approach using BERT based embeddings (2022). https://arxiv.org/abs/2205.07259
14. Sánchez-Franco, M.J., González Serrano, M.H., dos Santos, M.A., Moreno, F.C.: Modelling the structure of the sports management research field using the BERTopic approach. In: Retos: nuevas tendencias en educación física, deporte y recreación, pp. 648–663 (2023)

15. Alhaj, F., Al-Haj, A., Sharieh, A., Jabri, R.: Improving Arabic cognitive distortion classification in Twitter using BERTopic. IJACSA **13** (2022). https://doi.org/10.14569/IJACSA.2022.0130199
16. Hutama, L.B., Suhartono, D.: Indonesian hoax news classification with multilingual transformer model and BERTopic. IJCAI **46** (2022). https://doi.org/10.31449/inf.v46i8.4336
17. Reimers, N., Gurevych, I.: Sentence-BERT: sentence embeddings using siamese BERT-Networks (2019). https://doi.org/10.48550/arXiv.1908.10084
18. Radović, A.: Srtools (2021). https://pypi.org/project/srtools/
19. Ljubešić, N., Štefanec, V.: The CLASSLA-StanfordNLP model for lemmatisation of non-standard Serbian 1.1. Slovenian language resource repository CLARIN.SI (2020). https://doi.org/10.18653/v1/W19-3704
20. Dieng, A.B., Ruiz, F.J.R., Blei, D.M.: Topic modeling in embedding spaces. Trans. Assoc. Comput. Linguist. **8**, 439–453 (2020). https://doi.org/10.1162/tacl_a_00325
21. Newman, D., Lau, J.H., Grieser, K., Baldwin, T.: Automatic evaluation of topic coherence. In: HLT 2010: Human Language Technologies: The 2010 Annual Conference of the North American Chapter of the Association for Computational Linguistics, pp. 100–108. Association for Computational Linguistics (2010)
22. Bouma, G.: Normalized (pointwise) mutual information in collocation extraction. In: Proceedings of GSCL, pp. 31–40 (2009)
23. Ljubešić, N., Lauc, D.: BERTić - The transformer language model for Bosnian, Croatian, Montenegrin and Serbian (2021). https://doi.org/10.48550/arXiv.2104.09243

Evaluation of Head-Up Display
for Conditionally Automated Vehicles

Kristina Stojmenova Pečečnik[1]($\boxtimes$) (ID), Alexander Mirnig[2,3] (ID),
Alexander Meschtscherjakov[2] (ID), and Jaka Sodnik[1] (ID)

[1] Faculty of Electrical Engineering, University of Ljubljana, 1000 Ljubljana, Slovenia
kristina.stojmenova@fe.uni-lj.si
[2] Department of Artificial Intelligence and Human Interfaces, University of Salzburg,
Salzburg, Austria
[3] Center for Technology Experience, AIT Austrian Institute of Technology GmbH,
Vienna, Austria

Abstract. This paper describes the user-evaluation process of a head-up display (HUD) display for conditionally automated vehicle. The main goal of the presented HUD is to help drivers with the monitoring of the environment when operating a conditionally automated vehicle, and the transiting between different levels of automation. This paper presents the perceived usability, user experience, and acceptance of such an automotive telematics solution. The high scores on all of the observed aspects indicate that the proposed HUD could contribute to the acceptance, willingness to use and with that, fasted and more successful adaptation on conditionally automated vehicles on European roads and wider. The presented work was part of an international project with the main goal to investigate different innovative approaches and technologies for supporting new driver tasks and responsibilities in vehicles with different levels of automation. In addition to the assessment of the HUD within the project, this paper provides also an overview of the project, its goals, and how its innovations address European mobility needs.

Keywords: Automated Vehicles · Conditional Automation · Head-Up Display · User Experience · Usability · User Acceptance

1 Introduction

Driving involves continuous and simultaneous performance of multiple motor, sensory, and cognitive tasks. In addition to physically operating a vehicle, driving also involves interaction with other road users, following traffic rules, and adapting to weather and road conditions. In particular, monitoring the environment is a crucial aspect of driving, as it enhances situational awareness and enables drivers to make informed decisions for a safe and comfortable journey. As the population, urbanization, and number of vehicles continue to grow, particularly in developing countries, the demands and complexity of the driving task are becoming increasingly challenging. To address these challenges, the automotive industry has been focused on developing automated vehicles which can partially or fully handle the task of driving. The ultimate goal is creation of a fully autonomous vehicle that can take over all driving functions.

© The Author(s), under exclusive license to Springer Nature Switzerland AG 2024
M. Trajanovic et al. (Eds.): ICIST 2023, LNNS 872, pp. 174–182, 2024.
https://doi.org/10.1007/978-3-031-50755-7_17

To distinguish between the vehicle's capabilities of performing the driving task, the Society of Automotive Engineers (SAE) has defined 6 levels of automation, ranging from level 0, where the driver operates the vehicle entirely manually, to level 5, where the vehicle is fully autonomous (Table 1) [1]. For example, vehicles that are able to perform longitudinal and lateral control of the vehicle under specific conditions, are defined as conditionally automated vehicles (SAE level 3).

Table 1. Levels of vehicle automation 1

Level of automation	Vehicle and driver responsibilities
L0 -No Automation	Zero autonomy; the driver performs all driving tasks
L1 - Driver Assistance	Vehicle is controlled by the driver, but some driving assist features may be included in the vehicle design
	Vehicle has combined automated functions, acceleration and steering, but the driver must remain engaged with the task and monitor the environment at all times
L2 - Partial Automation	Vehicle has combined automated functions, acceleration and steering, but the driver must remain engaged with the task and monitor the environment at all times
L3 - Conditional Automation	Driver is necessary, but is not required to monitor the environment during automation. The driver must be ready to take control of the vehicle at all times with notice
L4 - High Automation	The vehicle is capable of performing all driving functions under certain conditions. The driver mays have the option to control the vehicle
L5 - Full automation	The vehicle is capable of performing all driving functions under all conditions. The driver may have the option to control the vehicle

Most research concerning vehicle automation deals with obstacles and questions regarding the achievement of automation from a technological perspective. However, the acceptance and use of such novel technologies is highly related to the user experience and their perceptions of usefulness, usability and safety. In that regard, user-centric approaches that take into consideration also the users of automated vehicles are a must in order to achieve successful introduction of such vehicle technologies on the road.

1.1 Motivation

Automated vehicles with higher levels of automation have sensors (e.g. radar, lidar, cameras, GPS systems) to detect and recognize the environment and assist drivers with the monitoring and supervising of the driving process. The in-vehicle information systems utilize these sensors' information in order to provide visual, auditory, or tactile feedback to the driver through corresponding user interfaces (HMIs). However, human driving behavior in conditionally automated vehicles may still hinder road safety despite the higher level of automation. Because of their reduced driving task involvement, drivers may engage in non-driving tasks, neglect paying attention to the environment, and sometimes forget even forget about the primary task of driving, even though they are required to remain engaged (SAE level 2) or be ready to take over control at all times (SAE level 3).

Intrigued by this issue and motivated to provide solutions to tackle it, the Holistic Approach for Driver Role integration and Automation Allocation for European Mobility needs (HADRIAN) [2] project was born. This project aims to create different novel HMI solutions to assist different type of drivers (average, elderly and professional) with the operation of vehicles with different levels of automation in different conditions [2]. In this paper, we primarily focus on the presentation of one of the HADRIAN innovations, namely a head-up display (HUD) that supports the driver when changing between automated and manual driving modes. The HUD was designed in a manner that allows the driver to fully utilize the vehicle's capabilities while ensuring driving safety and prompting high acceptance of advanced telematics technologies.

2   HADRIAN Project

The HADRIAN project is an international project comprised of 16 partners from industry, academia, small and medium-sized companies and research institutions from nine countries (Austria, Slovenia, Spain, Greece, Germany, France, the Netherlands, Turkey, and the United Kingdom). In this constellation, the different professional, research and academic collaborators cover the entire chain of design, development, implementation and performance evaluation of new integrated and realistic automotive systems, which can enable the use of automated driving on the public roads of Europe in the very near future.

As shown in Table 1, SAE [1] currently distinguishes five levels of driving automation systems. However, this classification only takes into account the vehicle's capabilities and does not cover the driver's needs or desires. The distinction between the levels is far from simple, and requires understanding of numerous vehicle technology features in order to be aware about the level of automation the vehicle is in. In principle, the driver does not want to worry about distinguishing between levels and does not want to find out what his role is at a given moment, especially not when transitioning between different levels of automation. The driver expects and wants a safe and comfortable ride, which can be provided by the chosen vehicle with optimal support and guidance.

The main goal of the HADRIAN project was therefore to develop a comprehensive system for a seamless and fluid interaction between the driver, the vehicle and the environment/infrastructure, which guides both normal and professional drivers through different levels of driving automation, which are also adapted to their needs. The system consists of a model for real-time monitoring of the psychophysical state of the driver and an adaptive user interface that adjusts the display of information and the level of automation of the vehicle according to the needs of the driver at any given moment. In this paper, we present some of the efforts put in the development of and evaluation of a visual user interface that aims to ensure safe, efficient and comfortable driving of conditionally automated vehicles.

3 Methodology

3.1 Prior Work

The review of relevant research in this field revealed that there are discrepancies in how to present information, and moreover, how much information to feature in a HUD to help the driver maintain appropriate levels of situational awareness when operating a conditionally automated vehicle. An overview of the available research was featured in several reviews of related work in the past years, demonstrating how different HUD designs have been designed and evaluated [3–6]. However, all of these studies were completed in different testing settings, used different driving environments and were evaluated with different testing participants, which do not allow for a comprehensive or reliable direct comparison of results. To get a more consistent understanding on this, we performed an exploratory study in which we tried to identify how and which information to present in a HUD [7, 8]. The results from that study were used to develop the HUD presented in this study.

3.2 Head-Up Display

The HUD used in this study features elements that have a fixed position on the windshield and elements that exploit different focal lengths and the whole area of the windshield, to create a more immersive augmented reality (AR) experience. Elements such as vehicle speed, current speed limit, available ADAS functions, etc. are presented throughout the whole drive, and are hence presented in a fixed position on the windshield, right above the steering wheel. Elements that appear only when necessary, such as GPS directions or highlight of important road participants, are displayed or highlighted directly in the environment. All features are presented in Table 2 and Fig. 1.

Fig. 1. Visual Head-up Display preview

Table 2. Visual head-up display information

Information presented on the Visual HUD	Icon example
Information presented during the whole trop	
Speed limit	
Vehicle speed	
Speeding	
Active ADASs	
Distance to vehicle in-front only when driving in TTC < 2s.	
Vehicle level of automation (example level 3)	
Display surrounding traffic/road signs (pedestrian crossing, traffic light, priority road, non-priority road, stop) 150 m before their location in the environment	
GPS directions	
Short messages/email previews	New SMS
Information presented during takeover request	
Speed limit	
Vehicle speed	
Active ADASs	
Level of automation	
Highlight of important participants that can affect the takeover maneuver	
Visual takeover notification with a timer counting down the reaming-ing time for takeover, with takeover of 15 seconds	15 (counting down to 0)
Auditory takeover notification with five seconds lead time	4000 Hz pure tone

3.3 Participants

30 participants (15 females and 15 males), aged between 21 and 57 (M = 30.17, SD = 10.60), participated in the study. Their driving experience ranged from 2 to 39 years (M = 11.78, SD = 10.12). Due to technical difficulties, two of the participants did not complete the study, and their results were excluded from the analysis.

3.4 Study Set-Up

The study was conducted in a compact motion-based driving simulator [9] with real car parts (seat, steering wheel and pedals) and a physical dashboard (Fig. 2). The simulated environment was displayed on three 49" curved TVs ensuring a 145° field of view of the driving environment. The driving scenario was developed in SCANeR Studio [10] and included 13 km of simulated route from a suburban area to a city center.

Fig. 2. Driving simulator set-up used in the study

The goal of the study was to identify if the new HUD interface provides sufficient information to help the driver with the monitoring of the environment during different automation driving modes, and to get an insight on how is this affects the driver's user experience, perceived usability and acceptance. In that regard, the three main dependent variables were user experience, perceived usability, and general acceptance of the HUD as a new ICT solution. User experience scores were collected with the User Experience Questionnaire – UEQ [11], perceived usability with the System Usability Scale – SUS [12], whereas the acceptance of the HUD was recorded through the Acceptance of Advanced Transport Telematics (AATT) questionnaire [13].

4 Results

4.1 User-Experience

User experience data was collected with UEQ. The data was analyzed using the UEQ tool, which provides benchmarks for six aspects of user experience (attractiveness, perspicuity, efficiency, dependability, stimulation and novelty) that have been derived from values from an international benchmark data set. The comparison of the results for the evaluated product with the data in the benchmark allows conclusions about the relative quality of the evaluated product compared to other products. The results of the UEQ for our HUD indicated that the evaluated HUD was rated above average compared to the benchmark for attractiveness (M = 1.713, SD = 0.61), perspicuity (M = 2.259, SD = 0.55), efficiency (M = 1.647, SD = 0.77), dependability (M = 1.621, SD = 0.77) and novelty (M = 1.224, SD = 0.64), suggesting that the HUD was designed well. Although still positive, the HUD only received an average score for stimulation (M = 0.578, SD = 0.91), which is believed to be due to the lack of encouragement for the user to actively look for additional information in the environment. All the results are presented in Fig. 3.

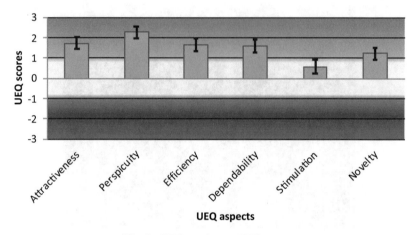

Fig. 3. UEQ results per UEQ aspect

4.2 Usability

Data about the perceived usability was collected with the SUS, which indicated that the HUD was also rated as above average for its perceived usability. The SUS results into scores ranging from 0–100. Despite its similarity of the scale, the final score does not represent a percentage. The score of 68 is set as a discriminatory limit – a score below 68 indicates below average, whereas a score above 68 indicates an above average perceived usability. The participants rated the HUD above average, with a mean score of 8.471 (SD = 9.88).

4.3 Acceptance

The user acceptance was assessed with the AATT questionnaire. The AATT scores indicated high acceptance of the HUD as a new ICT solution in transportation. The AATT scores can range from −2 to +2. The scores from all of the variables are used to calculate two aspects depicting user's perceived usefulness and satisfaction with the evaluated solution. The scores are then presented with a Cartesian coordinate system where the usability scores provide the ordinate values, and the satisfaction scores provide the abscissa values of the AATT per user. Except for one participant, all the AATT ratings from the remaining participants fell in the first quadrant of the coordinate system, with mean scores for perceived usefulness M = 1.291 (SD = 0.514), and mean scores on satisfaction M = 1.034 (SD = 0.501) (Fig. 4). These results are significant in the evaluation process, as acceptance of the technology is often seen as a direct indicator of future "willingness to use."

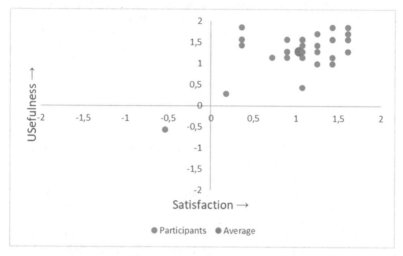

Fig. 4. AATT scores per test participant and overall average score of acceptance.

5 Discussion and Conclusions

Data on user experience, perceived usability and user acceptance provides an important insight about the performance of a solution from a user's point of view, as they can have a great influence on the adoption and willingness to use of such solution in real vehicles.

The results from this study show that user experience with the HUD was rated above average for all UEQ aspects except for Stimulation. The HUD provides a lot necessary information to help the driver with monitoring of the environment, which may result in lesser stimulation to engage in the task of driving that is concerning the monitoring part of it. The high scores on perspicuity provide important feedback about the clarity and precision of the presented information, which may have also affected the perceived efficiency and reliability of the HUD. The high score for novelty is somewhat expected, as although their value has been recognized in the research world, the implementation of HUDs in real vehicles is, mainly due to technological pitfalls, still a rare practice. In line with the high scores for user experience, the results from the SUS show that participants rated the HUD above average also for usability, which is closely related to the perceived efficiency and dependability. Lastly, when looking at the participants' acceptance of the Visual HUD Support System as an advanced telematics solution, the results revealed high levels of perceived satisfaction, and much higher scores on perceived usefulness.

The high scores on all aspects of user experience, usability, acceptance, and perceived usefulness and efficiency, indicate that the proposed HUD could contribute to the acceptance, willingness to use and with that, fasted and more successful adaptation on conditionally automated vehicles on European roads and wider.

Acknowledgement. The work presented in this paper was financially supported by the European Union's Horizon 2020 research and innovation program for the project HADRIAN, grant agreement no. 875597 and by the Slovenian Research Agency within the project Modelling driver's situational awareness, grant no. Z2–3204 and program ICT4QL, grant no. P2–0246. This document reflects only the authors view, the Innovation and Networks Executive Agency (INEA) is not responsible for any use that may be made of the information it contains.

References

1. SAE, T.: Definitions for terms related to driving automation systems for on-road motor vehicles. SAE Stand. J **2016**, 3016 (2016)
2. HADRIAN project: https://hadrianproject.eu/
3. Schömig, N., et al.: An augmented reality display for conditionally automated driving. In Adjunct Proceedings of the 10th International Conference on Automotive User Interfaces and Interactive Vehicular Applications, pp. 137–141 (2018)
4. Feierle, A., Beller, D., Bengler, K.: Head-up displays in urban partially automated driving: effects of using augmented reality. In: 2019 IEEE Intelligent Transportation Systems Conference (ITSC), pp. 1877–1882. IEEE (2019)
5. Riegler, A., Riener, A., Holzmann, C.: Augmented reality for future mobility: insights from a literature review and HCI workshop. I-Com **20**(3), 295–318 (2021)
6. Currano, R., Park, S.Y., Moore, D.J., Lyons, K., Sirkin, D.: Little road driving hud: heads-up display complexity influences drivers' perceptions of automated vehicles. In: Proceedings of the 2021 CHI Conference on Human Factors in Computing Systems, pp. 1–15 (2021)
7. Stojmenova, K., Tomažič, S., Sodnik, J.: Design of Head-Up Display Interfaces for Automated Vehicles. Available at SSRN 4182146
8. Stojmenova, K., Jakus, G., Tomažič, S., Sodnik, J.: Is less really more? A user study on visual in-vehicle information systems in automated vehicles from a user experience and usability perspective. In Proceedings of the 13th AHFE International Conference on Usability and User Experience, New York, USA, 24–28 July 2022. New York: AHFE Open Access (2022)
9. Vengust, M., Kaluža, B., Stojmenova, K., Sodnik, J.: NERVteh compact motion based driving simulator. In Proceedings of the 9th International Conference on Automotive User Interfaces and Interactive Vehicular Applications Adjunct, pp. 242–243 (2017)
10. AVSimulation. SCANeR studio. https://www.avsimulation.com/scanerstudio/
11. Laugwitz, B., Held, T., Schrepp, M.: Construction and evaluation of a user experience questionnaire. In: Holzinger, A. (ed.) USAB 2008. LNCS, vol. 5298, pp. 63–76. Springer, Heidelberg (2008). https://doi.org/10.1007/978-3-540-89350-9_6
12. Bangor, A., Kortum, P.T., Miller, J.T.: An empirical evaluation of the system usability scale. Intl. J. Hum.-Comput. Interact. **24**(6), 574–594 (2008)
13. Van Der Laan, J.D., Heino, A., De Waard, D.: A simple procedure for the assessment of acceptance of advanced transport telematics. Transp. Res. Part C: Emerg. Technol. **5**(1), 1–10 (1997)

Visual Programming Support
for the Explainable Artificial Intelligence

Mina Nikolić[(✉)] [ID], Aleksandar Stanimirović [ID], and Leonid Stoimenov [ID]

Faculty of Electronic Engineering, University of Niš, Niš, Serbia
mina.nikolic@elfak.ni.ac.rs

Abstract. The field of Explainable artificial intelligence is showing promising growth in recent years, thus giving researchers the option of deeply exploring the benefits and drawbacks of many different proposed models for solving the enigma of interpretability and explainability regarding machine learning models and their predictions. There are currently a number of techniques that can help to assist researchers in understanding the logic behind decisions made by various models, but the focus of this paper will mostly be on discussing and comparing two strong options, LIME (Local Interpretable Model-Agnostic Explanations) and SHAP (Shapley Additive Explanations). The proposed pipeline for the comparison will be given in a form of an Orange Data Mining workflow. Secondly, the paper aims to give a proposal of how a custom widget encapsulating the functionality of the LIME library can be integrated into the graphical interface, making its usability more appropriate towards less experienced users.

Keywords: Explainable artificial intelligence (XAI) · interpretability · explainability · LIME · SHAP · Orange Data Mining · Visual programming

1 Introduction

The rapid rise of machine learning and artificial intelligence-based techniques in recent years has provided many opportunities for enabling researchers from different branches to develop a variety of solutions for real world problems. These problems are distributed through various domains such as healthcare, education, business, scientific research and many more.

Some of the machine learning models are considered to have a black box approach which can be a problem when working with sensitive matters involving human lives. Black box models can't give much of the necessary explanations that would be greatly needed and beneficial for researchers. Knowing that, to make models and predictions more transparent and interpretable, we need to look for different approaches that would resolve those shortcomings of black box models. Entering the field of Explainable AI can give researchers the necessary interpretability and explainability of models and predictions made by them.

Terms such as interpretability and explainability are often used in the Explainable artificial intelligence (XAI) domain interchangeably and currently there is not a clear consensus for their meaning, thus they appear underspecified.

M. Trajanovic et al. (Eds.): ICIST 2023, LNNS 872, pp. 183–192, 2024.
https://doi.org/10.1007/978-3-031-50755-7_18

On the other hand, as argued in [1], these terms should be viewed on their own. The authors take the stance that interpretability is not enough by itself, rather that explainability is needed for people to trust black-box models. In their view, the goal of interpretability is "to describe the internals of a system in a way that is understandable to humans" [1, p.2].

They also add that "in order for a system to be interpretable, it must produce descriptions that are simple enough for a person to understand using a vocabulary that is meaningful to the user" [1, p.2].

As stated in [2, p.1], the desired characteristics for explainable models (explainers) are that they must be interpretable, i.e., "to provide qualitative understanding between the input variables and the response". The meaning behind this definition also acknowledges that explanations need to be easy to understand. It is also stated that interpretability is not a term that can be used in the same manner for different types of target audiences with various levels of expertise. To make a leap towards fair and ethical decision making, there is a need for discussing the transparency notion of interpretability as shown by Lipton in [3].

Regarding interpretability Lipton proposes that there are two categories in which techniques and model properties fall, first being transparency and the second post hoc explanations. The two terms are defined as "how does the model work" and "what else can a model tell me", respectively [3, p.4]. It is also stated that transparency is on the opposite side of the spectrum from the black-box approach mentioned earlier.

Knowing all of that, there was a need to develop solutions that would solve this problem and provide some additional explanations for the inner logic behind machine learning models and predictions. Several solutions were proposed, giving a variety of options to choose from. Linardatos et al. [4] give a comprehensive review of interpretability methods to explain any black-box model.

The focus of this paper is to give an understanding how two different models, LIME (Local Interpretable Model- Agnostic Explanations) [2] and SHAP (Shapley Additive explanations) [5] work in order of explaining the reasoning behind a machine learning model on a given dataset. Secondly, we propose an example of an alternative way of using LIME through Orange Data Mining as a custom Add-on widget.

We believe that this way of using LIME could make the library more accessible to a greater number of people that don't have the necessary technical skills, as Orange Data Mining provides visual programming experience.

2 Related Work

To understand the different approaches, present in the explainable AI domain, a taxonomy of XAI techniques has been proposed in [6]. The categorization into appropriate sections has put both LIME and SHAP libraries in the Feature- Based techniques section. Feature-Based techniques focus on finding the most relevant attributes for the specific model and provided data. Following the proposed taxonomy, LIME and SHAP are defined as model-agnostic techniques, meaning that they can be used with any type of machine learning model. Regarding the classification based on the scope of interpretation, LIME can be seen as a model that involves local interpretations, while SHAP can use both local, and global interpretations.

There are several research papers discussing the comparison of different Explainable AI models including LIME and SHAP among other models. Some of the work is directed towards raising the question of trustable Explainable AI [7], while on the other hand there are proposals on how a theoretical framework could be used to enable a comparison between different models [8]. A practical implementation of a framework for accountable and explainable AI has been shown in [9], with the demonstration being done regarding the real time Affective State Assessment Module of a robotic system. The implemented practical solution is based on a theoretical approach described in [10].

Explainable AI and the interpretability of machine learning models have the potential to be used in a variety of different scenarios regarding a vast number of fields, such as healthcare, economy, law, and many others.

The use of Explainable AI in the healthcare domain has been a topic of numerous papers [11–14] where LIME and SHAP were some of the models used.

Evaluation and comparison of these two models has also been done in the domain of credit risk management where authors give a side-by-side review of the results obtained by these models [15].

Using the power and capabilities of Explainable AI has also been shown in the domain of law, where a comparison of LIME and SHAP was given to explain the process of rational discovery and predictions gathered from a neural network [16].

The specific problem that this paper wanted to address is to make Explainable AI both more available and understandable to a larger audience that doesn't have the necessary technical skills to do complex scientific work in this field. We wanted to empower audiences from different domains by giving them an option of understanding the logic behind predictions made by various machine learning models.

To do such a thing, we utilized the benefits of the paradigm of visual programming that is present in the Orange Data Mining software. Visual programming can be defined as a sort of programming paradigm that allows users to illustrate processes. The tool used for implementing visual programming support for XAI tools is Orange Data Mining. Orange Data Mining enables users to construct a complete machine learning workflow by using a simple drag-and-drop interface that accommodates a vast number of useful widgets.

By proposing a custom-made widget that encapsulates some of the functionality offered by LIME, we give users the ability of seeing and visually comparing results of prediction explanations done by LIME and SHAP models.

3 Methodology

As previously stated, LIME and SHAP libraries are not the only tools used in order to implement explainable and interpretable AI solutions but are among the most commonly used.

In this section, the methodology for creating a comparison pipeline using custom widgets will be described. Firstly, it is important to acknowledge that Orange Data Mining supports an Explain Add-on in their widget catalog. It consists of Feature Importance, Explain Model, Explain Prediction, Explain Predictions and ICE widgets.

The inner core of Explain Prediction widget is the SHAP model of explainability, while the implementation shown in the paper provides an alternative using LIME. Both SHAP and LIME are widely used but they have some distinct strengths and weaknesses, so it is beneficial to use them both to be more certain of the given results.

3.1 LIME and SHAP Models for Explainable AI

LIME (Local Interpretable Model-Agnostic Explanations)
As stated in [2], LIME is an algorithm that can explain the prediction of any classifier or regressor in a faithful way, by approximating it locally with an interpretable model. An essential criterion for explanations is that they must be interpretable, meaning that they provide qualitative understanding between the input variable and the response variable. To make an explanation meaningful, it must be at least locally faithful, which means that it must correspond to how the model behaves in the surroundings of the instance being predicted.

Another characteristic of an explainer is that it should be model agnostic, meaning that it can explain any model. The explanation produced by LIME is described by the following mathematical equation [2]:

$$\xi(x) = argmin_{g \in G} \mathcal{L}(f, g, \pi_x) + \Omega(g) \tag{1}$$

The first argument is a measure of how unfaithful g is in approximating f in the locality defined by πx. The second argument represents the measure of complexity. The key takeaway from this equation is that to ensure interpretability and local fidelity, the first argument needs to be minimized while the second needs to be low enough to be interpretable by humans.

SHAP (Shapley Additive Explanations)
As stated by Lundberg and Lee [5], SHAP values can be described as a unified measure of feature importance. The inner workings of the SHAP library are based on the game-theory approach involving the use of Shapley values. Calculations are done by assigning a value to each feature in the prediction by using coalition, which results in considering different combinations of features and measuring their impact on the prediction outcome. The process of computing SHAP values is challenging, so to make calculations faster, the values are approximated using additive feature attribution methods like Shapley sampling values and Kernel SHAP. Regarding the desired properties of Additive Feature Attributions, local accuracy, missingness and consistency were mentioned in the paper [5].

3.2 Dataset

The dataset used for analysis is the Pima Indians Diabetes dataset [17]. As described by the creators of the dataset, the objective is to predict whether a patient has diabetes based on certain diagnostic measurements. The dataset consists of several input variables, and an outcome variable. Input variables include the number of pregnancies the patient has had, their BMI, insulin level, age, glucose level, blood pressure, skin thickness and the diabetes pedigree function.

The number of analyzed instances is 786 and it is worth noting that the dataset is not particularly balanced, as it is a common occurrence while working with medical data. There were no missing values, and no preparation or preprocessing techniques were done on the dataset.

3.3 Proposed Workflow

Before discussing the proposed workflow for the comparison of results obtained by both LIME and SHAP, a feature importance analysis was done on the dataset, using the Feature Importance widget, included in the Orange Data Mining library. The Feature Importance widget, as the name suggests, uses the data to compute the contribution of each feature toward the prediction. The reasoning behind using this widget alongside LIME and SHAP was to have a reference point when comparing the results from these Explainable AI libraries.

In Fig. 1, the proposed implementation of a workflow comparing the results from both LIME and SHAP models can be seen. Orange Data Mining workflows are designed with modularity in mind, meaning that a workflow consists of numerous widgets.

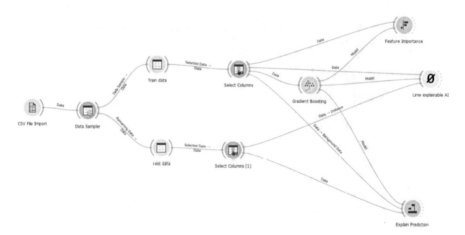

Fig. 1. Orange Data Mining workflow for the comparison of LIME and SHAP

The workflow in Fig. 1 consists of various widgets that are connected in a specific manner, as previously shown:

1. *CSV File import* widget that is used for loading the dataset of choice,
2. *Data Sampler* widget that divides data into parts for training and testing,
3. *Data Table* widget which is needed for the purpose of visualizing the dataset in a tabular form,
4. *Select Columns* widget used with the intent of choosing features (input columns) and the target variable (output column),

5. *Feature Importance* widget, used for calculating the most relevant features,
6. Machine learning model widget (*Gradient Boosting*) that is used for encapsulating different machine learning models,
7. *Explain prediction* widget that encapsulates some of the functionality of SHAP,
8. *Lime explainable AI,* which is a custom widget, designed to encapsulate the necessary functionality of LIME

The implementation of *Lime explainable AI* widget can be seen at [18]. The Lime explainable AI widget has the same inputs as the *Explain Prediction SHAP* widget provided by Orange Data Mining. Inputs include data from the training set, a machine learning model (in this case Gradient Boosting was used) and an instance for which the prediction explainability was done.

4   Results

The output from the Feature Importance widget suggests that the most relevant features for the provided model and dataset are Glucose, BMI, DiabetesPedigreeFunction, Age, Pregnancies, Insulin, BloodPressure and SkinThickness (in that order). The results are displayed in Fig. 2.

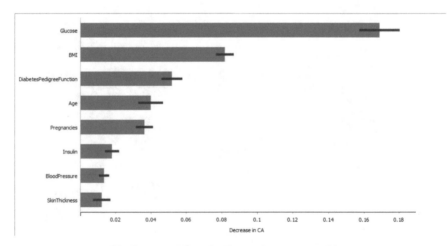

Fig. 2. Output from the Feature Importance widget

The instance used in order to explain predictions has the following values for attributes:

- Pregnanices: 0
- Glucose: 179
- BloodPressure: 50
- SkinThickness: 36
- Insulin: 159
- BMI: 37.8

- DiabetesPedigreeFunction: 0.455
- Age: 22

The output from the Explain Prediction widget that uses SHAP can be seen in Fig. 3. The red values on the diagram are those that push probabilities from the baseline probability toward probability 1.0, while the blue values do the opposite. For the previously described instance, Glucose levels, BMI and SkinThickness are the attributes that contribute the most towards an instance having diabetes. The prediction value using SHAP was 0.92 towards an instance having diabetes.

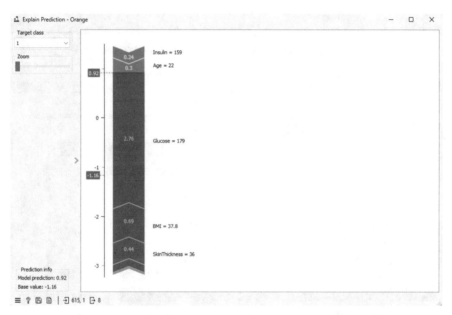

Fig. 3. Output from Explain Prediction widget (SHAP)

By using LIME, it can be seen that features that contributed the most towards this instance having diabetes were high Insulin and high Glucose levels. The prediction probability for this instance having diabetes was 0.94. On the left side of the widget, values for the concrete instance can be seen. Below, a graph type section allows for changing the way of visualization. The right, main area, is used to display predictions. The calculated results can be seen in Fig. 4 and Fig. 5, where different types of visualization were provided.

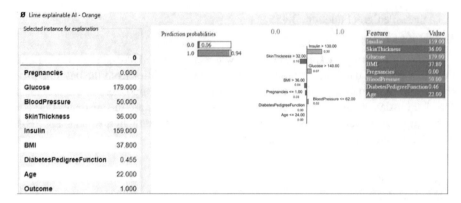

Fig. 4. Output from LIME Explainable AI widget (HTML view)

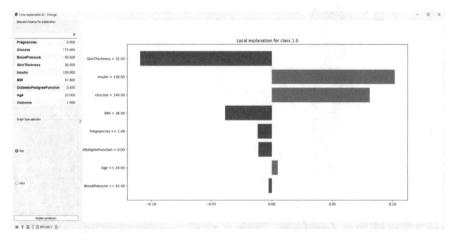

Fig. 5. Output from LIME Explainable AI widget (Bar plot view)

5 Conclusion

The proposal of such a way of using the benefits of Visual programming to explain complex predictions is more intuitive and straight forward than implementing traditional methods of programming. By encapsulating the functionality of LIME into an Orange Data Mining widget, the authors of this paper give a more user centric experience for the end user, thus enabling audiences from different backgrounds to equally participate in the discussion.

Implementing a workflow for comparing the outputs of two widely used Explainable AI models has the potential of giving much more useful insights than just relying on one of them. This is especially needed in situations when working with datasets regarding the medical and healthcare domain, as human lives are on the line. Knowing that, the proposed comparison and functionality of the custom widget was shown on a dataset from a medical field. But, as previously stated, the field of Explainable AI is meant to give a broad understanding of predictions coming from a vast number of different domains and fields.

It is important to acknowledge that the power of implementing Explainable AI tools is not limited only to the domain of tabular data, as previously shown. The vast expansion of different types of data is leading the way towards the explainability and interpretation of numerical, textual, image, video or even audio data.

One of the paths for further development of this implementation would need to include the support for explainability of those previously mentioned types of data, including a user interface that would differ based on the type of data used. On that note, the visual programming support would also need to be integrated in a way that would make working with complex data types more intuitive and accessible.

Further development can be focused towards implementing a more complex and detailed analysis, with a data set involving a larger number of attributes and values. In that case, a collaboration with domain experts would be needed for validating the obtained explanations. *Lime explainable AI* widget and the proposed workflow are intended to be the first step in analyzing the potential solutions, while domain specific questions should be discussed with experts in that field.

References

1. Gilpin, L.H., Bau, D., Yuan, B.Z., Bajwa, A., Specter, M., Kagal, L.: Explaining explanations: an overview of interpretability of machine learning. In: 2018 IEEE 5th International Conference on Data Science and Advanced Analytics (DSAA), pp. 80–89 (2018)
2. Ribeiro, M.T., Singh, S., Guestrin, C.: Why should I trust you?: Explaining the predictions of any classifier. In: Proceedings of the 22nd ACM SIGKDD International Conference on Knowledge Discovery and Data Mining (2016)
3. Lipton, Z.C.: The mythos of model interpretability: in machine learning, the concept of interpretability is both important and slippery. ACM Queue **16**(3), 31–57 (2018)
4. Linardatos, P., Papastefanopoulos, V., Kotsiantis, S.: Explainable AI: a review of machine learning interpretability methods. Entropy (Basel) **23**(1), 18 (2020)
5. Lundberg, S.M., Lee, S.-I.: A unified approach to interpreting model predictions. In: Proceedings of the 31st International Conference on Neural Information Processing Systems, pp. 4768–4777 (2017)
6. Rudesh, D., et al.: Explainable AI (XAI): core ideas, techniques and solutions. ACM Comput. Surv. **55**(9), 1–33 (2023). https://doi.org/10.1145/3561048
7. Ignatiev, A.: Towards trustable explainable AI. In: Proceedings of the Twenty-Ninth International Joint Conference on Artificial Intelligence (2020)
8. Palacio, S., Lucieri, A., Munir, M., Hees, J., Ahmed, S., Dengel, A.: XAI handbook: Towards a unified framework for explainable AI, pp. 3766–3775 (2021). arXiv:2105.06677. [cs.AI]
9. Vice, J., Khan, M.M.: Toward accountable and explainable artificial intelligence part two: the framework implementation. In: IEEE Access, vol. 10, pp. 36091–36105 (2022). https://doi.org/10.1109/ACCESS.2022.3163523
10. Khan, M.M., Vice, J.: Toward accountable and explainable artificial intelligence part one: theory and examples. In: IEEE Access, vol. 10, pp. 99686–99701 (2022). https://doi.org/10.1109/ACCESS.2022.3207812
11. Loh, H.W., Ooi, C.P., Seoni, S., Barua, P.D., Molinari, F., Acharya, U.R.: Application of explainable artificial intelligence for healthcare: a systematic review of the last decade (2011–2022). Comput. Methods Programs Biomed. **226**(107161), 107161 (2022)

12. Knapič, S., Malhi, A., Saluja, R., Främling, K.: Explainable artificial intelligence for human decision support system in the medical domain. Mach. Learn. Knowl. Extr. **3**(3), 740–770 (2021). https://doi.org/10.3390/make3030037
13. Duell, J., Fan, X., Burnett, B., Aarts, G., Zhou, S.-M.: A comparison of explanations given by explainable artificial intelligence methods on analysing electronic health records. In: 2021 IEEE EMBS International Conference on Biomedical and Health Informatics (BHI), pp. 1–4 (2021)
14. Dave, D., Naik, H., Singhal, S., Patel, P.: Explainable AI meets healthcare: A study on heart disease dataset (2020). arXiv:2011.03195. [cs.LG]
15. Misheva, B.H., Osterrieder, J, Hirsa, A, Kulkarni, O., Lin, S.F.: Explainable AI in credit risk management (2021). arXiv:2103.00949. [q-fin.RM]
16. Schweighofer, E.: Rationale discovery and explainable AI. In: Legal Knowledge and Information Systems: JURIX 2021: The Thirty-fourth Annual Conference, Vilnius, Lithuania, 8–10 December 2021, vol. 346, IOS Press (2022)
17. Smith, J.W., Everhart, J.E., Dickson, W.C., Knowler, W.C., Johannes, R.S.: Using the ADAP learning algorithm to forecast the onset of diabetes mellitus. In: Proceedings of the Symposium on Computer Applications and Medical Care, pp. 261–265. IEEE Computer Society Press (1988)
18. Github link for the widget implementation. https://github.com/minanikolic916/LIME_ML

Multi-criteria Decision Making in Turning Operations Using AHP, TOPSIS and WASPAS Methods

Rajko Turudija$^{(\boxtimes)}$ 🔾, Ljiljana Radović 🔾, Aleksandar Stanković 🔾, and Miloš Stojković 🔾

Faculty of Mechanical Engineering, University of Niš, 18000 Niš, Serbia
rajko.turudija@masfak.ni.ac.rs

Abstract. In machining shops, it is very important to be as productive as possible while achieving the required surface roughness values depicted in technical drawings. Adding to this the ever-growing importance of limiting the pollution of the environment, energy consumption has also become very important in industry. However, different cutting parameters influence these factors differently, so it is very hard to know which combination of parameters to use in each machining process. This paper focuses on the implementation of Multi-Criteria Decision Making (MCDM) methods to evaluate the combined influence of cutting parameters in turning processes depending on different importance ranking (weights) of criteria. The cutting parameters used in the study are cutting speed (Vc), cutting depth (ap) and feed rate (fn). Combining each of the parameters, multiple alternatives are achieved, with each alternative having a different influence on energy consumption (EC), surface roughness (SR) and productivity (MRR - material removal rate). As in machining industry most often EC and SR need to be at a minimum and MRR at a maximum, these are the criteria in presented multi-criteria decision-making problem: minimum EC, minimum SR and maximum MRR. For solving this problem, Analytic Hierarchy Process (AHP) method was used for determining the objective and constant weights of the criteria. TOPSIS and WASPAS methods were implemented for recommendations of the best alternative depending on previously determined weights of criteria.

Keywords: Multi-criteria decision making · AHP · TOPSIS · WASPAS · turning of AISI 1045

1 Introduction

During programing of turning operations for CNC (Computer Numerical Control) turning machines, it is possible to change various cutting parameters, with the aim of obtaining parts that comply with all given nominal measures and tolerances depicted on technical drawings. In addition, it is also important for the turning to be as productive as possible. Essentially, all machinists want to achieve the desired results in the shortest time possible. However, as society becomes more aware of the impact we humans have

© The Author(s), under exclusive license to Springer Nature Switzerland AG 2024
M. Trajanovic et al. (Eds.): ICIST 2023, LNNS 872, pp. 193–205, 2024.
https://doi.org/10.1007/978-3-031-50755-7_19

on our environment, all processes (including machining) must also be made energy efficient, so that there is no unnecessary pollution of the world in which we live. This puts the machinist in an unenviable situation, since productivity and efficient use of energy are often contradictory. Add to that the need to achieve the appropriate surface roughness, it is often very difficult to decide which combination of cutting parameters to choose. Because of this, the main motivation for the work is to facilitate the selection of combinations of different cutting parameters for machinists in machine shops depending on what importance maximum productivity (MRR), minimum energy consumption (EC) and minimum surface roughness (SR) have on a given turning process. This could also enhance production productivity through minimizing the time required for configuring input parameters and optimizing resource utilization, which is paramount to the manufacturing process [1, 2]. This study presents an implementation and a different approach for selecting optimal cutting parameters, aiming to determine the most favorable parameter combination for the machining process. To address this challenge, a hybrid multi-criteria decision-making (MCDM) methodology is employed to assess the combined influence of solution parameters in turning processes based on their ranking. In the initial phase of the MCDM, the Analytic Hierarchy Process (AHP) is utilized to establish objective constant weights for the criteria, while the Technique for Order of Preference by Similarity to Ideal Solution (TOPSIS) and the weighted aggregated sum product assessment (WASPAS) methods are subsequently applied for ranking the optimized alternatives. The global significance of MCDM methods is evidenced by their extensive application in solving real-world problems, as reflected in numerous studies. Additionally, this method is adaptable to various cutting processes due to its flexibility in accommodating changes in criterion weights [3, 4]. Determining the weighting coefficients proves to be a crucial factor in selecting the optimal alternative and ranking research outcomes, as observed in the works of [5] and [6]. Additionally, the works of [7] and [8] emphasize the necessity of retaining subjective perspectives during the definition of initial weights to yield relevant results applicable to concrete examples. The selection of criterion weights directly impacts the research outcomes, showcasing the interdependence between weight assignment and decision-making results. Thus, the present study employs multiple methods to enhance the concreteness of decision-making and ensure the relevance of the results [9–11]. Extensive research on multi-criteria decision-making highlights the recommendation by most authors to employ robust models for selecting optimal alternatives within the decision-making framework, as also supported by the works of [12] and [6]. This research aims to introduce a robust and hybrid multi-criteria decision-making approach to evaluate the combined impact of cutting parameters in turning processes. Subsequent sections of this paper will delve into the key components, as elaborated below.

In the introductory section of the paper, Sect. 1 presents the fundamental objective and main motivation behind the development of the idea. Section 2 provides a literature review. Section 3 proposes an approach based on multi-criteria decision-making methods. Section 4 presents a case study, encompassing the definition of all input parameters, robust alternative ranking, and a discussion of the optimality of the obtained solutions. Section 5 presents obtained results and discussion. The final section of the paper, titled

"Conclusion," highlights the experimental research results and the benefits derived from the utilization of the proposed methodology.

2 State of the Art

During the turning process, the main cutting parameters that can be changed and that have the greatest impact on the turning of parts are cutting depth (ap), cutting speed (Vc) and feed rate (fn). However, considering that each of these cutting parameters affects MRR, EC and SR in a different way, it is difficult to predict how exactly which combination of cutting parameters will affect the three previously mentioned values, and which combination is the best. Therefore, the paper wants to answer the question: depending on the importance that minimum EC, minimum SR and maximum MRR have on the given turning process, what is the best preset combination of cutting parameters (mainly ap, Vc and fn) for that specific process.

In the realm of milling operations, the advent of a multi-criteria decision-making framework for parameter selection, as introduced by the authors in [13], has sparked great interest. The presented case study serves as a validation of the reliability and effectiveness of these methods specifically in the context of milling parameter selection. The significance of choosing optimal cutting parameters cannot be overstated, given its direct impact on crucial factors such as process efficiency, productivity enhancement, and the minimization of machine operation time. To address this challenge, the TOPSIS method has been implemented and its application in selecting optimal alternatives for cutting parameters has been extensively explored in the research conducted by the authors [14]. Encouragingly, the dependability of this method has been further substantiated through the successful deployment of experimental results on real-world examples, lending credence to its efficacy and practicality. It is noteworthy that the TOPSIS method has emerged as one of the most frequently employed MCDM methods, as evidenced by its prominent standing in relevant research and review articles [15, 16]. This further augments the reliability and robustness of the investigations conducted in this domain.

Building upon the foundation laid by previous studies, the authors in [17, 18] have employed a composite approach, incorporating multiple methods and a selection of relevant criteria within the MCDM framework. This methodology allows for the ranking and exploration of optimal parameters that capture the combined influence of cutting parameters in turning processes, as witnessed in the comprehensive works presented in [19–21]. Moreover, a comparative study involving the AHP, WASPAS, and TOPSIS methods has been undertaken, focusing on their application in the machining process. This insightful exploration, chronicled in [22–26], not only sheds light on the effectiveness of these methods but also provides a valuable benchmark for further investigations and methodological advancements. Furthermore, the practical applicability of the WASPAS method in addressing decision-making challenges encountered in production and the selection of material processing parameters has been examined by the authors in [27, 28]. These investigations highlight the robustness and optimality of the proposed methodology. Additionally, the effectiveness of MCDM methods in turning processes has been extensively studied and documented in the works of the authors [28–34], underscoring the broader significance and impact of these methodologies in real-world machining applications.

A comprehensive review of the existing literature reveals several key observations. Notably, researchers have exhibited a remarkable inclination towards hybrid MCDM approaches in recent years, striving to combine diverse methods to address complex decision-making problems. These hybrid paradigms have been specifically designed to tackle challenges that defy resolution through classical MCDM approaches, particularly those involving the consideration of multiple aspects and diverse types of criteria [35, 36]. Moreover, hybrid approaches offer valuable insights into issues related to criterion evaluation and weight management [6], thereby enriching the decision-making process. Collectively, the emergence of hybrid MCDM approaches represents a significant stride forward in the field of decision-making, poised to unlock a multitude of possibilities and practical applications in real-world scenarios [37].

However, in the context of machining and MCDM methods, while previous research has demonstrated the effectiveness of MCDM techniques and even hybrid methodologies, a notable gap exists concerning the integration of multiple MCDM methods, namely AHP (Analytic Hierarchy Process), TOPSIS (Technique for Order of Preference by Similarity to Ideal Solution), and WASPAS (Weighted Aggregated Sum Product Assessment), in the domain of turning operations, specifically applied to AISI 1045 steel. Moreover, the scarcity of studies addressing the simultaneous optimization of energy consumption, productivity, and surface roughness in turning processes is apparent. The research seeks to bridge this void in the literature by combining these well-established MCDM techniques and applying them to the machining of AISI 1045 steel. By addressing the challenge of simultaneously minimizing energy consumption, maximizing productivity, and ensuring required surface quality, this study aims to contribute significantly to the machining industry's pursuit of efficiency and precision.

Moving forward, this paper will delve into presentation of a combination of several MCDM methods, shedding light on the combined influence of cutting parameters in turning processes while accounting for distinct rankings (weights) assigned to each criterion based on their relative importance. The next section of the paper will deal with a presentation of the employed methods.

3 Methodology

The objective of this study is to select the optimal combinations of cutting parameters that maximize productivity (MRR) while minimizing energy consumption (EC) and surface roughness (SR). To achieve this, a three-step approach was adopted. Firstly, the Analytic Hierarchy Process (AHP) method was utilized to establish objective constant weights for the evaluation criteria, considering the relative importance of MRR, EC, and SR. Subsequently, the Technique for Order of Preference by Similarity to Ideal Solution (TOPSIS) and the weighted aggregated sum product assessment (WASPAS) methods were applied to rank the optimized alternatives based on their performance across the defined criteria. This comprehensive methodology enables a systematic and efficient evaluation of the cutting parameter combinations, facilitating informed decision-making in turning operations. The novelty of our approach lies in the combining of AHP, TOPSIS, and WASPAS methods for MCDM within the specific context of turning operations on AISI 1045 steel. While previous studies have explored MCDM techniques in various machining applications and materials, the integration of these three methods in the realm of turning processes is a novel endeavor. Furthermore, the selection of criteria - minimizing energy consumption, maximizing productivity (material removal rate), and minimizing surface roughness - represents a unique combination of objectives that align closely with the essential goals of machinists in today's environmentally conscious and quality-focused machining industry. By fusing these methods and criteria, the research brings forth an innovative approach that addresses the multifaceted challenges faced by machinists, enabling them to make informed decisions that optimize their processes in terms of both efficiency and product quality. This approach's novelty is anticipated to make a significant contribution to the field of machining and Multi-Criteria Decision Making.

3.1 AHP Method

The Analytic Hierarchy Process (AHP) method, developed by Saaty [38, 39], is a powerful tool employed in multi-criteria decision-making (MCDM) that facilitates the determination of objective constant weights for evaluating criteria. AHP provides a structured approach to decision-making by decomposing complex problems into a hierarchical structure comprising a goal, criteria, and alternatives. In the context of our study, the goal is to optimize the cutting parameters for turning operations. The criteria include productivity (MRR), energy consumption (EC), and surface roughness (SR). AHP utilizes pairwise comparisons to assess the relative importance of these criteria. Expert opinions or data-driven techniques can be used to establish pairwise comparison matrices. By quantifying the preferences through a scale, such as Saaty's scale (Table 1), the matrices are populated with numerical values. Subsequently, through the calculation of eigenvectors and eigenvalues, the weights are derived, indicating the relative importance of the criteria. To ensure the consistency of the pairwise comparisons, the consistency ratio (CR) is calculated. A CR value close to 0 indicates a high level of consistency, whereas higher CR values indicate the need for further examination and adjustment of the comparisons. The flexibility of the AHP method allows its application to multiple processes, as the CR ensures the consistency of the criteria weights. This feature makes the hybrid method, which combines AHP with other MCDM methods, robust and reliable for decision-making in turning operations.

Table 1. Saaty's scale for AHP method

Saaty's scale	Meaning
1	Equally preferred
3	Moderate preference of one over the other
5	Strong preference of one over the other
7	Very strong preference of one over the other
9	Extreme preference of one over the other
2,4,6,8	Intermediate values, indicating the degree of preference between adjacent values

3.2 TOPSIS Method

The Technique for Order of Preference by Similarity to Ideal Solution (TOPSIS) method, a widely used MCDM approach, is employed in our study to rank the optimized alternatives for turning operations. TOPSIS is a distance-based method that evaluates alternatives based on their similarity to the ideal and anti-ideal solutions. The ideal solution represents the maximum achievement of the desired criteria (e.g., maximum productivity, minimum energy consumption, and minimum surface roughness), while the anti-ideal solution represents the minimum achievement. To calculate the TOPSIS rankings, the alternatives are represented as a decision matrix, where each row corresponds to an alternative and each column corresponds to a criterion. The decision matrix is normalized to eliminate the effects of scale differences among the criteria. Subsequently, the ideal and anti-ideal solutions are determined based on the maximum and minimum values for each criterion, respectively. Euclidean or other distance metrics are then utilized to calculate the proximity of each alternative to these reference solutions. The relative closeness of each alternative to the ideal solution is calculated using the concept of relative closeness coefficients. Finally, the alternatives are ranked based on their relative closeness coefficients. The TOPSIS method provides a robust and intuitive approach for ranking the alternatives, enabling the selection of the most favorable combination of cutting parameters. By incorporating the TOPSIS method into the hybrid MCDM framework, the overall decision-making process becomes more comprehensive and effective in optimizing turning operations.

3.3 WASPAS Method

The weighted aggregated sum product assessment (WASPAS) method is an integral part of our study's hybrid multi-criteria decision-making (MCDM) framework for ranking the optimized alternatives in turning operations. In our research, the AHP method was employed to establish objective constant weights for the criteria. These criteria weights, obtained through the AHP process, were subsequently utilized in the WASPAS method for evaluating and ranking the alternatives. The WASPAS method considers both the performance values of the alternatives and the pre-determined criteria weights. It aggregates the performance values using the weighted sum product approach to assess the alternatives comprehensively. By multiplying each alternative's performance values by the corresponding criteria weights and summing the results, the method generates a performance index for each alternative. In our research, the criteria weights derived from the AHP method offer a robust and reliable basis for the evaluation conducted by the WASPAS method. By utilizing these pre-determined weights, we ensure the consistency and objectivity of the evaluation process. This approach allows us to effectively rank the alternatives based on their performance indexes, providing valuable insights for selecting optimal cutting parameter combinations in turning operations. The integration of the WASPAS method within the hybrid MCDM framework, utilizing the criteria weights obtained from the AHP method, enhances the overall decision-making process. This approach offers flexibility and robustness, as it combines the strengths of both methods. The use of the AHP-derived weights ensures consistency and objectivity, while the WASPAS method enables a comprehensive evaluation of the alternatives based on the determined weights.

4 Case Study

In this case study, the aim was to determine the optimal combination of cutting parameters for turning AISI 1045 steel, considering the criteria of maximum productivity, minimum energy consumption and minimum surface roughness, using a hybrid MCDM approach with AHP, TOPSIS, and WASPAS methods. The data utilized for this study is derived from two previously conducted research papers, [40] and [41] In the first study [40], the focus was on optimizing the input parameters to achieve minimum energy consumption while maintaining the targeted productivity. From this study, we extracted data related to EC and MRR. The second study [41], specifically explored the influence of cutting parameters on surface roughness. We obtained data on SR from this study. Both studies employed the same set of input parameters:

- cutting depth (ap) values of $\{1, 1.75, 2.5\}$ mm,
- feed rate (fn) values of $\{0.1, 0.2, 0.3\}$ mm/rev,
- cutting speed (Vc) values of $\{240, 270, 300\}$ m/min.

The experiments were conducted on a Gildemeister NEF 520 lathe machine equipped with a 12 kW motor, a maximum spindle speed of 3000 rev/min, and Heidenhain Manual Plus 4110 control units. The cutting tool used was a Sandvik Coromant DCLNL 2020K 12 toolholder with a ZCC-CT CNMG120408-PM insert. In both studies, a full factorial design of experiments was employed to explore the effects of various combinations of cutting parameters. This approach allowed for a comprehensive analysis of the turning process. A total of 27 unique combinations were obtained, each representing different values of EC, MRR, and SR (Table 2).

The AHP method was used to determine the weight coefficients. The pairwise comparison matrix was formulated to establish the relative importance of the criteria. In this case, the matrix indicated that energy consumption (EC) was considered four times more important than surface roughness (SR), and twice as important as material removal rate (MRR), as can be seen in Table 3.

The weight coefficients obtained as a result of the AHP analysis are as follows·

- w1 (weight coefficient for EC) = 0.571428571;
- w2 (weight coefficient for SR) = 0.142857143;
- w3 (weight coefficient for MRR) = 0.285714286.

Furthermore, the consistency of criterion importance was verified, confirming that the criteria were adequately compared in terms of importance, with a consistency ratio (CR) of 0.

Table 2. Alternatives generated through full factorial design of experiments

#	ap	fn	Vc	#	ap	fn	Vc
1	1	0.1	240	15	1.75	0.2	300
2	1	0.1	270	16	1.75	0.3	240
3	1	0.1	300	17	1.75	0.3	270
4	1	0.2	240	18	1.75	0.3	300
5	1	0.2	270	19	2.5	0.1	240
6	1	0.2	300	20	2.5	0.1	270
7	1	0.3	240	21	2.5	0.1	300
8	1	0.3	270	22	2.5	0.2	240
9	1	0.3	300	23	2.5	0.2	270
10	1.75	0.1	240	24	2.5	0.2	300
11	1.75	0.1	270	25	2.5	0.3	240
12	1.75	0.1	300	26	2.5	0.3	270
13	1.75	0.2	240	27	2.5	0.3	300
14	1.75	0.2	270				

Table 3. Pairwise comparison matrix

	EC	SR	MRR
EC	1	4	2
SR	0.25	1	0.5
MRR	0.5	2	1

5   Results and Discussion

Determining the best combination of cutting parameters depending on the importance of the minimum EC, minimum SR and maximum MRR is multi-criteria decision-making problem, with minimum EC, minimum SR and maximum MRR being the criteria. The results of the study reveal interesting insights regarding the recommendations provided by the TOPSIS and WASPAS methods (Fig. 1).

These two methods, although aimed at achieving the same goal, presented divergent recommendations for the best alternatives. Notably, both methods concurred on identifying alternative number 21 as the worst option. However, when it comes to determining the best alternatives, the TOPSIS method suggested alternative number 16, while the WASPAS method favored alternative number 13. It is worth noting that these two alternatives exhibit a high degree of similarity, differing only in one parameter: the feed rate (fn), where alternative 13 has fn = 0.2 mm/rev and alternative 16 has fn = 0.3 mm/rev.

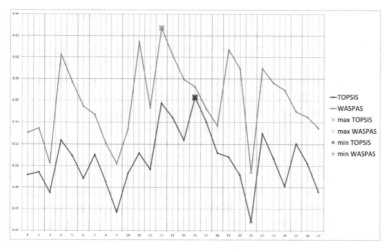

Fig. 1. Results of TOPSIS and WASPAS methods

Although the discrepancy between these recommendations may not be a major concern, further investigation is warranted to understand the underlying reasons for this discrepancy.

The obtained results from the TOPSIS and WASPAS methods provide a clear distinction in the ranking of the top five best and worst alternatives (Table 4).

Table 4. Top 5 best and worst alternatives

Best 5 TOPSIS	Best 5 WASPAS	Worst 5 TOPSIS	Worst 5 WASPAS
A16	A13	A21	A21
A13	A11	A9	A9
A14	A19	A3	A3
A17	A4	A27	A8
A22	A14	A24	A1

The analysis of the worst alternatives based on the TOPSIS and WASPAS methods reveals a significant overlap in the rankings. Both methods consistently identify alternatives A21, A9, and A3 as the worst performing options. This alignment in rankings suggests that these alternatives exhibit unfavorable characteristics across the evaluated criteria, making them less desirable choices in terms of energy consumption, productivity, and surface roughness.

According to the TOPSIS method, the top five alternatives in descending order of preference are A16, A13, A14, A17, and A22. Conversely, the WASPAS method ranks the alternatives as A13, A11, A19, A4, and A14. Interestingly, alternative A13 emerges as a consistent top choice according to both methods. This suggests that A13 possesses favorable characteristics across the evaluated criteria, making it a strong contender for the optimal combination of parameters in turning operations. On the other hand, alternative A14 appears in the top five rankings for both methods, indicating its prominence in terms of performance. The disparities in the remaining alternatives between the two methods warrant further investigation. While TOPSIS places alternatives A16, A17, and A22 in the top five, WASPAS includes A11, A19, and A4 instead. These discrepancies highlight the sensitivity of the decision-making process to the employed MCDM method and the various factors influencing the evaluation and ranking of alternatives. One possible explanation for the disparity in results could be attributed to the utilization of different normalization methods in the TOPSIS and WASPAS methods. As these methods employ distinct approaches, they inherently introduce variations in the way alternatives are evaluated and ranked. Expanding the decision-making process by incorporating more methods would not only provide a more comprehensive assessment but also shed light on the strengths and weaknesses of each approach and contribute to improved decision-making in the field of turning operations.

6 Conclusion

The main motivation for this work was to facilitate the selection of combinations of different cutting parameters for machinists in machine shops, taking into consideration the varying importance of maximum productivity (MRR), minimum energy consumption (EC), and minimum surface roughness (SR) in a given turning process. To achieve this objective, a hybrid multi-criteria decision-making (MCDM) method was employed, combining the Analytic Hierarchy Process (AHP), Technique for Order of Preference by Similarity to Ideal Solution (TOPSIS), and Weighted Aggregated Sum Product Assessment (WASPAS) methods. The AHP method played a crucial role in establishing objective constant weights for the criteria, ensuring consistency in the decision-making process. It provided the necessary foundation by determining the relative importance of EC, SR, and MRR. The TOPSIS and WASPAS methods were subsequently employed to rank the optimized alternatives based on the established criteria weights. For the case study, AISI 1045 steel was chosen as the material of interest. Data from two previous studies, namely [40] and [41], were utilized. These studies provided insights into the effects of cutting parameters on energy consumption, productivity, and surface roughness.

One of the most intriguing findings of this research was the discrepancy in the rankings of the top five alternatives between the TOPSIS and WASPAS methods. While both methods identified similar alternatives as the worst performers, they diverged in their recommendations for the best alternatives. This discrepancy highlights the influence of different normalization methods and the inherent variations between the TOPSIS and WASPAS techniques, which makes them very sensitive. Further investigation is necessary to better understand the underlying reasons for these differences and to refine the decision-making process. Incorporating additional MCDM methods could provide a more comprehensive understanding of the decision-making process. By including more methods, one can gain valuable insights into the robustness and reliability of the hybrid approach, and insight into what the different MCDM methods are good and bad at. Additionally, exploring the integration of other factors, such as tool life, cost, and environmental impact, could enhance the decision-making process and enable a more comprehensive evaluation of turning operations, further facilitating the work of the in-shop machinist.

References

1. Stanković, A., Petrović, G., Marković, D., Ćojbašić, Ž.: Solving flexible job shop scheduling problem with transportation time based on neuro-fuzzy suggested metaheuristics. Acta Polytechnica Hungarica **19**(4), 209–227 (2022)
2. Stanković, A., Petrović, G., Ćojbašić, Ž, Marković, D.: An application of metaheuristic optimization algorithms for solving the flexible job-shop scheduling problem. Oper. Res. Eng. Sci. Theor. Appl. **3**(3), 13–28 (2020)
3. Žižović, M., Miljković, B., Marinković, D.: Objective methods for determining criteria weight coefficients: a modification of the CRITIC method. Decis. Mak. Appl. Manag. Eng. **3**, 149–161 (2020)
4. Mukhametzyanov, I.: Specific character of objective methods for determining weights of criteria in MCDM problems: Entropy CRITIC and SD. Decis. Mak. Appl. Manag. Eng. **4**, 76–105 (2021)
5. Pamučar, D., Žižović, M., Biswas, S., Božanić, D.: A new logarithm methodology of additive weights. Application in logistics. Fact. Univ. Ser. Mech. Eng, LMAW) for multi-criteria decision-making (2021)
6. Petrović, G.S., Madić, M., Antucheviciene, J.: An approach for robust decision making rule generation: solving transport and logistics decision making problems. Expert Syst. Appl. **106**, 263–276 (2018)
7. Keshavarz-Ghorabaee, M., Amiri, M., Zavadskas, E., Turskis, Z., Antucheviciene, J.: Determination of objective weights using a new method based on the removal effects of criteria (MEREC). Symmetry **13**, 525 (2021)
8. Petrović, G., Pavlović, J., Madić, M., Marinković, D.: Optimal synthesis of loader drive mechanisms: a group robust decision-making rule generation approach. Machines **10**, 329 (2022)
9. Olabanji, O.M., Mpofu, K.: Appraisal of conceptual designs: coalescing fuzzy analytic hierarchy process (F-AHP) and fuzzy grey relational analysis (F-GRA). Results Eng. **9**, 100194 (2021)
10. Zavadskas, E.K., Turskis, Z., Stević, Ž, Mardani, A.: Modelling procedure for the selection of steel pipes supplier by applying fuzzy AHP method. Oper. Res. Eng. Sci. Theor. Appl. **3**, 39–53 (2020)
11. Stević, Ž., Vasiljević, M., Puska. A., Tanackov, I., Junevicius, R., Vesković, S.: Evaluation of suppliers under uncertainty: a multiphase approach based on Fuzzy AHP and Fuzzy EDAS. Transport **34**(1), 52–66 (2019)
12. Brauers, W.K.M., Zavadskas, E.K.: Robustness of MULTIMOORA: a method for multi-objective optimization. Informatica **23**, 1–25 (2012)
13. Wuyang, S., Yifei, Z., Ming, L., Zhao, Z., Dinghua, Z.: A multi-criteria decision-making system for selecting cutting parameters in milling process. J. Manuf. Syst. **65**, 498–509 (2022)
14. Chong, P., Hanheng, D., Warren, T.L.: A research on the cutting database system based on machining features and TOPSIS. Robot. Comput.-Integr. Manufact. **43**, 96–104 (2017)
15. Çelikbilek, Y., Tüysüz, F.: An in-depth review of theory of the TOPSIS method: an experimental analysis. J. Manag. Anal. **7**, 281–300 (2020)
16. Hwang, C.-L., Lai, Y.-J., Liu, T.-Y.: A new approach for multiple objective decision making. Comput. Oper. Res. **20**, 889–899 (1993)

17. Trung, D.D.: Application of TOPSIS and PIV methods for multi-criteria decision making in hard turning process. J. Mach. Eng. **21**(4), 57–71 (2021)
18. Trung, D.D., Thinh, H.X.: A multi-criteria decision-making in turning process using the MAIRCA, EAMR, MARCOS and TOPSIS methods: a comparative study. Adv. Prod. Eng. Manag. **16**(4), 443–456 (2021)
19. Divya, C., Raju, L.S., Singaravel, B.: Application of MCDM methods for process parameter optimization in turning process—a review. In: Narasimham, G.S.V.L., Babu, A.V., Reddy, S.S., Dhanasekaran, R. (eds.) Recent Trends in Mechanical Engineering. LNME, pp. 199–207. Springer, Singapore (2021). https://doi.org/10.1007/978-981-15-7557-0_18
20. Pathapalli, V.R., Basam, V.R., Gudimetta, S.K., Koppula, M.R.: Optimization of machining parameters using WASPAS and MOORA. World J. Eng. **17**(2), 237–246 (2020)
21. Singaravel, B., Selvaraj, T.: Optimization of machining parameters in turning operation using combined TOPSIS and AHP method. Tehnički vjesnik **22**(6), 1475–1480 (2015)
22. Nguyen, H.-Q., Le, X.-H., Nguyen, T.-T., Tran, Q.-H., Vu, N.-P.: A comparative study on multi-criteria decision-making in dressing process for internal grinding. Machines **10**, 303 (2022)
23. Varatharajulu, M., Duraiselvam, M., Kumar, M.B., Jayaprakash, G., Baskar, N.: Multi criteria decision making through TOPSIS and COPRAS on drilling parameters of magnesium AZ91. J. Magnes. Alloys **10**(10), 2857–2874 (2021)
24. Tran, Q.-P., Nguyen, V.-N., Huang, S.-C.: Drilling process on CFRP: multi-criteria decision-making with entropy weight using grey-TOPSIS method. Appl. Sci. **10**, 7207 (2020)
25. Kumar, G.V.A., Reddy, D.V., Nagaraju, N.: Multi-objective optimization of end milling process parameters in machining of EN 31 steel: application of AHP embedded with VIKOR and WASPAS methods. i-manager's J. Mech. Eng. **8**(4), 39–46 (2018)
26. Vats, P., Singh, T., Dubey, V., Sharma, A.K.: Optimization of machining parameters in turning of AISI 1040 steel using hybrid MCDM technique. Mater. Today Proc. **50**, 1758–1765 (2022)
27. Chakraborty, S., Zavadskas, E.K.: Applications of WASPAS method in manufacturing decision making. Informatica **25**(1), 1–20 (2014)
28. Zavadskas, E.K., Turskis, Z., Antucheviciene, J., Zakarevicius, A.: Optimization of weighted aggregated sum product assessment. Elektr. ir Elektrotech **122**, 3–6 (2012)
29. Singaravel, D., Shankar, P., Prasanna, L.: Application of MCDM method for the selection of optimum process parameters in turning process. Mater. Today Proc. **5**(5), 13464–13471 (2018)
30. Trung, D.D.: A combination method for multi-criteria decision making problem in turning process. Manufacturing Rev. **8**, 26 (2021)
31. Mane, S.S., Mulla, A.M.: Relevant optimization method selection in turning of AISI D2 steel using Crygenic cooling. Int. J. Creat. Res. Thoughts **8**, 803–812 (2020)
32. Pathapalli, V.R., Basam, V.R., Gudimetta, S.K., Koppula, M.R.: Optimization of machining parameters using WASPAS and MOORA. World J. Eng. **17**, 237–246 (2020)
33. Raigar, J., Sharma, V.S., Srivastava, S., Chand, R., Singh, J.: A decision support system for the selection of an additive manufacturing process using a new hybrid MCDM technique. Sādhanā **45**, 1–14 (2020)
34. Saha, A., Majumder, H.: Multi criteria selection of optimal machining parameter in turning operation using comprehensive grey complex proportional assessment method for ASTM A36. Int. J. Eng. Res. Africa **23**, 24–32 (2016)
35. Zavadskas, E.K., Turskis, Z., Kildienė, S.: State of the art surveys of overviews on MCDM/MADM methods. Technol. Econ. Dev. Econ. **20**(3), 533–572 (2014)
36. Figueira, J., Greco, S., Ehrgott, M. (Eds.). Multiple criteria decision analysis: state of the art surveys. Springer Science & Business Media (2013)
37. Wu, J., Liang, L., Zhang, H.: Hybrid multi-criteria decision making methods: a review of applications and studies. Int. J. Inf. Technol. Decis. Mak. **18**(5), 1315–1352 (2019)

38. Saaty, T.L.: A scaling method for priorities in hierarchical structures. J. Math. Psychol. **15**(3), 234–281 (1977)
39. Saaty, T.L.: The Analytic Hierarchy Process. McGraw-Hill Company, New York (1980)
40. Stojković, M, Trifunović, M., Madić, M., Turudija, R., Manić, M.: Partial effect of cutting parameters on engaged power and energy consumption: The case of external turning of an AISI1045 steel workpiece. Innovative Mechanical Engineering, Faculty of Mechanical Engineering, University of Niš, 1, 2, pp. 34–47, 2812–9229 (2022)
41. Turudija, R., Janković, P., Stojković, M., Madić, M., Ivanović, M.: Investigation of the Cutting Parameters Influence on Surface Roughness in Turning AISI 1045 Steel, Innovative Mechanical Engineering, Faculty of Mechanical Engineering, University of Niš, 1, 2, pp. 22–33, 2812–9229 (2022)

CNN - Based Object Detection for Robot Grasping in Cluttered Environment

Ivan Ćirić[1] ⓘ, Nikola Ivačko[1](✉) ⓘ, Stefan Lalić[1], Valentina Nejković[2] ⓘ,
Maša Milošević[1] ⓘ, Dušan Stojiljković[1] ⓘ, and Dušan Jevtić[3]

[1] Department of Mechatronics and Control, Faculty of Mechanical Engineering, University of
Niš, Aleksandra Medvedeva 14, 18000 Niš, Serbia
nikola.ivacko@masfak.ni.ac.rs
[2] Faculty of Electrical Engineering, University of Niš, Aleksandra Medvedeva 14,
18000 Niš, Serbia
[3] Fazi D.O.O., Novoprojektovana Bb, Crveni Krst, 18000 Niš, Serbia

Abstract. The objective of robot vision is to identify objects of interest within
the robot's workspace from digital images. Despite advances in the programming
and control of collaborative robots, object recognition and localization remain
a challenging task. Given that the performance of collaborative robots operating in cluttered environments is heavily reliant on robot vision, the implementation of an appropriate object recognition algorithm can significantly improve
its performance. In this paper, a robot vision algorithm is developed based on a
Convolutional Neural Network (CNN) classifier for object detection and spatial
localization for robot arm manipulation and grasping and tested in cluttered scene
in terms of its robustness to variations in light conditions, viewing angles, and
other uncertainties and noise. After successful object recognition, homography
is employed for object localization, which serves as input for the robot's manipulation control. The experimental setup involves a collaborative robot equipped
with a haptic gripper and a static camera placed above the manipulation area. The
results of the robot's manipulation and grasping based on the implementation of
developed algorithms are presented and discussed in final chapters of the paper.

Keywords: Robot Vision · CNN · Cobot · Robot Manipulation · Robot Grasping

1 Introduction

1.1 Motivation

The objective of this paper is to address the issue of robot guidance and grasping by
leveraging conventional and intelligent image processing techniques. Robot vision for
manipulation and grasping involves the use of static cameras to capture images of objects
in the robot's environment. These images are then analyzed using image processing
algorithms to identify and locate objects of interest. This process involves several steps,
including image acquisition, feature extraction, object recognition, and pose estimation.
Overall, the use of static cameras for robot vision provides a powerful tool for enabling

© The Author(s), under exclusive license to Springer Nature Switzerland AG 2024
M. Trajanovic et al. (Eds.): ICIST 2023, LNNS 872, pp. 206–217, 2024.
https://doi.org/10.1007/978-3-031-50755-7_20

robots to perform complex manipulation and grasping tasks in a wide range of applications. However, the development of effective algorithms and techniques for handling issues such as lighting variations and camera calibration is critical for achieving reliable and accurate performance. The authors' first results were presented in [12] were focused on conventional computer vision methods, while the research presented in this paper is further development of the robot vision algorithms with CNN implementation and improvements regarding grasp approaching angle determination.

1.2 Research Question

The primary challenge involves accurately recognizing the target object intended for manipulation and grasping, using a monocular camera and object recognition/classification algorithm. In order to determine adequate object recognition algorithm for specific task of robot grasping, we aim to compare the CNN-based pretrained classifier and conventional template matching algorithm regarding speed and accuracy for a scenario involving a static monocular camera system operating in diverse lighting conditions and a cluttered environment. The algorithm provides the 2D coordinates of the object's center of mass in the camera image, and the corresponding 3D coordinates in the real world, which are crucial for the robot's grasping task. The secondary challenge relates to the cluttered environment, where the objects of interest are intermingled with other objects that share similar features in terms of shape, size, color, and symmetry, resembling a stochastic working environment. Additionally, we seek to overcome the problem of illumination inconsistency that occurs under varying lighting conditions by developing an adaptive shutter adjustment algorithm. Finally, approaching angle positioning of the end-effector is determined based on the class of the detected object and object angular posture.

2 State of the Art

Over the past three decades, vision-based motion control has made significant advancements in both scientific research and technological applications [1]. What were once experimental techniques in laboratories have now become commonplace in various practical settings. These methods, involving visual perception and learning, have the potential to enhance the performance of industrial robots operating in unstructured environments.

One notable approach for object recognition, as described by M. Peña et al. [2], utilizes an Artificial Neural Network (ANN) architecture that takes a specific description vector called CFD&POSE as input. Numerous other recognition methods have been developed, including the two-step template matching method, which incorporates correlation coefficient and genetic algorithm techniques to accurately identify transformation parameters [3]. This algorithm offers high precision and robust rotation invariance.

Another crucial aspect of robot vision involves recognizing the geometry of the robot's workspace based on digital images. By comparing selected pixel coordinates in a two-dimensional image with their corresponding coordinates in a three-dimensional robot environment, the relationship between them can be determined [4]. The prevailing methodologies for visual recognition and vision-based robot motion control combine both object recognition and 3D reconstruction within a unified framework. These methods employ image morphing and intermediate virtual images to guide the manipulator toward its target. Noteworthy advantages of this proposed technique include its ability to recognize partially occluded and deformable shapes, enable controlled motion without complete calibration information, and eliminate the need for manual feature selection and correspondence [5–8].

Separate approach [9] highlights utilization of hashing methods and approximate nearest neighbors (ANN) voting, along with the proposed "Encoding + Selection" pattern and the use of label information, to improve the efficiency and accuracy of RGB-D-based object recognition in robot vision. The experiment results confirm the effectiveness of the mentioned voting recognition method along with the proposed bit selection method, demonstrating value of applying hashing learning with the "Encoding + Selection" pattern for enhancing recognition efficiency and discriminative binary representations in robot vision.

Further exploration reveals the challenges of applying deep learning-based object recognition models developed using ImageNet to robot vision. It highlights the differences between web images and images captured by robots, which often show objects from unusual angles and scales. The paper proposes a data augmentation layer that simulates the robot's visual experience by zooming in on the object of interest, resulting in up to a 7% increase in object recognition performance [10].

The effectiveness of the approach in bridging the gap between computer vision and robotics is discussed within the development of a computer vision system for automatic image labeling in robotics scenarios. The proposed method leverages weak supervision provided by a human demonstrator to enable realistic human-robot interaction and achieve automatic image labeling. The authors extend their previous method for ego-motion compensation and employ an object recognition framework to test the image labeling process [11].

3   Methodology

The robot vision algorithm used for robust grasping starts with the image preprocessing. The improved image is then used for object detection. The features of the detected object of interest, namely 3D coordinates and posture angle are determined in image, and then transformed in robot coordinate system and further used as an input for the robot control algorithm. The automatic exposure time adjustment program, as well as the programs that combine object recognition and robot grasping task were written in Python using the OpenCV library and YOLO v8.

3.1   Experimental Setup

The experimental configuration involves a black canvas with dimensions of 1000 mm × 1000 mm, placed on a flat work surface, onto which objects of interest, along with similar components regarding shape, size, and color, are scattered (shown in Fig. 1a). The sensory input of the system includes a Rapoo C260 web camera mounted on a Bosh profile, positioned directly 120 mm above the work surface. On the other side of the work desk, 500 mm from the black canvas, is a 5 degree of freedom robot arm xArm5 that comprises base and rotary joints, and an end-effector with a maximum gripper force of 30N (shown in Fig. 1b).

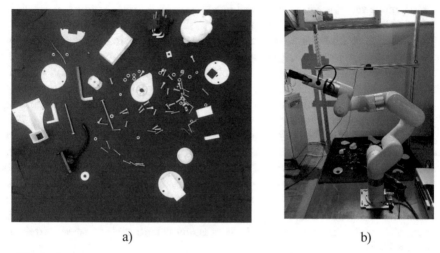

a)   b)

Fig. 1. Scattered parts on the work surface a) and U-Factory X-arm and camera setup b)

3.2 Image Preprocessing

In image preprocessing, varying lighting conditions can significantly influence the intensity distribution of a grayscale image matrix, leading to contrast variations. While methods like histogram equalization can be employed to address these variations, in this study, a different approach was taken. An automatic adjustment of the physical camera shutter speed was implemented to regulate the amount of light reaching the sensor, modulating the shutter speed based on the detected brightness level. Specifically, if an image is perceived as too dark, the exposure time is increased, ensuring optimal brightness. Preliminary tests showed that images processed with this method resulted in a noticeable improvement in object detection accuracy compared to non-processed images [12].

While the automatic adjustment of the camera shutter speed proves effective when curating a new dataset, it's recognized that this method might not be applicable to pre-existing datasets. In such scenarios, post-capture image processing techniques would be more suitable. The adaptive normalization method was considered but was not implemented in this study due to the effectiveness of the shutter speed adjustment based approach [12].

3.3 Object Recognition

After successful image preprocessing, CNN-based object recognition algorithms were implemented, with a particular focus on the YOLO (You Only Look Once) model. YOLO is a deep learning-based object detection algorithm that identifies objects in real-time by dividing the image into a grid and predicting bounding boxes and class probabilities simultaneously. This unique approach offers a balance between speed and accuracy, making it especially suitable for applications requiring real-time object detection, such as autonomous vehicles and video object recognition.

Alternative models such as Faster R-CNN, SSD, and RetinaNet also offer robust object detection capabilities. However, YOLO's efficiency in simultaneously detecting multiple objects sets it apart for specific application needs. It's important to note YOLO's limitations, including its demand for significant computing resources and potential performance degradation in scenarios with complex backgrounds or occlusions.

On the other hand, Template Matching used in previous research [12] is a simpler algorithm that compares a pre-defined template with a region of interest in the image. The algorithm calculates the similarity score between the template and the image patch, and the location with the highest score is considered the object's position. Template Matching is computationally less expensive and requires minimal training, making it suitable for applications with limited computing resources. However, Template Matching is sensitive to changes in scale, rotation, and illumination, and it may not be able to recognize objects with significant visual variations. As shown in authors previous research the best template matching method for robotic grasping is Normalized Cross- Correlation Matching Method (NCCM).

3.4 Data Preparation and Training of YoloV8

In order for our CNN based robot vision algorithm to successfully detect objects of interest it was necessary to additionally train YOLO classifier. For preparation of training data set for Yolo V8 the appropriate dataset was formed that contained images of the work surface in different settings, alongside with augmented images created using common augmentation techniques (random scaling, flipping, rotation, and adding noise). The total of 56 images was produced for the training dataset, with each of the 5 classes represented exactly once in each image. To bolster the dataset and introduce variability, common augmentation techniques such as random scaling, flipping, rotation, and noise addition were employed using the 'imgaug' Python library. Post-augmentation, each of the 56 images underwent five augmentations, resulting in a total of 280 images for the training dataset. The annotation process utilized CVAT (Computer Vision Annotation Tool) to define the object classes for the grasping task and apply bounding boxes to the images (Fig. 2). Additionally, the annotated dataset was exported in the YOLO 1.1 format. The dataset was split into a training set and a validation set, while the training parameters, network architecture, and the dataset's path were specified within the yaml file.

A scaled-down lightweight yolov8n model was employed to train a neural network for object detection. The network underwent training for 100 epochs, allowing it to iteratively learn and refine its ability to accurately detect objects in images (Fig. 3). The training outcomes, along with the interpretation of the obtained results, are presented in the Results and discussion chapter.

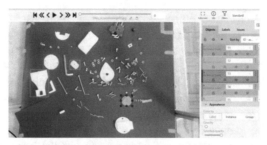

Fig. 2. Annotations are being carried using online CVAT engine (Computer Vision Annotation Tool).

Fig. 3. Training of CNN after loading the dataset.

3.5 Coordinate Transformation

To relate the camera and robot's coordinate systems, the camera is positioned in a fixed location above the robot's workspace, and calibration is performed using the calibration tip of the robot's end-effector. Three markers that are visible in the image are used as a reference or a measure, as a minimum of three points are required to define a plane uniquely. The calibration process involves aligning the end-effector's tip on specific markers and measuring the robot's coordinates at the marker's position, along with the marker's image coordinates.

The aim of the calibration is actually coordinate transformation by finding the transformation matrix due to a mismatch between the robot's x-y-z coordinates and the image's x_i-y_i-z_i coordinates. Their equivalence can be expressed by the equation:

$$H_{robot} = T * H_{image} \tag{1}$$

where, respectively, H_{robot} and H_{image} are the calibration configurations given in the robot and image coordinate frames. Matrix H_{image} represents a rotation around the z_i axis and translation along x_i and y_i axes of the image coordinate frame. The pose of the calibration pattern H_{robot} expressed in the robot coordinate frame x–y–z can be determined with the calibration tip at the robot end-effector and the calibration points marked on the calibration pattern. Finally, the transformation matrix is derived from the first equation as:

$$T = H_{robot} * H_{image}^{-1}. \tag{2}$$

3.6 Robot Grasping Task

Due to the inherent asymmetry of the objects and the varying characteristics of each class, it is crucial to consider class-dependent gripper orientation and positioning for effective grasping. Each class of objects necessitates a predefined grasp position to ensure successful manipulation. This entails determining the appropriate approach angle and orientation for the robot gripper to achieve a secure and stable grasp on the object.

To address the challenge of object orientation, the Python OpenCV library is utilized, specifically the "minAreaRect()" function. This function allows for the calculation of the minimum bounding rectangle that encloses the recognized object on the ground. The minimum bounding rectangle represents the tightest rectangular region that encompasses the object's boundary points. By leveraging the "minAreaRect()" function, it becomes possible to extract essential information about the angle at which the object is positioned on the floor. The resulting angle is a valuable parameter for achieving proper alignment and manipulation during the grasping process. The angle represents the orientation of the object, which is vital for instructing the robot gripper to rotate its final joint accordingly.

By integrating the extracted angle into the robotic grasping system, the gripper can align itself precisely with the object's orientation, enhancing the chances of a successful and stable grasp. This angle information, derived from the "minAreaRect()" function, acts as a key component in ensuring the robot's ability to adapt its gripper's configuration dynamically based on the recognized object's characteristics.

4   Results and Discussion

Training and validation results presented in this section shows that additionally trained YOLO classifier can accurately and reliably detect selected objects in various environments and scenarios.

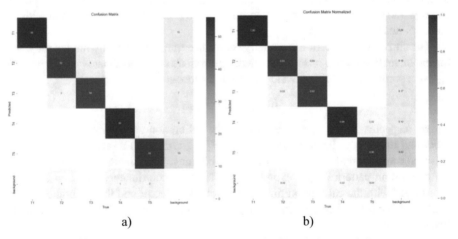

a)   b)

Fig. 4. Confusion matrix a) and normalized confusion matrix b)

In a confusion matrix (Fig. 4a), the rows represent the actual class labels, and the columns represent the predicted class labels. Each cell in the matrix shows the count or frequency of instances that belong to a particular combination of predicted and actual classes. A normalized confusion matrix (Fig. 4b), represents these counts as proportions or percentages, allowing for a more intuitive understanding of the classifier's performance. The values in the matrix are normalized by dividing each count by the total number of instances in the corresponding actual class, resulting in values between 0 and 1.

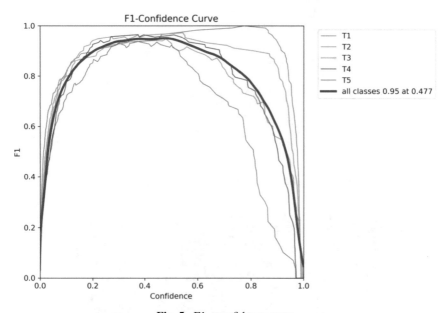

Fig. 5. F1 - confidence curve

The F1- confidence curve (Fig. 5) represents the relationship between the confidence level of predictions and the F1 score. The F1 score is a commonly used evaluation metric for binary classification problems that combines precision and recall into a single value.

The number of total instances of classes being recognized is shown in Fig. 6, as well as relative width and height, along with position of objects center of mass.

In object detection tasks, the training dataset (Fig. 7) consists of a collection of images with corresponding ground truth annotations.

Validation batch (Fig. 8) shows evaluated model's performance during training.

By leveraging the "minAreaRect()" function, it becomes possible to extract essential information about the angle at which the object is positioned on the floor (Fig. 9).

The experimental results (Table 1) are showing that although small training set was used, YOLO is superior to Normalized Squared Difference Method template matching regarding accuracy. With greater training set it is presumable that accuracy of YOLO would be even higher. However, for the objects of near-circular shape both algorithms are absolutely accurate while traditional template matching is much faster and therefore applicable for real time implementation. The results of both algorithms for specially prepared validation set of 6 scenarios are shown in Table 1.

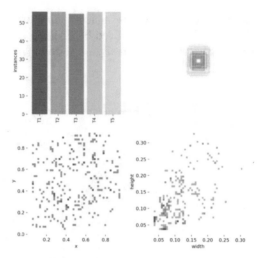

Fig. 6. Recognized instances of the class and class features

Fig. 7. Image sample from the training dataset

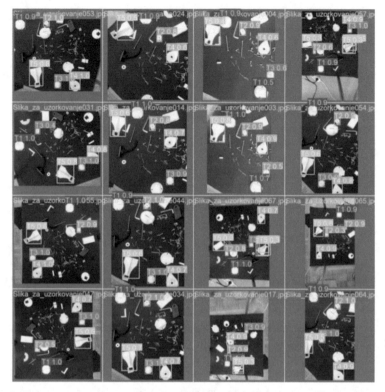

Fig. 8. Validation batch

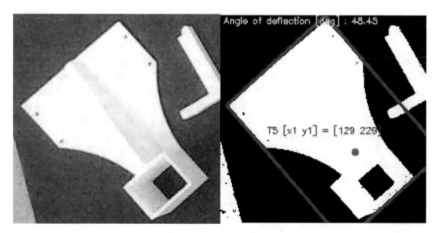

Fig. 9. Recognized object and determined angle

Table 1. Comparison of the experimental results showing the success rate of a algorithms recognizing 5 different objects on 6 diverse cluttered images, previously gone through shutter speed adjustment.

	T 1	T2	T3	T4	T5	Success Rate (%)
YOLO v8	6	6	6	6	6	100.00
NCCM	6	4	6	4	4	80.00

Conclusion

The CNN based solution for the robotic manipulation and grasping is developed and tested in the research presented in this paper. The presented methodology solves object localization using additionally trained YOLO v8 and gives reliable and accurate starting point for the object localization and posture angle determination. Experimental results showed the dominance of CNN based method for the object localization in terms of accuracy and reliability compared to conventional NCMM template matching method. However, the simplicity of the template matching method and calculation speed joined with high accuracy and reliability of detection of round objects are showing that NCMM is still usable in some cases. Further research could be focused on more complex path planning in cluttered environment and vision based grasping control, as well as integration of the soft touch end-effector in cobot system.

References

1. Corke, P.I., Hager, G.D.: Vision-based robot control. In: Siciliano, B., Valavanis, K.P. (eds.) Control Problems in Robotics and Automation. Lecture Notes in Control and Information Sciences, vol. 230. Springer, Heidelberg (1998). https://doi.org/10.1007/BFb0015083
2. Peña, M., López, I., Osorio, R.: Invariant object recognition robot vision system for assembly. In: Proceedings of the Electronics, Robotics and Automotive Mechanics Conference, CERMA 2006, vol. 1, pp. 30–36. IEEE Computer Society, Washington (2006)
3. Baek, G., Kim, S.: Two step template matching method with correlation coefficient and genetic algorithm. In: Huang, D.-S., Jo, K.-H., Lee, H.-H., Kang, H.-J., Bevilacqua, V. (eds.) ICIC 2009. LNCS (LNAI), vol. 5755, pp. 85–90. Springer, Heidelberg (2009). https://doi.org/10.1007/978-3-642-04020-7_10
4. Mihelj, M., et al.: Robotics, pp. 107–122. Springer, Cham (2019). https://doi.org/10.1007/BFb0015083
5. Singh, R., Voyles, R.M., Littau, D., Apanikolopoulos, N.P.P.: Shape recognition and vision-based robot control by shape morphing. In: Proceedings in 1999 International Conference on Information Intelligence and Systems, Bethesda, pp. 188–195 (1999)
6. Zhang, Z., Shang, H.: Low-cost solution for vision-based robotic grasping. In: Proceedings of the 2021 International Conference on Networking Systems of AI, INSAI 2021, Shanghai, pp. 54–61 (2021)

7. Ćirić, I., Pavlović, M., Banić, M., Simonović, M., Nikolić, V.: AI powered obstacle distance estimation for onboard autonomous train operation. Tehnicki Vjesnik - Technical Gazette, Slavonski Brod (2022)

8. Milojević, A., Linß, S., Ćojbašić, Ž., Handroos, H.: A novel simple, adaptive, and versatile soft-robotic compliant two-finger gripper with an inherently gentle touch. J. Mech. Robot. **13**(1), 011015 (2021)

9. Feng, L., Liu, Y., Li, Z., Zhang, M., Wang, F., Liu, S.: Discriminative bit selection hashing in RGB-D based object recognition for robot vision. Assem. Autom. **39**(4), 504–511 (2019)

10. D'Innocente, A., Carlucci, F.M., Colosi, M., Caputo, B.: Bridging between computer and robot vision through data augmentation: a case study on object recognition. In: Liu, M., Chen, H., Vincze, M. (eds.) ICVS 2017. LNCS, vol. 10528, pp. 384–393. Springer, Cham (2017). https://doi.org/10.1007/978-3-319-68345-4_34

11. Fanello, S.R., Ciliberto, C., Natale, L., Metta, G.: Weakly supervised strategies for natural object recognition in robotics. In: IEEE International Conference on Robotics and Automation (ICRA), Karlsruhe, Germany, 6–10 May 2013 (2013)

12. Ćirić, I., et al.: Comparison of simple object recognition algorithms for robot vision. In: Proceedings of the XVI International SAUM Conference on Systems, Automatic Control and Measurements, Niš, Serbia, November 17–18 (2022)

Performance Prediction of Bio-inspired Compliant Grippers Using Machine Learning Algorithms

Maša Milošević$^{(\boxtimes)}$ (ID), Dušan Stojiljković (ID), Ivan Ćirić (ID), Nenad T. Pavlović (ID),
Nikola Despenić (ID), Nikola Ivačko (ID), and Ana Kitić (ID)

Faculty of Mechanical Engineering, University of Niš, 18000 Niš, Serbia
masa.milosevic@masfak.ni.ac.rs

Abstract. The utilization of compliant grippers is rapidly growing in numerous fields due to the need for greater control and safety in grasping applications. Compliance grippers' design and evaluation process often encounter challenges such as prolonged duration and low success rate when traditional methods are employed. Additionally, the current process of designing and testing these grippers is time-consuming and often relies on a trial-and-error method or the application of finite element method (FEM) analysis. This study endeavors to enhance compliant grippers' design and testing stage by employing machine learning algorithms (MLA). Support Vector Regression (SVR), K-Nearest Neighbor (KNN) Regression, Decision Tree Regression, XGBoost Regression, and Gaussian Process Regression are used to generate rapid and precise predictions. Two datasets are utilized, with the first one comprising experimental measurements aimed at validating the results obtained by FEM analysis, as they are subsequently used to train MLAs. The primary objective was to assess the accuracy of these MLAs in predicting gripping linear displacement based on various design parameters. Ultimately, the research revealed the MLA that exhibited the highest performance, characterized by the lowest prediction error with selecting the optimal set of design parameters.

Keywords: Compliant Grippers · Finite Element Method · Machine Learning Algorithms · Performance Prediction · Curved Flexure Hinges

1 Introduction

In fields such as science, medicine, and industry, there is an increasing demand for improved control and safety in grasping applications, leading to the rise in the popularity of compliant grippers. However, the current process of designing and adapting compliant grippers for complex tasks is time-consuming. Traditionally, the analysis of their performance involves either using software based on finite element method (FEM) for motion simulation or relying on a trial-and-error approach, which entails extensive prototyping and experimental evaluation. Unfortunately, both methods suffer from long development cycles, delaying the transition from concept to the final product.

© The Author(s), under exclusive license to Springer Nature Switzerland AG 2024
M. Trajanovic et al. (Eds.): ICIST 2023, LNNS 872, pp. 218–228, 2024.
https://doi.org/10.1007/978-3-031-50755-7_21

To address this challenge, the integration of machine learning algorithms (MLA) offers an automated and efficient solution for predicting the behavior of compliant grippers. By leveraging MLA, the prediction process can be accelerated, reducing the time and effort required during the design and testing phase of the gripper counterpart model. Moreover, real-time predictions can be achieved, enabling immediate control of the gripper counterpart's motion in real-world applications, such as grasping and robotic manipulation.

The utilization of MLA for prediction purposes also provides an opportunity to gain a deeper understanding of the geometric and mechanical aspects of the gripper counterpart model, as well as the relationships between its adaptable parameters. Machine learning can effectively capture and analyze complex patterns in the gripper's behavior, facilitating improved design optimization and performance enhancement.

By combining the benefits of compliant grippers and MLAs, gripper counterparts' design and testing phases can be streamlined, resulting in faster and more accurate predictions. This advancement contributes to the development of advanced and effective gripper counterparts, with applications ranging from scientific research to medical procedures and industrial processes. Additionally, the insights gained from MLA-based predictions can inform the design of customized grippers for specific tasks by selecting the optimal set of design parameters, further expanding their potential impact in diverse domains.

The field of bio-inspired compliant grippers and machine learning algorithms has witnessed significant advancements, particularly in the design and control of bio-inspired soft grippers and the design and analysis of compliant robotic grippers.

Bio-inspired soft grippers, as discussed in the paper [1] by Manti et al., have gained attention due to their ability to adapt to object surfaces. These grippers utilize soft and flexible materials that can deform during interaction, enabling effective grasping. The integration of soft and rigid materials in a hybrid approach further enhances the gripper's capabilities. By leveraging the gripper's morphology and the mechanical properties of the soft materials, control tasks are partially achieved, reducing the complexity of control systems.

Compliant robotic grippers, on the other hand, have been explored as an alternative to traditional rigid-body grippers. The paper [2] addresses the limitations of rigid-body grippers and proposes a solution that improves performance, reduces costs, and simplifies assembly processes. By employing compliant mechanisms, which rely on the deflection of flexible members, issues such as backlash, wear, part count, weight, assembly cost, and maintenance are mitigated.

The paper [3] is the first to predict the fatigue life of a 2-DOF compliant mechanism through a combination of subtractive clustering, the adaptive neuro-fuzzy inference system (ANFIS), and the differential evolution algorithm. Le Chau et al. in [4] proposed a new evolutionary multi-objective optimization technique for a linear compliant mechanism used in a nanoindentation tester. This approach combines several techniques: The central composite design (CDD) for organizing experimental data, FEM for retrieving high-quality performances, the artificial neural network (ANN) acting as a black box to call pseudo-objective functions, and the multi-objective genetic algorithm (MOGA) for the optimization process. In [5] different machine learning techniques (artificial neural networks, support vector machines, and Gaussian progress regression) were developed to anticipate the displacement and stiffness of a compliant cross-axis joint by changing material, lattice orientation, volume fraction, lattice topology, and force. The research article [6] employed Gaussian process regression to accurately represent the hysteresis curve of the Piezo-actuated Micro-gripper. Additionally, the regression model was utilized to forecast the resulting displacement of the gripper. The authors of [7] proposed the use of unsupervised machine learning techniques K-means to reduce the design space dimensionality during the design parameterization stage of the conceptual design.

The focus in the papers varies from design and control to optimization and performance estimation, showcasing the potential of machine learning in advancing the capabilities and effectiveness of compliant gripper technologies.

The aim of this paper is to enhance the design and testing process of compliant grippers, which are used in various fields for safer and more controlled grasping applications. The paper proposes the use of MLAs to generate rapid and accurate predictions for gripper performance, with the aim of selecting the optimal set of design parameters. The final goal is to find the algorithm with the lowest prediction error, improving the design and optimization of compliant grippers.

2   Compliant Gripper Design

Universal grippers cannot meet today's requirements for grippers, such as the need to grip objects with irregular or unpredictable shapes. Likewise, universal graspers can cause harm to the delicate object, necessitating a control mechanism that would regulate contact forces. All of this influenced extensive research in the field of compliant mechanisms for the creation of more flexible mechanisms for grasping. Compliant grippers represent compliant mechanisms that are particularly created for the manipulation of fragile objects [8]. The term "compliant mechanisms" refers to moving, single-piece structures that can only convert movement due to elastic deformation. The function of elastic deformation in the case of compliant mechanisms is taken over by flexure hinges which, due to their geometry, perform the desired movement [9]. The design of the compliant gripper investigated in this paper is represented in Fig. 1. The compliant gripper consists of the passive and active (compliant finger) parts. In a nutshell, the passive part represents the thumb, and the active compliant finger is the index finger of a human hand (Fig. 1(a)). The parameters that are of interest in this research are shown in Fig. 1(b), and the remainder of the paper will go into further detail regarding them.

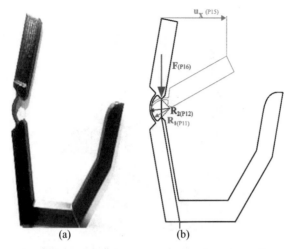

Fig. 1. (a) 3D printed compliant gripper (prototype), (b) Design of compliant gripper shown with input and output parameters of interest

3 Experimental Validation and Finite Element Analysis

In order to validate the relevance of our dataset for further testing, we conducted an experiment as a preliminary step before implementing FEA. Through a comparison of the experimentally obtained results with those from FEA, our goal was to evaluate if there are differences in significant load points spanning from 0 to 250g, as these loads are critical and serve as a boundary before entering the realm of plastic deformation.

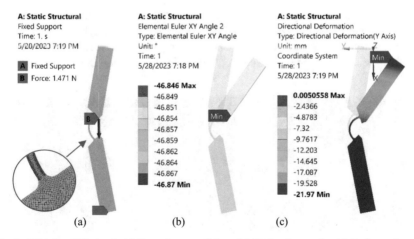

Fig. 2. FEA for 150g load: (a) Boundary condition, (b) Deflection angle, (c) Deformation u_x

Fig. 3. Experimental setup for deflected angle measuring

FEA was done in ANSYS version 19.2 [10] for the given design of a compliant finger. Figure 2 represents the results obtained by FEA that were taken from the ANSYS Workbench platform. In Fig. 2(a) boundary conditions are shown: Fixed support A; Force B equal to 150g of load and defined finite element mesh for an area of interest. The deflection angle of the inner part and directional deformation u_x of the compliant fingertip are represented in Fig. 2(b) and (c). As can be seen, the gray-colored compliant finger is defined in the initial, not deformed position. At the locations of the flexure hinge, a mesh is produced using the "Refinement" function with quadrilateral components of the size 0.2 mm. The resulting mesh of the FE model contained 24719 elements and 25551 nodes.

The experimental setup involved placing weights (force) on a prototype, suspended by a non-stretchable thread, while a camera was positioned on the right side to capture the process (Fig. 3). The results were obtained using computer vision techniques, specifically through a Python program that calculated the angle of deflection in degrees (Fig. 4).

Fig. 4. Results of deflected angle of compliant finger for 150g of applied load obtained by computer vision

Comparing the experimental results with those generated from FEM simulations revealed a significant level of agreement, reaching up to 95% at critical points, such as 0, 50, and 150g of applied load, demonstrating minor deviations of only a few degrees as can be seen in Table 1.

Table 1. Deflection angle comparison between FEM and experiment

Deflection Angle (degrees)		
Load (grams)	FEM	Experiment
50	28.677	26.85
100	37.3109	39.33
150	46.846	47.25
200	55.3407	58.12
250	63.834	66.04

4 Predictive Modeling Parameters in a Compliant Gripper Design

In numerous industrial applications, compliant grippers hold significant importance, necessitating precise dimension prediction for optimal design and performance. However, given the unique characteristics of the gripper, our primary focus in this study was to predict the displacement in the x-direction, denoted as P15. Thus, the objective of this research was to develop highly accurate predictive models capable of estimating the P15 dimension based on the provided input variables.

Table 2 presents the dataset utilized in this study, comprising numerical measurements. The input variables consist of P11 and P12, which correspond to R1 and R2 in [mm], while P16 indicates the force magnitude in [N]. The remaining variables, namely P5 (Total Deformation Maximum), P6 (Equivalent Elastic Strain Maximum), P7 (Equivalent Stress Maximum), P13 (Elemental Euler XY Angle Average), and P14 (Elemental Euler XY Angle 2 Minimum), are the output values of interest. The dataset consists of 2000 data points, which have been validated through simulation to ensure the reliability of the measurements.

We conducted correlation analysis using Spearman's correlation coefficients to understand the relationships among the variables. The heatmap (Fig. 5) reveals the substantial impact of geometry variables (P11 and P12) on the P15 dimension. Specifically, P12 exhibits a positive correlation with P15, indicating that increasing its value leads to a larger displacement. On the other hand, P11 displays a negative correlation with P15, implying that higher values of P11 result in a smaller displacement. Additionally, P16 (force) demonstrates a negative correlation with P15, suggesting that higher force intensities lead to reduced displacement.

Based on the correlation between inputs and outputs, as well as the design of the compliant mechanism, we have concluded that variables from P5 to P14 do not have a significant impact on the output variable. Therefore, these variables were dropped from

Table 2. Description of the dataset

	count	mean	std	min	0.25	0.5	0.75	max
P11	2339	9.4703	0.2610	8.87	9.27	9.47	9.67	9.995
P12	2339	10.77	0.2610	10.245	10.57	10.77	10.97	11.37
P16	2339	1.7634	1.0481	0.049	0.833	1.716	2.745	3.529
P5	2339	22.406	23.803	0.1017	4.9316	11.357	33.627	87.879
P6	2339	0.0197	0.0192	0.0002	0.0067	0.0122	0.0266	0.0987
P7	2339	6.3892	6.2463	0.0522	2.1707	3.9631	8.6311	32.008
P13	2339	−48.3496	24.366	−112.4	−61.97	−36.52	−30.19	−24.28
P14	2339	−46.5369	27.916	−122.6	−58.75	−33.54	−26.35	−20.89

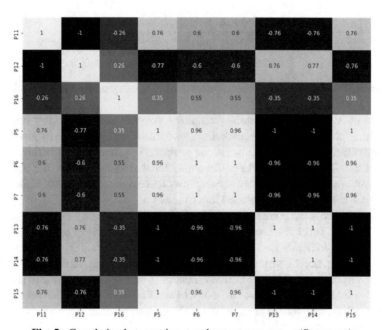

Fig. 5. Correlation between input and output parameters (Spearman)

the analysis. Since all input and output variables are positive, we applied scaling using the Min-Max Scaler method. We also conducted a distribution analysis of the input variables using a diverse dataset derived from FEM analysis. This dataset encompasses a wide range of scenarios and design parameter values, enabling us to capture the real-world complexity of compliant gripper design. The primary goal of our distribution analysis was to gain a deep understanding of the data's characteristics. Our analysis, based on histograms, revealed that the input variables generally follow nearly uniform distributions, although certain data points occur less frequently. To enhance the robustness of our analysis, we utilized boxplot diagrams, which visually illustrate the data distribution and highlight extreme values. Additionally we examined the distribution function of the output variable (P15) to gain insights into its range and variability (Fig. 6). Through this approach, we ensure a thorough and realistic evaluation of our predictive models, which is essential for capturing the intricacies of the design space and the associated parameter variations. This is crucial for a comprehensive understanding of our gripper's performance and design optimization.

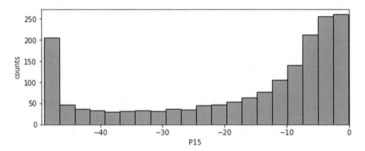

Fig. 6. Distribution of output parameter

Additionally, various tree-based algorithms for the predictive models were employed, including Decision Tree, KNeighborsRegressor, GaussianProcessRegressor, and XGBRegressor from the boost library.

Evaluation metrics such as Mean Absolute Error (MAE), Mean Squared Error (MSE), and R2 score were used to assess the accuracy and performance of the predictive models. Lower values of MAE and MSE indicate superior predictive performance, while an R2 score closer to 1 signifies a stronger correlation between predicted and actual values.

The results demonstrate the performance of each predictive model based on the evaluation metrics (Fig. 7). The XGBRegressor model exhibit lower MAE and MSE values and higher R2 scores compared to other models, indicating its superior performance in accurately predicting the P15 dimension. The KNeighborsRegressor model also performs reasonably well, although with slightly higher MAE and MSE values. However, the DecisionTreeRegressor and GaussianProcessRegressor models display comparatively higher MAE and MSE values, suggesting weaker predictive performance.

Accurate prediction of the P15 dimension is crucial for the design and performance optimization of compliant grippers. The correlation analysis highlights the significant influence of geometry variables (P11, P12) and force (P16) on the P15 dimension.

XGBRegressor demonstrates promising performance in accurately estimating the P15 dimension, offering potential enhancements for compliant gripper design and performance. Future research endeavors can delve into alternative algorithms and consider the incorporation of additional features to further refine and improve predictive models.

By employing predictive modeling techniques, we can gain valuable insights into the behavior of compliant grippers and optimize their performance for various applications. The utilization of MLAs not only streamlines the design process but also reduces the need for time-consuming and costly physical experiments. The proposed approach offers

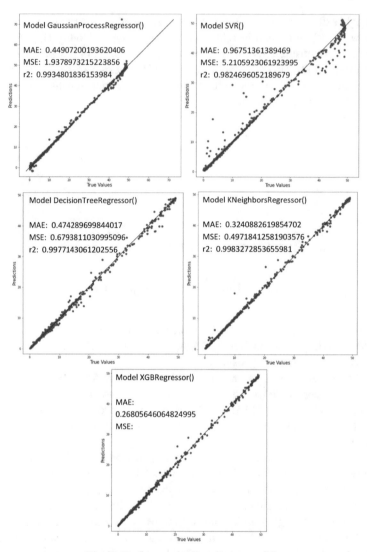

Fig. 7. Performance of predictive models

a systematic and efficient means of predicting the P15 dimension, ultimately leading to improved gripper designs that meet specific requirements.

Furthermore, the successful implementation of MLAs in this study sets the stage for future investigations and advancements in the field of compliant grippers. Looking ahead, it can be further leveraged to explore and optimize additional parameters in compliant gripper design. Parameters such as the length of rigid segments (L1 and L2), the arc length of the flexible segment (l), thickness (w) and height (H) of the compliant gripper, height of the flexible segment (h), curvature angle (α), load force (γ), the positional radius of the flexible segment (R), the radius of the flexible segment (r), and material properties (E - Young's modulus) can be incorporated into the predictive model. By expanding the scope of the algorithm, a comprehensive understanding of the interplay between these design parameters and gripper performance can be achieved, leading to further advancements in compliant gripper design and optimization (Fig. 8).

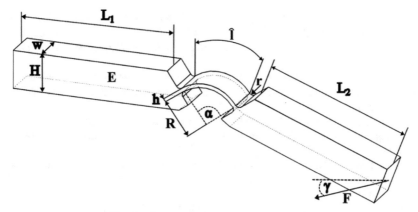

Fig. 8. Geometry of compliant mechanism

5 Conclusion

This study has presented a solution utilizing machine learning algorithms for predicting the performance of compliant grippers based on their design parameters. The implementation of these algorithms offers significant advantages over traditional manual methods, streamlining the design and testing phase of the gripper counterpart model. By automating the prediction process, valuable time and effort are saved, allowing for faster iterations and design improvements. In this context machine learning algorithms have proven to be highly efficient compared to time-consuming FEM analysis, which required several days to obtain results. On the other hand, the trained machine learning model can generate predictions within a matter of seconds and this remarkable improvement in speed enables designers and engineers to rapidly assess the performance of compliant grippers under various design scenarios, facilitating iterative design processes and accelerating time-to-market. Continued research and application of machine learning in this field hold promise for future advancements in compliant gripper technology.

Acknowledgements. This research was financially supported by the Ministry of Science, Technological Development and Innovation of the Republic of Serbia (Contract No. 451-03-47/2023-01/200109).

References

1. Manti, M., Hassan, T., Passetti, G., D'Elia, N., Laschi, C., Cianchetti, M.: A bioinspired soft robotic gripper for adaptable and effective grasping. Soft Rob. **2**(3), 107–116 (2015)

2. Krishnakumar, N., Guhan, M., Krishnan, R.V., Kaarthicsudhan, S.M.: Design and analysis of robotic gripper with complaint mechanism. Int. J. Innov. Sci. Res. Technol. **7**(11), 2016–2021 (2022)
3. Tran, N.T., Dao, T.-P., Nguyen-Trang, T., Ha, C.-N.: Prediction of fatigue life for a new 2-DOF compliant mechanism by clustering-based ANFIS approach. Math. Probl. Eng. **2021**, 1–14 (2021)
4. Le Chau, N., Dao, T.-P., Nguyen, V.T.T.: An efficient hybrid approach of finite element method, artificial neural network-based multiobjective genetic algorithm for computational optimization of a linear compliant mechanism of nanoindentation tester. Math. Probl. Eng. **2018**, 1–19 (2018)
5. Cáceres-C, C., Cuan-Urquizo, E., Alfaro-Ponce, M.: Compliant cross-axis joints: a tailoring displacement range approach via lattice flexures and machine learning. Appl. Sci. **12**(13), 6635 (2022)
6. Chen, F., Gao, Y., Dong, W., Du, Z.: Design and control of a passive compliant piezo-actuated micro-gripper with hybrid flexure hinges. IEEE Trans. Industr. Electron. **68**(11), 11168–11177 (2021)
7. Liu, K., Tovar, A., Nutwell, E., Detwiler, D.: Thin-walled compliant mechanism component design assisted by machine learning and multiple surrogates. SAE Technical Paper 2015-01-1369 (2015)
8. Pavlović, N.D., Pavlović, N.T.: Gipki Mehanizmi, Mašinski fakultet Univerziteta u Nišu, Niš, Serbia (2013). (in Serbian)
9. Howell, L.L.: Compliant mechanisms. In: McCarthy, J. (ed.) 21st Century Kinematics, pp. 189–216. Springer, London (2013). https://doi.org/10.1007/978-1-4471-4510-3_7
10. Ansys. https://www.ansys.com/. Accessed 25 May 2023

Digital Water

High-Performance Computing Based Decision-Support System for Remediation Works on a Dam

Boban Stojanović[1]([⊠]) [ORCID], Vladimir Milivojević[2] [ORCID], Maja Pavić[2] [ORCID], and Višnja Simić[1] [ORCID]

[1] Faculty of Science, University of Kragujevac, Radoja Domanovića 12, Kragujevac, Serbia
`boban.stojanovic@pmf.kg.ac.rs`
[2] Jaroslav Černi Water Institute, Jaroslava Černog 80, 11226 Belgrade, Serbia

Abstract. Even in the early years of the Višegrad Dam's operation, water leakage was noted. It was required to stop leaking by inserting granular materials into subsurface holes since further escalation of this phenomena may have had a number of negative implications. However, closing subterranean karst conduits is a very complicated procedure that needs ongoing oversight and management. To assist decision-makers plan their future actions and make the best choices during remediation, we have created a decision-support system that can assess the consequences of the applied technical procedures almost instantly. In order to determine piezometric levels, flow rates, and tracer concentrations for an assumed system design and under specified boundary conditions, the system uses numerical models of hydraulics and mass transport. To get the model states that best match reality, the measuring system's data were routinely and automatically integrated into a numerical model. The system is made to allow conducting all computations and data assimilation on a high-performance computing infrastructure in order to ensure that simulation results are timely provided.

Keywords: high-performance computing · genetic algorithm · decision-support tool

1 Introduction

A water leakage through the karst topography has been observed since the dam of the hydroelectric plant Višegrad was built; it increased steadily from 1.4 m^3/s in 1990 to 14.68 m^3/s in 2009. It was necessary to remediate the dam environment in order to decrease water and energy losses, stop additional erosion, and potentially prevent the dam from collapsing. A variety of different research and remedial operations were carried out, as stated in [1], in order to ascertain the as-yet-unknown geometry of the karst conduits, which serve as the primary channels of water leakage. A sophisticated computational decision-support system (DSS) was needed for the initial investigations as well as ongoing monitoring and control of the remediation process. This DSS had to be able to add an in silico perspective to the problem on top of the in situ experimental investigations as well as combine these two perspectives to create an augmented and comprehensive picture of the issue.

© The Author(s), under exclusive license to Springer Nature Switzerland AG 2024
M. Trajanovic et al. (Eds.): ICIST 2023, LNNS 872, pp. 231–243, 2024.
https://doi.org/10.1007/978-3-031-50755-7_22

Many publications that deal with some elements of situations that are comparable have been published in the recent past, although the majority of them are only recommendations for the use of particular methodologies [2, 3]. Several methods have been used to learn about karst development and leakage studies, as can be seen from an overview of these documents: (a) regional and dam site geological mapping [4–9]; (b) boreholes and galleries [4, 8–16]; and (c) geophysical methods [10, 14, 16–18]. To the best of the authors' knowledge, there aren't any papers regarding computational platforms that integrate terrain research, in-situ data, and computational models in an automated, holistic way.

The primary goal of the DSS for the Višegrad dam seepage investigation and remediation was to provide an automated and continuous estimation of the spatial distribution of karst faults, their geometric parameters, and hydraulic properties before and during the remediation process, based on the measurements obtained by the installed monitoring system and deterministic and empirical physical laws. The model was created as a collection of elements for simulating phenomena like sinking upstream the dam, the velocity of the flow at the downstream springs, sodium fluorescein dye tracing, sodium chloride tracing, tomography, etc. that would otherwise only be particularly observed using a system for continuous monitoring or measured through periodic experiments.

The discrepancy between measured and estimated values is a good indicator of how accurately the model captures the behavior of the real system. We may get alternative model findings that diverge less or more from observed data by changing the values of the conduit parameters (the diameters of karst fractures and frictions along the conduits). Given the complexity of the models under consideration, it is obvious that the construction of suitable algorithms is necessary in order to carry out an iterative process this demanding. This issue may be viewed as a mathematical optimization problem that seeks to minimize the discrepancy between calculated and measured values by determining the size and hydraulic characteristics of the conduits within the specified network. Due to the genetic algorithm's (GA) intrinsic universality and resilience, we used it to optimize the highly nonlinear and multi-objective optimization problem. A DSS that automates the process of determining the best suitable model parameters utilizing the concepts and methodologies of high-performance computing (HPC) is used to accomplish the whole simulation-based optimization using GA. Using the DSS throughout the entire remediation process, a series of automated simulations of granulate injection, transport, and deposition were run on a regular basis to determine the ideal granulate mixture and injection speed for sealing cracks.

2   The Concept of Decision-Support System

Figure 1 depicts the high-level idea of the HPC-based computational platform for the analysis and repair of seepage under the Višegrad dam. The foundation of the platform is a hydraulic model developed in accordance with research on the morphology of karst fissures.

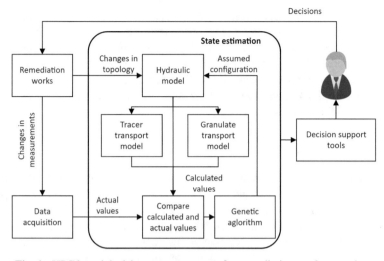

Fig. 1. HPC based decision-support system for remediation works on a dam.

Since the conduits' exact geometrical and hydraulic qualities are unknown, they are first presumed within realistically predicted ranges. The computed flow rates are then utilized to run the tracer transport model, which produces tracer dynamics over time, following the hydraulic simulation on the presumptive model configuration. Piezometric levels, flow rates, and tracer concentrations are calculated and utilized as indications of the quality of the assumed configuration by comparison with measured values. The GA develops ever-better configuration assumptions, performs simulations, assesses their fitness, and chooses the best for the following generation in an iterative process. The amount of the deposited material is determined in each GA iteration based on the previously determined flow rates in order to account for the impacts of the granulated material injection on the hydraulic model. Once the most suitable system configuration has been found, it is made available to the remediation operators through a specially created user application, enabling them to watch the repair process in almost real-time and take the necessary steps.

The DSS is made to run GA and all required computations on high-performance computing infrastructure in order to give continuous estimation of the system status within a reasonable amount of time. The DSS also includes a data management system with automatic quality control, ensuring accurate and timely data delivery for computations. The program for monitoring remediation is a 3D graphical tool that offers all the indications crucial for making decisions in numerical, tabular, and graphical form.

2.1 Hydraulics

Given the intricacy of the seepage under the Višegrad dam, it is obvious that it consists of a few distinct physical processes that are tightly connected and call for an all-encompassing strategy. Fluid movement via subsurface fissures created by stone faulting and erosion over millions of years is the fundamental underlying mechanism. We used a

1D hydraulic model, assuming relatively long conduit lengths compared to their transverse dimensions, to avoid the danger of missing any significant hydraulic events. On the other hand, large cross-sections and flow rates result in high Reynolds numbers, which indicate a turbulent flow regime. The Finite Element Method (FEM) was employed to solve equations relating to turbulent flow. Piezometric levels at the characteristic sites of the system and the fluid velocity and flow rates within every conduit are outcomes of the simulation of the hydraulic model.

2.2 Tracer Transport

We do stationary hydraulic calculations and do not take into account the temporal dimension of the flow since changes in boundary conditions are relatively slow in comparison to flow hydraulics. Tracers are introduced into the critical locations of the systems during tests to learn more about the velocities. We created a suitable computational model of solution transport to simulate the dynamics of the tracer through the system based on the hydraulic calculation findings. The temporal path of tracer concentrations at significant locations along the subterranean network is the outcome of Finite Difference Method (FDM) modeling.

2.3 Granulated Material Transport and Sedimentation

The main sinkholes and many boreholes were filled with inert granular material as part of the dam's remediation. Using the installed monitoring tools, the effects of the granulate injection on the piezometric levels and velocities were observed. However, we had to create a computer model of the movement and sedimentation of granulated material in order to determine the position and volume of the deposited material. The sedimentation model is based on the computation of critical velocities and tangent stresses for the deposition of the granulate within the sloping conduits, using the velocities and flow rates derived from hydraulic simulations as inputs. Critical velocities and tangential strains are computed for each conduit for a specific granulate dimension; these values are then utilized to compute the probability of material deposition within each conduit. Each class of granulate in a granulate mix is calculated separately to determine how much of each class was deposited during the period under consideration.

A hydraulic calculation is made to determine the velocities and flow rates in conduits at the start of each simulation time step. We determine the rate of granulate deposition using the computed velocities and the transportation and sedimentation model, and then we determine changes in conduit diameters and frictions. The computed hydraulic and sedimentation variables are recorded in output files at the conclusion of the phase, and the entire procedure is repeated for the following time step. Up until the simulation time period is exceeded or the system is completely sealed, the iterative process is repeated.

2.4 Estimation of Conduits' Parameters

Prior to starting the remediation, it was required to estimate the geometry and hydraulic properties of the karst conduits using measurements and the output of computational models. Additionally, the sedimentation of the injected granular material changes the size and frictions of conduits during remediation, leading to a variety of various hydraulic system topologies. It is important to perform the hydraulic simulation and compare the outcomes with the measurements in order to compare piezometric levels and water velocities in the expected and actual system design for the identical upstream and downstream levels. Tracer transport is computed, and the estimated concentration time course is then compared with the concentrations acquired by the experiment, in order to determine how well the model complies with tracer dynamics.

3   Data Management

Within the DSS depicted in Fig. 1 a data management system has been built in order to supply accurate and timely data for computations. The system is used to track variables important to the remediation process; while most measures are automated, some are still done manually. The system is made up of an acquisition server and a central database server, both of which include services that have been especially created for them. It is housed at the HPP Višegrad server room.

3.1   Data Acquisition

Data from monitoring systems are gathered via the data acquisition procedure, whether they are supplied in file form (CSV, TXT, XLS, etc.) or manually input by monitoring employees. It is based on the widely used Extraction, Transformation, and Loading (ETL) method, in which data is taken from the original source, converted into a standard data model, and loaded into a data storage.

Groundwater levels, headwater and tailwater levels, as well as water velocity and NaCl sensors at significant karst system exits (underwater springs), are all automatically acquired from the monitoring system on the Višegrad dam. For direct monitoring of levels in karst conduits beneath the dam, various water level and NaCl sensors have also been placed in upstream deep boreholes.

Weir flows, groundwater levels, well levels, and uplift pressures are all measured manually. Throughout the remediation procedure, quantities of inert granules and grouting materials are also tracked and acquired. Control measures, borehole logs, salt tracing studies, tomography experiments, etc. also have manually collected data.

At the acquisition server, acquired data is locally kept before being routinely sent to the central database server. The standard data model for time series, which is implemented in a central database on a relational database system, transforms all measurement data. Together with metadata on measurements, investigations, and remedial operations, additional data is also kept in the central database. The setup of computational models is handled by a specific section of the central database. This section is essential for the real-time application of calculations because it includes information on the most recent state of the computational model that decision support tools use to analyze and forecast the impacts of remedial activity.

3.2 Data Quality Control

To better comprehend the consequences of remediation efforts, all collected data is later employed in computations and optimizations. Even if gathered data may be easily accessible, it is crucial to perform quality control of the data in order to deliver accurate input for computations. The purpose of the quality control process is to assess each measured value's level of reliability in comparison to user-defined criteria, which may use both past values for the same variable and values from a different variable. The criteria is often established based on the properties of the measuring apparatus (such as the measurement range), as well as trend values (such as the moving average) and values of other relevant variables (real or averaged). Data quality mark, which may be any number between 0 and 1, with 0 designating faulty data and 1 indicating trustworthy data, is the outcome of the quality control process.

3.3 Data Processing

The central database server can handle data processing on demand in addition to delivering raw data from the database. This is done using a specialized data processing service that may be set up to produce engineering values based on raw data (for example, the water level determined by the frequency of vibrating wire in a piezometer). Analytical formulae for data processing and aggregation of values across time (minimum, maximum, or average values) are also available. When redundancy is crucial, a special sort of processing may be employed to combine two or more time series from different measurement equipment into a single time series.

3.4 Data Distribution

For the purpose of distributing data to automated software components and decision support tools, a dedicated service is offered on a central database server. Data access through this service requires user authentication, and the data sets to which a user has access are determined by user permission rules. When a user or software component asks a data distribution service for specific data, the service responds with processed or raw data from a database, depending on the request. The request also specifies an acceptable quality level (often 0.5), which can be used to pick data for the response and to analyze it, particularly when aggregating data. This allows for the use of only high-quality data, which is essential for estimating the impacts of remediation efforts.

4 Computational Model

As stated in [1], a 3D geological model is developed based on the known geological structure in order to construct a fictitious network of karst conduits connecting sinkholes and the zone of springs. The comprehensive examination of all accessible geological exploration objects (boreholes), their segmentation, and monitoring of the drilling procedure itself led to the development of the fault distribution in three dimensions. By logically linking the majority of potential fault paths and employing tectonics as the most important component in the early stages of karst fissure formation, the network of conduits is designed.

We established a potential network of karst conduits after positioning all the karst components in three dimensions, which served as the basic topology for mathematical modeling of subterranean flow. Since it represents the collection of finite elements with certain unknown parameters that are generated during the parameter estimation procedure, the fictitious 3D network of karst conduits served as the foundation for all simulations. The proximity of an element to a certain fault structure allowed for the reasonably accurate determination of the approximate ranges of its geometric properties.

4.1 Boundary Conditions

The primary sinkholes are represented by the exterior nodes of the finite element hydraulic model that are located upstream of the dam. The downstream springs are represented by the downstream external nodes, on the other hand. As a result, piezometric levels at nodes that represent sinkholes are equivalent to the upstream water level, whereas levels at nodes that represent springs are equivalent to the downstream water level.

4.2 Tracer Transport Model

Hydraulic calculations may be used to determine the flow rates for an assumed system geometry and specified boundary conditions, after which the dynamics of the tracer injected into certain system sites can be computed. On the other hand, experimental results are also obtained for the kinetics of the tracer insertion and the tracer concentration time course. The parameter estimate method finds the most likely system configuration by comparing the computed and measured tracer concentrations, among other criteria.

5 Data Assimilation

The following indications are used to assess how well an assumed system architecture and associated hydraulic model reflect the actual condition of the subsurface network of fissures:

– matching between the actual and estimated piezometric values;
– matching between the actual and estimated water velocities values;
– matching between the estimated tracer dynamics and the actual concentration time.

The quality of each configuration is assessed by comparing calculated piezometric levels, velocities, and tracer dynamics with measured values for the same boundary conditions (upstream and downstream water levels). Unknown parameters of each configuration (equivalent diameters and frictions of conduits) are determined using genetic algorithm (GA).

5.1 Evolutionary Algorithms

Multiobjective GA is used, as previously described, to estimate the subterranean conduits' unknown characteristics. The first production of parameter sets (individuals) is where the process of parameter estimates starts. The hydraulic model is simulated based on the parameters specified by each individual, producing piezometric levels and flow rates. Tracer transfer is also computed based on the flow rates. The quality of the considered parameter set (the individual fitness) is determined by comparing the hydraulics and tracer dynamics to the observed values. GA selects the top solutions to be subsequently exposed to processes of selection, cross-over, and mutation, resulting in a new generation of individuals, based on the evaluation of all individuals. Once the computed findings and observed values are sufficiently matched, the preceding process is repeated iteratively, and the parameters are then accepted as the most likely.

5.2 HPC Implementation

We have created WoBinGO [19], a software framework for tackling optimization issues over heterogeneous resources, including HPC clusters and Globus-based Grids, in order to enable speedup of the algorithm execution and to minimize the optimization period to a manageable timeframe. The framework was created to achieve the following objectives: (1) accelerating the optimization process through parallelization of GA over the Grid; (2) relieving the researcher's burden of locating Grid resources and interacting with various Grid middlewares; (3) enabling quick allocation of Grid jobs to avoid waiting until requests for computing resources are processed by Grid middleware; and (4) providing flexible worker job allocation in accordance with the dynamics of the users' requirements. It enables parallel GA population assessment using a master-slave parallelization approach.

This framework includes the Work Binder (WB) [20], which gives client applications interactive access to Grid resources practically immediately. The WB integration into the framework frees the programmer from having to deal about specific Grid computing features and allows them to concentrate entirely on the optimization problem. Additionally, WB offers automatic occupancy flexibility depending on recent and current client behavior, which enhances the usage of the Grid infrastructure. Evaluation of individuals is carried out on Grid computing elements (CEs) with the use of the master-slave parallelization paradigm and WB.

Figure 2 depicts the framework's fundamental structure. The distributed evaluation system based on WB and the optimization master make up the framework. The evaluation pool and the WB subsystem make up the distributed evaluation system. The core evolutionary loop is carried out by the master, while the grid-based assessment procedures are executed by the evaluation system. It should be emphasized that the pool of ready jobs developed and maintained by WB itself is NOT the same as the evaluation pool.

Up to the moment where a generation needs to be evaluated, the master completes the primary evolutionary loop. The master then directs everyone in a generation to the evaluation pool. The master continues the process until the halting criterion is met after each generation's individuals have been assessed and given an objective function value.

An intermediary layer between the master and the WB service is the evaluation pool. It offers generational appraisal of people in asynchronous parallel. The evaluation pool gets people from the master and enqueues them each time a generation has to be reviewed. For each individual in the queue, the evaluation pool launches the WB client.

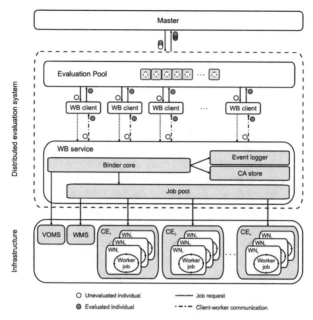

Fig. 2. Structure of WoBinGO framework [19].

When a individual is assessed, the evaluation pool gets the outcome from a WB client and then gives the associated individual an objective function value.

The client, worker, and WB service are the three tiers in which the software components that make up the WB environment are delivered. The WB service's principal function is to keep a pool of available worker tasks on the Grid and connect them to clients that need evaluations. In order to load enough worker jobs into the pool for incoming requests, it submits jobs to the Grid CEs. The client connects to the WB service and makes a worker request. Following a successful coupling between the client and worker, the WB service serves as a relay for traffic between them. The client transmits an individual to a worker for the purpose of distributed evaluation, who then computes fitness value and delivers it back over WBproxy. Following evaluation, the worker re-connects to the WB service and requests further work within the WB-configured job time constraints. The maximum work lifetime imposed by the local Grid site administrator using MDS cannot be exceeded.

6 Decision Support Tools

In order to properly define and coordinate subsequent activities, decision-makers had to move quickly in response to the different obstacles provided by the investigation and remediation of Višegrad dam seepage. To help decision makers make well-informed choices and consider the effects of their actions, specifically developed decision support tools were created. One tool is made especially for tracking remediation efforts in real-time, while the other is made for analysis and planning. Both tools are dependent on calculations made on mathematical models that are a component of the HPC platform, as well as observed data and metadata.

6.1 Real-Time Tracking of Remediation

This tool's primary focus is on current measurements because it is intended to help decisions in real-time, but it also offers insight into the outcomes of calculations. This tool also gives data from continuous calculations done on an HPC platform, in addition to the fact that recent values of measured variables may be utilized for evaluating the impacts of remediation activity. By injecting both inert and grouting material into karst conduits, remediation seeks to lower the hydraulic conductivity and effective cross-section of conduits while minimizing seepage beneath the dam. Model parameters had been estimated before the start of the remediation procedure, and a preliminary model configuration had been saved in a central database. Using the starting model and the current boundary conditions, an automatic computation was run continuously during the remediation procedure. In this method, reference values that match to seepage undisturbed by remediation efforts have been supplied. The impacts of remediation on seepage beneath the dam are seen by comparing these reference values to the most recent measurements. On Fig. 3a, reference water velocity at spring DS1 is depicted as the red line, which is essentially constant, as contrast to observed water velocity, which is depicted as the blue line and is significantly decreasing with time due to remediation efforts.

Along with additional data from a central database, such as injection logs for both grouting material and inert granular material, tomography charts, and other data, the tool also offers 2D and 3D representations of measurements, model results, and configurations. Individual injection logs can be reported or the total volume over time can be deduced by data processing. Users may observe the volume of material injected every batch, the volume for each granulation, the borehole, and the depth at which the material was injected in the detailed logs.

6.2 Analysis and Planning

Different scenarios of inert material injection have been routinely simulated during the cleanup efforts. The best parameters for the injection procedure have been determined based on study and are then utilized for planning.

Daily and weekly planning has employed applications for analysis and planning to assist decisions. Inert material injection has been simulated using an updated state of hydraulic model, and boundary conditions have been determined using data on headwater and tailwater levels. Additionally, user-defined parameters included the rate and position of injection as well as the ratio of different grain sizes, which were also examined. The simulation produced findings for conduit flow rates and velocities, piezometric heads, as well as variations in conduit volumes brought on by the deposition of inert material. In this manner, the effects of injection or grouting could be evaluated, and preparations could be made for more work, often lasting up to seven days. In Fig. 3b, conduits with effective sizes and consequent flows are shown in three dimensions.

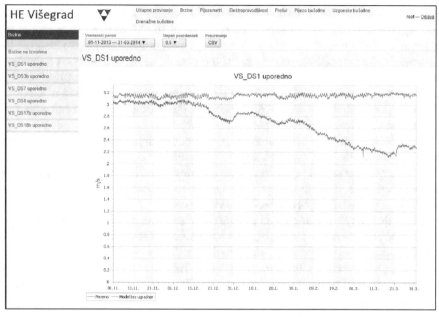

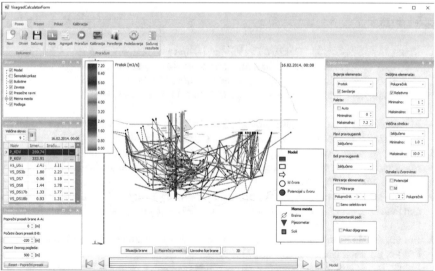

Fig. 3. a) Real-time tracking of remediation (above), b) Tool for analysis and planning (below).

7 Conclusion

We described the HPC infrastructure for Višegrad dam remediation decision-support system in this work. The DSS's main building blocks include numerical models of hydraulics, tracer transport, and the transit and sedimentation of granular materials. The key component of the DSS was the method for parameter estimation based on genetic algorithm because some of the hydraulic model parameters were known roughly within the ranges determined by geological study. In order to ensure that the results of the simulations best reflect the in situ observations, the algorithm was able to estimate the geometric and hydraulic characteristics of the postulated karst conduits, producing a model configuration that accurately depicts the subterranean network of fissures. The computations were performed in a distributed computing environment as a result of the parallelization method that was used, allowing for daily decision support and parameter estimate within a reasonable time period.

The DSS met every criteria established prior to and during remediation, and it allowed for near real-time monitoring of the impacts of granular material injection as well as the forecasting of hypothetical injection situations during remediation planning. The results' satisfactory degree of reliability was demonstrated by a comparison between the major variables' projected changes and their observed values.

Given that the DSS was created as a modular software-hardware system, the algorithms used are inherently robust, and the parallelization strategy is effective, the DSS is easily adaptable and can be turned into a very potent decision support system for projects of a similar nature. Implementing current HPC techniques, Cloud technologies, and perhaps AI approaches that might aid in modeling poorly-defined physical processes are possible routes for platform upgrades.

References

1. Divac, D., Mikavica, D., Dankov, Z.: Višegrad dam remediation – case study. In: Divac D., Milivojević N., Kostić S. (eds.) Contemporary Water Management: Challenges and Research Directions, Proceedings of the International Scientific Conference in the Honour of 75 years of the Jaroslav Černi Water Institute, pp. 125–142. Jaroslav Černi Water Institute, Belgrade (2022)
2. Mohammadi, Z., Raeisi, E., Bakalowicz, M.: Method of leakage study at the Karst dam site. A case study: Khersan 3 Dam, Iran. Environ. Geol. **52**, 1053–1065 (2007)
3. Chuang-Bing, Z., Yi-Feng, C., Ran, H., Zhibing, Y.: Groundwater flow through fractured rocks and seepage control in geotechnical engineering: theories and practices. J. Rock Mech. Geotech. Eng. **15**(1), 1–36 (2023)
4. Redwine, J.C.: Logan Martin dam deep grouting program: hydrogeologic framework in folded and faulted appalachian karst. In: Applied Karst Geology, pp. 243–253. Balkema, Rotterdam (1993)
5. Ertunc, A.: The geological problems of the large dams constructed on the Euphrates River (Turkey). J. Eng. Geol. **51**, 167–182 (1999)
6. Pollak, D.: Engineering–geological investigations of karst terrain. In: Proceeding of Present State and Future Trends of Karst Studies, Marmaries, Turkey, vol. 1, pp. 295–305 (2000)
7. Altug, S., Saticioglu, Z.: Breke arch dam, Turkey: hydrogeology, karstification and treatment of limestone foundation. In: Proceeding of Present State and Future Trends of Karst Studies, Marmaries, Turkey, vol. Z.1, pp. 315–323 (2000)
8. Satti, M.E.: Engineering karst study in Kowsar dam site. In: Proceeding of Present State and Future Trends of Karst Studies, Marmaries, Turkey, vol. 1, pp. 325–329 (2000)
9. Turkmen, S., Ozguler, E., Taga, H., Karaogullarindan, T.: Seepage problems in the karstic limestone foundation of the Kalecik Dam (South of Turkey). J. Eng. Geol. **63**, 247–257 (2002)
10. Xu, R., Yan, F.: Karst geology and engineering treatment in the Geheyan project on the Qingjiang River, China. J. Eng. Geol. **76**, 155–164 (2004)

11. Bhattacharya, S.B.: Dams on karstic foundation. Ir Power **46**(3), 37–46 (1989)
12. Degirmenci, M.: Karstification at Beskonak dam site and reservoir area, Southern Turkey. Environ. Geol. **22**, 111–120 (1993)
13. Merritt, A.H.: Geotechnical aspects of the design and construction of dams and pressure tunnel in soluble rocks. In: Karst GeoHazards, pp. 3–7. Balkema, Rotterdam (1995)
14. Iqbal, M.A.: Engineering experience with limestone. In: Karst Geo Hazards, pp. 463–468. Balkema, Rotterdam (1995)
15. Karimi, H., Raeisi, E., Zare, M.: Physicochemical time series of karst spring as a tool to differentiate the source of spring water. Carbon Evaporate **20**(2), 138–147 (2005)
16. Guangcheng, L., Zhicheng, H.: Principal engineering geological problems in the Shisanling pumped storage powerstation, China. J. Eng. Geol. **76**, 165–176 (2004)
17. Al-Saigh, N.H., Mohammed, Z.S., Dahham, M.S.: Detection of water leakage from dams by self-potential method. J. Eng. Geol. **37**, 115–121 (1994)
18. Milanovic, P.T.: Water Resources Engineering in Karst, p. 312. CRC (2004)
19. Ivanovic, M., Simic, V., Stojanovic, B., Kaplarevic-Malisic, A., Marovic, B.: Elastic grid resource provisioning with WoBinGO: a parallel framework for genetic algorithm based optimization. Fut. Gener. Comput. Syst. **42**, 44–54 (2015)
20. Marović, B., Potočnik, M., Čukanović, B.: Multi-application bag of jobs for interactive and on-demand computing. Scalable Comput. Pract. Exp. **10**, 413–418 (2001)

Analysis of Seepage Through an Embankment Dam Within the Framework of Hydropower System Resilience Assessment

Dragan Rakić[1]([✉]) [iD], Miroslav Živković[1] [iD], Milan Bojović[1] [iD], and Vukašin Ćirović[2] [iD]

[1] Faculty of Engineering, University of Kragujevac, 34000 Kragujevac, Serbia
drakic@kg.ac.rs
[2] Jaroslav Černi Water Institute, 11226 Pinosava, Belgrade, Serbia

Abstract. The paper presents a numerical analysis of the influence of the partial loss of the water-retaining function of the embankment dam elements, such as the grout curtain and the clay core, on the seepage through the dam, as well as on the dam stability. The mentioned structural elements are divided into several regions, which enables the distributed management of the filtration coefficients of individual zones, i.e. simulation of possible damage to these elements. By increasing the filtration coefficient of a certain region, the flow of water through and under the dam body increases, depending on the damage region. This analysis was conducted at different water levels in the reservoir for each of the potentially damaged region and the total water flow through the dam body and through the grout curtain was measured. In addition to the filtration analysis in which the impact of damage on the increase in flow through the dam was considered, an analysis of the partial and global stability of the dam as a function of damage on these elements was also conducted. Partial stability was analyzed through the reduction of the remaining load-bearing capacity of individual dam regions, which was followed by the reduction of the stress point distance from the failure surface. The impact of damage on the global stability of the structure was analyzed through the analysis of the global factor of safety of the structure. The presented analysis was carried out as part of the assessment of the resilience of the hydropower system. The Zavoj dam, which is a part of the Pirot hydropower system in the Republic of Serbia, was used as a case of study.

Keywords: Dam safety · Filtration analysis · Resilience · FEA · Partial stability

1 Introduction

The global stability of geotechnical structures is analyzed by determining the global factor of safety (*FoS*) [1–3]. However, the fact that a building structure has a greater global *FoS* than prescribed does not always indicate the structure is safe, especially when it comes to complex systems such as hydropower system [4, 5]. Based on the above, the conclusion is that in addition to the analysis of global safety, it is necessary to

© The Author(s), under exclusive license to Springer Nature Switzerland AG 2024
M. Trajanovic et al. (Eds.): ICIST 2023, LNNS 872, pp. 244–253, 2024.
https://doi.org/10.1007/978-3-031-50755-7_23

conduct an analysis of the impact of processes that do not threaten the structural stability in the short term, but in the long term. In order to overcome this deficiency in the safety analysis of the multipurpose water management systems, such as dams and reservoirs, in addition to the global stability analysis, it is necessary to conduct an analysis of the functionality of the individual system elements exposed to the loads whose intensity is lower than the loads that threaten the global stability analysis of the structure. This paper analyzes the impact of damage to certain regions of the clay core and the grout curtain on the increase in seepage through the dam body and under the dam body. Water seepage causes an increase in the filtration rate, which in the long term can cause washing out of soil material particles. This particle washing accelerates over time which would increase the porosity and thus reduce structural safety. The analysis of filtration processes in the model, as well as the analysis of stability, were conducted using the finite element-based program PAK [6, 7]. Determination of the global FoS is carried out using the shear strength reduction (SSR) method [8, 9]. The Zavoj dam, as part of the HPP Pirot in the Republic of Serbia, was used as a case of study. The presented analysis is part of a study carried out as part of the assessment of the resilience of the hydropower system [10].

2 Theoretical Basics

2.1 Basic Equations of Flow Through a Porous Medium

Water in the soil flows through a complex system of interconnected pores. However, it is not possible to observe the flow through a porous medium in this way, rather, the flow through a porous medium is observed as a flow through a continuum and is described by applying Darcy's law [11–13]:

$$\mathbf{q} = -\mathbf{ki} \tag{1}$$

where the quantity $\mathbf{k}$ represents a permeability matrix while $\mathbf{i}$ represents a hydraulic gradient calculated using the following expression:

$$\mathbf{i} = \nabla\varphi \tag{2}$$

The total potential φ can be calculated as:

$$\varphi = \frac{p}{\gamma} + h \tag{3}$$

where p represents pore pressure, γ is the unit weight of the water, while h represents the distance from the reference level.

The equation of steady-state fluid flow through a porous medium is derived based on the continuity equation and Darcy's law and has the following form:

$$\nabla^T(\mathbf{k}\nabla\varphi) = -Q \tag{4}$$

where the quantity Q represents the flow per unit volume.

2.2   Basic Equations of Strength Analysis

In strength analysis, stress and strain as well as other relevant quantities are calculated using displacement **u** which is known quantity [7, 14]. In the static analysis it is necessary to satisfy the following equilibrium equation at each integration point:

$$\nabla^T \sigma + \mathbf{F}^V = 0 \tag{5}$$

where σ represents the stress tensor, while $\mathbf{F}^V$ is the body force vector. The numerical solution of this equation system is performed using known boundary conditions. By using the previous equation and the corresponding boundary conditions, the equilibrium equation can be written using the virtual work principle of a deformable body in the following form:

$$\int_V \sigma \delta \varepsilon dV = \int_V \mathbf{F}^v \delta \mathbf{u} dV + \int_{S^\sigma} \mathbf{F}^s \delta \mathbf{u} dS \tag{6}$$

In addition to the equilibrium equation, the strength analysis requires the use of appropriate constitutive models for the analysis of nonlinear soil mechanics problems. For this strength analysis, the Mohr-Coulomb constitutive model was used to describe the mechanical behavior of the dam and the surrounding rock mass [15, 16]. The failure surface of this model is a function of stress invariants and has the following form:

$$f = \frac{I_1}{3} \sin \phi + \sqrt{J_{2D}} \left(\cos \theta - \frac{1}{\sqrt{3}} \sin \theta \sin \phi \right) - c \cos \phi \tag{7}$$

where quantities I_1 and J_{2D} represents the first stress invariant and second deviatoric stress invariant, respectively, while θ represents Lode's angle. Quantities c and ϕ represents constitutive model parameters, cohesion and internal friction angle, respectively.

2.3   Shear Strength Reduction

To determine the global *FoS* of the dam and the surrounding rock mass the *SSR* method was used. Global *FoS* of the structure represents a ratio between available shear strength and the shear stress occurring in the material [17, 18]:

$$FoS = \frac{\tau_f}{\tau} \tag{8}$$

In the *SSR* method, the *FoS* of the structure represents the maximum value of the *SSR* factor for which the structure is in equilibrium, or the value for which there is convergence of numerical solutions.

2.4   Remaining Load-Bearing Capacity

In order to obtain partial stability of the structure, the remaining load-bearing capacity vector is introduced. This vector represents the distance of the stress point from the

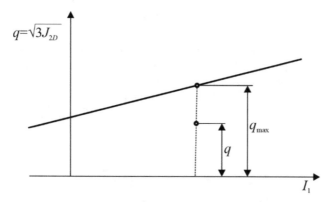

Fig. 1. Remaining load-bearing capacity

failure surface, as shown in the Fig. 1. The remaining load-bearing capacity is expressed in relation to limit value of the stress at which failure occurs. In other words, field of remaining load-bearing capacity shows the distribution of the unused strength of the material.

The remaining load-bearing capacity can be expressed for any constitutive model according to the following equation:

$$RP = \left(1 - \frac{q}{q_{max}}\right) \cdot 100\% \tag{9}$$

where q_{max} represents the distance of the failure surface for a known stress state (known value of I_1), while quantity q is the distance of the stress point for the same stress state (for the same value of I_1).

The distance of the failure surface q_{max}, for a specific stress state, can be calculated by using second deviatoric stress invariant:

$$q_{max} - \sqrt{3J_{2D}} \tag{10}$$

In the case of Mohr-Coulomb model the value of J_{2D} for specific stress state can be calculated using failure surface Eq. (7):

$$\sqrt{J_{2D}} = \frac{c \cos\phi - \frac{1}{3}I_1 \sin\phi}{\cos\theta - \frac{1}{\sqrt{3}}\sin\theta \sin\phi} \tag{11}$$

The distance to the stress point q can also be calculated by applying the second invariant of deviatoric stress [19] according to:

$$q = \sqrt{3J_{2D}^*} \tag{12}$$

where the second invariant of deviatoric stress is calculated as:

$$J_{2D}^* = I_2 + \frac{1}{3}I_1^2 \tag{13}$$

where the I_1 and I_2 represent the first and second stress invariants, respectively.

3   Numerical Model of the Zavoj Dam

The Zavoj Dam forms a reservoir on the Visočica River near the town of Pirot in the south-east of the Republic of Serbia [20]. The upstream and downstream dam body are formed by stone embankments between which is a clay core with appropriate filtration layers. The structure of the dam is 86 m high and 250 m long at the dam crest. The surrounding rock mass at the dam site consists of alternating layers of sandstone and sandy clay.

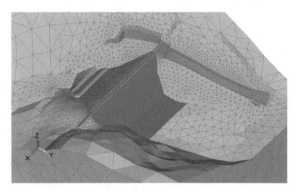

Fig. 2. Finite element model of the Zavoj dam

However, due to the lack of the quasi-homogeneous zones distribution data in this analysis was assumed that the surrounding rock mass is homogeneous. The three-dimensional model of the Zavoj dam was formed based on the available data, which mainly consists of two-dimensional drawings [21]. The formed three-dimensional model was the basis for developing the finite element model that includes all the basic elements of the structure. Numerical analysis of filtration and the dam stability were performed using program PAK [6]. The finite element model of the dam with the associated rock mass is shown in Fig. 2, while the structural elements of the dam whose potential damage was analyzed are shown in the Fig. 3.

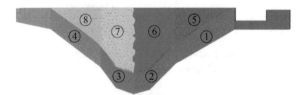

Fig. 3. Clay core and grout curtain regions

In order to conduct a numerical analysis of the impact of the individual dam elements damage on the dam operating, the identification of the potentially structure weak elements was performed. As structural elements whose potential loss of functionality would affect to the dam operation, the impact of the grout curtain damage on leakage under the dam, as well as the damage of the clay core was analyzed [10]. For numerical analysis of hypothetical scenarios such as loss of water resistance function of the grout curtain and clay core, grout curtain is divided into 6 regions and the clay core is divided into 4 regions, so it is possible to independently manage with appropriate filtration coefficients.

4 Numerical Analysis of the Dam

The fields of filtration quantities were obtained within the conducted numerical analyzes of the dam and the surrounding rock mass for different reservoir water levels. Grout curtain damages are, as mentioned earlier, simulated by increasing the filtration coefficient of the corresponding regions (Fig. 3). The dependence of the flow through the contour of the grout curtain as a function of the reservoir water level, for different regions damage is shown in Fig. 4. It can be seen that the damage of some regions of the growth curtain has a less impact on the total flow through this area. This primarily refers to the zones of the grout curtain, which are located at the deeper layers of the rock mass (regions 2 and 3). Possible damage of regions 1 and 4 has the greatest impact on increasing of the total flow through the grout curtain. The reason for this behavior is that the deeper regions of the grout curtain are wedged into the rock mass, so their damage has less impact on the total flow below the dam.

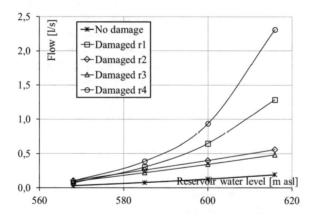

Fig. 4. Flow through growth curtain for different damaged zones

Potential damage of the grout curtain regions also causes an increase in the filtration rate on the upstream dam face, as shown in Fig. 5. Increased filtration rates in these narrow regions would cause increased washing of material from the dam body and thus the occurrence of suffusion and loss of stability of the dam. The effect of this damage could be easily observed visually because it occurs on the upstream face of the dam just below the water surface, near the left and right side of the dam.

The results of the analysis of potential damage to certain regions of the clay core are presented below. The dependence of the total flow through the clay core as a function

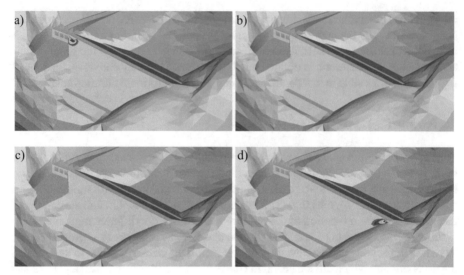

Fig. 5. Filtration rate – upstream view: a) damaged r1, b) damaged r2, c) damaged r3, d) damaged r4

of the reservoir water level for the case of potential damage to different regions of the clay core, is shown in Fig. 6. Unlike the previously analyzed case of potential damage to the grout curtain, when the damage to the regions closer to the left and right banks had a greater impact on the total flow under the dam, in the case of the clay core analysis, the hypothetical damage to the central regions has the greatest impact on the total flow. However, the total value of the flow through the damaged clay core is much higher compared to the value of the total flow through the damaged grout curtain, so it can be concluded that damage to the clay core would have a much greater impact on the long-term stability of the dam compared to damage to the grout curtain.

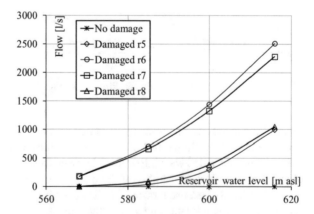

Fig. 6. Flow through clay core in case of failure

It is clear that potential damage to the clay core would cause a significant increase in the rate of filtration through the dam body. This would cause significant damage to the supporting elements of the dam body through the occurrence of suffusion, which would lead to the loss of its stability. The effects of damage to the clay core would be visible on the surface of the dam, that is, they would be visible on the upstream and downstream faces of the dam body (Fig. 7).

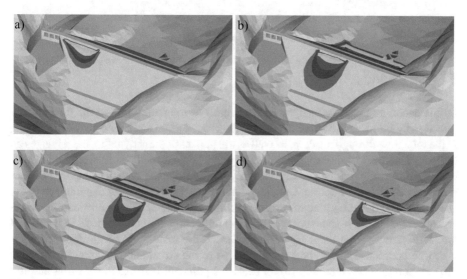

Fig. 7. Filtration rate – upstream view: a) damaged r5, b) damaged r6, c) damaged r7, d) damaged r8

As a result of this analysis, in addition to the field of filtration quantities shown in the previous part, the pore pressures field, i.e. filtration forces in nodes of finite elements, which represent one of the loads in the structural stability analysis, were calculated. The analysis of stability came to the conclusion that these damages would not significantly affect the short-term stability of the structure, i.e. they do not significantly affect the change in the *FoS* of the dam, however, it should be borne in mind that such damage leads to the washing of material from the dam body, which leads to the appearance of suffusion, and thus to a change in the strength of the supporting elements of the dam and loss of its stability.

In addition to the global stability analysis by determining the global *FoS* of the dam, a partial stability analysis was performed based on the remaining load-bearing capacity field (Fig. 8), from which it can be concluded that this field does not change significantly with the introduction of damage to the grout curtain and the clay core.

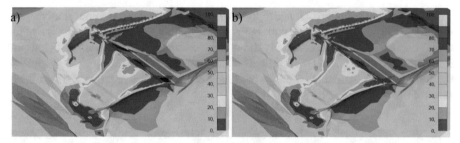

Fig. 8. Remaining load-bearing capacity: a) no damage, b) damaged r6

5 Conclusions

In this paper, a numerical analysis of the impact of the partial loss of the dam's water retention function, as a result of damage to the grout curtain and clay core, on the change in the total flow through the dam, as well as the impact of these hypothetical damages on the global stability of the dam, was carried out. The presented results were obtained as part of the dynamic resilience analysis of the Pirot hydropower system in the southeastern part of the Republic of Serbia. Conducted numerical analyzes show that damage to the grout curtain and the clay core would not significantly affect the short-term stability of the structure, that is, these damages would not significantly affect the change in the global factor of safety of the dam. However, based on the results of the filtration analysis, it can be concluded that hypothetical damage to the structural elements that perform the function of water retention would cause a significant increase in the filtration rates through the dam body. As a consequence of the increased filtration rates, the washing of the material would increase, which would lead to a local reduction in the porosity of the material, that is, to the occurrence of suffusion, which would reduce the strength of the material, and thereby impair the stability of the structure. The short-term consequences of damage to the grout curtain and the clay core would be visible to the naked eye on the upstream and downstream dam faces in the form of an increased filtration rate, so the appearance of leachate on the downstream dam face would represent a reliable signal that these elements have been damaged and that urgent measures need to be taken how the collapse of the dam was prevented.

In the following, it is planned to update the surrounding rock mass of the finite element model of the dam using real geology and conduct a transient dynamic analysis using previous seismograms at the dam site.

References

1. Yu, Y., Xie, L., Zhang, B.: Stability of earth–rockfill dams: influence of geometry on the three-dimensional effect. Comput. Geotech. **32**(5), 326–339 (2005)
2. Huang, M., Jia, C.Q.: Strength reduction FEM in stability analysis of soil slopes subjected to transient unsaturated seepage. Comput. Geotech. **36**(1–2), 93–101 (2009)
3. Ziccarelli, M., Rosone, M.: Stability of embankments resting on foundation soils with a weak layer. Geosciences **11**(86), 1–16 (2011)

4. Escuder-Bueno, I., Mazzà, G., Morales-Torres, A., Castillo-Rodríguez, J.: Computational aspects of dam risk analysis: findings and challenges. Engineering **2**(3), 319–324 (2016)
5. Wu, Z., Xu, B., Gu, C., Li, Z.: Comprehensive evaluation methods for dam service status. Sci. Chin. Technol. Sci. **55**(8), 2300–2312 (2012)
6. Živković, M., Vulović, S., Rakić, D., Dunić V, Grujović N.: Software for solving coupled problems - PAK Multiphysics. University of Kragujevac, Faculty of Mechanical Engineering, Kragujevac (2019)
7. Kojić, M., Slavković, R., Živković, M., Grujović, N.: Finite element method, Linear analysis (in Serbian). University of Kragujevac, Faculty of Mechanical Engineering, Kragujevac (1998)
8. Rakić, D., Živković, M., Milivojević, N., Divac, D.: Stability analysis of a concrete gravity dam using shear strength reduction method. Ann. Fac. Eng. Hunedoara Int. J. Eng. **15**(2), 117–122 (2017)
9. Dawson, E.M., Roth, W.H., Drescher, A.: Slope stability analysis by strength reduction. Géotechnique **49**(6), 835–840 (1999)
10. Rakić, D., Stojković, M., Ivetić, D., Živković, M., Milivojević, N.: Failure assessment of embankment dam elements: case study of the pirot reservoir system. Appl. Sci. **12**(2), 558 (2022)
11. Schrefler, B., Zhan, X.: A fully coupled model for water flow and airflow in deformable porous media. Water Resour. Res. **29**(1), 155–167 (1993)
12. Lewis, R.W., Schrefler, B.A.: The Finite Element Method in the Static and Dynamic Deformation and Consolidation of Porous Media. Wiley, West Sussex, England (1998)
13. Wolfgang, K.: Groundwater Modelling: An Introduction with Sample Programs in Basic. Elsevier, NY (1986)
14. Bathe, K.J.: Finite Element Procedures. Massachusetts Institute of Technology, USA (1996)
15. Hoek, E.: Strength of rock and rock masses. ISRM News J. **2**(2), 4–16 (1994)
16. Hoek, E., Carranza-Torres, C., Corkum, B.: Hoek-Brown failure criterion. In: 2002 Proceedings of the 5th North American Rock Mechanics Symposium and the 17th Tunnelling Association of Canada, NARMS-TAC 2002, Toronto, Canada (2002)
17. Nian, T., Jiang, J., Wan, S., Luan, M.: Strength reduction FE analysis of the stability of bank slopes subjected to transient unsaturated seepage. EJGE **16**, 165–177 (2011)
18. Rakić, D., Živković, M., Vulović, S., Divac, D., Slavković, R., Milivojević N.: Embankment dam stability analysis using FEM. In: 3rd South-East European Conference on Computational Mechanicsan ECCOMAS and IACM Special Interest Conference, Kos Island, Greece (2013)
19. Potts, D., Zdravković, L.: Finite Element Analysis in Geotechnical Engineering: Finite Element Analysis in Geotechnical Engineering: Theory (Vol. 1). Thomas Telford Publishing, London (1999)
20. PE Electric Power Industry of Serbia, HP Pirot; Energoprojekt—Hidroinženjering AD, Belgrade, Built Design, General Report (In Serbian: Projekat Izvedenog Objekta, Opšti Izveštaj), Electric Power Industry of Serbia, Belgrade (2005)
21. Jaroslav Černi Water Institute: Innovative project of technical monitoring of the dam "Zavoj" (in Serbian), Jaroslav Černi Water Institute, Belgrade (2020)

Statistical Method for the Depth-Duration-Frequency Curves Estimation Under Changing Climate: Case Study of the Južna Morava River (Serbia)

Luka Vinokić[1]([✉]) [ID], Milan Stojković[2] [ID], and Slobodan Kolaković[1] [ID]

[1] Faculty of Technical Sciences, Trg Dositeja Obradovića 6, 21102 Novi Sad, Serbia
lukavinokic@yahoo.com, kolakovic.s@uns.ac.rs
[2] The Institute for Artificial Intelligence Research and Development of Serbia, Fruškogorska 1,
21000 Novi Sad, Serbia
milan.stojkovic@ivi.ac.rs

Abstract. This paper addresses the influence of climate change on flood occurrence within the Veternica river stream, a tributary of the Južna Morava River. Depth-Duration-Frequency (DDF) curves serve as the indicators of climate change impacts. The Generalized Extreme Value (GEV) distribution is selected as the theoretical distribution function, and the Modified Method proposed by the Indian Meteorological Department (MIMD method) is employed to downscale daily precipitation data to sub-daily levels. DDF curves are established using both the standard and Koutsoyiannis' methods, respectively. In addition, the HEC-HMS model will be utilized for the computation of flood hydrographs, and a comparative analysis will subsequently be carried out. Under climate change conditions, the hydro-graph peak for a 100-year return period, based on the standard method, exhibits variation ranging from a decrease of 17% to an increase of 2% compared to the reference period. However, the peak based on Koutsoyiannis' method shows variation from a decrease of 4% to an increase of 12%, thereby indicating it as a more conservative approach.

Keywords: Climate Change · DDF · GEV · Precipitation · MIMD · Hydrograph

1 Introduction

Climate change is one of the emerging issues, and its impact goes beyond extreme air temperature, as precipitation patterns are also changing. Climate change is intensifying the water cycle, and it brings more intense rainfall and associated flooding, as well as more severe drought [1]. Understanding the changes in climate is of utmost importance for flood management and mitigation.

More than 30 people have died during the most recent floods in Serbia. In May 2014, the recorded flows within the Kolubara river exceeded the previous records by several times [2]. Please note that all-time precipitation maximum sums were recorded

© The Author(s), under exclusive license to Springer Nature Switzerland AG 2024
M. Trajanovic et al. (Eds.): ICIST 2023, LNNS 872, pp. 254–266, 2024.
https://doi.org/10.1007/978-3-031-50755-7_24

in Belgrade, Loznica, and Valjevo (Serbia). People were evacuated, homes were lost, and financial loss added up to over 1 billion euros. With the enormous impact that these problems cause, climate change is clearly becoming an important topic.

With the impact of climate change and urbanization, flood events could be even more severe. Also, with the current passive flood management plan, floods will become an upcoming problem and active flood management measures need to prevail (e.g. existing and planned reservoir systems).

The assessment of flood potential is generally based on Depth-Duration-Frequency (DDF) curves, which are fundamental tools for determining the rainfall volume for a specific return period. DDF curves are used to describe the relationship between the depth, duration, and frequency of precipitation events. There has been a significant amount of research on developing appropriate methods for generating DDF and Intensity-Duration-Frequency (IDF) curves, comparing different sampling techniques, and examining the effect of climate change on precipitation values [3]. Some studies have used Bayesian analysis or statistical downscaling techniques to estimate future changes in precipitation patterns, while others have used weather generators or regional climate models [4]. These studies have generally found that climate change is likely to result in changes to the intensity, duration, and frequency of precipitation events, although the specific changes are difficult to predict due to the complexity of the climate system and the limitations of the available data and modeling techniques [3]. A deep understanding of the nature of these curves under the influence of changing climate can aid in predicting flood events and implementing better strategies for flood management.

Considering the previously mentioned, this study focuses on the Veternica river basin. The study area is selected due to the existence of the Barje dam, a multi-purpose reservoir used for flood mitigation, water supply, and environmental conditions improvements.

In this research, a method developed for the estimation of the consistent DDF (IDF) curves is used, with a theoretical distribution function that best fits observed data [5]. The three-parameter General Extreme Value (GEV) distribution has shown to be the best fit for observed data, which is in accordance with several studies which show that GEV distribution to be one of the most reliable theoretical distributions that gives an excellent fit to extreme rainfall under the impacts of climate change by many authors [2, 3].

Precipitation records are often an issue for certain regions, where data are limited in space and time, including the area of interest of this research. Therefore, it is important to select an adequate method for downscaling daily rainfall into sub-daily rainfall. There are a great number of techniques for precipitation transformation from daily into sub-daily rainfall. In some previously performed research [6], the formula proposed by Hershfield and Bell (1969) has been used, in others, more recent studies, sub-daily rainfall calculations are based on ratios provided by World Meteorological Organization [4], or a method based on regional parameter estimation [7–9]. The modified version of the method proposed by the Indian Meteorological Department is used in this research, where the coefficient is adjusted to southeast Europe [10].

The hydrologic modeling system Hec-HMS is utilized for calculating flood hydrographs for the study area under current and future climate change scenarios. A comparative analysis is conducted on these hydrographs to study the influence of climate change on flood events.

The paper structure envelops the characteristics of the study area and data used (Sect. 2), methodology (Sect. 3), and results obtained (Sect. 4), as well as conclusions (Sect. 5).

2 Study Area and Data Collection

2.1 Study Area

The Veternica river basin, located in Southern Serbia near Leskovac, is used as the study area. This area basin has been subjected to several flood events, thereby emphasizing the necessity for a comprehensive understanding of how these incidents might alter under the influence of climate change. The inflow point of this basin is selected to be the Barje Dam, considering its crucial role in flood prevention and mitigation (Fig. 1).

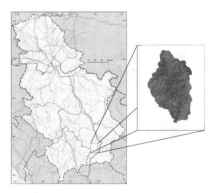

Fig. 1. Location of Veternica river basin

A Digital Elevation Model (DEM) used in this research is extracted from South-East Europe DEM with a resolution of 25×25 m [11]. The basin is largely composed of agricultural lands, forests, and scattered settlements, with its hydrological responses significantly affected by both land-use practices and climatic variations. In addition, the CORINE Land Cover is used to help estimate hydrological model parameters [12].

2.2 Data Collection

For the purpose of this research, historical daily precipitation data has been collected from the Leskovac meteorological station for an available time period from 1950 to 2005.

For the assessment of future rainfall data under the influence of climate change, daily precipitation data has been collected from the regional climate model CCLM4-8-17, extending up to 2100. The climate model, developed under the Coordinated Regional Climate Downscaling Experiment (CORDEX) initiative, provides high-resolution climate projections which are highly reliable for regional impact studies [13]. Given data is separated into two groups based on the projected period: the reference period; and the future period, which consists of two Representative Concentration Pathways (RCPs) scenarios (RCP 4.5 and RCP 8.5) [13].

3 Methodology

The proposed methodology consists of three major parts: (1) finding the best distribution fit for the precipitation events, (2) downscaling precipitation data via the MIMD method, and (3) constructing DDF curves with standard and Koutsoyiannis' method. The methodology is visually interpreted in Fig. 2.

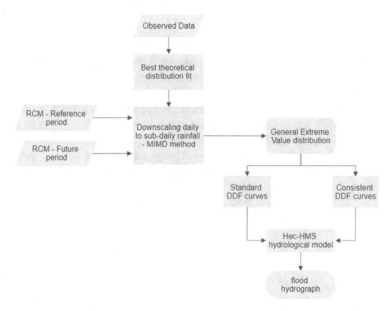

Fig. 2. Flowchart of the research methodology used in this study

3.1 Theoretical Distribution Fit

The initial step involves the analysis of the observed daily precipitation data acquired from the Main Meteorological Station of Leskovac City. The data, spanning from 1950 to 2005, is analyzed to find the best theoretical distribution function fit. This step is crucial as it sets the foundation for estimating the DDF curves under various climate change scenarios.

Extreme events are often analyzed using two methods: peaks-over threshold (POT) and block maxima (BM). POT applies a generalized Pareto distribution to events exceeding a set threshold, while BM uses a theoretical distribution to analyze blocks of yearly maximum precipitation [14]. Both these methods are commonly used in engineering practices to estimate the design floods. However, implementing the POT method requires certain subjective decisions, such as setting the threshold, managing threshold exceedances, and addressing the annual cycle [14]. For this reason, the daily recorded precipitation is converted into yearly extremes. The maximum annual precipitation sums are often aligned with the theoretical distributions of maximums, such as GEV, Gumbel, Fréchet, or Weibull distributions.

A set of 56 annual extreme precipitation sums are used for finding the best theoretical distribution fit based on the Standard Error of fit (SE) [15]:

$$SE = \sqrt{\frac{\sum_{i=1}^{n} \left(P_i^{empirical} - P_i^{theoretical} \right)^2}{n-k}} \tag{1}$$

where $\sum_{i=1}^{n} (P_i^{empirical} - P_i^{theoretical})^2$ represents a sum of all square differences between the empirical and theoretical distribution of the same probability of exceedance within a dataset, n denotes the number of data within a dataset, while k represents the number of parameters of a theoretical distribution.

Amongst the above-mentioned distributions, General Extreme Value (GEV) distribution function has led to the lowest standard error compared to the empirical distribution function. Therefore, GEV is chosen as a theoretical distribution function, where the inverse cumulative distribution function (inverse CDF) is used to calculate the precipitation for the given exceedance probability, given by the following equation [16]:

$$F^{-1}(T) = \xi + \frac{\lambda}{\kappa} \left[\left(-\ln\left(1 - \frac{1}{T}\right) \right)^{-\kappa} - 1 \right] \tag{2}$$

where T represents the return period (years), while ξ, λ and κ are parameters for GEV distribution.

The selection of the GEV distribution function is in accordance with many studies performed in flood management, as extreme precipitation data tend to follow this distribution [17]. The next step involves the application of theoretical distribution to the climate change data by fitting the climate change data to the GEV distribution, which allows for the estimation of extreme future rainfall events [18].

3.2 Downscaling Daily to Sub-Daily Rainfall

Considering the lack of sub-daily precipitation data due to the difficulties associated with data collection at the such fine temporal resolution, a daily to sub-daily rainfall downscaling method is incorporated. This technique is particularly relevant in order to accurately estimate the DDF curves and consequently the flood hydrographs.

The method is based on the use of a modified version of the formula proposed by the Indian Meteorological Department (MIMD):

$$P_t = P_{daily} \left(\frac{t}{24} \right)^n \tag{3}$$

where P_t represents estimated sub-daily rainfall (mm), P_{daily} represents daily precipitation (mm), t denotes rainfall duration (h) and n is the regional coefficient with the value of $n = 0.29$ for South-East Europe [10].

3.3 DDF Curves Estimation with Quasi-Standard Method

The standard procedure of DDF curve estimation consists of an independent analysis of precipitation for different rainfall durations and combining the results into DDF or Intensity-Duration-Frequency (IDF) curves [19].

Since only the daily precipitation data are available, the standard method of defining the DDF curves cannot be used. Therefore, the MIMD downscaling method is implemented. Firstly, daily precipitation depth quantiles are calculated for annual exceedance probability $(0.1\%, 1\%, 2\%, 5\%, 10\%, 20\%, 50\%, 80\%)$ in accordance with GEV distribution. Subsequently, the MIMD formula is used on the bases of daily precipitation, thereby downscaling it to a sub-daily time scale. Finally, by connecting the depth quantiles of different rainfall durations and the same return period, the DDF curve is constructed [19].

3.4 Consistent DDF Curves Estimation - Koutsoyiannis

The consistent DDF curve, on the other hand, is estimated using a method proposed by Koutsoyiannis [5]:

$$i = \frac{a(T)}{b(t)} \qquad (4)$$

where i represents the intensity (mm), with two separated functional dependences of return period and duration, respectively: $a(T)$ and $b(t)$.

Since the method uses an IDF relationship, it is necessary to transform it into a DDF relationship, which is achieved simply by multiplying the intensity with the rainfall duration to obtain the rainfall depth (mm) [19]:

$$P(t, T) = \frac{a(T)}{b(t)} * t \qquad (5)$$

The function $a(T)$ takes into consideration the theoretical distribution of datasets. The essential difference compared to a standard DDF curve formation is the assumption that the whole dataset has the same distribution, regardless of rainfall duration [5]. In this case, the mentioned expression is given through GEV inverse CDF:

$$a(T) = F^{-1}(T) = \xi + \frac{\lambda}{\kappa}\left[\left(-\ln\left(1 - \frac{1}{T}\right)\right)^{-\kappa} - 1\right] \qquad (6)$$

Note that the parameters of GEV distribution are now a function of the precipitation data of all rainfall durations $(10, 20, 30, \ldots, 90, 100, 200, 300, \ldots, 900, 1000, 1400\ \text{min})$ calculated through MIMD.

The denominator $b(t)$ serves as a part of the equation which takes rainfall duration into account for intensity calculation, and is given through the following expression [5]:

$$b(t) = (t + \theta)^{\eta} \qquad (7)$$

where θ and η are parameters to be estimated, defined in the following domains: $\theta > 0$ and $0 < \eta < 1$.

Parameter Estimation. The estimation of three parameters of GEV distribution ξ, λ and κ is done using the L-moment method, while parameters θ and η are estimated in the following process. The process is composed of several steps, the first being assigning ranks r_{ji} to all of the m data values of intensity, where m represents the overall number of data values [5]:

$$m = \sum_{j=1}^{r} n_j \tag{8}$$

where n_j represents the number of data for each j out of r rainfall durations. The second step is computing the average rank for each group of n_j values of the same group, after which the Kruskal-Wallis statistic is calculated and implemented [5]:

$$k_{KW} = \frac{12}{m(m+1)} \sum_{j=1}^{r} n_j \left(\bar{r}_j - \frac{m+1}{2} \right)^2 \tag{9}$$

Finally, the estimation is completed with an iterative process of the trial-and-error, where the aim is to minimize k_{KW}:

$$k_{KW} = f(\theta, \eta) \rightarrow min \tag{10}$$

4 Results and Discussion

4.1 Standard Method

This chapter provides the outputs obtained from the study and a discussion of these findings. The results are focused on the formation of DDF curves through the application of the GEV distribution (Fig. 3, 4, and 5). The DDF curves themselves don't provide much visual information. Nonetheless, this information is used in the hydrological model to create a better and somewhat more comparable visual output.

The Hec-HMS hydrological model simulation clearly shows that precipitation of climate model RCP-4.5 produces significantly lower hydrographs compared to the reference period, which is not expected, while RCP-8.5 is just above the reference hydrograph. This can be explained by at least two uncertainties: one is the lack of bias correction of the RCMs, and the other is the application of the quasi-standard method. Most likely, that represents the combination of the two mentioned uncertainties.

Hydrographs in Fig. 6 show that climate change has altered the peak discharge varying from -16% to $+2\%$.

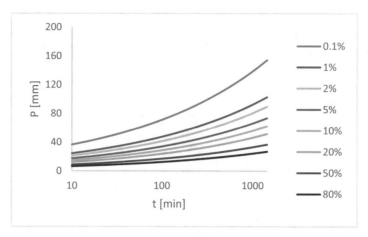

Fig. 3. DDF curves for reference period using the standard method

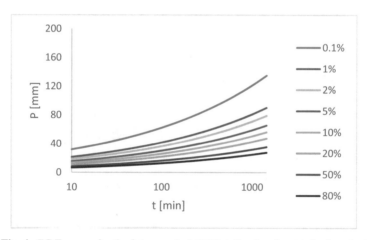

Fig. 4. DDF curves for the future period (RCP-4.5) using the standard method

4.2 Koutsoyiannis' Method

Parameters needed for the application of this method are estimated using techniques described in the previous chapter. The values of parameters for $b(t)$ are estimated through the iterative process, and are as follows: $\theta = 0$, $\eta = 0.71$ for the reference period; $\theta = 0.5$, $\eta = 0.713$ for RCP-4.5; and $\theta = 0.275$, $\eta = 0.711$ for RCP-8.5. The values of parameters for $a(T)$ are obtained by L-moments and are as follows: $\xi = 4.14$, $\lambda = 1.82$ and $\kappa = 0.07$ for the reference period; $\xi = 4.87$, $\lambda = 1.62$ and $\kappa = 0.08$ for RCP-4.5; and $\xi = 5.57$, $\lambda = 1.65$ and $\kappa = 0.11$ for RCP-8.5.

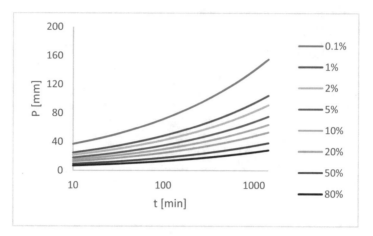

Fig. 5. DDF curves for the future period (RCP-8.5) using the standard method

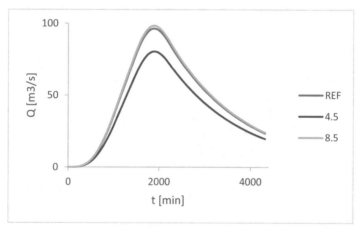

Fig. 6. Hydrographs for the projected precipitation of a return period of 100 years using precipitation data acquired with the standard method.

The DDF curves provided in Fig. 7, 8, and 9 are similar to the curves estimated in the previous section, with higher values of rainfall depth on average. However, this method takes into account the variability of rainfall data under different climate change scenarios and provides more accurate estimates in comparison to the standard DDF curves.

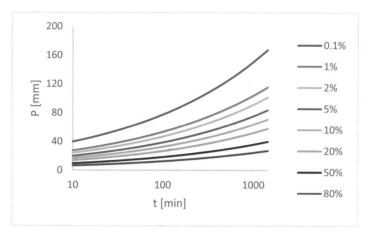

Fig. 7. DDF curves for reference period using Koutsoyiannis' method

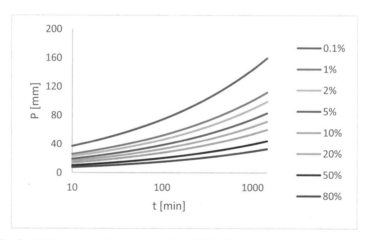

Fig. 8. DDF curves for the future period (RCP-4.5) using Koutsoyiannis' method

Compared to the previous hydrological model simulations, the DDF curves using Koutsoyiannis' method produce higher peaks, presenting a more conservative approach. Hydrographs in Fig. 10 show that climate change has altered the peak discharge varying from −4% to +12%.

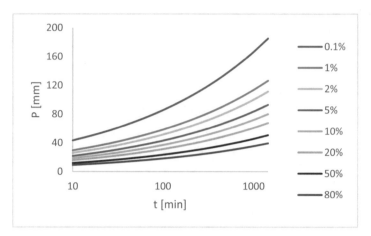

Fig. 9. DDF curves for the future period (RCP-8.5) using Koutsoyiannis' method

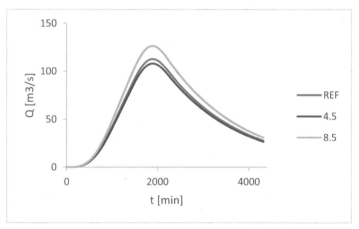

Fig. 10. Hydrographs for the projected precipitation of a return period of 100 years using precipitation data acquired with Koutsoyannis' method.

5   Conclusions

This research has significantly contributed to understanding how DDF curves form and how they change under varying climatic conditions. The use of the MIMD rainfall down-scaling method, GEV distribution, and Koutsoyiannis' method to establish DDF curves led to more reliable and conservative outcomes. This demonstrates the effectiveness of this approach in reflecting the influence of climate change, supporting its broader applicability in hydrological studies.

The research offers an understanding of how climate change may increase flood-related risks in the Veternica river basin. By emphasizing the importance of the application of the consistent DDF curve forming method and highlighting an increase in precipitation due to climate change, it underlines the urgency for adapting flood management strategies. While the findings are specific to the Veternica river basin, the proposed approach can be applied to river basins with different climate and hydrology.

While this research provides valuable insights, there are inherent limitations related to the availability and resolution of the data used. Additionally, the authors of this paper acknowledge the lack of uncertainty analysis has led to certain errors. Therefore, future research should quantify uncertainty to provide more accurate results, which can further be used in decision-making process in flood management. Also, extending the investigation to other climatic factors, such as temperature and evaporation rates, would provide a more comprehensive understanding of their collective impact on flood events.

Acknowledgment. This research is supported by the Science Fund of the Republic of Serbia, Grant No. 6707, REmote WAter Quality monitoRing and INtelliGence, REWARDING.

References

1. IPCC Homepage. https://www.ipcc.ch. Accessed 15 May 2023
2. Stojković, M., Plavšić, J., Prohaska, S.: Annual and seasonal discharge prediction in the middle Danube River basin based on a modified TIPS (Tendency, Intermittency, Periodicity, Stochasticity) methodology. J. Hydrol. Hydromech. **62**(2), 165–174 (2017)
3. Solaiman, T., Simonović, S.: Development of Probability Based Intensity-Duration-Frequency Curves under ClimateChange. Department of Civil and Environmental Engineering, The University of Western Ontario, Canada, Report No: 072, 1913-3219 (2011)
4. Simonović, S.: Bringing future climatic change into water resources management practice today. Water Resour. Manage **31**, 2933–2950 (2017)
5. Koutsoyiannis, D., et al.: A mathematical framework for studying rainfall intensity-duration-frequency relationships. J. Hydrol. **206**, 118–135 (1998)
6. Mauriño, M.: Generalized rainfall-duration-frequency relationships: applicability in different climatic regions of Argentina. J. Hydrol. Eng. **9**, 269–274 (2004)
7. AlHassoun, S.: Developing an empirical formulae to estimate rainfall intensity in Riyadh region. J. King Saud Univ. Eng. Sci. **23**, 81–88 (2011)
8. Kawara, A., Elsebaie, I.: Development of rainfall intensity, duration and frequency relationship on a daily and sub-daily basis (case study: Yalamlam Area, Saudi Arabia). Water **14**, 897 (2022)
9. Wheater, H., Larentis, P., Hamilton, G.: Design rainfall characteristics for south-west Saudi Arabia. Proc. Inst. Civil Eng. Part 87, Part 2, 517–538 (1989)
10. Galoie, M., et al.: Converting daily rainfall data to sub-daily—introducing the MIMD method. Water Resour. Manage. **35**, 3861–3871 (2021)
11. Copernicus EU-DEM v1.1. https://land.copernicus.eu/imagery-in-situ/eu-dem/eu-dem-v1.1. Accessed 19 May 2023
12. Copernicus CORINE Land Cover. https://land.copernicus.eu/pan-european/corine-land-cover. Accessed 19 May 2023
13. Climate Information Homepage. https://climateinformation.org. Accessed 19 May 2023
14. Silva, D., et al.: Introducing non-stationarity into the development of intensity-duration-frequency curves under a changing climate. Water **13**, 1008 (2021)
15. Scielo. https://shorturl.at/acGH5. Accessed 30 May 2023

16. Silva, D., Simonović, S.: Development of nonstationary rainfall Intensity Duration Frequency curves for future Climate conditions. Department of Civil and Environmental Engineering, The University of Western Ontario, Canada, Report No: 106, (2020)
17. Markiewicz, I.: Depth–duration–frequency relationship model of extreme precipitation in flood risk assessment in the upper Vistula basin. Water **13**, 3439 (2021)
18. Millington, N., Das, S., Simonović, S.: The Comparison of GEV, Log-Pearson Type 3 and Gumbel Distributions in the Upper Thames River Watershed under Global Climate Models. Department of Civil and Environmental Engineering, The University of Western Ontario, Canada, Report No: 077, 1913-3219 (2011)
19. Plavšić, J., et al.: Konsistentno određivanje racunskih kiša [Consistent assessment of computational rainfall]. Vodoprivreda **4**, 151–159 (2015)

Technology Platform for Hydroinformatics Systems

Nikola Milivojević[(✉)] [ID], Vukašin Ćirović [ID], Luka Stojadinović [ID],
Jovana Radovanović [ID], and Vladimir Milivojević [ID]

Jaroslav Černi Water Institute, Jaroslava Černog 80, 11226 Belgrade, Serbia
`nikola.milivojevic@jcerni.rs`

Abstract. This paper sheds light on the authors' extensive involvement in the development and use of hydroinformatics systems in Serbia and the surrounding region. The definition of the hydroinformatics system, a list of its potential applications, and a summary of current best practices worldwide are all provided in the introduction. An overview of the general platform that was developed using knowledge gained during the creation of various systems is also presented. We have discussed the features built into a single software platform that uses computational services and mathematical models. With the specifics of implementation in line with the goal and traits of the researched systems, a variety of real-world examples of application by the Institute are presented.

Keywords: hydroinformatics · hydro-information system · software platform · hydrological modelling · hydraulic modelling · decision-support tool

1 Introduction

To address the serious issues of equitable and effective water use for various purposes, hydroinformatics, an interdisciplinary field of technology has emerged from computational hydraulics [1]. Hydroinformatics focuses on the integration of information and communication technologies with hydrology, hydraulics, environmental science, and engineering. Many cutting-edge uses of contemporary information technology are included in hydroinformatics for the management and decision-making of water resources. The most recent IT advancements in artificial intelligence (including knowledge-based systems, machine learning, evolutionary algorithms, and artificial neural networks), artificial life, cellular and finite state automata, as well as other, previously unrelated sciences and technologies, are also utilized.

In its widest meaning, hydroinformatics refers to the use of information technology in the water industry. One of them is hydroinformatics systems (HIS), which have emerged as a way to aid in the optimal management of water resources as well as to resolve current and future disputes within a specific basin or in a particular region in relation to conflicts of interest or development projects that exist in different states, local communities, individual companies, and other legal or physical bodies. In contrast

M. Trajanovic et al. (Eds.): ICIST 2023, LNNS 872, pp. 267–275, 2024.
https://doi.org/10.1007/978-3-031-50755-7_25

to hydroinformatics systems, which are broader systems that include electricity generation, water supply, irrigation, flood and drought risk assessment, water quality and other artificial activities within the system, hydrological information systems are only a component that are common to most HIS systems and are typically characterized by a single-purpose application.

The creation and implementation of the hydroinformatics system requires the creation of a sophisticated software platform with an emphasis on the user (user applications), real-time system execution (services), and specialized system administration (administrator tools). This system includes software components for quick and safe data access by users through specialized applications designed to support system management. It also offers comprehensive and precise data retrieval methods from multiple sources, their validation, processing, and archiving. This group of HIS data management components enables the creation of software for "real-time" formation of an up-to-date computational state of the system, bringing HIS as an IT platform for management support closer to daily operational use with the least amount of human resources required.

2   Hydroinformatics Systems in Practice

Modern hydroinformatics systems are evolving in a way that suggests more flexibility in applying various models. The Open Modelling Interface (OpenMI), which focuses on integrating models and tools in the environmental domain with an emphasis on water, was introduced at the end of 2005. Due to the accessibility of computer resources and the widespread use of deep learning (DL) algorithms in numerous water resource data analyses and hydrological activities, DL research has experienced substantial growth.

Deep learning models [2], together with increased data [3], image synthesis [4], and web-based modeling [5], have been employed more and more in hydrological research in recent years. In the form of serious games, simulation techniques that take participant input into account are also suitable for the management of water resources [6]. The present trend toward "digitalization" of water [7] is particularly significant since it should provide the most comprehensive picture of the health of water resources in real time by fusing data and models with risk management.

Because of large amount of data accessible in contemporary information systems in the subject of hydroinformatics, processing them with standard techniques may be challenging. As a result, it is crucial to take into account the use of Big data approaches in hydroinformatics [8].

The functions for data and model management, data assimilation in hydrological and hydraulic models, as well as a description of the functionality of decision support tools, are some of the ways in which this paper contributes to the knowledge and idea of HIS.

3   Hydroinformatics System Technology Platform

The development and deployment of HIS for a particular system entails the implementation of the fundamental set of features. Since data management is the fundamental function of every HIS, specific software components relating to data gathering, archiving, and processing must be included. In addition, regardless of HIS's intended use, the functionality of organizing mathematical models and calculations is implied. Individual implementations can only be separated in terms of the management support capabilities that depends on the HIS's intended use, as not all HISs are designed to support decision-making at both the operational and strategic management levels. The authors, who are Jaroslav Černi Institute experts, determined the general structure of HIS based on their experience in the construction of numerous HISs, as indicated in the following figure (Fig. 1).

The background-running software components are logically divided into three categories: acquisition, central, and compute server. User tools are divided into four categories based on their intended use: data analysis, system maintenance, operational management support, and strategic planning support. The diagram clearly shows the data flows across the aforementioned components, highlighting the transformation of data from various sources into validated data and finally data with quality control used in the models to get computational values. The HIS is implemented in accordance with the concepts of service-oriented architecture (SOA) [9], and contemporary web technologies are used in accordance with the principles of scalability and robustness.

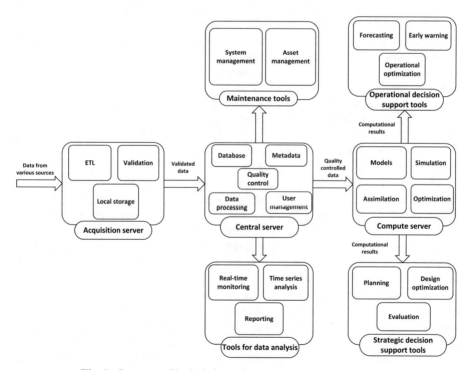

Fig. 1. Structure of hydroinformatics system technology platform.

3.1 Data Management

Users of the platform have access to user and administrator tools where they access data, while processes linked to data management in the HIS take place on the acquisition and central server. The major objective of data management in HIS is to enable data processing and retrieval with quality control information as well as to automatically extract and translate data from diverse sources into data and metadata models defined in HIS. All components involved in data management operations must be effective and dependable for the data to be immediately downloaded and used for calculations with the necessary level of accuracy when it is needed for decision-making.

The acquisition server's role within the system is to gather measurement data as well as pertinent data regarding the system's status. The system can accommodate any number of acquisition servers. Each acquisition server has a data transmission component that connects it to a central server. The foundation of automatic data gathering is the ETL (Extraction, Transformation, and Loading) principle, which enables the blending of data from numerous sources. The ETL process is implemented in a modular and configurable way that makes it simple to adapt to system changes that might arise.

The data validation process is an essential component of the acquisition server, which is a part of the data receiving layer. This procedure indicates that each item of data received is checked before being registered in the local storage. This check yields a data status that may be accurate or inaccurate. This is how the acquisition server's data validation procedure can filter the input data so that only the right data is used by the rest of the system.

All data is downloaded, verified, and then kept in the local data storage. The storage of the data in the local storage is time-limited once it has been transmitted from the local storage to the central server. This means that the data is deleted from the local storage once the configuration-defined amount of time has passed. This is how the local storage provides the central server's acquired data with time-limited redundancy.

The central server, as a key subsystem of the system, plays a central role in the operation of the HIS. It serves as the core component of the data management process. All system elements interact with the central server, which is responsible for ensuring seamless access to its resources, with the central database being the primary resource.

On the central server, the data receiving service is implemented, acting as the interface through which system components can write data to the central database. This ensures centralized control and consistency in data management within the system.

The Technical Data Quality Service is an important component that determines the technical quality of the data in the system. This service utilizes evaluators, which are created by the system administrator based on the characteristics and intended uses of the data. These evaluators assess the value and reliability of the data in the system. To ensure reliability, evaluators need to be applied, and they only require data from the relevant data series for computation purposes. Additionally, the service allows for judging the data's reliability by comparing it with values from other time series.

Users and automated software components can apply their own criteria for interpreting data based on quality evaluation, such as whether the data should be used regardless of quality or only if its quality evaluation is higher than a certain threshold.

Any request to the central server from any component of the system must pass through the access management service in order to confirm the necessary access rights. The system's access privileges have been implemented at several levels. The username and password serve as the initial and most fundamental level of access to the system and, consequently, to the data. This degree of security is present in every user application. Determining registered users' access rights to specific system components is the next level of protection. Determining the right of access to data in the system is the third degree of protection. The definition of rights to data in the central database constitutes the final and fourth level of protection. The measured data may be updated by authorized professionals (experts) and saved with a specific system note in a central database.

All components and supporting systems that use information from the central database have access through the data processing and provisioning service. The data processing and provision service handles all requests for data. Only the original data are stored in the database, and data processing is carried out as requested by the client. In the processing process, arithmetic expressions and data aggregations are permitted, and data quality information must be taken into account. The administrator tools allow for the configuration of all processing parameters.

The central database serves as a repository for all HIS data since it compiles all time series obtained through observation in the system, as well as information on the observation system, documents, and HIS functionality. Graphical layers, time charts, and user accounts are among the functionality information found in the central database. It is possible to track the history of modifications made while using the system because all data are time-referenced.

Observation time series data make up the majority of the core database. The central database enables the preservation of pertinent measurements into a system that is intended to collect and deliver to the user the data captured at various places and systems in the most efficient manner possible.

The central database contains information on the observation and data collecting system, including spatial relationships, measuring equipment features, the state of the measuring equipment, etc. It is possible to track the state of the system across time, from the addition of entities to the system, through changes in attributes, to the archiving of entities, because observation system data are recorded in a central database with a time reference.

The central database also contains the following information in addition to the data listed above: user notes, user data kinds, groups, documents, etc. This makes it easier to evaluate time series data and potential modifications to the measuring device that can result in abrupt changes in observed values or breaks in time series.

The metadata database holds data that accompanies subject data in addition to the fundamental data that is kept in the central database, making it possible to comprehend the data in the central database better. The administration of metadata in the system has been greatly facilitated by the implementation of a metadata model in HIS that enables inheritance via hierarchy.

Data management techniques, whose goal is to provide access to data in a consistent way and with the use of system content made by administrators, serve as the foundation for user applications for data analysis. Particularly, layers and diagrams for reviewing historical data are created, as well as panels for reviewing real-time data. It is also feasible to create reports using information from the central database and information gleaned through their processing.

Tools used by administrators to manage databases and metadata as well as configure software components make up a distinct group. These technologies also make it possible to maintain data on measuring apparatus, which greatly simplifies the upkeep of the monitoring system by users.

3.2 Model and Simulation Process Management

The HIS supports a variety of mathematical models. Although hydrological models are most frequently seen, they are frequently integrated with hydropower and 1D and 2D hydraulic models. It is also possible to use models for the movement of water in materials, groundwater flow, and many other natural phenomena. The nature of the studied system makes the introduction of additional models frequent (e.g., a hydraulic model of flow through karst canals for karst basins). In order for HIS to function, models must be able to communicate with one another and with data collected by the system.

By combining the numerical solver's executable files and model files into a wrapper, it is possible to easily perform calculations with user data. It implements an interface for setting up initial values and conducting computations. Through a wrapper, it is feasible to control the calculation and keep track of the data from the original solver. Reading the results is also viable through the interface, where various post-processing may be carried out before the results are shown in the tools.

The use of algorithms for optimization and assimilation of measured data is made possible using wrappers for conducting calculations in mathematical models. To handle multi-criteria optimisation issues that arise in the management of water resources, the HIS specifically offers the use of a platform for parallel evolutionary algorithms. In order for the optimization results to be accessible at the operational level, it is required to match the division of the hydro potential of the Serbian and Romanian sides with the limitations set out in the regulation on usage of the sophisticated system, as in the case of HIS Iron Gate.

On the same platform, it is also possible to implement algorithms for assimilation of measured data into model states in addition to optimization problems for decision support. This will improve the match between the values of the reconstructed model state and the actual values, which will improve forecasting of the state and output from the model. The service offered by Flood Early Warning System (FEWS) "Kolubara" is one of the most recent instances of assimilation in use. This implementation uses information gathered from the observation network on air temperatures, rainfall, and river course levels to update the coupled hydrological-hydraulic model of the basin in order to provide early warnings and flood warnings.

The compute server, another essential component of HIS, is where all the aforementioned functions for managing models and computations are put into practice. Dedicated high-performance hardware is frequently found inside the compute server, where it is employed to carry out calculations and address optimization issues. If the purchase and upkeep of the High-Performance Computing (HPC) cluster proves to be too expensive for the HIS user, the option of using cloud resources is also contemplated.

3.3 Decision Support Tool

In general, HIS includes user-based decision support tools that allow analysis of measurable data with quality control and computation outcomes in mathematical models to be made in order to make judgments at the short- or long-term levels [10]. Depending on the user's needs and the specific HIS installation, it could only contain tools for operational management or only have capabilities for strategic planning.

Real-time model state reconstruction methods and value forecasts will soon be used in HIS to assist operational management. Short-term planning, expert evaluations, and the creation of notifications on impending events may all be quickly verified by users using tools designed for this purpose. Since it is frequently impossible to measure the values that are important for making decisions directly, these methods may be used to estimate the values, allowing for the informed making of decisions. An up-to-date state of the model may be created using the automatic assimilation process that takes place on the compute server, which allows for the creation of more precise forecasts. The daily production plan at Hydropower Plant (HPP) Iron Gate 1 and HPP Iron Gate 2 is therefore checked in HIS Iron Gate, for example, using these forecast data to check the plan. If the plan needs to be modified, the tool can provide substitute plans or allow the plan to be verified as needed. In the early warning and flood warning systems, another use of the forecast values is the comparison of the predicted values of the level with the threshold or, more accurately, with the actual geometry of the bank and embankment for even more specific alerts (Fig. 2).

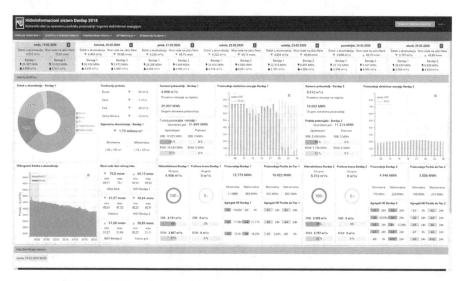

Fig. 2. HIS Iron Gates decision-support tool for operative management.

User tools enable analysis to be implemented over extended periods of time in order to examine the consequences of investments or changes in management rules, which supports strategic planning, or decision making at the long-term level. With the use of

HIS tools, statistical data processing based on lengthy time series may be done, and pertinent hyetographs and hydrographs can be formed in order to produce fictitious situations. As an illustration, evaluate design options using water hydrographs from the last 100 and 1000 years. The tool for analyzing the impact of floods on the roadway and nearshore to help the planning and construction of the Morava Corridor is an example of one of these tools. The program makes it possible to launch a hydrological model, the output of which serves as an input for a 2D hydraulic model. They were utilized to examine the risk areas along the Zapadna Morava riverbanks as well as any possible hazards with the planned road grade (Fig. 3).

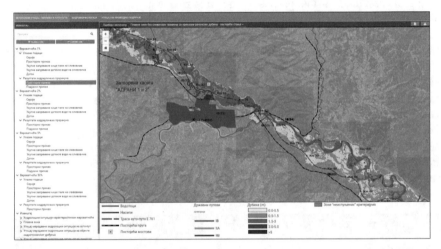

Fig. 3. Morava Corridor decision-support tool. Shades of blue represent the depth of flood waters, while red represents potential hazards in risk areas.

In order to calculate the consequences of investments across decades, mostly hydrological models are used. This is accomplished by accounting for the outcomes of climate models for various scenarios of climate change. Hydropower models then utilize the outcomes of hydrological models to evaluate the effectiveness of various technological solutions. The strategic planning tool in HIS Drina is an illustration of one such instrument [11]. It is feasible to calculate the yearly generation at potential hydroelectric power facilities in the Drina River basin while taking other water users (water supply, industry, etc.) into consideration. The use of models and optimisation algorithms for determining the effectiveness of the management rules for the multi-purpose reservoir is demonstrated by the example of the tool for periodic update of the HIS Prvonek management rules. This tool accounts for changes in anticipated freeboard as a result of potential climate change as well as changes in water usage.

4 Conclusion

Application of a general approach is required in the development of HISs in order to lower costs and increase standardization in implementation across various systems. This is due to the growing need for decision support tools in the management of water resources and the ongoing development of numerical procedures and hardware platforms. It can be argued that the HIS idea described in this work is a good foundation for further development and expansion to additional areas of application in water management based on the presented implementations in real systems. Main areas of application may include advanced flood early warning systems with integrated remote sensing, optimal operational management tools for hydropower and utility companies, and various resilience analyses [12].

References

1. Abbott, M.B.: Hydroinformatics: Information Technology and The Aquatic Environment: Avebury Technical. Aldershot, UK (1991). p. 145
2. Sit, M., Demiray, B.Z., Xiang, Z., Ewing, G.J., Sermet, Y., Demir, I.: A comprehensive review of deep learning applications in hydrology and water resources. Water Sci. Technol. **82**(12), 2635–2670 (2020)
3. Demir, I., Xiang, Z., Demiray, B. Z., Sit, M.: WaterBench: a large-scale benchmark dataset for data-driven streamflow forecasting. EarthArxiv. (2021)
4. Gautam, A., Sit, M., Demir, I.: Realistic river image synthesis using deep generative adversarial networks. Front. Water **4**, 784441 (2020). https://doi.org/10.31223/osf.io/n5b7h
5. Ramirez, C.E., Sermet, Y., Molkenthin, F., Demir, I.: HydroLang: an opensource web-based programming framework for hydrological sciences. EarthArxiv. (2021)
6. Van der Wal, M., de Kraker, J., Kroeze, C., Kirschner, P.A., Valkering, P.: Can computer models be used for social learning? A serious game in water management. Environ Model Softw. **75**, 119–132 (2016)
7. Savić, D.: Digital water developments and lessons learned from automation in the car and aircraft industries. Engineering **9**, 35–41 (2022)
8. Gohil, J., Patel, J., Chopra, J., et al.: Advent of Big Data technology in environment and water management sector. Environ. Sci. Pollut. Res. **28**, 64084–64102 (2021)
9. Milivojević, V., Divac, D., Grujović, N., Dubajic, Z., Simić, Z.: Open software architecture for distributed hydro-meteorological and hydropower data acquisition, simulation and design support. J. Serb. Soc. Comput. Mech. **3**(1), 347–372 (2009)
10. Divac, D., Grujović, N., Milivojević, N., Stojanović, Z., Simić, Z.: Hydro-information systems and management of hydropower resources in Serbia. J. Serb. Soc. Comput. Mech. **3**(1), 1–37 (2009)
11. Milivojević, V., Milivojević, N., Stojković, M., Ćirović, V., Divac, D.: Development of distributed hydro-information system for the Drina river basin. In: Zdravković, M., Trajanović, M., Konjović, Z. (eds), Proceedings of the 4th International Conference on Information Society and Technology - ICIST 2014, vol. 1, pp. 50–55. Society for Information Systems and Computer Networks, Kopaonik (2014)
12. Ignjatović, L., Stojković, M., Ivetić, D., Milašinović, M., Milivojević, N.: Quantifying multi-parameter dynamic resilience for complex reservoir systems using failure simulations: case study of the pirot reservoir system. Water **13**(22), 3157 (2021)

Computer Based Learning

Novel Approach for Education in Biomedical Engineering Based on Atomic Learning

Nikola Vitković[1](✉) (iD), Miroslav Trajanović[1] (iD), Milica Barać[1] (iD),
Milan Trifunović[1] (iD), Jelena Stojković[1] (iD), Sergiu-Dan Stan[2] (iD),
and Razvan Pacurar[2] (iD)

[1] Faculty of Mechanical Engineering in Niš, University of Niš, Niš, Serbia
{nikola.vitkovic,miroslav.trajanovic,milica.barac,
jelena.stojkovic}@masfak.ni.ac.rs,
milan.trifunovic@masfak.mni.ac.rs
[2] Technical University of Cluj-Napoca, Cluj-Napoca, Romania
Sergiu.Stan@mdm.utcluj.ro, Razvan.Pacurar@tcm.utcluj.ro

Abstract. Biomedical engineers are often responsible for preserving and improving the quality of life for ailing patients. Furthermore, engineering and medical students and practitioners must be constantly informed of newly developed methods in the specific domains of interest, properly educated, and widely connected to exchange knowledge and best practices. To meet these requirements, the authors propose improvement and application of the Novel Educational Methodology (NEM), and dynamic collaboration network - DCN of knowledge triangle elements created by the authors of this research. The NEM addition includes the implementation of atomic learning using graph relations and connections and personalised sequence-based learning. The DCN network is represented by atomic structure and collaboration is enabled by different node connections. The enhanced NEM and DCN will be implemented through open-source e-portals dedicated to education and collaboration in biomedical engineering.

1 Introduction

Biomedical engineers are often responsible for preserving and improving the quality of life for ailing patients so that the work can be quite fulfilling. However, because of the complex and demanding nature of the work, interested individuals should possess certain character traits and complete their engineering education before beginning to work in the field. Furthermore, engineering and medical students and practitioners must be constantly informed of newly developed methods in the specific domains of interest, properly educated, and widely connected to exchange knowledge and best practices [1, 2]. These goals define three pillars of proper education and practice in the field of biomedical engineering: Just in time introduction to the novel methods applicable in healthcare; Proper education and training; Connection and collaboration between scientists, engineers, students and medical practitioners. These three pillars are essential to properly acquire knowledge and implement and develop new engineering methods and medical techniques. The application of novel educational methodologies supported by

M. Trajanovic et al. (Eds.): ICIST 2023, LNNS 872, pp. 279–287, 2024.
https://doi.org/10.1007/978-3-031-50755-7_26

machine learning, augmented reality, simulation, modelling, e-learning, m-learning, and distance and blended approaches are the focus and the backbone of the enhancement of medical education [3, 4], and should provide support for the defined pillars. The main shortcomings of today's approaches in eLearning are missing or inadequate feedback from the students, system adaptation not focused on learning context, course presentation (learning material) limited to one teaching style and standardised learning by using standard software solutions. Besides stated advantages and disadvantages of eLearning methods, other shortcomings are necessary to address: they are weekly adoptable to students with disabilities; they are not suitable for groups with different knowledge backgrounds and cognitive capabilities; they are poorly customisable to immersive business demands [5]. The research outcome aims to thoroughly improve the teaching process of selected branches/disciplines of medical and engineering sciences, by answering the stated issues using ICT and related methodologies and technologies.

The main research goal of this study is to improve Novel Educational Methodology - NEM [6] to make it applicable in complex fields like biomedical engineering and thus create a better learning methodology. To accomplish the main goal, the following research objectives are addressed in this paper:

- The objectives stated in the original NEM: A learning system that will always provide modern content following the requirements of society, universities and companies; A learning platform that will be able to adapt to the specific needs of educational institutions, companies, public institutions and organisations; Work-based learning will allow students to learn using different types of courses developed by SMEs and enterprises, which can be done online using a web platform or learning at a physical company site.
- The application of additional methods will help improve NEM, and enable better learning procedures and outcomes, through the developed E-COOL platform. The enhanced NEM includes adding atomic learning represented by graph relations and connections. Atomic learning is based on the "First principle" method of learning by understanding basic knowledge [7], and on sequence-based learning methodology for the creation of sequential learning [8].
- To define a novel virtual centers network (Dynamic Collaboration Network - DCN) of knowledge triangle elements (innovation, academia, and business), defined in a new innovative way and represented by a developed DCN web application. The basis for the DCN creation is NEM. The atoms from NEM are used as the foundation for defining centers as atomic structures. The DCN formed in the presented way is required to enable the formation of a database of contemporary knowledge to be used in NEM.

The combination of upgraded NEM and support provided by DCN should enable the fulfillment of the research goal, and, therefore constant upgrade of knowledge database, knowledge sharing and distribution, and capability to make a strong impact in the eLearning world.

2   The Methodology and Its Application

It should be clear that two dependent methodologies are described in this paper. First is NEM, and the second is DCN foundation methodology. The DCN provides resources (people, knowledge atoms, experience) which enable building the NEM knowledge database and improving the NEM application and verification in the real world. Two connected applications are developed, one for NEM (E-COOL) and one for DCN. These applications can use appropriate integrated services to share their resources and form a unique and complex eLearning framework.

NEM combines traditional and eLearning concepts tailored to the needs or requirements of the student, student group, and teacher. This methodology enables the creation or application of standard eLearning content, like context-based learning, feedback-based learning and flexible learning. It introduces new learning methods and approaches in both universities and business educational processes, e.g. work-based education, life-long learning and long-term learning. The main improvement to NEM methodology is introducing knowledge graphs to represent its structure.

The NEM relations and objects are not an exact copy of knowledge graph structure. Yet, they represent an adapted variant in which atoms reflect nodes, connections between atoms are edges, and labels are atoms main descriptions. An edge (connector) defines the relationship between the nodes/atoms. Knowledge graphs [9] are typically made up of datasets from various sources, which frequently differ in structure. Schemas, identities and context work together to provide structure to diverse data. Schemas provide the framework for the knowledge graph, identities classify the underlying nodes appropriately, and the context determines the setting in which that knowledge exists. This reflects atom definition and atoms connections. The different context can reflect atoms belonging to different molecules (different courses or science fields). Knowledge graphs that are fueled by machine learning utilise natural language processing (NLP) [10] to construct a comprehensive view of nodes, edges, and labels through a process called semantic enrichment. When data is ingested, this process allows knowledge graphs to identify individual objects and understand the relationships between different objects. This working knowledge defined in NEM courses is compared and integrated with other datasets from different sources, like DCN, which are relevant and similar in nature. When a complete knowledge graph structure is defined, it will allow answering a query to find a knowledge atom or a group of atoms that will produce required educational material, i.e. it will be some type of recommender learning system.

In the current stage of the NEM methodology, atomic basic and atomic graph based learning is used to create learning material by using graphs to connect basic elements of knowledge (atoms), enabling the learner and trainer to personalise the course material. In NEM, atoms are defined as a unit of knowledge and can be represented by a different format like textual, video or audio. The graph connections enable the creation of different paths and multivariant combinations of atoms and their representations, resulting in different and personalised online courses. This means that each path define specific context or course. Furthermore, this approach can adapt courses to various students' requirements, including students with specific disabilities or learning issues.

Graphs will provide additional resources for classifying (machine learning method) the learning material and forming an intelligent tutoring system in future work. One of the most important things that can help NEM to form user-adapted courses is knowledge graph capabilities to support the creation of new knowledge, establishing connections between data points that may not have been realised before. The example of one course defined in knowledge graph/NEM context is presented in Fig. 1.

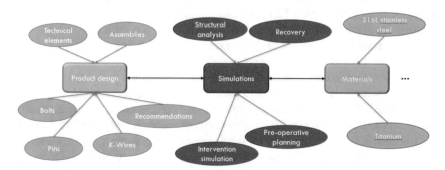

Fig. 1. Graph-based atomic learning in medical engineering

Figure 1 presents three molecules of knowledge or complex educational entities: product design and simulations. Both contain individual atoms of knowledge. For example, technical elements for product design can refer to CAD software technical features for 3D modelling; bolts can be defined as standardised elements used for assemblies, etc. Each knowledge atom can be connected to others by different relationships and connections. The graph approach should allow the connection of elements into educational entities, i.e. courses with different orders, which depends on the stated, customised or sequential learning requirements. In the presented example, the course is composed of three molecules, product design, simulation and material, and the course should be conducted sequentially. Still, if different relations are imposed, then graph can define complex routes to the realisation of this specific course and other potentially tailored courses that can be made from these atomic elements. Atomic elements are initially defined as self-contained, but they can be part of any course, which is one of the benefits of the NEM application. The e-platform (E-COOL), presented in Fig. 2a, is used to create courses by using atomic elements as a base and course as a representation of the combination and order of the atoms of learning, i.e., graph nodes and connections.

A DCN will be formed by implementing the atomic principle through a collaborative e-platform, thus, enabling creation and implementation of open personalised courses and courses with customised content, for education, innovation, and business (knowledge triangle). Furthermore, this platform enables registration and collaboration between scientist, engineers, students and other participants from the private and public sectors to create current healthcare-required courses. Connection and cooperation between portal participants are essential to properly acquire required knowledge and implement and develop new engineering methods and medical techniques, like one presented in this research.

Considering everything stated before, DCN as a virtual network of knowledge triangle centers is and will be connected through existing internet structure. Each center will be defined as node in a network. The main components of each node are virtual

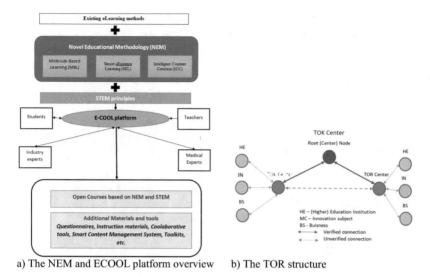

a) The NEM and ECOOL platform overview   b) The TOR structure

Fig. 2. The NEM and ECOOL platform overview

avatars of knowledge triangle entities (academia, innovation, business). These three avatars will form a (Triangle of Knowledge - TOK). Virtual avatars are currently in the creation process and represent each partner in TOK. Virtual avatars can be considered optimised digital tweens, where only selected parts of behavior and characteristics are transferred into the digital world. TOK currently forms a basic connection which enables bi-directional transfer of knowledge, to help improve education, research, and innovation. The network nodes (centers) will be connected by using proposed inverse tree diagram. This diagram uses one center node as root and connects it to other nodes using branches (Fig. 2b). Each branch has the root node as start node, and end node (individual participant center). In this way one node can be connected directly to root node. This is important because every piece of information from individual centers must be compared and verified before it is sent to others. Direct communication between centers can be done, but final verification must be done by a selected body that works as part of the main center.

The system conforms to the fluctuational staff management [11] and information system management rules. This means that other nodes can take over the main center's function in the case of some interrupts, like unavailability. Each node doesn't need to have a complete TOK triangle. If some of the components are missing, they can be acquired and used from another node. This approach will eliminate a communication bottleneck, which is why network is very important. It will bring resources from around the Europe just click away. The direct communication between nodes in network, and inside node (between triangle components), will be based on web services, and it will use new XML specification for knowledge transfer, which will be developed in future work. In the current version classic SOAP messages are implemented. The one of very important elements of future work is to enable indirect communication between avatars by using AI agents, which will collect data (new knowledge in education and research) from each partner in a node and/or whole network, and alert main center (root node) about detected updates. After verification, main center will propagate updates to each node and new knowledge will be available to all included partners. To show practical example of DCN network application in Fig. 3 is presented a use case of personalised implant application.

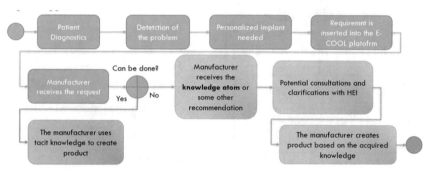

Fig. 3. E-COOL Smart Content Management System Example

The presented use case refers to the classical clinical case in orthopedic for the treatment of bone trauma. The initial step is patient diagnostic procedure done in a medical clinic. The diagnosis is defined, and the conclusion states that a personalised (not standard) implant is required. The personalised implant as one of the atoms in the E-COOL platform is selected, and description (requirement) is inserted. Based on the connection with DCN application the possible manufacturer(s) receive the request to produce the personalised implant. If this specific implant can be manufactured, it is made, and process is finished. Suppose production is not possible because of a lack of required knowledge. In that case, the E-COOL platform can provide instructions from its database and enable collaboration with academia or other businesses for additional consultation using the DCN network resources. When all required knowledge is acquired the production is performed and the plate implant is transferred to the medical clinic.

To enable realisation of the previous use case, two main connected web applications (platforms) are developed and presented in Fig. 4a and 4b, for E-COOL platform and for iCenters (DCN) application respectively. The first one is E-COOL application intended for NEM application and knowledge sharing, while DCN application is oriented for centers forming and their connections through established network. Both applications are in beta development, but all initial and previously described functionalities are integrated. The communication between applications is currently done using web services and messages, but in the final version, one whole application encompassing both platforms will be formed.

The main entities of the E-COOL platform are: data atoms, users, candidates, companies, and surveys. The data atoms are main entities, while others are supportive. Companies can contribute by adding data atoms, and courses can be created by manually connecting atoms or by semi-automatic using graphs. Surveys are used for course validation and also users (candidates) testing.

a) E-COOL platform data elements (knowledge atoms), current link - https://supre-num.com/callmeed/

b) DCN platform with Center cores displayed, current link - https://callmeportaleu.supre-num.com/icenter/cores

Fig. 4. E-COOL and DCN platforms

The main entities of the DCN platform are: cores, fields (science, industry), lay-ers (specific fields, or layer with fields combination, e.g. for some project), people (belonging to several fields and layers), organisations (academia, industry, etc.), organi-sation_service used for communication with E_COOL platform (with companies). Com-panies will be replaced with organisations in the E-COOL platform to have a unique connection. The Fig. 5 logic schema diagram of the DCN platform is presented to understand the centers' data model better.

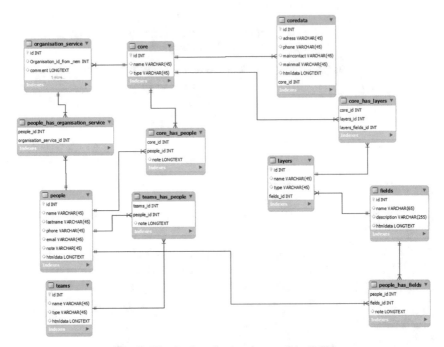

Fig. 5. The database/logic schema of the DCN

3 Conclusion

This study shows how Novel Educational Methodology can be used to create external resources like the Knowledge Triangle Center represented through Dynamic Collaboration Network. The NEM should be a tool for cooperation, knowledge exchange and improvement, skills and competencies upscaling and improving educational processes in higher educational institutions. It can be used as an educational paradigm for improving lifelong, project- and work-based learning and a tool for academia and industry education.

The presented solution for education in biomedical engineering is innovative, includes different actors from private and public sectors and is represented by an open-source e-platforms integrating various modern e-Learning technologies and methods, together with networking. Therefore, it offers complete and contemporary solutions for education in specific fields and the capability for implementation in other complex educational and scientific areas.

Important future output will be open e-platform for collaboration and knowledge exchange, which will enable the application of Novel Educational Methodology (NEM), molecular network structure of knowledge triangle elements (business, academia and innovation), enhancement of existing Higher Education curriculums, creation of innovative patient-oriented products (hardware and software).

Acknowledgement. Research funded by ERASMUS+ "Collaborative e-platform for innovation and educational enhancement in medical engineering - CALLME", No. 2022-1-RO01-KA220-HED-000087703, and Ministry of science, technological development and innovation of the Republic of Serbia.

References

1. Remi, K., Jones, L.: Effect of e-learning on health sciences education: a protocol for systematic review and meta-analysis. High. Educ. Pedag. **6**(1), 22–36 (2021). https://doi.org/10.1080/23752696.2021.1883459
2. Vekli, G.S., Çalik, M.: The effect of web-based biology learning environment on academic performance: a meta-analysis study. J. Sci. Educ. Technol. **32**(3), 365–378 (2023)
3. Paranjape, K., Schinkel, M., Nannan, P.R., Car, J., Nanayakkara, P.: Introducing artificial intelligence training in medical education. JMIR Med. Educ. **5**(2), e16048 (2016). https://doi.org/10.2196/16048
4. Grunhut, J., Wyatt, A.T., Marques, O.: Educating future physicians in artificial intelligence (AI): an integrative review and proposed changes. J. Med. Educ. Curric. Dev. **8**, 23821205211036836 (2016). https://doi.org/10.1177/23821205211036836
5. Murad, H., Yang, L.: Personalised E-learning recommender system using multimedia data. Int. J. Adv. Comput. Sci. Appl. **9**(9) (2018). https://doi.org/10.14569/IJACSA.2018.090971
6. Vitković, N., et al.: Novel educational methodology for personalized massive open online courses. In: ICIST 2020 Proceedings, vol. 1, pp. 5–9. ISBN 978-86-85525-24-7
7. Shukla, A.: First Principles Thinking: Building winning products using first principles thinking, 95 pages, Paperback Published 9 January 2022. ISBN 979-87-91280-35-0
8. Qian, L., Yi, Z., Bo, X.,: Temporal-sequential learning with a brain-inspired spiking neural network and its application to musical memory. Front. Comput. Neurosci. **14** (2020). https://doi.org/10.3389/fncom.2020.00051
9. Gutierrez, C., Sequeda, J.F.: Knowledge graphs. Commun. ACM **64**(3), 96–104 (2021). https://doi.org/10.1145/3418294
10. Dessì, D., Osborne, F., Recupero, D.R., Buscaldi, D., Motta, E.: Generating knowledge graphs by employing Natural Language Processing and Machine Learning techniques within the scholarly domain. Futur. Gener. Comput. Syst. **116**, 253–264 (2021). https://doi.org/10.1016/j.future.2020.10.026
11. Ruso, J., Glogovac, M., Filipović, J., Jeremić, V.: Employee fluctuation in quality management profession: exploiting social professional network data. Eng. Manag. J. **34**(4), 511–525 (2022). https://doi.org/10.1080/10429247.2021.1952022

Ontology Development Approach Adopting Analogy and Competency Questions

Valentina Nejković[(⊠)] [iD] and Nenad Petrović [iD]

Faculty of Electronic Engineering, University of Niš, Aleksandra Medvedeva 14, 18000 Niš, Serbia

{valentina.nejkovic,nenad.petrovic}@elfak.ni.ac.rs

Abstract. Ontologies provide a useful format for structuring arbitrary knowledge domains. However, their development is a time-consuming and costly process, that requires not only domain expertise, but also knowledge and skills for mapping the domain knowledge to the ontology language of classes, instances and relations. In this paper, we propose a novel method for developing ontologies that exploits analogies between different but related knowledge domains represented by ontologies. Our approach processes two ontologies for related problems or knowledge domains, where one of them is complete, while another needs further development; it then generates a series of competency questions for the domain expert in order to extend the new ontology which will be generated as outcome. The proposed method is evaluated in case of ontology from wireless networking domain – creation of ZigBee ontology starting from the one for WiFi. The main implications of our method are twofold: 1) reduces the time need for an ontology engineer from the ontology development process, and it reuses the domain expertise already encoded in the original ontology making the process faster and less costly 2) provides further cognitive links between different knowledge domain. Our approach also opens some new directions in the knowledge discovery and management research regarding the role of machine-learning algorithms in ontology engineering.

Keywords: Ontology · Semantics · Knowledge Engineering

1 Introduction

Ontologies have been widely adopted for knowledge representation across many domains, ranging from healthcare, biology and Internet of Things (IoT) to enterprise information systems, enabling different usage scenarios, such as data integration, device interoperability, reasoning against large amount of information and code generation based on specific parameters [1]. They represent formalizations defining the crucial concepts and their relations within a specific (domain ontologies) or all domains (upper ontologies) in form of *(subject, predicate, object)* triplets [2]. The so-called semantic data, structured with respect to ontology-based definition is stored within triple stores, forming a knowledge graph, represented usually using RDF format. When it comes to querying and retrieval of semantic data from triple store, SPARQL query language is used. Additionally, reasoning mechanisms can be applied on the knowledge within the semantic graph in order to infer new facts based on the existing information.

M. Trajanovic et al. (Eds.): ICIST 2023, LNNS 872, pp. 288–297, 2024.
https://doi.org/10.1007/978-3-031-50755-7_27

However, involvement of humans in the process of ontology engineering represents a bottleneck in ontology development [3]. One common approach to addressing the problem is knowledge reuse as a possible solution for such problem, where ontologies are already designed and where the encoded knowledge can be reused in different applications. On the other side, in case of developing new ontologies from scratch, development of such solution can not much help.

The challenge we addressed in this paper is how we continuously make new ontology development more effective as previously encoded knowledge accumulates. In this paper, we propose algorithm for development of new ontology by using an existing ontology under the assumption that the two knowledge domains are analogous. We believe that this approach is reasonable because analogy by definition relates different concepts with similar meaning. As outcome, the proposed framework is implemented in Java programming language and evaluated in realistic scenario when it comes to wireless networking domain – mapping WiFi to ZigBee ontology.

2 Background and Related Work

In general, most methodologies for developing ontologies identify common steps when it comes to ontologies development process, such as identifying ontology purpose and scope, building, evaluation, and maintaining. Ontology building includes ontology capture or conceptualization, coding and existing ontologies integration. Ontology capture is identification of the key concepts and relationships in the domain of interest, defining concept and relationships, identification of terms referred to such concepts and relationships. Ontology coding is explicit representation of the conceptualization in formal language. There is a constant pressure in capture and coding steps to give answer on the question related to use all or part of existing ontologies.

This represents a challenge in developing a comprehensive methodology for ontologies development. It is easy enough to identify synonyms and to extend an ontology where no concepts readily exist. When there are obviously similar concepts defined in existing ontologies it is rarely clear how and whether such concepts can be adapted and reused. But we envisage that analogy can be used to respond this challenge, too.

For all those steps, the costs related to the required steps can be estimated, for example efforts in person months or performed activities duration. We envisage that involvement of reusing of source ontologies into new ontology development or to extend some existing ontology will bring to final time-consuming reduction for development.

In order to form analogy, we need to map elements and relations between them from one knowledge domain referred as target discourse onto elements and their relations of other knowledge domain referred as source discourse [4]. Analogy is built between two different knowledge domains or within the same knowledge domain but with elements from different situations. The target discourse represents new input to the system. A relation among two elements denotes how the elements are linked in the discourse [4]. Elements have properties that are called features. When two elements are analogue, they can be mapped. Building analogy among two discourses represents a process of retrieving a set of mappings among their elements. Set of mappings between elements in two discourses is known as alignment. There is a difference between feature (attribute) and relation- based alignments [5]. Feature-based alignment represents attributional similarity, while relation-based alignment represents relational similarity or analogy [4, 5]. An algorithm of analogical reasoning that build alignments using attributional and relational similarities we envisage as promising for analogical comparison of two discourses. Please note that we construct the set of presuppositions for MQs from the source ontology, which considerably can help to expert to give answers on received set of questions.

On the other side, the so-called competency questions [6] are already proven effective in ontology development processes and used for ontology requirements capturing. Their structure is reflected to natural language sentences following certain patterns of answers that people expect to get from an ontology. During competency questions design their answerability is very important [7]. A competency question that is answerable is known as meaningful question (MQ). Each MQ is assigned a certain set of presuppositions that tells it apart from other non-meaningful competency questions.

In the process of ontology development there are approaches that used analogies [8, 9] as well as approaches that used competency questions [5], but to the best of our knowledge there are no approaches that combine analogy computing with MQs. In this paper, we present original approach with analogy computing and MQs for ontology extending, building upon our previous work presented in [3]. We base our method on assumption that analogies are discovered relying on contextual structure-based approach [3], including the involvement of domain expert with aim to decide how much certain features and relations with other elements contribute to analogy.

Table 1 provides summary including state-of-art works leveraging analogy and competence question approaches to ontology development across various domains.

Table 1. Experiments and results.

Domain	Description
Food processing requirements [10]	Analogy-based merging of various food production process-related aspects within a unified knowledge base by making use of terminology alignments between various dialects
EU criminal procedural rights in judicial cooperation [11]	Analogy-driven reasoning approach to building unified European Union (EU) legal framework in case of international cooperation
SPARQL query generation [12]	Approach which translates natural language competency questions into SPARQL queries with goal to aid domain experts in testing ontologies
Drafting ontologies from competence questions in area of biology and geography [13]	Approach to automated domain ontology drafting relying on competence questions

3 Methodology

3.1 Formalization of Abstract Analogy Ontology

We consider that the problem space is composed by n ontologies that model domains. Ontology in our system is represented (on abstract level) as a set of micro-structures. In this subsection, we introduce an abstract version of the Analogy Ontology data model.

Assume there is an infinite set U (resources) and an infinite set of blank nodes $B = \{N_j: j \in N\}$. Blank nodes are anonymous objects (both subject and object). A triple $(s, p, o) \in (U \cup B) \times U \times (U \cup B)$ is called an RDF triple. In such a triple, s is called the subject, p the predicate and o the object. We often denote by UB the union of the sets U and B.

Let us consider two segments of a same or two different ontologies that are supposed to be analogue. Moreover, we assume there is an infinite set U (resources or entities or nodes). A triple (s, p, o) is referred to as *micro-structure*, where s is called the subject, p the predicate and o the object. Such notation is used in order to put emphasis on establishing analogy between ontology elements on abstract level regardless of underlying representation or data format. In this context, an *Ontology relational structure* represents a set of micro-structures.

Definition 1. (Ontology relational structure): **Relational structure** *is a model of micro-structures constituted by relationships between entities in a domain of interest (discourse).*

Let N be the set of all nodes $ND = \{ND_1, ND_2, \dots, ND_n\}$, where n is the number of all nodes of a domain. Nodes are entities. Entities are concepts, individual names and roles. Concepts are sets of individuals, and individual names are single individuals in domain.

For example, an ontology modelling the domain of *University* and its individuals' relationships might use concepts such as *Professor* to represent the set of all professors and *Student* to represent the set of all student individuals, roles such as *teacherOf* to represent the binary relationship between teachers and their students, and individual names such as Nikola and Marko to represent the individuals Nikola and Marko, respectively.

Roles (relations) are associated with predicates. Let $MS = \{MS_1, MS_2, \dots, MS_n\}$ be the set of all micro-structures of an ontology or it segment. A *micro-structure $MS_i \in MS$* is composed by the pair of nodes $ND = \{ND_l, ND_k\}$ and relation $R_{l,k}$ among that nodes pair. Simply, micro-structure is composed by concepts and relation among concepts. Mapped to RDF knowledge representation we can say that micro-structure corresponds to RDF triple. Triple $(s, p, o) \in (U \cup B) \times U \times (U \cup B)$, where U is an infinite set of resources and $B = \{N_j: j \in N\}$ is an infinite set of blank nodes. Blank nodes are anonymous objects both subject and object). In such a triple, s is called the subject, p the predicate and o the object. In our example of the domain of university one first-order micro-structure is *teacherOf(Nikola,Marko)*.

We define *first-order* and *second-order micro-structures*. First-order micro-structure is a relation between two concept nodes represented by a single role node. Graphically, each of the concept nodes is connected to the role node by an undirected edge. The role node identifies the micro-structure. Second-order micro-structure is relation between two role nodes and one is relation where two roles are viewed as nodes with relation among them which constitutes micro-structure (Fig. 1).

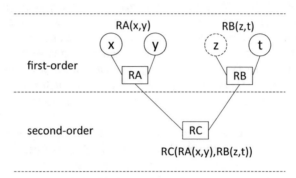

Fig. 1. First and second-order micro-structures.

Let us consider example of parent (source) and child (target) ontology segments. Figure 2 shows that their relational structures are the same for the pair of relations PR1 and CR1, PR2 and CR2 followed by pairs of nodes which participate in pair of relations PN1 and CN1, PN2 and CN2, PN3 and CN3.

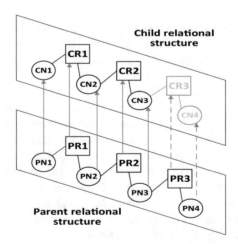

Fig. 2. Analogy mapping of ontology segments with relation contextual structure.

We can say that relational structures of parent and child are analogue, where parent relational structure mapped child relational structure. It can be noted that relation PR3 doesn't has its pair companion in child relational structure. The logical next step in reasoning is to extend child relational structure mapping PR3 on CR3 as well as PN4 to CN4. Two questions raise in this example: 1) Does relational structures of parent and child are analogues? 2) Does child relational structure can be extended by CR3 and CN4 as analog companions of PR3 and PN4, respectively?

3.2 Algorithm

The benefit of the proposed approach is that expert does not have to develop ontology, since she receives set of questions, which could reduce efforts and time made during usual development phase. The approach increases reliability of outcome ontology, because special attention is given to process where are questions, known as competency questions, carefully designed and formed. In Table 2, the algorithm for analogy-based ontology generation leveraging competency questions in synergy with analogies is given. The implementation was done in Java programming language relying on Open TaaSOR API [14] for ontology-related management operations, triplet insertion and SPARQL query execution.

Table 2. Analogy-based ontology generation algorithm.

Input: Base onotology - *BO*
Output: To be done (target) ontology - *TD*
Steps:
  1. Establish equivalent in the *TD* terminology.
      For each class *C* in Base ontology
         Set of rules generate questions: What is the equivalent concept from *C* in the *TD*?
            Example: *What concept is equivalent concept in TD to concept C?*
         Establish the subclass relationships if they exist in the Base ontology.
  2. All relations that are not *part_of* relations should be established.
  3. For each class *C* from the TD, establish its attributes.
      Example: *What are the attributes of class C?*
  4. For each class *C* from TD generate its individuals.
      To generate individuals, you can ask any the following question:
         Provide examples or instances of class *C*
            Example: *VisaCard is an instance of class CreditCard*
  5. Validate new ontology.
     5.1. If initial set of rules is not empty
         5.1.1 Generate equivalent rule to target ontology
         5.1.2 Run that rule in the target ontology to confirm it holds (i.e. it is true)
     5.2. If initial set of rules is empty, you can just confirm that the standard axioms hold
  6. End

4 Case Study

We define *first-order* and *second-order micro-structures*. First-order micro-structure is a relation between two concept nodes represented by a single role node. Graphically, each of the concept nodes is connected to the role node by an undirected edge. The role node identifies the micro-structure. Second-order micro-structure is relation between two role nodes and one is relation where two roles are viewed as nodes with relation among them which constitutes micro-structure. Complexity of IoT wireless networks is increased by growing requirements for simultaneous use of different radio access technologies such as ZigBee, Bluetooth, LTE and Wi-Fi. In this case study, based on [9], our goal is to create ZigBee ontology starting from one related to WiFi. Illustration of analogy-based ontology creation process for ZigBee ontology using the existing OMN Wireless Ontology is given in Fig. 3. Moreover, the set of extracted rules is summarized in Table 3.

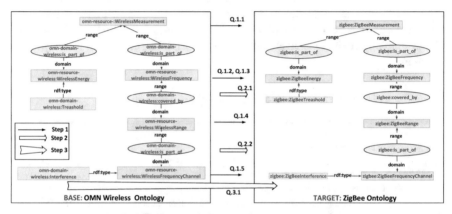

Fig. 3. Wireless to ZigBee ontology – rules extraction.

Table 3. Set of rules for ZigBee ontology creation.

Aspect	Example
Base domain (Bdom)	Wireless
Base ontology (Bont)	WirelessOnt
Target ontology (Tont)	ZigBeeOnt
Rb	{Rint}
Rint	*Measurement (?f, ?e)^exceedTreashold^coveredBy(?r, ?f)^part of(?c, ?r)-> Interference(?c)*
Set of rules	*covered_by: WirelessRange-> WirelessFrequency* *part_of: WirelessFrequency-> WirelessMeasurement* *part_of: WirelessEnergy-> WirelessMeasurement* *part_of: WirelessChannel-> WirelessRange*

The whole procedure of ZigBee ontology creation starting from WiFi ontology using the proposed approach described in Table 1 is given in Fig. 4

In Table 4, an overview of experimental results is given. The processing time required for various steps within the proposed approach is given and compared to manual ontology

Step1. Establish equivalent in the ZigBee domain terminology. For each class C set of rules generate questions (What is the equivalent concept from C in the ZigBee domain. Example: What concept is equivalent concept in ZigBee domain to concept C?).

    Q.1.1
            Intro:
            Question: What concept is analogue concept in ZigBee domain to concept WirelessMeasurement?
    Q.1.2
            Intro:
            Question: What concept is analogue concept in ZigBee domain to concept WirelessFrequency?
    Q.1.3
            Intro:
            Question: What concept is analogue concept in ZigBee domain to concept WirelessRange?
    Q.1.4
            Intro:
            Question: What concept is analogue concept in ZigBee domain to concept WirelessEnergy?
    Q.1.5
            Intro:
            Question: What concept is analogue concept in ZigBee domain to concept WirelessChannel?

Q.1.1-Answer: A11 - ZigBeeMeasurement
Q.1.2-Answer: A12 - ZigBeeFrequency
Q.1.3-Answer: A13 - ZigBeeRange
Q.1.4-Answer: A14 - ZigBeeEnergy
Q.1.5-Answer: A15 - ZigBeeChannel

Step 2: All relations that are not part_of relations should be established.
        covered_by: WirelessRange-> WirelessFrequency
        part_of: WirelessFrequency-> WirelessMeasurement
        part_of: WirelessEnergy-> WirelessMeasurement
        part_of: WirelessChannel-> WirelessRange

    Q.2.1
    Analogy: WirelessRange is cover by WirelessFrequency. ZigBeeRange is analogous to WirelessRange. ZigBeeFrequency is analogous to WirelessFrequency.
    Question: Is ZigBeeRange covered by ZigBeeFrequency?
    Q.2.2
    Analogy: WirelessFrequency is part of WirelessMeasurement. WirelessFrequency is analogous to ZigBeeFrequency. WirelessMeasurement is analogous to ZigBeeMeasurement.
    Question: Is ZigBeeFrequency part of ZigBeeMeasurement?
    Q.2.3
    Analogy: WirelessEnergy is part of WirelessMeasurement. WirelessEnergy is analogous to ZigBeeEnergy. WirelessMeasurement is analogous to ZigBeeMeasurement.
    Question: Is ZigbeeEnergy part of ZigbeeMeasurement?
    Q.2.4
    Analogy: WirelessChannel is part of WirelessRange. WirelessChannel is analogous to ZigBeeChannel. WirelessRange is analogous to ZigBeeRange.
    Question: Is ZigbeeChannel part of ZigbeeRange?
Q.2.1-Answer: Yes
Q.2.2-Answer: Yes
Q.2.3-Answer: Yes
Q.2.4-Answer: Yes

Step 3: For each class C from the ZigBee domain, establish its attributes. Example: What are the attributes the class C?
    Q.3.1
    Analogy: WirelessChannel is analogous to ZigBeeChannel. WirelessChannel attributes are: Interference, ...
    Question: What are the attributes of the class ZigBeeChannel?
Q.3.1-Answer: StartFrequency, StopFrequency, CentralFrequency, ChannelWidth, ChannelNumber, Interference (better term is maybe Occupied)

Fig. 4. ZigBee ontology creation – step by step automated using procedure.

creation procedure. As it can be seen, manual procedure lasts around 12 times longer compared to our method. The step that requires user interaction still lasts several minutes, while the entirely automatic procedure does not exceed 1 s.

Table 4. Experiments and results.

Aspect	Time required [s]
Class template generation	0.43
Answering competency questions	125
Ontology generation	0.83
Manual procedure	1512

5 Conclusion and Future Work

According to the achieved results, the approach seems promising as it speeds up the ontology creation significantly, by more than one order of magnitude. Furthermore, it emphasizes knowledge re-usage, which eliminates overhead and amount of effort required for further developments and extensions.

In future, it is planned to adopt machine learning enabled techniques, such as trending ChatGPT service in order to further accelerate creation of domain ontologies by recognizing the key concepts of particular domain and their relations based on huge amount of textual data, even in scenario when base ontology is not available. On the other side, we would also leverage ChatGPT in order to semantically extract raw data with respect to given ontology.

References

1. Petrovic, N., Tosic, M.: SMADA-Fog: semantic model driven approach to deployment and adaptivity in fog computing. Simul. Model. Pract. Theory **101**(102033), 1–25 (2020). https://doi.org/10.1016/j.simpat.2019.102033
2. Gruber, T.: Toward principles for the design of ontologies used for knowledge sharing. Int. J. Hum. Comput. Stud. **43**(5–6), 907–928 (1995)
3. Nejkovic, V., Tosic, M.: Using analogy computing for ontology development. In: 2016 IEEE 32nd International Conference on Data Engineering Workshops (ICDEW), Helsinki, Finland, pp. 115–120 (2016). https://doi.org/10.1109/ICDEW.2016.7495628
4. Gentner, D., Smith, L.: Analogical reasoning. In: Ramachandran, V.S. (ed.) Encyclopedia of Human Behavior, 2nd edn., pp. 130–136. Elsevier, Oxford (2012)
5. Presutti, V., Blomqvist, E., Daga, E., Gangemi, A.: Pattern-based ontology design. In: Suárez-Figueroa, M.C., Gómez-Pérez, A., Motta, E., Gangemi, A. (eds.) Ontology Engineering in a Networked World, pp. 35–64. Springer, Heidelberg (2012). https://doi.org/10.1007/978-3-642-24794-1_3
6. Tahani, A., Parsia, B., Sattler, U.: Mining Ontologies for Analogy Questions: A Similarity-based Approach. http://ceur-ws.org/Vol-849/paper_32.pdf
7. Reynolds, J., Pease, A., Li, J.: Analogy and deduction for knowledge discovery. In: IKE, pp. 39–48 (2004)
8. Stojkovic, M., Trifunovic, M., Misic, D., Manic, M.: Towards analogy-based reasoning in semantic network. Comput. Sci. Inf. Syst. **12**(3), 979–1008 (2015)
9. Nejkovic, V., Jelenkovic, F., Makris, N., Passas, V., Korakis, T., Tosic, M.: Semantic coordination on the edge of heterogeneous ultra dense networks. J. Netw. Syst. Manage. **29**(2), 1–28 (2021). https://doi.org/10.1007/s10922-020-09576-3
10. Dooley, D., et al.: Food process ontology requirements. Semantic Web, preprint, pp. 1–32. IOS Press (2022). https://doi.org/10.3233/SW-223096
11. Audrito, D., Sulis, E., Humphreys, L., Di Caro, L.: Analogical lightweight ontology of EU criminal procedural rights in judicial cooperation. Artif. Intell. Law **31**, 629–652 (2023). https://doi.org/10.1007/s10506-022-09332-9
12. Benhocine, K., Hansali, A., Zemmouchi-Ghomari, A., Reda Ghomari, A.: Towards an automatic SPARQL query generation from ontology competency questions. Int. J. Comput. Appl. **44**(10), 971–980 (2022). https://doi.org/10.1080/1206212X.2022.2031722

13. Gangemi, A., Lippolis, A.S., Lodi, G., Nuzzolese, A.G.: automatically drafting ontologies from competency questions with FrODO. In: Studies on the Semantic Web Volume 55: Towards a Knowledge-Aware AI, pp. 107–121. IOS Press (2022). https://doi.org/10.3233/SSW220014

14. Tosic, M., Seskar, I., Jelenkovic, F.: TaaSOR – Testbed-as-a-service ontology repository. In: Korakis, T., Zink, M., Ott, M. (eds.) TridentCom. LNICSSITE, vol. 44, pp. 419–420. Springer, Heidelberg (2012). https://doi.org/10.1007/978-3-642-35576-9_49

Educational System for Demonstrating Remote Attacks on Android Devices

Mihajlo Ogrizović$^{(\boxtimes)}$ ⓘ, Pavle Vuletić ⓘ, and Žarko Stanisavljević ⓘ

The School of Electrical Engineering, King Alexander Blvd. 73, Belgrade, Serbia
{ogrizovic,pavle.vuletic,zarko.stanisavljevic}@etf.bg.ac.rs

Abstract. Android is one of the most popular operating systems for smartphones. It's based on a modified version of the Linux kernel. Due to the popularity of this operating system, there have been many attackers that have tried to exploit the weaknesses of this system. In the CVE database, as of 2023, there have been over 8000 exploits discovered on different devices with the Android operating system. With every update of the operating system, Android developers try to fix the discovered bugs that lead to attacks. However, by unawareness of the average Android user, many smartphones aren't updated to use the newest version of the Android operating system, which leads to them being targets of many attacks. Research has shown there haven't been many attempts to implement an educational tool to inform of the different kinds of attacks on mobile devices, especially ones that focus on the specifics of the Android devices. In this paper we present a system for education that strives to focus on attacks that could happen on Android devices, with a special focus on remote attacks. This will be shown with a couple of different attack methods, that exploit different weaknesses of the Android operating system. The final product of this research is an educational tool that is ready for use in the course Advanced Network and System Security in the following semester.

Keywords: Android · mobile security · software development · computer science education

1 Introduction

During the last two decades, mobile phones have become the most used personal devices. It has been estimated that in 2022, over 1.2 billion smartphones have been sold [1]. The most popular type of mobile phones is smartphones. Smartphones contain numerous conveniences and functionalities to their users, such as making calls, sending messages, access to the internet and maps.

Android is one of the most popular operating systems for smartphones. It's based on a modified version of the Linux kernel. Research has shown that Android has around 68.61% of the market share [2].

However, due to the popularity of this operating system, there have been many attempts to exploit the vulnerabilities that exist in this system. In the CVE database, as of May 2023, there have been over 8000 exploits discovered on different devices

© The Author(s), under exclusive license to Springer Nature Switzerland AG 2024
M. Trajanovic et al. (Eds.): ICIST 2023, LNNS 872, pp. 298–306, 2024.
https://doi.org/10.1007/978-3-031-50755-7_28

with the Android operating system [3]. With every update of the operating system, Android developers try to fix the discovered bugs that lead to attacks. However, due to the unawareness of the average Android users, many smartphones aren't updated to use the newest version of the Android operating system, which leads to them being targets of many attacks. Statistics have shown that, as of April 2023, only 20% of Android devices have the latest stable version of the operating system (Android 13.0) [4].

To prevent future attacks, both awareness campaigns towards all the Android users and the education of professionals lead to the conclusion that there needs to be a way of educating people of the possible danger their smartphones could be in.

Therefore, this paper shows a new environment for demonstrating different kinds of attacks on Android devices to better educate Android users of said dangers. It allows users to access virtual machines, where they have the necessary tools to execute attacks on Android devices, which are emulated with the help of Android Virtual Devices.

The following section delves into the problem that arose during the implementation of the environment and explains the details of the attacks that are going to be demonstrated. The third section explores the intricacies of the laboratory environment as well as the results of the attacks on different devices. The last section goes into the planned improvements of the implemented environment.

2 Problem Analysis and Related Work

Research has shown that there are some attempts to develop different tools and environments used for security education. Some of the developed tools include SREG [5], which is an education game that helps players get an understanding of security attacks and vulnerabilities. Another is a tool which enables a fully customizable platform for digital object-oriented teaching called Drag&Fly [6].

However, there haven't been many attempts to implement an educational tool to inform of the different kinds of attacks on mobile devices, especially ones that focus on the specifics of the Android devices. There have been more educational tools for learning more universal mobile security concepts that apply to any smartphone.

One of these examples is Labware, that tries to educate smartphone users of common attack methods that lead to attackers gaining sensitive information [7]. This security labware is implemented by separating the attacks into seven self-contained modules. Every module covers an important threat related to mobile-device security and privacy, mobile-app security, and mobile network and communication security. One of the module examples is based on exploring Mobile SMS security.

Another educational tool is SACH [8], which assists users to write more secure Android applications. This is done by scanning Android applications and identifying possible security vulnerabilities, based upon CERT Java secure coding rules for Android.

There have been also a few attempts at designing educational tools by developing a purposefully vulnerable Android application in order to demonstrate insecure coding practices. These tools include OWASP GoatDroid Project [9] and DIVA [10].

Our system for education strives to focus more on attacks that could happen on Android devices, with a special focus on remote attacks. This is shown with a couple of different attack methods, that exploit different weaknesses of the Android operating system. Some are due to the usage of an old Android operating system version, while others are due to leaving an open port on the device that allows the attacker to communicate with the device, which are explained in more detail in the subsequent subsections. To better understand these attacks, firstly there will be a quick overview of the way Android applications work.

2.1 Android Applications

Applications are run on Android devices in such a way that they are isolated from the rest of the system as well as mutually exclusive [11]. This concept is commonly known as a sandbox. The Android OS assigns to every application a unique identifier and when run, every application is run as a separate process.

The kernel enforces security between applications and the system itself by using well-known concepts used on the Linux OS such as group and user identifiers. Consequently, if an application tries to access a well-protected resource or tries to do something that it should not do, the operating system will not allow it, because it does not have the appropriate user permissions. Every Android application is run on a separate instance of a virtual machine (ART virtual machine on Android devices with the operating system version over 4.4 and the Dalvik virtual machine for devices with an older version).

Having the permission concept in mind, applications can request specific permissions to function properly [12] (e.g., resources for using messages, the camera...), which is done by listing the needed permissions in the `AndroidManifest.xml` file, which is in the root of every Android application, and serves as the main configurational file for an application. During the installation process, the user is informed which permissions the application requests. In Fig. 1, an example of the `AndroidManifest.xml` file can be seen, where the permission to use the internet is requested by the application.

2.2 Remote Attacks Overview

This subsection will be dedicated to explaining a couple of specific types of remote attacks which are demonstrated in the laboratory environment.

Creating a Reverse Shell. This attack is based on a universal attack pattern which is based on the victim device unknowingly starting a connection (usually a TCP connection) with the attacker's device. This method is usually combined with other types of attacks, where the reverse shell essentially represents a backdoor to the victim's device.

One common way a reverse shell is implemented on an Android device is by disguising it as a seemingly normal application, which has malicious intentions. It is since users won't notice the requested permissions when installing the application, which therefore leads to the fact that the malicious application can use all permissions.

Once started, the applications start an Android service that establishes a TCP connection with the attacker's device. Now the attacker can communicate via a shell with the victim, since the service always runs in the background (even if the application that started it is closed). The attacker can send commands to the victim's device to do whatever he needs which can lead to the leakage of sensitive information (images, messages and perhaps even credit card numbers).

```
<manifest xmlns:android="http://schemas.android.com/apk/res/android"
    package="com.example.androidlab"
    android:versionCode="1"
    android:versionName="1.0" >
    <uses-permission android:name="android.permission.INTERNET" />
    <application
        android:allowBackup="true"
        android:dataExtractionRules="@xml/data_extraction_rules"
        android:fullBackupContent="@xml/backup_rules"
        android:icon="@mipmap/ic_launcher"
        android:label="AndroidLab"
        android:roundIcon="@mipmap/ic_launcher_round"
        android:supportsRtl="true"
        android:theme="@style/Theme.AndroidLab"
        tools:targetApi="31" >
        <activity
            android:name=".MainActivity"
            android:exported="true"
            android:label="@string/app_name" >
            <intent-filter>
                <action android:name="android.intent.action.MAIN" />
                <category android:name=
                "android.intent.category.LAUNCHER" />
            </intent-filter>
        </activity>
    </application>
</manifest>
```

Fig. 1. Example of the `AndroidManifest.xml` file

Exploiting the Weaknesses of the `WebView` Component. One commonly used component in Android applications is the `WebView` component, which allows embedding of web pages into the application.

The included web pages can include JavaScript code. The `WebView` component allows interaction between the application code and the loaded page by allowing the passing of objects defined in the application source code to the web page. This can be done using the `WebView` component's method `addJavaScriptInterface`.

The problem here lies in the fact that the loaded page could be one with malicious JavaScript code that exploits the passed object by accessing methods of the `Runtime` object, which allow the malicious code to invoke various system calls on the victim's device. One way this can be used is by loading and running a malicious application that starts a reverse shell onto the victim's device.

This exploit has been deprecated on newer versions of the Android operating systems (versions over 4.2) by having application programmers state explicitly which methods of the passed object can be used in the JavaScript code. This is done with the annotation `@JavascriptInterface`.

Code Tampering. As stated, one common attack vector is by tricking users into installing and running applications that seem harmless and useful. However, they contain malicious code that can, for example, start a reverse shell with the attacker's server. This can be done in multiple ways, but one of the more common ways is by changing the application code of well-known applications. This is called code tampering.

Code tampering can be done manually or automatically. When done manually, the attacker decompiles a well-known application, and in the decompiled code manually adds malicious code that, when the app is ran, starts a service that establishes a reverse shell with the attacker's device. Then, after successfully compiling the newly created application, distributes the new executable Android application file. This can be also done automatically, by using various tools which do the same steps which are done manually.

Attacks Using adb. Android contains a useful tool for application programmers that allows communication between the Android device and the device on which the application is being programmed on. This tool is the Android Debug Bridge (`adb`), which enables simple installation, testing and debugging of a developed Android application [13].

Communication using this tool can be done with a wired or a wireless connection. The wireless communication is enabled by opening port 5555 on the Android device on which the application is being tested on. This allows programmers to install any application on the device with this port opened, which can lead to these devices being a target for attackers.

If an attacker is on the same network as the victim, and notices the opened port 5555, this can lead to the attacker installing any kind of malicious application on to the device, like the already mentioned one which starts a reverse shell with the attacker's device.

3   Realization of the Environment

The main problem was the implementation of an environment where users could access the mobile phones and execute the attacks. There are a couple of different methods of integrating the attack scenarios in an operational environment that have been analyzed for this paper. The one that is used is an already existing laboratory environment used in the course Advanced Network and System Security, with the similar idea of demonstrating different kinds of attacks, which was used due to its proven reliability and modularity on the course, where it's easy to implement a new attack scenario.

3.1   An Overview of the Used Laboratory Environment

The environment consists of four mutually interconnected virtual machines, where three of them have the Linux Ubuntu 22 operating system installed, and the fourth one has the `Pfsense` firewall set up as can be seen in Fig. 2.

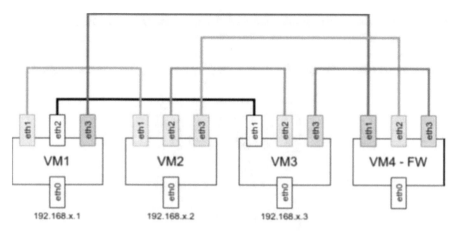

Fig. 2. Laboratory environment overview

The interface `eth0` allows access to the virtual machines themselves through VPN (Virtual private network), while others allow interconnection between the machines.

This laboratory environment is already successfully used by students of the course Advanced Network and System security, by allowing them to execute different types of attacks. Students are allowed `ssh`, `ftp` and `vnc` access to the machines, by using tools like Putty, WinSCP and TightVNC for using said protocols. There they can execute attacks such as ARP spoofing, DNS spoofing or setting up IP filters on the machines.

3.2 Adjustment of the Laboratory Environment for Android Attacks

For the uses of demonstrating attacks on Android devices, only one of the mentioned four virtual machines is used, specifically VM1, which has all the necessary tools installed. The mentioned tools include all the Android devices, as well as the attacker's server, the Android application used to demonstrate attacks and the tools which need to be run to demonstrate attacks.

Android Devices. To demonstrate the importance of using the latest version of the Android operating system, three Android devices with varying versions of the operating system have been installed, shown in Table 1.

Table 1. Overview of used Android devices

Android Device	Operating system version	API level
Google Pixel (AVD1)	Android 4.1.1	API 16
Samsung Nexus 6 (AVD2)	Android 8.0	API 26
Google Pixel 5 (AVD3)	Android 11.0	API 30

The mentioned instances of devices (Android Virtual Devices) are installed on VM1 using the `emulator` command that is integrated in Android SDK [14].

Victim Application. The application used to demonstrate attacks in our laboratory environment essentially represents a simplified web browser. It consists of a single Android activity which presents a `WebView` component that loads an URL of the Page which the user inputs through an `EditText` (Fig. 3).

Fig. 3. Appearance of the victim application

The only required permission for this application to function correctly is the permission to use the internet. An object from the source code of the application is passed to the `WebView` component to demonstrate the weakness of that component. The victim application will be used also to demonstrate code tampering and reverse shell attacks.

Used Tools. To demonstrate these attacks a variety of different tools have been used, described below.

- `Metasploit` – One of the most popular frameworks for hacking, used to exploit weaknesses of different kinds of systems, among them the Android operating system. Here it is used to set up the attacker's server, as well as the `WebView` component weakness, where it sets up the web page with the malicious JavaScript code as well.
- `Msfvenom` – This tool is used to inject malicious code into existing executables. In this case, it is used to inject code which sets up the reverse shell with the attacker's server into the already mentioned victim application.
- `Ghost` – Once the attacker's server is connected to the victim using the open port 5555, tools (such as this one) allow easier exploiting of the victim's device, by installing a reverse shell with the attacker's server.

3.3 Results

Once the attacks have been tried out on the three emulated devices, the outcomes can be seen in the following table (Table 2).

Table 2. Outcomes of the attacks on the Android virtual devices

Attack	AVD1	AVD2	AVD3
Exploiting the WebView component	Success	Failure	Failure
Code tampering	Success	Success	Success
Exploiting the adb tool	Success	Success	Success

As it can be seen in the table, the different versions of the Android operating system appear to make a difference when using the WebView component exploit. This is due to the already mentioned fact that newer versions fixed this exploit. Attacks such as code tampering and exploiting the adb tool are less based on the Android operating system, and more on deceiving ignorant users, therefore a difference in the operating system version didn't change the outcome of the attack.

Since the proposed addition to the course Advanced Network and System security laboratory environment is planned to be added the following semester, students haven't had the chance to try out the modified environment. In the previous semester, over 80 students have studied this course, therefore, the laboratory environment is going to be thoroughly tested in the following semester.

4 Conclusion

The final product of this research is an educational system that is ready for use in the course Advanced Network and System Security in the following semester, that serves as an education tool for which security precautions a person should have as an Android user but as well as an Android programmer. This educational system is fully functional in the existing laboratory environment, and therefore will work from any computer with access to virtual machines present in the environment. The system will contain attack scenarios that exploit weaknesses in Andorid's WebView component, the adb tool, code tampering and as well as reverse shell, which are described in detail in the paper.

In addition to possible problems that may arise after more massive testing of the system, further directions of research are primarily based on adding other attack scenarios. There will be more scenarios that exploit obsolete versions of the Android operating system, as well as scenarios that demonstrate privilege escalation attacks. There will also be more of a focus on exploits based on Android applications, by exploiting improper usages of the Android API.

Newer versions of the laboratory environment will also test software that detects suspicious code with a static or dynamic analysis of the application.

References

1. Global Smartphone Market Share. https://www.counterpointresearch.com/global-smartp hone-share/. Accessed 10 May 2023
2. Mobile Operating System Market Share Worldwide. https://gs.statcounter.com/os-market-share/mobile/worldwide. Accessed 20 May 2023
3. CVE database. https://cve.mitre.org/cgi-bin/cvekey.cgi?keyword=android. Accessed 28 May 2023
4. Mobile Android Version Market Share Worldwide. https://gs.statcounter.com/android-ver sion-market-share/mobile/worldwide. Accessed 10 May 2023
5. Yasin, A., Liu, L., Li, T., Wang, J., Zowghi, D.: Design and preliminary evaluation of a cyber Security Requirements Education Game (SREG). Inf. Softw. Technol. (2017)
6. Barpi, F., Dalmazzo, D., De Blasio, A., Vinci, F.: Hacking higher education: rethinking the EduHack course. Educ. Sci. **11**, 40 (2021). https://doi.org/10.3390/educsci11020040
7. Qian, K., Dan Lo, C.-T., Guo, M., Bhattacharya, P., Yang, L.: Mobile security labware with smart devices for cybersecurity education. In: IEEE 2nd Integrated STEM Education Conference, Ewing, NJ, USA, pp. 1–3 (2012). https://doi.org/10.1109/ISECon.2012.6204180
8. Abernathy, A., Yuan, X., Hill, E., Xu, J., Bryant, K., Williams, K.: SACH: a tool for assisting Secure Android application development. In: SoutheastCon 2017 (2017). https://doi.org/10. 1109/secon.2017.7925374
9. OWASP-GoatDroid-Project. https://www.openhub.net/p/owasp-goatdroid-project. Accessed 10 May 2023
10. DIVA Andorid. https://payatu.com/blog/damn-insecure-and-vulnerable-app/. Accessed 28 May 2023
11. Application Sandbox. https://source.android.com/docs/security/app-sandbox. Accessed 01 May 2023
12. Chell, D., Erasmus, T., Colley, S., Whitehouse, O.: Understanding Permissions, The Mobile Application Hacker's Handbook. Wiley, Indianapolis
13. Android Debug Bridge (adb). https://developer.android.com/studio/command-line/adb. Accessed 07 May 2023
14. Start the emulator from the command line. https://developer.android.com/studio/run/emu lator-commandline. Accessed 10 May 2023

CRETE – Code REview Tool for Education

Milos Obradovic[(✉)] ⓘ, Drazen Draskovic ⓘ, Tamara Sekularac ⓘ,
Mihajlo Ogrizovic ⓘ, and Dragan Bojic ⓘ

School of Electrical Engineering, University of Belgrade, Belgrade, Serbia
milos.obradovic@etf.bg.ac.rs

Abstract. At the School of Electrical Engineering at the University of Belgrade, the course Principles of Software Engineering was formed to make future software engineers familiar with the basic concepts of the field of Software Development. The core of this course presents a team project, where students engage in the complete software development life cycle using version control tools. A significant challenge in organizing such a practical team project is finding a suitable tool that combines code review techniques with version control. This tool should be suitable for educational purposes while enabling progress monitoring and evaluation of students' work. The difficulty lies in that the requirements for educational code review differ from those of tools designed for instructional development, which most existing tools are tailored for. This paper focuses on adapting the Gerrit platform for utilization at the School of Electrical Engineering, resulting in a tool that can be employed in the current semester.

Keywords: code review · software development · computer science education

1 Introduction

The course Principles of Software Engineering was established at the School of Electrical Engineering, University of Belgrade, with the aim of familiarizing future software engineers with fundamental concepts in the field of software development. This course features a team project that guides students through all stages of software system development, encouraging teamwork and the creation of functional software products throughout the semester.

Currently, during the project's feature addition phase, there is a lack of code review [1] among team members. Consequently, the teachers wish to introduce the practice of code review in software development. By implementing code review, students can receive valuable feedback on the quality of their program code, promoting knowledge sharing and enhancing programming techniques through exposure to different solutions and feedback from their peers.

Code review technique is highly recommended in software development [2]. However, the implementation of this technique lacks standardization and varies based on project requirements [3]. Furthermore, the utilization of code review techniques in

M. Trajanovic et al. (Eds.): ICIST 2023, LNNS 872, pp. 307–317, 2024.
https://doi.org/10.1007/978-3-031-50755-7_29

university courses remains limited [4]. Therefore, this paper aims to establish a foundation for implementing code review practices in courses at the School of Electrical Engineering.

To ensure the success of course activities, it is crucial to identify suitable software tools that are accessible to the School of Electrical Engineering. These tools should primarily support version control and code review capabilities. While various open-source version control tools are available and well-suited for course purposes, finding a suitable code review tool that is publicly accessible, motivates student participation, and minimizes the risk of team members avoiding their responsibilities poses a greater challenge.

To identify an appropriate tool that meets our requirements, a research study was conducted [5]. The findings suggest that the most effective approach is to develop an independent tool, as no existing tool fully satisfies all the identified requirements. However, creating a tool from scratch is deemed unnecessary. Instead, the research recommends leveraging the flexibility and open-source nature of the Gerrit platform [6], which can be easily customized to align with the specific needs of the project.

The paper is divided into four sections. The second section highlights the main problem of adapting the Gerrit platform for university use, with subsections addressing specific challenges arising from the differences between code review needs in education and open-source projects. The following section provides solutions to these problems, including utilizing Gerrit platform mechanisms and introducing additional modifications. The final section concludes the paper, summarizing the work done and suggesting future research directions.

2 Problems with Using Code Review in Education

This section introduces the primary research questions. Each question highlights a specific issue arising from the lack of adaptation of code review tools for educational purposes. The section is organized into several subsections, each focusing on a distinct problem that needs to be addressed. The subsequent section delves into each research question and presents solutions through the adaptation of the Gerrit platform.

2.1 Access Rights

To address the problem of setting access rights for students, a brief case study is conducted to explore the differences between software development in open-source projects (for which the Gerrit platform was originally designed) and the use of code review tools in teaching. Table 1 provides a comparison between the access rights commonly encountered in open-source software development projects and the access rights that are better suited to meet the specific requirements of a university setting.

A significant distinction between open-source projects and educational settings lies in the transparency of program code. While open-source projects require public availability, this may not be desirable in an educational context. Making student solutions publicly accessible often leads to increased instances of plagiarism. Consequently, for the purpose

Table 1. Comparison between access rights in open-source and university projects

Access right	Open-source projects	University projects
Reading project repository	all platform users	students working on a project
Commenting on new change	all registered users	students working on a project
Adding new change	all registered users	students working on a project
Accepting or rejecting change	project owners	students working on a project

of using the tool in university courses, it was chosen that students should only have access to the projects they are directly involved in.

Another notable difference arises in the code review process and the acceptance of new solutions. In open-source projects, all platform participants have the opportunity to propose solutions and provide feedback on added changes. However, ultimately, a smaller group of users, typically project owners, decide which changes are accepted and which are rejected. On the other hand, in small projects within a university course, the objective is to grant all students equal access rights and enable each of them to approve changes that are added to the project.

To address these differences, Subsect. 3.1 discusses the default access rights in Gerrit platform and provides guidelines for customizing them. These modifications aim to ensure every team member can access the project, add changes to the project and approve the changes added by other team members. Furthermore, the access rights should guarantee that every commit undergoes code review and cannot be directly added to the project repository.

2.2 Accepting New Changes

By assigning the appropriate access rights, students can access the project, contribute changes, comment on and approve changes made by other team members. Additionally, access rights can ensure that every commit is sent to code review. However, this alone does not guarantee active student participation in the code review process, as a student could potentially approve their own change without consulting other team members and incorporate it into the project.

In open-source projects, this problem is addressed by selecting a small group of individuals as project owners [7], who have the ultimate decision-making authority regarding the acceptance or rejection of changes. However, implementing this solution in university courses is not appropriate since all students should have the opportunity to participate in code review and contribute changes to the project once specific criteria are met.

The objective is to allow students to approve their own changes, but only after another individual has reviewed the added program code. Although this may appear as a mere formality, it significantly impacts the actual code review process. Some of the advantages of this approach include:

- The person conducting the code review does not bear the responsibility of accepting the change.
- Once one person provides a positive rating, it becomes possible (but not obligatory) for someone else to review the code without the initial reviewer's involvement in the code review process again.
- The individual conducting the code review can leave comments for the person who added the change and provide a positive code review label. The person who added the change can then read the comment and have responsibility for accepting the change if they agree with the feedback.

Hence, the question arises: How can we ensure that at least one team member reviews the program code before it is added to the project repository? Subsection 3.2 addresses this question by defining the necessary criteria and project setup to tackle this challenge effectively.

2.3 Automating Project Creation

When setting up a new project, the following steps need to be followed:

1. Create a group for the students involved in the project and add the respective students to the group.
2. Create a new project with the appropriate initial configurations.
3. Configure the access rights for both students and teachers, ensuring appropriate permissions are granted.

Given that the Principles of Software Engineering course enrolls a significant number of students, and projects are typically developed in teams of 3–4 individuals, a substantial number of projects are created each year. Manually performing these steps for each project is impractical and time-consuming. Moreover, manual adjustments increase the risk of errors, potentially jeopardizing the entire project setup.

To address this challenge, a script has been developed to automate all three steps involved in creating and configuring a new project. Automating these processes not only saves time but also minimizes the likelihood of errors, ensuring consistent and reliable project creation.

3 Adapting Gerrit Platform for Use in Education

This section focuses on the adaptation of the Gerrit platform for educational purposes, addressing the primary questions discussed in the previous section. The section consists of four subsections, with the first three dedicated to solving the primary problems identified in the previous section. Each of these subsections discusses the specific mechanisms within the Gerrit platform that can contribute to solving the problem and explains how these mechanisms are utilized to derive the CRETE platform. The final subsection outlines a few additional modifications made to enhance the completeness and readiness of the CRETE platform for production use.

3.1 Access Rights

In the Gerrit platform, users access rights are assigned based on the groups they belong to [8]. Additionally, within a project, access rights can be granted at the reference namespace level, and a project hierarchy determines the inheritance of access rights. Specifically, this paper focuses on the following namespaces:

- "refs": This namespace contains all references related to the project's code. Granting access rights to this namespace provides access to the entire project repository.
- "refs/heads": Within this namespace, all references used for software development are located. Adding a change to any of these references follows the conventional Git workflow, where the change is directly added to the project without undergoing the code review process.
- "refs/for/refs/heads": This namespace is specific to the Gerrit platform. Adding a change to a reference under "refs/for/refs/heads/x" indicates that a code review is required for the change. After the change is accepted, it is then added to the "refs/heads/x" branch. Figure 1 illustrates an example of adding changes to these references.

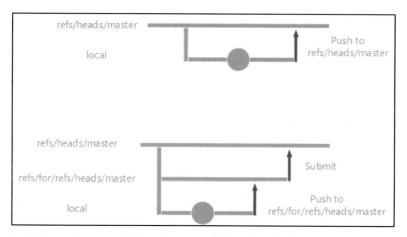

Fig. 1. Difference between adding a change to refs/heads and refs/for/refs/head namespaces.

There are various types of access rights that can be assigned to a group on a reference namespace [8]. This paper focuses on a subset of access rights, including:

- Ownership: Grants full control over the namespace.
- Reading (Read): Enables viewing the content of the namespace.
- Adding a change (Push): Allows pushing changes to the namespace.

- Change review (Label Code-Review): Permits reviewing and providing feedback on changes.
- Change acceptance (Submit): Allows accepting changes and merging them into the project repository.

To ensure that each student has access only to the project they are assigned to and cannot access other projects on the platform, account groups [8] are created for each team of students. Each user belonging to an account group is granted access rights assigned to that group.

Every account group has an owner group responsible for managing the group. The group owner can modify the group's name, description, add or remove members. To prevent misuse of adding unauthorized members, a separate group consisting of course teachers is designated as the owner of all student groups.

For each team project, a project is created where students collaborate throughout the semester. Within these projects, student groups have the following access rights:

- Read access to all project code.
- Ability to provide code review labels for changes under review.
- Submission and addition of changes to the main project code after meeting the required criteria.

To ensure that every commit undergoes code review and cannot be directly added to the project repository, students are granted push access only to the "refs/for/refs/heads" namespace, not the "refs/heads" namespace. This restricts their ability to push changes directly.

To prevent students from modifying project settings, a teachers group is assigned as the project owner. Table 2 provides an overview of all access rights granted to students for their assigned projects.

Table 2. Table captions should be placed above the tables.

Access right	Namespace	Access
Read	refs/heads	✓
Push	refs/for/refs/heads	✓
Label Code-Review	refs/heads	$[-2, 2]$
Submit	refs/heads	✓

It is important to note that upon installing the Gerrit platform, it comes with predefined access rights commonly used in open-source development (as discussed in Subsect. 2.1). However, keeping those access rights alongside the access rights described in this subsection could undermine the purpose of setting access rights specifically for students. Therefore, after the initial launch of the Gerrit platform, all access rights that are granted to unregistered and registered users should be removed.

3.2 Accepting New Changes

The Gerrit platform allows the definition of criteria for accepting changes into the project repository. Each criterion corresponds to a category, and a change must satisfy the requirements of each category to be accepted. Categories also specify the range of ratings that can be assigned. To be accepted, a change must receive at least one maximum rating for each category and must not have any minimum ratings, which would constitute a veto vote.

By default, the Gerrit platform includes two categories: Code Review and Verification [9]. The Code Review category is typically used to indicate that the code has been reviewed and is considered correct. On the other hand, the Verification category is automatically applied to check if the code compiles and passes basic tests. These categories can be customized using additional rules [10] written in the Prolog programming language [11].

In the Principles of Software Engineering course, the objective is to ensure that students cannot add and approve their own changes without involving other team members. To address this, a Non-Author Approval category is implemented, which mandates that at least one team member, who is not the author of the added code, confirms the validity of the change. Additionally, the Verification category is not used in the current version of the course and should be removed.

Modifying the criteria for change submission can be done by writing a submit rule or a submit filter. The key distinction is that a submit rule is applied directly to a project, while a submit filter is utilized across all parent projects in the hierarchy. Considering that these rules should be same for every project in the CRETE platform, a submit filter is implemented and added to root project. This approach simplifies the configuration process for new projects, ensuring uniformity in the criteria for change submission. Submit filter used in CRETE platform is shown in program code 1.

```
submit_filter(In, Out) :-
    In =.. [submit | Ls],
    remove_verified_category(Ls, R1),
    add_non_author_approval(R1, R),
    Out =.. [submit | R].

add_non_author_approval(S1, S2) :-
    gerrit:commit_author(A),
    gerrit:commit_label(label('Code-Review', 2), R),
    R \= A, !,
    S2 = [label('Non-Author-Code-Review', ok(R)) | S1].
add_non_author_approval(S1,
    [label('Non-Author-Code-Review', need(_)) | S1]).

remove_verified_category([], []).
remove_verified_category([label('Verified', _) | T], R)
    :- remove_verified_category(T, R), !.
remove_verified_category([H|T], [H|R])
    :- remove_verified_category(T, R).
```

Prog. code 1. Modifying criteria for change submission.

To apply the submit filter, it needs to be added to the All-Projects settings. Each project can have its own default rules, which are then passed as input parameters to the submit_filter. The filter removes the verification category and adds the non-author code review category, ensuring that someone other than the author has approved the code.

In the latest version of the Gerrit platform, the recommended approach for specifying criteria for accepting change is through submit requirements [12] rather than writing Prolog submit rules. However, at the time of this research, writing Prolog submit rules was a valid method to address the problem. The change mainly pertains to the formality and format of the rules, rather than the underlying concept and functionality of the rules themselves.

3.3 Automating Project Creation

As mentioned in Subsect. 2.3, automating the three steps for creating and setting up a new project involves:

1. Creating an account group for students.
2. Creating a new project with initial configurations.
3. Configuring access rights for students on the project.

The first two steps can be easily accomplished using the Gerrit platform's interface, which offers various commands such as `gerrit create-group` and `gerrit create-project`. However, solving the third problem of configuring access rights requires a different approach.

There is no specific command for setting access rights on the project. Instead, the solution leverages the feature of the Gerrit platform where project settings are stored in "refs/meta/config", which can be managed like any other reference in the platform. Program code 2 demonstrates how this problem can be solved:

```
# params:
# 1 - hostname
# 2 - ssh port
# 3 - gerrit admin name
# 4 - project name
# 5 - group name
change_access_rights() {
    # get config file
    mkdir $4
    cd $4
    git init
    git remote add origin "ssh://$3@$1:$2/$4"
    git fetch origin refs/meta/config:config
    git checkout config
    # change config file
    # 1 - remove all accces rights for students
    grep -v $5 project.config > tmpfile
      && mv tmpfile project.config
    # 2 - add access rights for students
    echo "[access \"refs/for/refs/heads/*\"]"
      >> project.config
    echo -e "\tpush = group $5" >> project.config
    echo "[access \"refs/heads/*\"]" >> project.config
    echo -e "\tlabel-Code-Review = -2..+2 group $5"
      >> project.config
    echo -e "\tsubmit = group $5" >> project.config
    echo -e "\tread = group $5" >> project.config
    # push new config file
    git add project.config
    git commit -m "new access rights"
    git push origin "HEAD:refs/meta/config"
}
```

Prog. code 2. Changing access rights for students on project.

During the project creation process, the ownership of the project is initially set for both students and teachers, allowing both groups to collaborate on the project. To modify their access rights, the ownership information for students is first removed from the configuration file. Then, the necessary access rights for students, as described in Subsect. 3.1, are added to the configuration file. Once the access rights file is modified, the updated version is pushed to the Gerrit platform.

3.4 A Few Additional Modifications

To ensure the completeness and readiness of the CRETE platform, several additional modifications need to be implemented. This subsection discusses two key aspects: student authentication and the tracking of platform usage.

For student authentication, the LDAP [13] method is chosen due to its compatibility with the existing infrastructure used in the School of Electrical Engineering. LDAP authentication is already employed for project assignments and laboratory exercises, making it a suitable choice. By leveraging the LDAP server, students can use their existing credentials without the need to create new accounts. This method ensures that only students can access the tool and simplifies their identification. Additionally, using LDAP enables the creation of projects through the script described in Subsect. 3.3 before students sign in to the platform for the first time.

Gerrit platform does not provide built-in functionality for statistics tracking. Therefore, external plugins are employed. After testing various plugins, the Gerrit Stats plugin [14] is selected for CRETE. This plugin offers an intuitive interface with two commands. The first command pulls Gerrit stats from the server into a JSON object, while the second command generates a web page with a series of graphs and diagrams presenting the platform usage. These visual representations allow course teachers to easily track and evaluate students' work throughout the semester. The Gerrit Stats plugin has an official demo [15] showcasing its usefulness and capabilities.

4 Conclusion

This paper has presented the CRETE platform, an adaptation of the Gerrit platform specifically designed for educational purposes. The platform has been successfully tested among hundreds of students and has proven to be effective. It is compatible with various operating systems, ensuring accessibility for all users.

Throughout the spring semester, the Principles of Software Engineering course successfully utilized all the features described in the third section. Over fifty projects were created using the script outlined in Subsect. 3.3. The configurations described in Subsects. 3.1 and 3.2 functioned smoothly without any issues. The teachers were able to have a comprehensive overview of the students' work using the plugins discussed in Subsect. 3.4. These tools and functionalities greatly facilitated the teaching and learning process in the course.

Moving forward, future research directions focus on enhancing the platform by incorporating additional features. Some ideas include integrating static code analyzers to help students identify basic mistakes more efficiently and improving support for reviewing non-program code files and documents.

Furthermore, there is a desire to introduce code reviews of projects that students are not directly involved in. This can be achieved by configuring additional access rights and implementing anonymous code reviews to encourage serious evaluations.

The platform's flexibility also allows for potential curriculum reforms, enabling further adaptation of the platform to meet evolving educational needs. As the platform continues to be utilized, new ideas for improvements are likely to emerge, further enhancing the learning experience.

References

1. Ackerman, A.F., Buchwald, L.S., Lewski, F.H.: Software inspections: an effective verification process. IEEE Softw. **6**(3), 31–36 (1989)
2. Duncan, S.: Modern software review: techniques and technologies. Softw. Qual. Prof. **9**(3), 46 (2007)
3. Winters, T., Manshreck, T., Wright, H.: Software Engineering at Google: Lessons Learned from Programming Over Time. O'Reilly Media, Sebastopol (2020)
4. Garousi, V.: Applying peer reviews in software engineering education: an experiment and lessons learned. IEEE Trans. Educ. **53**(2), 182–193 (2009)
5. Obradović, M., Kostić, M., Knežević, B., Drašković, D.: An overview of software code review tools and the possibility of their application in teaching at the School of Electrical Engineering in Belgrade. In: Proceedings of IX International Conference on Electrical, Electronic and Computing Engineering, IcEtran 2022, pp. 636–641. ETRAN Society, Belgrade (2022)
6. Gerrit Code Review. https://www.gerritcodereview.com/. Accessed 20 May 2023
7. Jin, L., Robey, D., Boudreau, M.-C.: The nature of hybrid community: an exploratory study of open source software user groups. J. Commun. Inform. **11**(1) (2015)
8. Gerrit Code Review – Access Control. https://gerrit-review.googlesource.com/Documentation/access-control.html. Accessed 20 May 2023
9. Gerrit Code Review – Review Labels. https://gerrit-review.googlesource.com/. Accessed 20 May 2023
10. Gerrit Code Review - Prolog Submit Rules. https://gerrit-review.googlesource.com/Documentation/prolog-cookbook.html. Accessed 20 May 2023
11. Prolog programming language. https://en.wikipedia.org/wiki/Prolog. Accessed 20 May 2023
12. Gerrit Code Review - Submit Requirements. https://gerrit-review.googlesource.com/Documentation/config-submit-requirements.html. Accessed 20 May 2023
13. Harrison, R.: Lightweight Directory Access Protocol (LDAP): Authentication Methods and Security Mechanisms. Technical report (2006)
14. Gerrit Stats. https://github.com/holmari/gerritstats. Accessed 20 May 2023
15. Gerrit Stats – Demo. https://gerritstats-demo.firebaseapp.com/. Accessed 20 May 2023

The Architecture of Citizen Science Open Data Repository Based on Version Control Platforms

Dušan Nikolić(✉) ⬤ and Dragan Ivanović ⬤

Faculty of Technical Sciences, University of Novi Sad, Trg Dositeja Obradovića 6, Novi Sad, Serbia
{nikolic.dusan,dragan.ivanovic}@uns.ac.rs

Abstract. There is an increasing number of citizen science projects, all with the potential to generate new data at a lower cost and arguably greater value than data generated by expert knowledge. However, access to citizen science data is still mostly limited to case-by-case project maintainers and in different data formats. This study proposes a novel approach to accessing open data by designing a hosted open data citizen science repository. We then propose an approach to bootstrap and maintain that data on a version-controlled platform. As a result, hosted open access data has the potential for better exchange and further reuse of data.

Keywords: Citizen Science · Open Data Repository · Open Data · Version Control

1 Introduction

Citizen science platforms are constantly evolving and are becoming more important in providing support to citizen science activities, both nationally and internationally [1].

The Horizon Europe (2021–2017) research and innovation programme encourages the development of EU's scientific and technological excellence and aims to strengthen the EU's research area. This includes but is not limited to boosting EU's innovation update, competitiveness and tackling policy priorities, such as green and digital transitions and sustainable development [2]. To maximize the impact that EU citizen science research has on society, there is a need for strong visibility of citizen science data. At present, there is an increasing number of citizen science projects, all with the potential to generate new data at a lower cost and with arguably greater value than data generated by expert knowledge [3]. However, citizens data access, re-use and interoperability across numerous citizens science platforms is limited due to the fact that those projects host their data and datasets in different formats and on a case-by-case basis. To achieve strong visibility and interoperability of datasets, this paper proposes an approach to storing and versioning citizen science data on a version-controlled platform which will allow access to datasets in various machine-readable formats.

The paper is organized in the following sections. In the next section of this study, we briefly describe existing citizen science solutions, data accessibility and format support across a sample of citizen science platforms and related work. In Sect. 3 we propose an

© The Author(s), under exclusive license to Springer Nature Switzerland AG 2024
M. Trajanovic et al. (Eds.): ICIST 2023, LNNS 872, pp. 318–325, 2024.
https://doi.org/10.1007/978-3-031-50755-7_30

approach to overcome data accessibility limitations by designing an open data repository based on version control. In Sect. 4 an open data repository architecture based on version control platforms has been presented and discussed. The paper is ending with conclusions and directions for further research and development.

2 Related Work

Over the past decade, a variety of citizen science platforms and infrastructures have been developed and verified with overlapping goals and at varying scopes. In this section, some of the most common types of infrastructures will be briefly described.

According to analysis made in [4], current types of citizen science platforms can be divided into five broad categories:

- Commercial platforms for citizen science initiatives,
- Citizen science platforms for specific projects,
- Citizen science platforms for specific scientific topics,
- National citizen science platforms,
- EU citizen science platforms

Commercial platforms for citizen science initiatives aim to provide paid professional services as well as data handling for institutions, citizen science project leaders and other stakeholders interested in citizen science research on a global scale. These platforms, such as SPOTTERON [5], represent a general solution with customizable web apps, add-ons, and other specific feature sets. These types of platforms are highly versatile and facilitate valuable citizen science research. However, in our study we focused on analyzing platforms with open and easily available datasets.

Specialized citizen science platforms aim to provide tools, information and interact with participants about a specific project or topic. An example is Zooniverse [6], one of the world's largest and most popular research platforms, for which we analyzed citizen science project data access. The results of our analysis are shown in Table 1. Moreover, the platform hosts dozens of active projects that allow volunteers to participate in crowdsourced scientific research. Another notable platform is the CitSci [7] project, which is a mature citizen science platform as a service founded in 2007. Its purpose is aimed at creating, collecting, analyzing, and managing citizen science data for a variety of community driven observable projects, such as ecological and environmental observations.

National citizen science platforms have been developed to facilitate citizen science projects on a variety of citizen science topics. They unite different stakeholders, such as institutions, scientists, media, and policymakers with research participants within the respective countries. A keynote of these types of platforms is the use of their respective languages when communicating with interested users and stakeholders. A notable national platform for which we have analyzed data access is the German's Bürger schaffen Wissen [8].

Moreover, there are citizen science platforms which aim to disseminate and promote knowledge, as well as provide guidelines with citizen science best practices to institutions and researchers across the EU. Two notable examples are the EU-Citizen.Science [9] and

EC's Joint Research Centre (JRC) [10] platforms, funded by the European Commission Horizon 2020 program. Our analysis of data access and data re-use aspect of the EU-Citizen.Science platform is discussed in Sect. 1.2. Both platforms provide general access to projects from many scientific disciplines and share the vision of being a knowledge hub for citizen science projects and research.

Regarding data quality, the authors in [11] argue that citizen science can produce high-quality data that is comparable to professional data. They assert that successful projects use methods like iterated project design, volunteer training, expert validation, replication through volunteers, and statistical modeling to ensure accuracy and address biases. Additionally, the authors stress the need to evaluate citizen science data based on project-specific factors instead of dismissing them due to volunteer involvement. The authors conclude by emphasizing the importance of evaluating data quality practices, addressing challenges, and suggesting future directions for ensuring high-quality data in citizen science. However, almost all currently active citizen science projects host their data and datasets in different formats and on a case-by-case basis. A sample analysis was performed on four well-known citizen science platforms in order to analyze data accessibility offered by different citizen science platforms. The results are provided in Table 1.

Table 1. Characteristics of data access on four citizen science platforms.

Platform domain	Open access to dataset and data?	Open dataset format support
https://eu-citizen.science/	Data hosted on individual citizen science projects domain and can be private	N/A
https://www.zooniverse.org/	Dataset and data hosted on individual citizen science projects domain	Multiple, depends on project
https://citsci.org/	Dataset and data accessible to members of specific project	As datasheet, CSV
https://www.buergerschaffenwissen.de/	Dataset and data hosted on individual citizen science projects domain	Multiple, depends on project

3 An Approach to Overcome Citizens Science Data Accessibility Limitations

As shown in Table 1, access to data is still mostly limited to individual case-by-case project maintainers, suggesting that data access and data usability can be further improved. The case-by-case, tailor-made management of datasets might serve the intended purpose, but is likely to reduce the exchange and further use of data. Interoperability, which enables the seamless reuse of resources (in this case, data, and processing) across different systems, can be achieved by applying community-wide agreements [3].

The primary goal of this study is to overcome open data limitations listed in Table 1, by proposing an easily accessible open data repository for citizen science projects. Certain citizen science infrastructures rely on project maintainers for hosting their research results, while others generate datasheets and comma separated value (CSV) spreadsheets. Even though this approach has been shown to serve its original purpose of publishing data results, it has simultaneously reduced data interoperability and data re-use. Therefore, this paper wishes to improve project data access by proposing a solution for curating datasets on a version-controlled platform of choice. A significant advantage of having open data hosted on a versioned system is the aspect of dataset iteration, continuous dataset versioning and metadata provided by the version-controlled platform.

4 Architecture and Discussion

To be able to host citizen science repositories, we have decided to create and maintain data in citizen science projects on a version-controlled repository. For these purposes we chose a popular community hosting service - GitHub. Citizen science projects can be hosted either individually on a user's account repository or within an organization. As such, the following three prerequisites should be met: creating an organization, creating a project repository, and maintaining data submissions. Forming an organization represents a manual step completed by each newly registered research group or individual upon request. However, it is important to note that repository access is managed by the citizen science project maintainer. In the case of private repositories with limited access, the server keeps a project access token in a relational database. With an access token in place, the server can use git to retrieve as well as maintain the repository datasets. From a system's perspective, datasets are referred to as collections of user data submissions exported in machine and human readable formats. To transform versioned user submission data into datasets, we implemented a CI/CD (Continuous Delivery/Continuous Integration) process which is described in the current section below.

To maintain user submissions, the project maintainer creates a dynamic submission form which represents the user submission data. Types of data submissions from project participants should become easily extensible and configurable. From a data reuse perspective, submission types are most understood as HTML form submission data types, including all the following: an image, plain-text, number, location, time, or date. These data types would be moderated by the project manager and versioned in an open data repository. Dynamic form is expected to be created during project initialization. However, it is possible for the user submission form to be adjusted over time. An example would be an adjustment which omits or adds new types of submission data to the submission form.

Figure 1 shows the architecture represented as a UML diagram through which the user submits new data. The following components are described as acronyms: an open data server is represented as ODS, an open data repository as ODR and an Exporter component.

The ODS component is implemented as a Python HTTP server. The purpose of the server is twofold. It accepts user submitted data represented as JSON over REST API and stores the request content in a relational database. Its second task is to validate user submissions and pass forward the data to the next component in the system. ODS should forward relevant user metadata to the Exporter component.

The Exporter component is a Python module which serves as a data submission filter. Its role is to process the user submissions and apply any transformations necessary to publish user submitted data to the open data repository.

The ODR component represents a remote repository within an organization or under a personal account hosted on GitHub. ODR stores raw data submissions represented as JSON as well as any project configuration files. The ODR contains datasets in machine-readable formats. Moreover, datasets are versioned by utilizing git commits. An organization or a user can manage multiple citizen science projects, which we enumerate as "1...n" in the diagram.

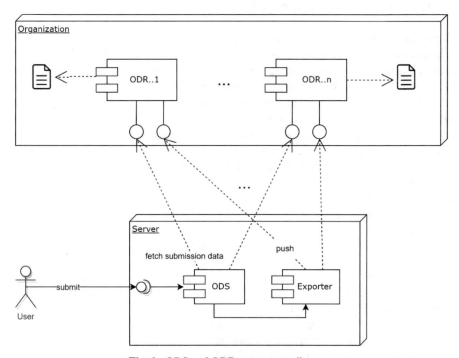

Fig. 1. ODS and ODR component diagram.

We will now describe the user data submission process. As discussed in this section and shown in Fig. 1, the format of the data submissions depends on the structure of the submission form. ODS manages submissions by fetching the latest submission form represented as a configuration file from the remote ODR. It then validates the user submitted data against the expected configuration structure. If the user submission data structure is valid, the submission is passed to the Exporter component for preprocessing.

During preprocessing, the Exporter should omit and redact any user metadata that should not be exported into the ODR. Some examples of sensitive user data are addresses, submission geo-location, or other device data which is not considered relevant to the citizen science dataset. At present, the current system does not store any user metadata. Therefore, the Exporter component is a simple module that publishes user submission data using git. After the preprocessing step, user submitted data is published to remote ODR.

After the data has been submitted to the ODR, it is possible to track history of data submissions by utilizing git commits. Each user submission should correspond to a single git commit. Since ODR tracks repository state and history through commits as snapshots of data, project manager has access to user submitted data at any point in time. The dataset generation and supplementation are an automatic process triggered by the Exporter. The process generates datasets in formats supported by the ODS. Machine readable formats change over time, and formats supported by ODS should be configurable. Currently supported formats are JSON and CSV. To maximize the impact of data interoperability and reuse, further work should be invested in supporting XML, semantic types such as RDF, JSON-LD, and any new machine-readable and open standards.

By default, the dataset within an ODR is considered work in progress. At any point, the project manager can preview the dataset by accessing the latest commit, as well as edit the submission form. However, once a dataset meets an expected number of user submissions, or a certain project milestone is reached, the project manager can label the current snapshot of the dataset as a "release". By release we can refer to GitHub's naming nomenclature, which states that releases are "deployable software iterations you can package and make available for a wider audience to download and use" [12]. Releases are usually highlighted on the platform and are easily downloadable. To release a dataset, we utilized the git tagging system. A project manager can tag the current commit, as well as add a release note, to start the dataset release process. Once the ODR is tagged, we use a CI/CD script to export the datasets to GitHub's Release page along with the custom note. The CI/CD script is shown in the following Listing 1.

```
name: Export datasets

on:
  push:
    tags:
      - "v*.*.*"

jobs:
  build:
    runs-on: ubuntu-latest
    steps:
      - name: Checkout
        uses: actions/checkout@v3
      - name: Display datasets
        run: tail -n +1 -- datasets/*
      - name: Release
        uses: softprops/action-gh-release@v1
        if: startsWith(github.ref, 'refs/tags/')
        with:
          files: |
            datasets/submissions.json
            datasets/submissions.csv
          token: ${{ secrets.CUSTOM_GITHUB_TOKEN }}
```

Listing 1. Dataset release script using GitHub Actions.

At present, released datasets can be downloaded from the GitHub Release page. One drawback of our current approach is that project managers need to be familiar with git tooling. In the future, a user-friendly UI for dataset export should be implemented.

5   Conclusion and Future Work

Extensive efforts have been made to enhance and further improve data access. There are abundant open data initiatives and solutions aimed at disseminating machine readable open access data. For citizen science, which aims to collect data submitted by many different citizens and participants, data submission and data acquisition represents an initial and crucial step in citizen science research. However, as shown in Table 1, data acquisition and data transformations are still on its case-by-case basis, and there is room for data re-use improvement. We proposed that an extendable ODR (open data repository) hosts citizen science user data submissions. Advantages of the proposed architecture are:

- Ability to preview, compare and track user submission history on ODR using git tooling or by user-friendly GitHub UI.
- ODR offers a configurable submission form. Project managers can examine the dataset and modify the user submission form depending on specific project submission requirements and domain.

– Data re-use and interoperability has significantly increased. GitHub has a well-documented and standardized REST API [13]. Dataset retrieval, processing and re-use should be faster and less programmatically complex.

The repository currently offers datasets in JSON and CSV. Our plan in the future is to improve dataset interoperability and provide support for other types of media, such as image and video submissions. Dataset interoperability of the ODR with other systems is improved by introducing custom schemas for user data submissions and metadata. These schemas enable data conversions into standardized open formats, such as XML, RDF, JSON-LD, GEO-JSON, DCAT AP, and others. As for storing images and other types of media on the ODR, we plan to explore the use of git LFS [14].

Moreover, it is also worth exploring the capability of automatic integration of open data repository datasets on mature open-source frameworks, such as CKAN [15], which serve to publish and manage existing collections of datasets.

References

1. Liu, H.-Y., Dörler, D., Heigl, F., Grossberndt, S.: Citizen science platforms. In: Vohland, K., et al. (eds.) The Science of Citizen Science, pp. 439–459. Springer, Cham (2021). https://doi.org/10.1007/978-3-030-58278-4_22
2. European Commission Horizon Europe page. https://research-and-innovation.ec.europa.eu/funding/funding-opportunities/funding-programmes-and-open-calls/horizon-europe_en. Accessed 22 May 2023
3. Williams, J., et al.: Maximising the Impact and Reuse of Citizen Science Data. UCL Press, London (2018)
4. Vohland, K., et al.: The Science of Citizen Science, p. 529. Springer, Cham (2021). https://doi.org/10.1007/978-3-030-58278-4
5. SPOTTERON Homepage. https://www.spotteron.net/. Accessed 22 May 2023
6. Zooniverse Homepage. https://www.zooniverse.org/. Accessed 22 May 2023
7. CitSci Homepage. https://citsci.org/. Accessed 22 May 2023
8. Bürger schaffen Wissen Homepage. https://www.buergerschaffenwissen.de/. Accessed 22 May 2023
9. EU-Citizen.science Homepage. https://eu-citizen.science/. Accessed 22 May 2023
10. JRC EU Science Hub. https://joint-research-centre.ec.europa.eu/index_en. Accessed 22 May 2023
11. Kosmala, M., Wiggins, A., Swanson, A., Simmons, B.: Assessing data quality in citizen science. Front. Ecol. Environ. **14**(10), 551–560 (2016)
12. GitHub About Releases page. https://docs.github.com/en/repositories/releasing-projects-on-github/about-releases. Accessed 22 May 2023
13. GitHub REST API Documentation. https://docs.github.com/en/rest. Accessed 22 May 2023
14. Git LFS Homepage. https://git-lfs.com/. Accessed 22 May 2023
15. CKAN Homepage. https://ckan.org/. Accessed 22 May 2023

Automatic Population of Educational Ontology from Course Materials

Milan Segedinac[1(✉)], Nevena Rokvić[2], Milan Vidaković[1], Radoslav Dutina[3], and Goran Savić[1]

[1] Faculty of Technical Sciences, University of Novi Sad, Novi Sad, Serbia
{milansegedinac,minja,savicg}@uns.ac.rs
[2] Vega IT, Novi Sad, Serbia
[3] Rationale, Novi Sad, Serbia
radoslav@rationaletech.com

Abstract. In this paper, we present a procedure for automatic ontology population with the data from MIT OpenCourseWare course materials. The data was collected for ten Computer Science domains, by scraping the text from PowerPoint presentation and preprocessing it. In the next step, a word2vec neural network was trained for each domain. The obtained words with their embeddings were next used for populating the ontology. The procedure is evaluated by analyzing the clusters of word embeddings and comparing the affiliation of embeddings to clusters with their original class.

Keywords: Ontologies · Ontology Population · Neural Networks · Education · OpenCourseWare

1 Introduction

With the advancements of educational technologies, the repositories of educational materials keep getting bigger, and searching them keeps getting more complicating. The goal of this research is to build and populate an ontology of course materials that would simplify searching and managing the publicly available course repositories.

One such public repository is OpenCourseWare, that contains courses from MIT, and this research focuses on the ontology suitable for that repository. The course materials available in this repository are mostly represented by unstructured text. Therefore, the ontology population relies on the Machine Learning techniques that allow the extraction of structural representations from unstructured text. While the procedure proposed in this research is designed for populating OpenCourseWare ontology, with slight modifications it can be utilized for arbitrary course repository.

OpenCourseware initiative results with a large amount of publicly available course materials. These materials are mostly unstructured making their utilization impractical. As stated in the previous section, an ontology of OpenCourseware course materials would facilitate their management.

M. Trajanovic et al. (Eds.): ICIST 2023, LNNS 872, pp. 326–336, 2024.
https://doi.org/10.1007/978-3-031-50755-7_31

An application built on top of such an ontology might help students to personalize their learning paths, e.g. by suggesting appropriate modules in study programmes. It might also help them to follow the curriculum, by showing them the interconnections among the domains that are being taught typically in an isolated manner.

Educators might benefit from such an application because it would help them to prepare both curriculum and educational materials. The ontology would give them insight in the deep relations that their subjects have with other content taught in the study programme.

While the initial structure of the ontology is created manually, manual population of the ontology would be highly impractical because of the vast and dynamic domain. Therefore, the main research of question that we deal with in this paper is to develop an automatic procedure for populating the OpenCourseware ontology. The procedure is based on Machine Learning techniques.

2 Related Work

Ontology population has been an interesting topic from the first appearance of Semantic Web technologies. According to [1], the most prominent types of ontology population systems are rule-based, machine learning, hybrid, and statistical systems.

Rule-based ontology population systems use a set of predefined rules to extract instances of concepts and relations from unstructured or semi-structured data sources. These rules are created manually by domain experts and are based on the syntactic and semantic patterns that are commonly found in the data sources. The rules are applied to the data sources to identify instances of concepts and relations that match the patterns defined in the rules. Once the instances are identified, they are mapped to the corresponding classes in the ontology.

Rule-based systems are relatively simple to implement and can be effective in extracting instances from data sources that have a limited vocabulary and a well-defined structure. However, they tend to have low recall due to the limited coverage of the extraction rules, and building these rules can be an expensive and time-consuming process.

An example of a rule-based ontology population system is the SOBA system [2], which populates an ontology with information extracted from unstructured text resources and semi-structured resources, such as tables in web pages. SOBA uses an extended version of the SProUT rule-based information extractor and adds to it the rules to extract soccer-specific entities and events from two soccer websites [3]. The system employs manually crafted heuristics to extract precise instances of concepts from the text, without relying on extensive linguistic analysis. The information extractor can extract relation instances of any type, as long as a corresponding set of extraction rules is accessible. In order to address redundancy issues, SOBA utilizes mapping rules that rely on queries sent to the knowledge base, allowing for the reuse of existing entities.

Machine learning-based ontology population systems use statistical models to learn patterns and relationships from data sources. These systems require a large amount of training data to build the models, which are used to classify instances of concepts and relations in the data sources. The training data is manually annotated by domain experts to provide the correct classification of instances. Once the models are built, they are applied to the data sources to identify instances of concepts and relations that match the learned patterns. The identified instances are then mapped to the corresponding classes in the ontology. Machine learning-based systems can achieve high recall and precision rates, but they require a significant amount of training data and can be computationally expensive.

An early example of a machine learning-based ontology population system is the Adaptiva [4], which leverages a blend of natural language processing and machine learning techniques to extract concepts and relations from text sources. The system utilizes predefined templates to extract instances of concepts and relations from the text, subsequently employing these instances to train a statistical model. This model is then employed to classify instances within the text sources, aligning them with their corresponding classes in the ontology. Additionally, the system incorporates a collection of heuristics to address noisy instances and enhance the accuracy of the classification process.

In recent years, most of machine learning ontology population systems rely on deep learning techniques. One example of such a system is OntoEnricher [5].

OntoEnricher is a tool designed to populate a seed ontology with concepts, relations, and instances extracted from unstructured text. The approach encompasses four steps: Dataset Creation, Corpus Creation, Training, and Testing. In the first step, a training dataset is generated by extracting relevant terms from DBpedia that correspond to the ontology's concepts. The second step involves creating a domain-specific training corpus by parsing a Wikipedia dump while employing various filtering techniques. In the third step, OntoEnricher is trained for relation classification between term pairs using the training dataset and corpus. Finally, in the fourth step, the approach is tested by enriching the ontology with information obtained from domain-specific web pages.

Even though this approach can be applied to any domain of interest, the implementation of OntoEnricher focuses on an information security ontology. The dataset is utilized to train a bidirectional LSTM model within the proposed ontology enrichment approach. This trained model is then utilized to enrich the information security ontology by extracting and incorporating concepts, relations, and instances from unstructured text sourced from the internet. The system is tested in a standard manner, with a subset of the training dataset (10%) by removing terms from the ontology and employing unstructured text from web pages. The testing phase yields an average accuracy of 80%, surpassing the performance of current state-of-the-art approaches.

Statistical ontology population systems use similarity measures and fitness functions to calculate the similarity between the extracted instances and instances in the ontology. These systems are typically domain-independent and extract only concept instances.

The system proposed by Yoon [6] serves as an illustrative example of a statistical ontology population system. It operates in a semi-automatic manner, focusing on extracting concept instances exclusively from structured documents. The system employs a similarity measure to determine the appropriate class assignments within the ontology for these instances. To function, the system requires a seed ontology and a set of training documents as inputs. The system extracts concept instances from the training documents and compares their similarity to instances already present in the ontology. Based on the similarity scores, the instances are assigned to their corresponding classes. Additionally, the system incorporates a set of heuristics to handle instances that may contain noise and improve the accuracy of the classification process. The system is designed to be applicable across various domains, and it primarily addresses redundancy issues when mapping extractions into the ontology concepts.

Hybrid ontology population systems leverage the advantages of both rule-based and machine learning-based approaches. These systems employ a blend of predefined rules and statistical models to extract instances of concepts and relations from various data sources. The predefined rules facilitate the extraction of instances that conform to specified syntactic and semantic patterns, while the statistical models learn patterns and relationships directly from the data sources. The extracted instances are subsequently mapped to their corresponding classes within the ontology. Thanks to the fact that they are taking the best from both worlds, hybrid systems excel in achieving both high recall and precision rates, and they possess the capability to handle noisy instances and complex relationships. However, they necessitate a substantial amount of training data and can be computationally demanding.

An illustrative instance of a hybrid ontology population system is the BOEMIE project [7]. It embodies a semi-automatic approach to hybrid ontology population and evolution. The system harnesses a combination of machine learning and rule-based techniques to extract instances of concepts and relations from heterogeneous documents encompassing diverse resources like text, image, video, and audio. Resource analyzers are employed to extract precise concept instances, while machine learning methods facilitate the extraction of mid-level relation instances, encompassing taxonomic and non-taxonomic relations, from multiple types of data. The extracted instances serve as a basis for reasoning, guided by manually constructed inference rules. To ensure the ontology's consistency, the system incorporates techniques such as utilizing a similarity measure between the extracted instances and the existing ontology instances. By employing clustering, the system effectively avoids redundancy and facilitates the identification of instances representing the same real object extracted from different resources. Furthermore, the system promotes the reuse of existing ontology instances, enabling enhanced efficiency in ontology management.

From this overview it can be concluded that ML based techniques alone can achieve high both precision and recall if enough training data is available. Thanks to the fact that there are plenty of publicly available course materials and that they are already annotated by being classified by their educational institutions, we opt for an ML technique for ontology population. If the approach should be applied to a more specific domains with fewer educational materials, the proposed solutions can be modified and built as a part of a more complex hybrid ontology population system.

The system that we propose is applied to educational resources available through OpenCourseWare. OpenCourseWare (OCW) was originally conceived and implemented at the Massachusetts Institute of Technology (MIT) in 2001 as a model for providing Open Educational Resources (OER) [8]. Its primary goal is to grant access to high-quality learning materials and promote the idea of education as a universal right. By utilizing an open license (CC-BY-NC-SA), OCW allows learners worldwide to benefit from these resources for self-directed learning, while educators and curriculum developers can adapt and reuse them to suit their own contexts and communities.

It has to be mentioned that, although OCW emphasizes self-directed learning and self-assessment, it does not facilitate interaction with instructors or other learners. MIT academics have created comprehensive collections of excellent educational resources, which are made freely available to the public through OCW, following a highly structured approach and with extensive media production support.

The success of OCW has generated significant interest in similar Open Educational Practices (OEP), leading to the term OER being coined and attracting numerous university leaders worldwide who seek to replicate this model. To facilitate and promote the adoption of the OCW model, the OCW Consortium was established in 2006, along with regional associate consortiums like the Universia-OCW Consortium in the Iberoamerican region (including Latin American countries, Portugal, and Spain), the Japan OCW Consortium, the Korea OCW Consortium, the Taiwan OpenCourseWare Consortium, and the Turkish OpenCourseWare Consortium. The achievements of OCW have also spurred the development of other OER initiatives, including OpenStax and LibreTexts, which specifically focus on creating and adopting open textbooks.

In this research, we have focused on the educational materials available through MIT OpenCourseWare website [9]. The next step would be to extend the prototypical implementation to support multiple sources of educational resources, and, in that way, facilitate building personal learning paths that involve multiple educational institutions.

The approach proposed in this paper differs from others reviewed in this section in terms of its hybrid nature tailored for the specific domain. It focuses on the population of educational ontologies from the learning materials, and utilizes both machine learning techniques and the declarative processing of the seed ontology in order to achieve both high precision and recall.

3 Methodology

The first step in the population process was building the dataset. The dataset is formed from the educational materials collected from OpenCourseWare, the webpage that contains all the publicly available educational materials from MIT. Even though Open-Courseware consists of educational materials in multiple formats, such as test questions, images, video materials, etc., for this project only PowerPoint presentations were taken. While for this proof of concept only PowerPoint presentations were utilized, the fact that whole process of ontology population relies on the text extracted from the presentations allows for easy extension of the system to different formats. Each other format would just require a converter that transforms it into plain text. For example, if the system is to support video presentations, speech to text converter should be implemented that would extract plain text from the video recordings. It should also be mentioned that all the materials were in English, which greatly facilitated the procedure.

The scraped data were persisted into separate textual files classified by their domain. Standard techniques of preprocessing text were then applied, such as special characters and stop words removal and lemmatization. Thus, obtained data were used for training the neural network in the next step.

For populating the ontology, shallow neural network word2vec [2] was trained on the data. Since the data were already categorized by domain, a separate model was trained for each domain. The resulting models allowed word embeddings characteristic for the domains in which they were trained. The next step was to utilize the models for populating the ontology, and it consisted of two phases. The first phase was manually identifying the basic concepts (such as Topic Content, Study Material, Lecture Note, etc.). These concepts were mapped to the classes in the ontology and in that way the seed ontology was created.

In the second phase, the ontology is being populated by the concepts obtained from word2vec and hierarchical relations among them are being established. The process described above is shown in Fig. 1.

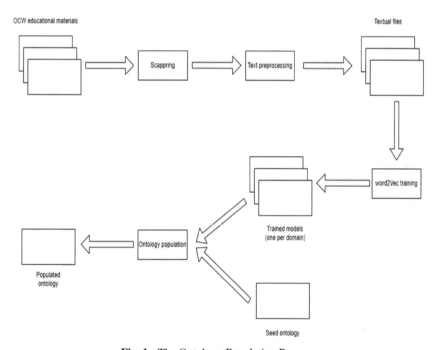

Fig. 1. The Ontology Population Process

The overall structure of the populated ontology is shown in Fig. 2.

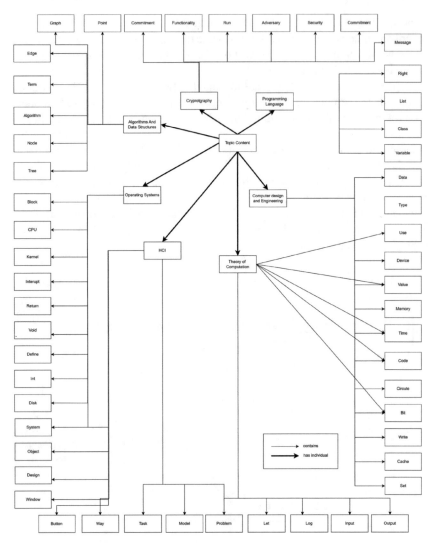

Fig. 2. The overall structure of the ontology

4   Evaluation

The evaluation of the solution's quality consisted of analyzing the clusters of word embeddings and comparing the affiliation of embeddings to clusters with their original classes (areas).

In Fig. 3, the distribution of individual areas and the proportion of clusters within them are depicted. The clustering results indicate that most areas predominantly belong to a single cluster. For instance, the Algorithms and Data Structures area is primarily associated with clusters 0 and 5, while Operating Systems is predominantly linked to cluster 5. This similarity in concepts between the two areas likely accounts for their clustering pattern. In the domain of Computer Design and Engineering, cluster 6 exhibits a significantly stronger affiliation compared to the other six clusters, indicating its clear association with this field. On the other hand, the Operating Systems area demonstrates the highest degree of term dispersion, as it is distributed across multiple clusters with similar percentages. The analysis of the clusters suggests that there are similarities between the word embedding clusters and the original manually assigned labels, implying the potential utility of this solution for constructing ontologies in the computing domain.

The detailed analysis is given in Table 1.

Table 1. The Percentage of the Domain Areas assigned to Clusters.

Cluster	Algorithms and Data Structures	Software Engineering	Cryptography	HCI	Operating Systems	Programming Languages	Software Design	Computation-al Theory
0 (%)	60,7	1,1	3,7	6,9	9,2	1,5	2,3	0,9
1 (%)	1,3	3,2	68,5	57,2	35,9	2,4	4,9	1,6
2 (%)	1,5	2,8	0	8,7	0,7	1,3	3,8	26,7
3 (%)	1,7	2,6	1,9	20,8	8,5	88,5	3,2	7,4
4 (%)	1,3	1,2	7,4	1,7	17,6	1,3	75,1	2,8
5 (%)	31,6	1,8	1,9	2,9	24,8	0,9	2,7	1,7
6 (%)	1	84,9	16,7	1,7	0,7	2,2	3,5	1,3
7 (%)	0,9	2,4	0	0	2,6	2	4,5	57,7

Table 1 shows the domain areas and the percentage of each area belonging to individual clusters. The highest percentages of belonging to clusters are marked in red. Some of the areas have a clear majority belonging to one cluster, and as mentioned earlier, some have smaller differences (more scattered affiliations). Examples of such fields are Computer Design and Engineering (84.9% belonging to cluster 6), Programming Languages (88.5% belonging to cluster 3) and Software Design and Engineering (75.1% belonging to cluster 4).

It is also interesting that none of the areas belong to the cluster 5, and that the highest percentage of belonging to this cluster is in the areas of Algorithms and data structures (31.6%) and Operating systems (24.8%), and that is a rather high percentage. This fact can be explained by the fact that a large number of words coincide within those two areas, since they are two essential areas in the field of computer science that are usually found in the lower years of study.

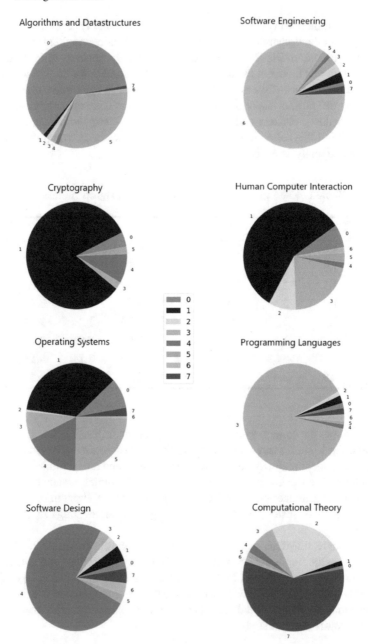

Fig. 3. The share of the clusters by domain areas

5 Conclusion

In this paper we propose a Machine Learning based population of an ontology of educational materials. The proposed research was conducted on the OpenCourseWare PowerPoint presentations made publicly available by MIT.

While developing this solution, we faced several challenges. The most important of them relates to the data itself. It is not possible to fully automatize the process of collecting and preprocessing the data since some of the steps required the intervention of the domain experts that had to decide whether the data is relevant for the ontology. The format of educational materials, i.e., are they textbooks, presentations, or lecture notes, greatly influences the possibility of automatic identification of the patterns in them and the overall quality of the ontology. Since this research utilizes only PowerPoint presentations, which typically contain only shortened explanations of lectures, the quality varies among the domains. The parts of the ontology for the domains with more elaborated presentations turn out to be superior in quality. Currently, we met this problem by involving domain experts in the process of selection of relevant educational materials.

Further research should follow several tracks. One of them would be populating the parts of ontology with metadata about the students and study programmes (number of students enrolled in a study programme, planned number of working hours, data about the exams, etc.). In that way a more precise picture of the educational process would be given. Most of this data can be obtained from the accreditation documentation.

Another track of further research will be the improvement of the existing part of the ontology by covering other formats of educational materials (e.g., video materials, lecture notes, exam examples, etc.). This will deepen the versatility of the proposed solution and allow the semantic enrichment of the existing ontology.

References

1. Lubani, M., Shahrul, N., Rohana, M.: Ontology population: approaches and design aspects. J. Inf. Sci. **45**(4), 502–515 (2019)
2. Buitelaar, P., Cimiano, P., Racioppa, S., Siegel, M.: Ontology-based information extraction with SOBA. In: Proceedings of the International Conference on Language Resources and Evaluation (LREC) (2006)
3. Piskorski, J., Schäfer, U., Feiyu, X.: Shallow processing with unification and typed feature structures–foundations and applications. Knstliche Intelligenz **1**(1), 17–23 (2004)
4. Brewster, C., Ciravegna, F., Wilks, Y.: User-centred ontology learning for knowledge management. In: Andersson, B., Bergholtz, M., Johannesson, P. (eds.) Natural Language Processing and Information Systems. NLDB 2002, Stockholm, Sweden, 27–28 June 2002, Revised Papers 6. LNCS, vol. 2553, pp. 203–207. Springer, Heidelberg (2002). https://doi.org/10.1007/3-540-36271-1_18
5. Sanagavarapu, L.M., Iyer, V., Reddy, Y.R.: OntoEnricher: A Deep Learning Approach for Ontology Enrichment from Unstructured Text. Cybersecurity and High-Performance Computing Environments, pp. 261–284. Chapman and Hall/CRC (2022)
6. Yoon, H.G., Han, Y.J., Park, S.B., Park, S.Y.: Ontology population from unstructured and semi-structured texts. In: Sixth International Conference on Advanced Language Processing and Web Information Technology (ALPIT 2007). IEEE (2007)

7. Castano, S., et al.: Multimedia interpretation for dynamic ontology evolution. J. Log. Comput. **19**(5), 859–897 (2009)
8. Marín, V.I., Villar-Onrubia, D.: Online Infrastructures for Open Educational Resources. Handbook of Open, Distance and Digital Education, pp. 1–20. Springer, Singapore (2022). https://doi.org/10.1007/978-981-19-0351-9_18-1
9. MIT. MIT OpenCourseWare. https://ocw.mit.edu/
10. GitHub - dav/word2vec: An efficient implementation of the continuous bag-of-words and skip-gram architectures for computing vector representations of words. GitHub. https://github.com/dav/word2vec

Visual Question Answering – VizWiz Challenge

Tamara Ranković[(✉)] [ID], Eva Janković[ID], and Jelena Slivka[ID]

Faculty of Technical Sciences, Novi Sad, Serbia
tamara.rankovic@uns.ac.rs

Abstract. The purpose of a Visual Question Answering (VQA) system is to answer questions related to a given picture. VQA systems have various applications, such as automatically generating picture descriptions and conducting searches using pictures as queries. This paper presents a VQA system designed specifically to provide automated assistance to blind individuals. The research utilizes the VizWiz dataset, the largest VQA dataset consisting of pictures taken by blind or visually impaired people. Pictures are processed by a Faster R-CNN model, which extracts regions of interest relevant to the posed questions. The output is a feature vector for each region. The questions are fed into a BERT model, which generates a single feature vector representing the entire question. The outputs from both models are combined using an attention layer. The output component of the system is a multi-label classifier that produces a list of probable answers. The system underperforms in comparison to the state-of-the-art. This can be attributed to the lack of fine-tuning for each individual component of the system, as well as shorter training times due to hardware limitations. Through error analysis, it was determined that the system tends to make more mistakes with blurry photos and images containing text. This represents potential directions for further research.

Keywords: Visual Question Answering · Computer Vision · Natural Language Processing · Deep Neural Networks

1 Introduction

The continuous advancements in the fields of computer vision and natural language processing (NLP) have opened up possibilities for tackling complex challenges. One such challenge is Visual Question Answering (VQA), which involves generating automated answers to questions in the context of a given image. To address this problem, a VQA system needs to effectively recognize the image content, comprehend the question expressed in natural language, and generate a meaningful answer based on both. VQA finds numerous applications in various domains, including:

- Human-computer interaction: VQA enables individuals to engage in more interactive communication by facilitating question-and-answer exchanges.
- Image corpus search: Users can leverage VQA to request images that fulfill specific queries, enhancing the efficiency and precision of image search tasks.

M. Trajanovic et al. (Eds.): ICIST 2023, LNNS 872, pp. 337–348, 2024.
https://doi.org/10.1007/978-3-031-50755-7_32

- Assistance to visually impaired and blind individuals in their daily activities: VQA systems can greatly support the visually impaired and blind population by providing them with valuable information and aiding them in their day-to-day tasks.

This paper introduces a VQA system specifically designed to provide assistance to individuals with visual impairments or blindness. The system is capable of automatically generating text-based answers to questions asked about a photo. To achieve this, the system leverages computer vision and natural language processing models that have been trained using the VizWiz dataset [1]. VizWiz is a specialized dataset consisting of photos and corresponding questions that have been generated through the daily use of an application designed to aid individuals with visual impairments. This dataset serves as a valuable resource as it closely reflects the real-world usage scenarios of the system described in this paper and stands as the largest publicly available dataset of its kind.

The system architecture is inspired by the previous state-of-the-art architectures described in [2]. In order to further improve accuracy, a BERT transformer was used to extract the feature vector of the question.

According to the VQA metric, the proposed architecture achieved a result of 35.63. Although this result falls below the state-of-the-art for the VizWiz dataset, it surpasses the baseline [3]. The difference in results can be attributed to the utilization of pre-trained models without further fine-tuning, which was due to hardware limitations. Furthermore, the preprocessing stage involved simplifying the response set, which could potentially impact the system's overall performance.

The advantages of the proposed solution are reflected in the application of the architecture on a real set of data for the target problem, unlike previous solutions. By overcoming hardware limitations and fine-tuning the components, the results could be further improved.

2   Related Work

In this section, we discuss proposed model architectures for solving the VQA and similar problems relevant to our work.

There are two predominant architecture types for solving the VQA problem [2]:

- Applying a CNN on the image and an RNN on the question to extract feature vectors of inputs; concatenating the two vectors and forwarding the result vector as the input into a classifier or an RNN that generates answers,
- Detecting interesting regions in the image, usually objects, and extracting feature vectors for each region using a CNN, while question's feature vector is generated using an RNN; using the attention layer to determine the weight of each region in the context of the question; forwarding result vectors as the input of a classifier or an RNN that generates answers

Majority of papers reviewed in this section use standard VQA datasets that consist of high-quality images, which don't necessarily reflect real-world use cases, for example helping blind and visually impaired, as is the case with our work. Our objective is to determine how effective are frequently chosen architectures when applied to a dataset containing realistic photographs.

Authors of [4, 5] propose models for solving the VQA problem that use a CNN to extract image feature vectors and a long short-term memory (LSTM) network for generating the feature vector of the question. Both models include a question-guided attention layer that determines the relevance of image segments for the question. Improvements the attention layer introduces are visible in both papers. Another useful component is a network for image region proposal, described in paper [6]. After regions are detected, feature vectors are generated for all of them.

A model similar to the previous ones, but trained on a VizWiz dataset, is presented in paper [7]. It generates question feature vectors using a Gated Recurrent Unit (GRU) network and image feature vectors for each image region with the help of the ResNet CNN. It also includes a question-guided attention layer and has a linear classifier as the end segment. The proposed architecture serves as the basis for our own model's architecture.

Transformer models currently perform highly when it comes to NLP tasks. BERT model is presented in paper [8], its architecture, training process and possible applications are discussed. Authors demonstrate how pretrained BERT can be fine-tuned and used for feature extraction of the input text and how that result can serve as input for other tasks. Aside from this approach, feature extraction can be done without fine-tuning. The benefit of the second approach is that this step can be performed in the preprocessing phase which leads to more efficient training process compared to LSTM, for example, which would have to be trained with the rest of the model.

Paper [9] discusses available VQA datasets, algorithms and main obstacles. Additionally, authors compared evaluation metrics for this problem. The most frequently used ones are accuracy, WUPS [10], a consensus-based metric, and manual evaluation. It was concluded that there is no universal metric used for VQA and that the choice is highly dependent on the dataset. Also, the results can differ because classification and text generation can provide differently structured answers. Based on guidelines provided in this paper, we opted for accuracy and VQA accuracy, a consensus-based metric.

3 Dataset

The dataset we used for solving the VQA problem was made publicly available as part of the VizWiz challenge [1, 11]. The challenge comprises a wide range of computer vision tasks. They include image captioning, classification, quality issues, object localization, reasons for answer differences, as well as three VQA-related tasks, each assigned a different dataset:

- VQA - Providing a natural language answer to a question about an image, which is the focus of our work
- Answer grounding for VQA - Providing the region in the image used to generate the answer
- Vision skills for VQA - Identifying the vision skills needed to answer the question (object recognition, text recognition, counting etc.)

Q: What type of chips are this? A: blue corn tortilla chips

Q: Can you tell me what is in this can please? A: coca cola

Q: What color is this? A: red

Q: Is my - is my light on? A: yes

Q: How many pink candles are in the picture? A: 2

Q: What is this? A: unanswerable

Fig. 1. Samples of images and questions from the VizWiz dataset

3.1 Structure

The dataset consists of photographs taken by the blind individuals and transcriptions of questions from audio recordings. The advantage of data collected this way is that questions and images, regarding their type and quality, resemble queries that real-world VQA services for helping blind and visually impaired people may encounter. Such queries frequently include low-quality photos, be it because they are blurry, not focused, taken in low light or something else. Each image-question tuple is paired with ten answers, gathered via crowdsourcing. Every answer has an *answer_confidence* property, with possible values *yes* and *no*, signaling whether the annotator was certain about their choice. Additionally, there was an option to label the question as *unanswerable*. If that was the most popular answer, the entire image-question pair was marked as unanswerable, using the property *Answerable - yes/no*.

Figure 1 emphasizes diversity of questions, ranging from the ones about object type, color and count to the ones requiring the VQA system to be able to recognize text or combine multiple mentioned capabilities. Authors of the dataset decided to classify possible answer types into four categories: *Yes/No, Number, Other* and *Unanswerable*. This information is stored in the *answer_type* property.

The dataset is divided into training, validation and test sets. Each set consists of a JSON file containing annotations and a set of related images. Dataset sizes are as follows:

- Training - 20523 image-question pairs and 205230 total answers
- Validation - 4319 image-question pairs and 43190 total answers
- Test - 8000 image-question pairs and no answers

Images and annotations can be downloaded separately, as images take up considerable amount of disk space. JSON files contain lists of objects where each of them has the reference to an image, which is its title, question, list of annotator-provided answers, *answer_type* and *unanswerable* properties. As the test set had no answers, we excluded it from further stages of our research.

3.2 Preprocessing

By analyzing the data, we noticed a high percentage of samples requiting the application of optical character recognition (OCR) techniques. Some questions falling into this category are:

- "What is the captcha on this screenshot?"
- "Hi, what can you see in this page that I have scanned with my phone camera?"
- "What book is this?"

As OCR is out of scope of this work, it was necessary to identify all image-question pairs that must be left out of the dataset. This filtering would require separate annotations and model to be trained. For that reason, we decided to use the Vision skills for VQA dataset mentioned earlier, which already contained vision skills required for answering the questions. The main advantage of this dataset is that it contains overlapping image-question pairs to the ones in our dataset. It also has a similar structure. Training, validation and test sets contain 14259, 2248 and 5719 pairs, respectively. Vision skills can be: *Color Recognition, Object Recognition* and *Counting*. For every pair, each skill is assigned a value from zero to five. Annotations are publicly available only for the training set [2].

Firstly, we aggregated training and validation subsets of our dataset. After that, we found its intersection with the annotated Visual Skills for VQA training set and discarded all pairs with *Text Recognition* value higher than 1. In the next step, we divided the resulting set into training and test sets, with ratio 90:10. In order to have a balanced number of representatives from all answer types in both sets, ratios of pairs were similar for each answer type. The number of pairs for each type in training and tests sets is shown in Fig. 2. After all steps, the training set contained 5614 pairs and the test set contained 624.

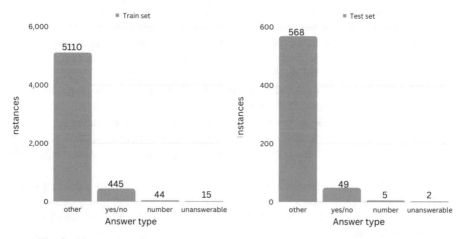

Fig. 2. The number of samples by answer type in training and test sets respectively

4 Methodology

In this section, we present the architecture of our model, shown in Fig. 3, and discuss the training process. In order to provide a correct answer, a VQA model has to be able to understand the question and image contents. Because the question is often related to only particular segments of the image, and not the entire image, it is useful to discover regions of interest (ROI) [6]. After ROI and question feature vectors are created, we use a question-guided attention layer to determine relevance of each ROI to the posed question, by assigning them weights. What we end up with is a set of weighted ROI feature vectors that we combine into a single vector representing the whole image. Derived image vector and question feature vector are merged into one vector, which carries information about both inputs and relationships between them. This is then an input to a multi-label classifier. Classifier output size is equal to the answer dictionary size, generated from answers in the training set. We opted for the multi-label classifier because:

- Annotators sometimes provided synonymous answers (e.g. TV and television)
- There were multiple correct answers (e.g. the object is both red and blue and some answers only say red or only blue or both)

4.1 Image Processing

The first step of image processing was finding ROIs, or more precisely objects. For that task, we used a Faster R-CNN network for region proposition [13] that uses ResNet CNN [14] with 50 layers as a backbone. The model was pretrained on the Microsoft COCO dataset [15]. Output of the model is an array of region coordinates, each paired with a value ranging from 0 to 1 that specifies the certainty of the prediction.

Out of all regions, we selected only the ones potentially useful for answering the question. A higher number of regions means a higher probability the question is regarding one of them, but low photo quality caused many regions to carry little relevant information. For that reason, we chose only the predictions with certainty score higher than 0.5. This threshold was determined empirically by observing the extracted regions. After this step, the number of regions varied from 0 to 100. As only 0.89% of images had more than 15 regions, we decided to take up to 15 regions for each image, sorted by their certainty score, and discard the others.

We encountered a problem of many images, precisely 28% in the training set, without any ROIs detected. As those images had no input for the feature extraction phase, we assigned them one region that represents the entire image. In the end, we had 2.52 ROIs per image.

Feature extraction for each ROI was done with the ResNet network with 18 layers trained on the ImageNet dataset [16]. Before sending it to the network, regions were cut and scaled to 224×224 size. The output of the encoder of size 512 was selected to be the ROI's feature vector.

As models mentioned used for image processing were not fine-tuned, we performed all steps in the preprocessing phase to make our model's training more efficient. We argue that fine-tuning of models for ROI proposition and feature extraction would not bring benefits as image contents are very diverse and can't be associated with any particular domain.

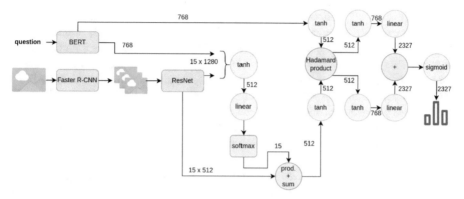

Fig. 3. Proposed model architecture

4.2 Question Processing

Feature vectors of questions were extracted using the BERT transformer model with 12 encoder layers, trained on the BooksCorpus dataset [17] and English Wikipedia [18]. BERT was chosen as the model because of its previous success in solving Question-Answering problems [19, 20]. Additionally, authors of the paper [21] highlight BERT's strengths in solving VQA problems over LSTM.

WordPiece tokenization [22] is performed on sentences before they are sent to BERT. It separates text into words or segments of words that are present in the dictionary. The first token is always a special [CLS] token, followed by original text tokens. Questions in our case are always one sentence, hence the [SEP] separator token is not used. Maximum number of tokens is 512. As all questions are much shorter than this, there was no need for their additional transformation. The last hidden state value corresponding to the [CLS] token was selected as the question feature vector, as it aggregates semantics of the entire sentence [8]. Its length is 768. This step was also performed in the preprocessing step, without fine-tuning, for the same reasons discussed in Sect. 4.1.

4.3 Answer Processing

As each image-question pair can have up to ten different answers, the size of the dictionary can become too large for us to produce any valuable results, given the limited size of the training set. To mitigate the problem, we applied multiple strategies. Because the answers were given in free form, there are cases where people provided similar, but still different answers. In order to shrink the dictionary size and discard answers that don't have distinct semantics, we performed the following preprocessing steps:

- Transforming Unicode characters into ASCII
- Removing redundant white spaces
- Removing special characters
- Transforming upper-case to lower-case letters
- Representing numerical values in answers with digits

Additionally, we noticed semantically similar answers that differ in the level of detail. For example, the question "What is this?" had answers: "coin", "ancient coin", "gold coin" and "prayer coin". We wanted to keep the most succinct version of the answer so we employed part-of-speech (POS) tagging mechanism in order to remove unnecessary adjectives and adverbs. Nevertheless, we abandoned this procedure in the end as it provided us with only minor improvements.

Another principle we adopted is to take only the answers the annotators were certain about. Also, the answer had to be provided by at least two annotators for it to be accepted. Questions left without answers after applying these criteria were labeled as unanswerable.

The dictionary of 2327 different one-hot encoded answers was created, based on the training set. We applied the same technique for filtering out answers from the test set as we did for the training set. After that, we used the dictionary to encode the remaining answers. 68 of them had no corresponding element in the dictionary and were represented with zero vectors.

4.4 Question-Guided Attention Layer

Each RIO's relevance to the question is quantified with the help of this layer, by assigning ROI a weight. The input to the layer is the image feature vector of size 15×512 and the question feature vector of size 1×768. Each row of the image feature vector should correspond to one ROI, or more precisely, it is ROI's feature vector. As there are cases when the number of regions is less than 15, rows that don't correspond to any ROI are zero vectors. This doesn't affect the end result as zero vector multiplied by any weight is still a zero vector.

Inputs of the first layer are created by concatenating image feature vector row and question feature vector. Exceptions are earlier mentioned zero vectors that are concatenated with a zero vector of size 1×768. As those rows aren't really regions, we want to neutralize their effect.

Rows of the newly formed vector are one by one sent to a nonlinear layer of size 512 with *tanh* activation function. The output of each row is fed into the linear layer which produces a scalar value. This architecture is inspired by the paper [7]. After all rows have gone through the both layers, we get a vector of size 1×15. We apply a *softmax* function to it, so that we get a relevance share of each ROI for the question. The result we get is the weight vector. Image feature vector is then combined with the weight vector by multiplication of each row with its corresponding scalar value from the weight vector. After that, image feature vector rows are summed to produce a final 1×512 image feature vector.

4.5 Classifier

The question feature vector and weighted image vector, which is the output of the attention layer, go through nonlinear layers with *tanh* activation function which produce outputs of length 512. Hadamard product is performed on the two outputs, as in the paper [7].

After that, the network forms two branches, both having a nonlinear layer followed by a linear layer, with dimensions displayed in Fig. 3. The output from both branches is the size of the dictionary, 2327.

After the addition of outputs from the branches, the result is forwarded to the sigmoid layer. We chose the sigmoid activation function because we are performing multi-label classification.

The final layer also uses a dropout mechanism [23], with dropout probability of 0.67. Motivation for hyperparameter's value stems from the fact that we have three answer types, excluding the *unanswerable* type that should produce zero output. We argue that parts of the network might specialize for certain answer types during the process of training.

4.6 Training Process

The network was trained with AdaDelta algorithm because it requires minimal or no manual setting of the learning rate over time. The initial learning rate was set to 1, as in the paper where the algorithm was originally published [24]. The loss function was binary cross-entropy, while the batch size was 30.

To avoid overfitting, we created a validation set by selecting 10% of the data from our training set. We used it to halt the training process if categorical accuracy didn't increase for more than 50 epochs. We saved only the model's state when categorical accuracy over the validation set was the highest.

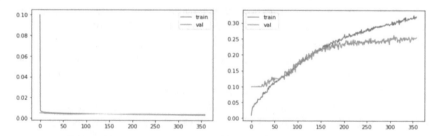

Fig. 4. Loss and accuracy by epochs during training

5 Results

VizWiz challenge imposes the VQA metric, originating from the VQA challenge [25], for testing performance of models. We used the same metric for our model. It is defined in the following way:

$$accuracy = \min\left(\frac{humans\ that\ provided\ that\ answer}{3}, 1\right)$$

Here we encountered a mismatch between our multi-label approach and the challenge requiring only one predicted answer. To be able to compare our results to the ones of other contestants, we chose only one predicted answer, the one having the highest score.

Results we managed to achieve are shown in Table 1. They are divided by the answer type and the total result on the entire test set is displayed too. The number of questions by answer type is shown in Fig. 2. The model performed the best for questions with *Yes/No* answers, with VQA score of 63.95, while score is the lowest for numerical answers, 6.67. Unanswerable questions had a VQA score 50, while questions falling into the category *Other* had 33.39. VQA score for the entire dataset is 35.63.

Table 1. VQA metric results by answer type

	Answer types				Total
	Yes/No	Number	Other	Unanswerable	
VQA metric	63.95	6.67	33.39	50	35.63

6 Discussion

Model's performance varies depending on the answer type, which can be explained by the fact that there are different requirements for different answer types. The model needs to have a more or less complete understanding of questions and images. Questions with *Yes/No* answers are usually less demanding and require less knowledge of details in the image. On the other hand, questions with *Number* answer type may require many different visual skills of the model, from object detection and classification to text recognition. We argue this is the reason we observed the lowest score in this category.

Another obstacle we encountered is the limited size of training and test sets, especially for more complex questions. For example, we only had 5 *Number* questions in the test set and also not enough of them in the training set for the model to be trained sufficiently. High amplitude in the question complexity can lead to model overfitting for certain types of questions and underfitting for others. However, Fig. 4, showing loss and categorical accuracy changes by epochs in the training process, tells us this wasn't the case. At the end of the training, the loss was still decreasing on the validation set, while the accuracy was increasing. Therefore, a bigger training set would most likely aid in better learning of answers to more complex questions.

Low quality of photographs generally affected predictions for all answer types. The main problems were poor lighting, blurriness and very low contrast, all shown in Fig. 5. The indicator of poor quality is also a very high percent, precisely 28%, of images in which the model for region proposal found no candidates. It is infeasible to generate meaningful feature vectors from such images. Image quality improvements in the pre-processing phase could bring benefits as they would make objects and their properties in the images clearer, thus easier to detect.

Hardware constraints also played a role during decision making process regarding model design. One example is choosing ResNet with 18 layers instead of ones with 50 or 152 layers, which produce longer and semantically richer feature vectors. Moreover, the total number of trainable weights was low compared to the state-of-the-art models. Such seemingly minor decisions could have a massive impact on model's ability to answer questions requiring a deep understanding of image contents.

Fig. 5. Examples of poor-quality images from the training set

7 Conclusion

In this paper, we presented a system architecture to address the VQA problem. Our main objective was to determine to what extent state-of-the-art architectures are applicable to datasets with low-quality images. The solution relies on transformer, CNN and Faster R-CNN models and attention layers. Results varied significantly, depending on the image quality and required vision skills.

With further improvements in results, our model can be integrated into a larger system, where it would receive input questions through Speech-To-Text components and provide answers to users through Text-To-Speech components. The integration would provide a comprehensive solution for blind and visually impaired individuals.

References

1. Gurari, D., et al.: Vizwiz grand challenge: answering visual questions from blind people. In: Proceedings of the IEEE Conference on Computer Vision and Pattern Recognition (2018)
2. Wu, Q., et al.: Visual question answering: a survey of methods and datasets. Comput. Vis. Image Underst. **163**, 21–40 (2017)
3. https://eval.ai/web/challenges/challenge-page/743/leaderboard/2020. Accessed 30 May 2023
4. Fukui, A., et al.: Multimodal compact bilinear pooling for visual question answering and visual grounding. arXiv preprint arXiv:1606.01847 (2016)
5. Xu, H., Saenko, K.: Ask, attend and answer: Exploring question-guided spatial attention for visual question answering. In: Leibe, B., Matas, J., Sebe, N., Welling, M. (eds.) ECCV 2016. LNCS, vol. 9911, pp. 451–466. Springer, Cham (2016). https://doi.org/10.1007/978-3-319-46478-7_28
6. Shih, K.J., Singh, S., Hoiem, D.: Where to look: focus regions for visual question answering. In: Proceedings of the IEEE Conference on Computer Vision and Pattern Recognition (2016)
7. Teney, D., et al.: Tips and tricks for visual question answering: learnings from the 2017 challenge. In: Proceedings of the IEEE Conference on Computer Vision and Pattern Recognition (2018)
8. Devlin, J., et al.: BERT: pre-training of deep bidirectional transformers for language understanding. arXiv preprint arXiv:1810.04805 (2018)

9. Kafle, K., Kanan, C.: Visual question answering: datasets, algorithms, and future challenges. Comput. Vis. Image Underst. **163**, 3–20 (2017)
10. Wu, Z., Palmer, M.: Verbs semantics and lexical selection. In: Proceedings of the 32nd Annual Meeting of Association for Computational Linguistics, Las Cruces, New Mexico (1994)
11. https://vizwiz.org/. Accessed 30 May 2023
12. https://github.com/chiutaiyin/Vision-Skills/blob/master/csv/vizwiz_skill_typ_train.csv. Accessed 30 May 2023
13. Ren, S., et al.: Faster R-CNN: towards real-time object detection with region proposal networks. In: Advances in Neural Information Processing Systems, vol. 28 (2015)
14. He, K., et al.: Deep residual learning for image recognition. In: Proceedings of the IEEE Conference on Computer Vision and Pattern Recognition (2016)
15. Lin, T.-Y., et al.: Microsoft COCO: common objects in context. In: Fleet, D., Pajdla, T., Schiele, B., Tuytelaars, T. (eds.) ECCV 2014. LNCS, vol. 8693, pp. 740–755. Springer, Cham (2014). https://doi.org/10.1007/978-3-319-10602-1_48
16. Deng, J., et al.: ImageNet: a large-scale hierarchical image database. In: 2009 IEEE Conference on Computer Vision and Pattern Recognition. IEEE (2009)
17. Zhu, Y., et al.: Aligning books and movies: towards story-like visual explanations by watching movies and reading books. In: Proceedings of the IEEE International Conference on Computer Vision (2015)
18. https://www.wikipedia.org/. Accessed 30 May 2023
19. Wang, Z., et al.: Multi-passage BERT: a globally normalized BERT model for open-domain question answering. arXiv preprint arXiv:1908.08167 (2019)
20. Yang, W., et al.: End-to-end open-domain question answering with BERTserini. arXiv preprint arXiv:1902.01718 (2019)
21. Yang, Z., et al.: BERT representations for video question answering. In: Proceedings of the IEEE/CVF Winter Conference on Applications of Computer Vision (2020)
22. Wu, Y., et al.: Google's neural machine translation system: bridging the gap between human and machine translation. arXiv preprint arXiv:1609.08144 (2016)
23. Srivastava, N., et al.: Dropout: a simple way to prevent neural networks from overfitting. J. Mach. Learn. Res. **15**(1), 1929–1958 (2014)
24. Zeiler, M.D.: AdaDelta: an adaptive learning rate method. arXiv preprint arXiv:1212.5701 (2012)
25. https://visualqa.org/challenge.html. Accessed 30 May 2023

Digitalization in Energy Sector

Hybrid Approach in Thermal Demand Forecasting of a Building

Anđela Marković[1,2(✉)] ⓘ, Marko Batić[1] ⓘ, and Katarina Stanković[1,2] ⓘ

[1] School of Electrical Engineering, University of Belgrade, 11120 Belgrade, Serbia
{marko.batic,katarina.stankovic}@pupin.rs
[2] Institute Mihajlo Pupin, University of Belgrade, 11060 Belgrade, Serbia
andjela.markovic@pupin.rs

Abstract. Energy consumption plays a significant role in contributing to the greenhouse effect, environmental degradation, and pollution. Developing energy-efficient buildings presents a promising approach to reduce dependency on non-renewable energy sources. However, transitioning to renewable energy sources poses unique challenges, as their utilization is highly influenced by weather conditions. Consequently, the implementation of demand forecasters becomes crucial in facilitating the integration of renewable energy sources. This paper aims to address the gap in research by investigating the modelling of demand forecasters in the absence of an implemented control strategy. The study focuses on a specific building located in Italy, providing a comprehensive analysis of the effectiveness and performance of the developed demand forecasting model. Through rigorous examination and analysis, the findings of this study demonstrate that the modelled demand forecaster achieved considerable results in accurately predicting energy demand. The implications of these findings suggest that the utilization of demand forecasters, even in the absence of an implemented control strategy, can contribute significantly to enhancing energy efficiency and promoting the integration of renewable energy sources. This research serves as a foundation for future studies aiming to optimize energy management strategies and advance sustainable practices in the built environment.

Keywords: HVAC · thermal model · grey box · ML · demand forecaster · negative feedback

1 Introduction

The excessive and often irrational utilization of traditional non-renewable energy sources, such as coal, natural gas, oil, and nuclear energy, has been a prevalent practice throughout the past century. This overdependence on finite resources presents a substantial challenge for humanity, as these sources cannot be replenished once depleted. Consequently, the world finds itself confronted with an urgent need to explore alternative energy solutions that are sustainable in the long term. Renewable energy sources, such as wind and solar energy, hold significant promise in this regard, offering an abundant and virtually unlimited supply of clean energy.

© The Author(s), under exclusive license to Springer Nature Switzerland AG 2024
M. Trajanovic et al. (Eds.): ICIST 2023, LNNS 872, pp. 351–361, 2024.
https://doi.org/10.1007/978-3-031-50755-7_33

However, the transition from non-renewable to renewable energy sources encounters several obstacles. One of the key challenges lies in accurately forecasting energy demand, particularly due to the high dependency of energy production from alternative sources on prevailing weather conditions. In this context, heating, ventilation, and air conditioning (HVAC) systems play a pivotal role, as statistics indicate that they contribute significantly to overall building energy consumption. Consequently, extensive research has been dedicated to the development of predictive thermal models for houses, which serve as valuable tools for studying energy efficiency.

Within the existing literature, a research project [1] has been conducted to explore the application of data-driven approaches utilizing deep neural networks for HVAC system parameter estimation. It is widely recognized that deep-learning algorithms exhibit their full potential when applied to large datasets, as their performance is contingent upon the availability of a substantial amount of data. However, in scenarios where the dataset is limited, alternative approaches are necessary to overcome this challenge. One viable solution involves leveraging machine learning (ML) algorithms, such as Random Forest Regressor, Decision Tree Regressor, and AdaBoost Regressor, to address the issue of small dataset sizes.

The focus of this research paper is to present a comprehensive study on modelling a predictive thermal model for energy-efficient buildings. By investigating and analysing the thermal behaviour of houses, this research aims to contribute to the development of sustainable energy practices and inform decision-making processes in the field. Emphasizing the importance of energy-efficient buildings, this study delves into the multitude of advantages they offer, including the reduction of greenhouse gas emissions, mitigating environmental degradation, minimizing the consumption of natural resources, reducing reliance on external energy sources, and mitigating environmental damage and pollution.

By addressing these critical aspects, this research paper seeks to foster a deeper understanding of the challenges and opportunities associated with sustainable energy transition. Ultimately, it aims to provide valuable insights and guidance for policymakers, researchers, and industry professionals in their endeavours to create a more sustainable and energy-efficient future.

2   Methods

2.1   House Building Structure

The pilot building serves as the proposed novel solution. It functions as an experimental facility rather than a conventional structure for daily requirements and features a predominantly south-facing main facade. Situated on a raised embankment approximately 1.6 m above ground level, the building exhibits a shape that closely resembles a cube. Its roof is flat and constructed using a combination of materials, including a 3 cm layer of wood, a waterproof membrane, and an 8 cm finishing layer of gravel. Internally, the pilot building comprises three rooms, two of which are relatively narrow, measuring 60 cm in width. These rooms serve as guardian spaces, specifically designed to mitigate solar gain originating from the east and west directions. The walls of the building consist of sandwich panels composed of a 10 cm layer of foam insulation and a metal coating. As no insulation has been incorporated, the roof may experience a maximum heat flux of approximately 30 $\frac{W}{m^2}$ during the summer season. The flooring follows a similar construction approach using sandwich panels and is placed atop a 20 cm thick concrete slab. Table 1 presents the walls' structure and corresponding layer thicknesses of the building.

The dimensions and layout of the flooring permit the potential installation of a heating/cooling system beneath the floor surface. Owing to the implementation of varying types of insulation materials, a dwelling will be conceptualized as a structure comprising a roof component and four external walls.

Table 1. Wall Structure and layers thickness

Wall Structure and layers thickness [mm]	
External Wall Layers	Sheet Metal (1) +. Polyurethane (118) + Sheet Metal (1)
Ceiling Layers	Expanded Polystyrene (60) + Fir Wood (30) + Bitumen (5) + Gravel (30)

2.2 Heating System

The architectural configuration of the prototype incorporates a comprehensive heat pump system, featuring a reversible heat pump capable of harnessing diverse thermal sources and sinks such as solar energy derived from CPC-PVT panels and PVT systems, ambient air facilitated by an air heat exchanger (AHX), and subsurface reservoirs accessed via a geothermal closed loop (GHX). Consequently, the heat pump (HP) is chosen as a conventional water-to-water technology, propelled by electrical means to leverage the photovoltaic (PV) power, enabling both heating and cooling functionalities. Additionally, the system integrates phase change materials (PCMs) for thermal storage, utilizing specific technologies integrated into the user-side (e.g., heating/cooling floor) and source-side (e.g., GHX, PVT) arrangements.

The system predominantly comprises three distinct loops with designated functionalities:

1. **PVT loop** (operating within a temperature range of −5 to +80 °C) primarily relies on the PVT device as its heat source. Heat energy generated by the PVT device is subsequently stored within the buffer tank labelled as BF$_1$.
2. **Air-ground loop** (operating within a temperature range of −5 to +50 °C) functions by utilizing either the surrounding air or the ground as potential sources or sinks for the heat pump (HP) system. This loop interacts with the buffer tank BF$_1$ of the PVT loop. In cases when the HP system is deactivated, this loop can be employed for regulating the temperature of the PVT device or storing energy within the ground for subsequent utilization.
3. **User loop** (operating within a temperature range of +10 to +50 °C) is responsible for heating or cooling operations and primarily serves to warm or cool the buffer tank labelled as BF$_2$, which is interconnected with radiant floor systems or fan coil systems.

The utilization of two tanks within the system serves as buffer tanks, enabling effective control over the thermal inertia of the overall system. Furthermore, these tanks facilitate hydraulic separation, ensuring seamless operation of the heat pump upon activation. The inclusion of primary and secondary loops further enhances the functioning of the heat pump. In Fig. 1 the distinct sides of the heat pumps are visually delineated using different colours: the source side and the user side, alongside the HP group.

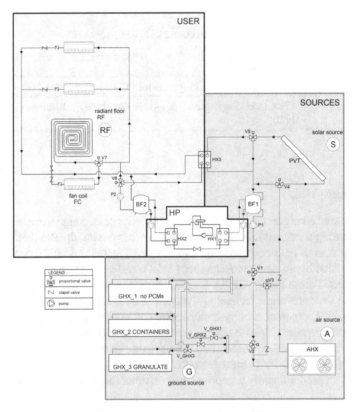

Fig. 1. Heat Pump system layout [2]

Initially, the supply of heat to the house was reliant on fan coils and radiant floor systems. However, several issues arose during the monitoring process:

1. The installed probes failed to collect crucial data pertaining to the fan coils, including flow rates and temperature differentials.
2. Additionally, the specifics of the control system in place remained unknown, presenting a challenge in comprehensively understanding and analysing the heating process within the system.

In response to the challenges, the modelled system has been adapted to rely solely on a water-based radiant floor as the primary heat source. The water-based radiant floor system functions as a direct supplier of heat within the system. Prior to distribution, the heated liquid is stored in a buffer tank, wherein the temperature is regulated through the operation of a heat pump.

2.3 RC Circuit Model of a House

The RC Circuit Model of the building, depicting the thermal behaviour and dynamics, is visually represented in Fig. 2. To begin, a state-space dynamic thermal model of a building is developed using electro-thermal analogies where heat flow is modelled as a current, temperature as a voltage source and thermal storage as a capacitance [3]. The equivalent thermal resistance of a layer can be calculated using Eq. (1), where λ represents thermal conductivity of the layer material. This model serves as a valuable framework for understanding the heat transfer processes within the building and capturing the transient responses to various inputs.

In the RC Circuit Model, each component of the building, such as walls, floors, roofs, and windows, is represented by an appropriate combination of thermal resistances (R) and capacitances (C). These elements reflect the thermal characteristics and interactions between the building components, enabling a comprehensive analysis of the heat flow and temperature distribution.

The thermal resistances (R) within the RC Circuit Model quantify the resistance to heat transfer through each building component, considering factors such as material properties, thickness, and surface area. On the other hand, the thermal capacitances (C) represent the ability of the building components and spaces to store and release thermal energy, considering their thermal mass and volume.

By incorporating the RC Circuit Model, the complex thermal dynamics of the building can be effectively captured and simulated. The model facilitates the evaluation of energy consumption, temperature profiles, and the impact of various heating and cooling strategies. Additionally, it provides valuable insights for optimizing energy efficiency, identifying potential areas of improvement, and guiding decision-making processes related to building design and retrofitting.

The inclusion of the RC Circuit Model in this study further enhances the understanding of the thermal behaviour of the building and complements the machine learning-based predictions and closed-loop simulations. Mentioned analogy is a crucial step because heat losses of a house caused by outside temperature depend on equivalent resistance of the entire building [4]. This integration of modelling approaches allows for a more comprehensive analysis of the thermal system, facilitating the development of effective control strategies and the optimization of energy management in buildings.

$$R_{eq_{layer}} = \sum_{layer=1}^{N_{layer}} \frac{thickness_{layer}}{\lambda_{layer} \cdot S_{layer}} \tag{1}$$

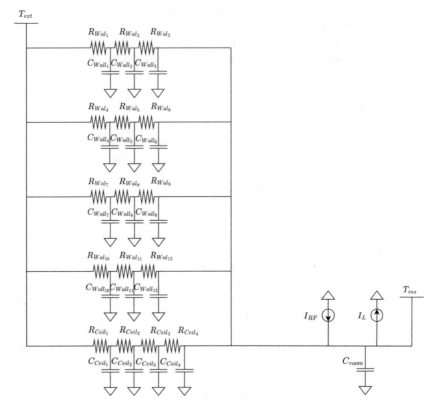

Fig. 2. RC Circuit of a building

2.4 White Box Model

The entire model is constructed as a negative feedback system, which is a fundamental concept in control theory. This system is depicted in Fig. 3, where the input is a *Set Point Temperature* predetermined by the user. The schematic representation of each subsystem is depicted in Fig. 4, Fig. 5 and Fig. 6. The negative feedback mechanism is essential for maintaining the desired temperature within the building by continuously adjusting the control variables.

Each subsystem within the model except *hysteresis* represents a distinct thermodynamic process and is characterized by linear and differential equations. These equations capture the fundamental principles of heat transfer, such as conduction, convection, and radiation, as well as the energy balance equations governing the flow of heat within the building [4]. By employing linear equations, the model assumes that the thermal behaviour of the building components and zones can be adequately described within a certain range of operating conditions. These linear equations simplify the mathematical representation of the processes, allowing for efficient calculations and simulations. The inclusion of differential equations accounts for the dynamic nature of heat transfer processes, considering the time-dependent changes in temperature and energy flows. These equations describe the rates of change of temperature within the building elements and rooms, reflecting the transient behaviour of the thermal system.

Through the integration of linear and differential equations, the model captures the complex interactions and dependencies between various components and zones of the building. This comprehensive approach enables a thorough analysis of the thermal behaviour and performance of the building under different operating scenarios, facilitating informed decision-making for energy optimization and control strategies.

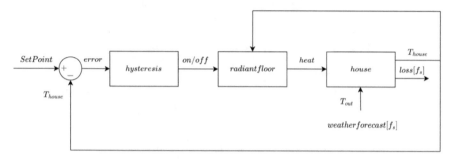

Fig. 3. White Box model of a Building

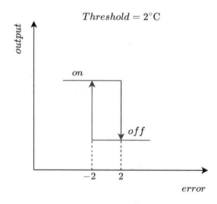

Fig. 4. Input-Output function relationship (hysteresis)

2.5 Grey Box Model

Within the Radiant Floor subsystem shown on Fig. 5, there exists a parameter ($T_{water} = T_{buffer\ tank\ 2}$, $T_{buffer\ tank\ 2}$ is unknown and controlled by heat pump) of unknown, non-deterministic value that requires prediction. Through the introduction of an predictor (*Black Box*), the originally characterized white box system transitions into a grey box system depicted on Fig. 7.

Within the system, there is a Black Box subsystem illustrated in Fig. 7, which employs a supervised machine learning (ML) approach. This approach utilizes Decision Tree

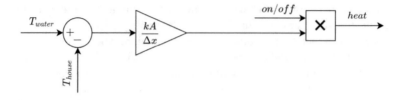

Fig. 5. Block Diagram of Radiant Floor Subsystem

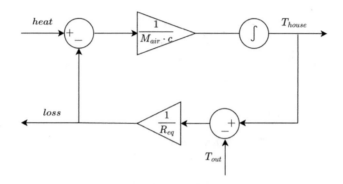

Fig. 6. Block Diagram of House Subsystem

Regressor to predict the unknown parameter - T_{water}. The objective is to approximate the mapping function that relates the inputs T_{out} (outdoor temperature) and T_{house} (indoor temperature) to the target variable T_{water}.

The dataset we're examining is fundamentally a time series. However, a noteworthy characteristic is the inconsistency in the time intervals between successive samples. Such discrepancies in temporal spacing add a layer of complexity to our analytical endeavors. Consequently, during the model training phase, we chose to diverge from the conventional approach to time series, wherein there's an intrinsic assumption of temporal interdependence among data points. This deviation was prompted by the marked irregularities in our dataset, especially the aforementioned variations in time intervals.

Given these intricacies, our attention was directed towards the Decision Tree Regressor. This method offers a structured, yet flexible approach to data interpretation, enabling us to discern patterns without being unduly constrained by the temporal relationships typically emphasized in time series data. In light of the distinctive attributes of our dataset, the Decision Tree Regressor presented itself as a fitting tool, allowing us to maintain both scientific rigor and adaptability.

To achieve accurate predictions, the *Black Box* subsystem undergoes hyperparameter tuning and model evaluation. Adequate hyperparameters for each ML model are determined using a 5-fold cross-validation score [5]. It was decided to use 5-fold cross-validation due to the small size of the available training dataset, which will be discussed in detail later. This technique divides the available data into five subsets, with each subset serving as a validation set in turn. The scoring metric utilized during cross-validation, which provides insights into the model's performance, is Mean Squared Error (MSE).

The use of cross-validation enables the estimation of how well the ML models will generalize to an independent dataset. By evaluating the models on different subsets of the available data, it becomes possible to assess their performance across various data distributions.

From a computational perspective, the continuous-time state space system representing the thermal dynamics of the building is approximated as a discrete-time system. This discretization is achieved through the utilization of a sampler with a sampling time of f_s = 1000 Hz. This discretization process allows for the translation of the continuous-time equations and variables into a discrete-time framework, which facilitates computational simulations and analysis.

The discrete-time approximation enables efficient numerical computations and simulations, as discrete-time systems are typically easier to handle and analyse than their continuous-time counterparts. By discretizing the state space system, the model can effectively capture and simulate the time-dependent behaviour of the thermal processes within the building, providing valuable insights into the energy consumption, temperature profiles, and overall thermal performance.

The choice of a sampling time of f_s = 1000 Hz in f_s the discretization process strikes a balance between computational accuracy and efficiency. A higher sampling rate ensures that the discrete-time approximation closely represents the continuous-time system, capturing the dynamics of the thermal processes with sufficient detail. However, it is important to consider computational resources and practical limitations when determining an appropriate sampling time.

Through the discretization of the continuous-time state space system, the model can effectively analyse and predict the thermal behaviour of the building, considering the discrete-time nature of computational simulations and control algorithms. This approach allows for a comprehensive understanding of the building's thermal dynamics, aiding in the optimization of energy management strategies and the improvement of overall thermal comfort.

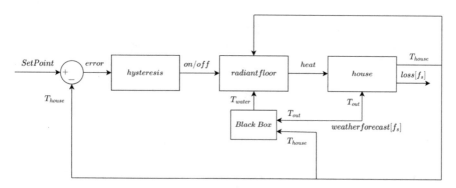

Fig. 7. Grey Box model of a Building

3 Solution/Discussion

After determining the optimal hyperparameters through cross-validation, the Decision Tree Regressor was trained using a split of 95% for training and 5% for testing. It's pivotal to recognize that the dataset used in this study is quite small, encompassing only 150 rows specifically tailored for the heating mode. In spite of the dataset's constrained size, it constitutes a significant foundation for gauging the model's performance. The Decision Tree Regressor, when subjected to 5-fold cross-validation, achieved a Mean Squared Error (MSE) of 2.022.

The Decision Tree Regressor was chosen for its ability to accurately predict the unknown parameter in this setting. The model was good at finding patterns between the input data and the target. Its performance shows it can handle the dataset's specifics and make reliable predictions.

Building upon the successful prediction of the unknown parameter, further investigation was conducted to evaluate the closed-loop control performance of the system. Utilizing prior knowledge and insights gained from the machine learning predictions, a closed-loop simulation was performed in Python, without the explicit utilization of control system knowledge. This approach demonstrates the potential of machine learning techniques in providing valuable information for closed-loop control applications.

In the specific case depicted in Fig. 8, the simulation scenario involved a Set Point Temperature of 16 °C, an outside temperature that follows a sinusoidal pattern over time, and an initial house temperature of 15 °C. Despite relying on regression-based predictions for T_{water}, the closed-loop control system successfully achieved and maintained a constant Set Point Temperature throughout the day. This outcome highlights the robustness and reliability of the machine learning predictions, which effectively guided the control system to ensure the desired temperature regulation.

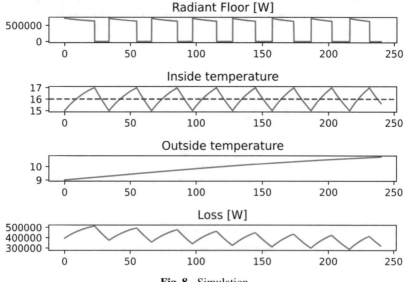

Fig. 8. Simulation

These compelling findings not only underscore the efficacy of the Decision Tree Regressor in approximating the unknown parameter but also emphasize the potential of machine learning techniques for closed-loop control in thermal management systems. By leveraging predictive capabilities and previous knowledge, machine learning models contribute to the optimization and automation of control strategies in building environments, facilitating energy efficiency and occupant comfort.

In summary, the utilization of machine learning models, particularly the Decision Tree Regressor, enables accurate prediction of unknown parameters within the thermal system. The successful integration of machine learning predictions into closed-loop simulations showcases their potential for real-time control applications, even without explicit knowledge of the underlying control system. These advancements have significant implications for the design and implementation of intelligent and adaptive control systems in the field of thermal management.

Acknowledgement. The research presented in this paper is partly financed by the European Union (H2020 IDEAS project, Grant Agreement No.: 815271 and SINERGY project, Grant Agreement No.: 952140) and the Ministry of Education, Science and Technological Development of the Republic of Serbia.

References

1. Kosarac, A., et al.: Thermal behavior modeling based on BP neural network in Keras framework for motorized machine tool spindles. Materials **15**(21), Article no. 21 (2022). https://doi.org/10.3390/ma15217782
2. Trinity College Dublin (Dublin University). D3.4 - ITES-MES prototype at small scale [Deliverable]. In: Novel Building Integration Designs for Increased Efficiencies in Advanced Climatically Tunable Renewable Energy Systems (IDEAS Project No. 815271, Work Package WP3: Heat Pump and Underfloor Heating/Cooling). Horizon 2020 Program: H2020-LC-SC3-2018–2019–2020. Stage: Final (2020)
3. Bastida, H., Ugalde-Loo, C.E., Abcysckera, M., Qadrdan, M., Wu, J.: Thermal dynamic modelling and temperature controller design for a house. Energy Procedia **158**, 2800–2805 (2019). https://doi.org/10.1016/j.egypro.2019.02.041
4. Jarmuda, T.: Thermal model of an intelligent house. https://core.ac.uk/display/295559358?utm_source=pdf&utm_medium=banner&utm_campaign=pdf-decoration-v1
5. Refaeilzadeh, P., Tang, L., Liu, H.: Cross-validation. In: Liu, L., Özsu, M.T. (eds.) Encyclopedia of Database Systems, pp. 532–538. Springer, Cham (2009). https://doi.org/10.1007/978-0-387-39940-9_565

Digital Solution to Estimate Solar Power Potential of Rooftops in City of Belgrade

Ana Dodig$^{(\boxtimes)}$ ⓘ and Vladimir Djapic ⓘ

The Institute for Artificial Intelligence Research and Development of Serbia,
11000 Belgrade, Serbia
ana.dodig@ivi.ac.rs

Abstract. Global warming is one of the main issues of today. The work to slow it down or even better stop it is getting more and more intense all over the world. Many countries are moving towards replacing the use of traditional fossil fuels with clean energy alternatives. Solar energy is a sustainable alternative to fossil fuels and has a low impact on the environment. It has been increasingly used recently because the price of solar panels has decreased, and the cost of electricity has increased significantly. In Serbia, rooftop solar power plants were a rarity until 2022, but the adoption of an appropriate regulatory framework enabled citizens to produce green energy for self-consumption. In order to increase the use of solar energy in Serbia, a solar 3D urban model of the part of the municipality in Belgrade was developed for the calculation and visualization of the solar energy potential of buildings rooftops. We received LiDAR data from the city of Belgrade, on the basis of which we created a 3D model and calculations. We have used Digital Surface Model (DSM) and Building Footprints layer to create solar radiation layer which takes into account the position of the sun throughout the year and at different times of day. The model accounts for the obstacles, such as nearby trees and buildings that may block sunlight, as well as the slope and orientation of the rooftops. Suitable roofs in terms of surface area, slope and orientation were selected and it was obtained that this part of the municipality in Belgrade can produce almost 20000 MWh per year.

Keywords: Solar Energy · PV · LiDAR · Renewable Energy · Solar Radiation

1 Introduction

Global warming and air pollution has a great impact on the health of both humans and animals. Air pollution is the sixth-leading cause of death, causing over 2.4 million premature deaths worldwide [1]. Global warming enhances multiple phenomena in nature which can in future lead to disasters [2]. On the other hand, energy insecurity and rising prices of traditional fossil fuels are also major threats to economic and political stability. One of the possibilities to reduce or possibly solve the above-mentioned problems is the use of solar energy which is a sustainable alternative to fossil fuels and has a low impact on the environment. Also, solar energy is the most abundant energy source of

M. Trajanovic et al. (Eds.): ICIST 2023, LNNS 872, pp. 362–374, 2024.
https://doi.org/10.1007/978-3-031-50755-7_34

renewable energy [3]. Certain studies showed that global energy demand can be satisfied by using solar energy as it is plentiful in nature and freely available at no cost [4]. Urban environments demand significant amounts of energy. In order to deploy photovoltaic (PV) systems appropriately in such an environment, it is necessary to assess the local PV potential. Solar radiation distribution and solar radiation intensity are two key elements which determine efficiency of PV systems [5]. The city of Belgrade is located in Southeastern Europe, on the Balkan Peninsula in a continental climate region. The summer season (July and August) brings the largest amount of solar energy. On the other side the lowest amount of solar energy is received during winter (in January and December). According to the available data, the territory of the city belongs to the areas relatively rich in solar energy, with an average solar radiation of 1446.8 kWh/m^2 per year [6]. Additionally, the construction of buildings has been increasing in recent years. Therefore, the deficit of energy is more noticeable, which encourages the development of self-sustainable buildings that produce their own energy. The objective of this study was to estimate how much energy can be produced using PV systems on a building rooftops in a part of a municipality in Belgrade. First, we needed to calculate solar radiation over the building rooftops in the region of interest. There are several solar irradiation models which are used for that. Well known examples are the GRASS [7], QGIS and its plugins (Urban Multi-Scale Environmental Predictor—UMEP Tools [8], Solar Energy on Building Envelopes—SEBE Tool [9]), CityGML Generation Tool [10], SAGA GIS [11] and the ArcGis Solar Analyst [12] which we used in this paper. A brief comparison of the mentioned tools can be seen in Fig. 1. We selected ArcGis tool for several reasons. Its precision, stemming from its advanced complexity, is a primary factor. Additionally, it offers a comprehensive feature set without the need for additional installations of extensions or plugins. Moreover, it benefits from a robust online community and an abundance of official tutorials for user guidance. This tool calculates diffuse and direct solar irradiance for all given points over any period from several days to the whole year. It also takes into account obstacles that may block sunlight such as buildings or nearby trees and the slope and orientation of the surface. After that we choose suitable rooftops according to size, orientation, slope and amount of solar energy that reaches rooftop surface. At the end we were able to estimate electrical energy potential by calculating the area covered by its suitable cells (in m^2) for solar panel mounting and converting the usable solar radiation values to electric power. The amount of power that solar panels can produce also depends on the solar panels' efficiency and the installation's performance ratio. The United States Environmental Protection Agency (EPA) [13] provides a conservative best estimate of 16 percent efficiency and 86 percent performance ratio and we have used these numbers to calculate electric power potential. Other researchers have also used these tools or similar ones and built on them. Some works, in addition to calculating the solar potential on the roofs, also calculated the solar potential of the facades, but showed that the solar potential of the facades is significantly lower [14, 15]. In [16] they used combination of QGIS tools and CityGML tool to calculate the solar potential and to generate and visualize an urban 3D model. When it comes to Serbia and the surrounding area [17] used GRASS tool to calculate solar irradiation over the city of Indjija and another 7 European cities. [18] used SAGA GIS tool to produce grids of potential solar radiation and duration in order to determine the location of potential solar power plants in Montenegro. Most of those works were performed based on data obtained with LiDAR technology. The novelty in this work is that the solar potential of this part of Belgrade and in general the whole of Belgrade has not yet been calculated so we used LiDAR data and ArcGIS Solar Analyst tool to calculate it. Additionally, we hope that this will help the further development of the city in terms of the use of solar energy.

	ArcGIS	GRASS	QGIS	SEBE	CityGML	SAGA GIS
Accuracy	Known for its accuracy and reliability, particularly when handling spatial data. It offers advanced geoprocessing tools for precise solar potential estimation.	Offers robust spatial analysis capabilities, can be suitable for accurate solar potential estimation	Highly depends on the quality of data and tools used. It can provide accurate results when configured correctly which requires great experience	Designed for accurate solar energy estimation in urban environments, focusing on building interactions	Accuracy highly depends on the quality of the 3D models generated	Offers accurate terrain analysis capabilities, which can be beneficial for solar potential estimation
Complexity	Powerful and complex GIS software. It has a wide range of features and capabilities especially for LiDAR data analysis. It may require training to use effectively	Sophisticated, open-source GIS platform, which can be complex dependig on usage.	Less complex and is often more user-friendly, making it accessible to users with varying levels of GIS expertise	Specialized for urban solar analysis and may be less complex to use for this specific task compared to general GIS software	The complexity depends on familiarity with 3D modeling. Generating CityGML files from LiDAR data can be challenging for non experienced users	Moderately complex, requires more manual setup compared to commercial alternatives
Plugins	Supports a wide range of extensions and plugins. Provides an all-in-one GIS solution with extensive capabilities	Has a collection of modules and plugins, many of which are open-source	Offers a variety of free and open-source plugins which can be additionally installed	Typically used as a standalone tool without traditional GIS plugins	Specific plugins for 3D modeling tasks. Should be used in conjunction with other software for comprehensive solar power potential estimation	Offers its own set of modules and extensions but the range of available plugins is not as extensive as what is found in ArcGIS
Licensing	Commercial software with licensing costs	Mostly open-source and free to use	Free and open-source software	Availability and licensing may vary	Licensing depends on the specific tool or software used for 3D modeling	Free and open-source software
Expertise	Suitable for users with advanced GIS expertise	Requires expertise in geospatial analysis and command-line operation	Accessible to users with varying levels of GIS expertise	Specialized for urban solar analysis and may require expertise in solar energy modeling	Requires expertise in 3D modeling	Moderate expertise needed for LiDAR data processing
Education	Abundant tutorials and training resources available	The availability of tutorials may not be as extensive as for some other GIS software.	Many online tutorials	Tutorials may be limited compared to more widely-used GIS software	Availability of tutorials may depend on the specific 3D modeling software used	Tutorials and documentation are available but may not be as extensive as some commercial GIS software

Fig. 1. Comparison of different methods for solar power potential estimation

1.1 Context and Justification

Currently, more than half of the world's population lives in urban areas and more than 40% of the total energy consumption in the European Union corresponds to buildings [16]. This situation initiated the EU to promote the use of renewable energy resources, especially solar energy, in order to increase the energy efficiency of buildings. Recently, numerous efforts are being done to develop systems that automatically calculate the solar energy potential of an area and show building rooftops which are the most suitable to place PV systems. In this sense, Serbia has also begun to develop rapidly in recent years. In order to motivate the usage of solar energy even further, in this work, we introduced system which calculates the solar energy potential of building rooftops based on LiDAR data. This system can help decide where to put a solar panel and whether it is even cost effective to put it on a particular rooftop. Hopefully, this will help Serbia to develop more in terms of using renewable energy sources, specifically solar and eventually reduce pollution.

2 Methodology

Estimation of the solar potential and the energy capacity of PV solar energy systems of a region requires that we have data of it. We have LiDAR point cloud data for part of the Zvezdara municipality for which we performed the estimations. Our aim of this work was to find suitable roofs of buildings on which solar panels could be installed and to estimate how much electricity could be produced from them in total per year. In order to achieve this, we took the following steps (see Fig. 2):

- Classification of LiDAR point cloud dataset to ground, low, medium and high vegetation and buildings. After that, points classified as building are used for building footprint layer extraction which is later used to form a solar radiation layer only on the rooftops of buildings. If we had not separated the points representing the buildings, then the solar radiation layer would have been created for the whole region, which would have been much more computationally demanding. We did not need that because we only consider rooftops and their solar potential in our work.
- Extraction of the digital surface model (DSM) layer from Lidar point cloud dataset which shows the elevation of the ground and features on the ground, such as buildings and trees. This is the second and final layer we need for a solar radiation layer formation.
- Creation of the solar radiation layer from building footprint and DSM layer with Area Solar Radiation tool which is part of the ArcGIS Spatial Analyst extension.
- Removing roof surfaces which are not suitable for solar panel installation based on size, slope, orientation and solar radiation.
- Converting solar radiation to electric power potential which also depends on solar panel's efficiency and the installation's performance ratio.

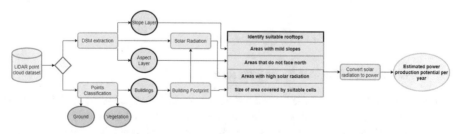

Fig. 2. Workflow for the estimation of electric power potential of the building's rooftops from LiDAR point cloud dataset

The following seven sections describe in detail previously mentioned steps and illustrates results after each.

2.1 LiDAR Point Cloud Dataset

Light Detection and Ranging (LiDAR) is a remote sensing technique that light from a laser to measure out the elevation like buildings, ground and forest. LiDAR point cloud data is a collection of accurate 3D (X,Y,Z) points with other attributes (intensity, classification of features and GPS time). LiDAR point cloud data is commonly used in Geospatial and Earth observations applications. The LAS file format is a standard binary file format to store 3D point cloud data [19]. For the benefit of this work, we received 8 LAS files from the city of Belgrade representing part of the Zvezdara municipality. We performed an analysis on them, which is explained in the following sections. Figure 3 shows our dataset in the way that the color of the points represents the height (blue - low, red - high).

Fig. 3. LiDAR point cloud representation. Color of the points represents the height (blue - low, red - high)

2.2 Classification of LiDAR Point Cloud Dataset

In order to be able to determine the solar potential of building rooftops, we first had to extract buildings from our data set, and therefore also roofs. We have used classification tools in ArcGIS to classify buildings and also ground and vegetation points. Within the points belonging to the vegetation, they are divided into low, medium and high vegetation. Points that do not belong to any of the mentioned classes remain unassigned. Figure 4a show classified points and Fig. 4b shows the colors of the classes. Ground is represented with brown color, vegetation with different shades of green and buildings with light red. By classifying the points, we also have the statistics of how many points belong to which class, and Table 1 represents those statistics. We can see that almost 15% of the points belong to buildings.

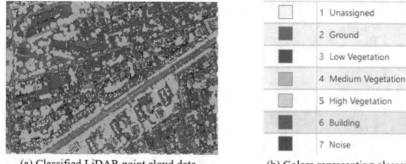

(a) Classified LiDAR point cloud data   (b) Colors representing classes

Fig. 4. Lidar point data classification

Table 1. Classified points statistics

Category	Number of points	%
Ground	159630219	34.21
Buildings	69172396	14.83
Low Vegetation	104950582	22.49
Medium Vegetation	130710242	28.02
High Vegetation	2053036	0.44
Unassigned	23412	0.01
Noise (Low)	13203	0.00

2.3 Building Footprints

In order to create solar radiation layer only for building's rooftops and in that way, we save the time needed for that compared to calculating a layer for the entire region, we had to extract building footprints layer. Building Footprints are polygon features representing captured building outlines. We extracted building footprints from LiDAR points which are classified as building with several ArcGIS tool. Firstly, we created a raster file from building points, and later converted it to polygon features to have separate data for every polygon/building. During the process, we performed some kind of regularization so that there are no holes in the polygons, and they look more regular. The final touch was manual polygon correction and separation in cases where two buildings are very close. Figure 5 shows extracted building footprint layer.

Fig. 5. Representation of building footprints

2.4   Digital Surface Model DSM

As mentioned earlier DSM represents the elevation of the ground and features on the ground, such as buildings and trees. It means that DSM captures both the natural and built features of the environment which is of essential importance to us. Using ArcGIS tool to extract DSM we have extracted DSM layer from our LiDAR pointcloud data set file. Figure 6 shows the DSM layer for our region. From the DSM layer, using surface parameters tool we have also created slope and aspect layer in order to later determine suitable rooftops. The slope layer for each cell has a slope value ranging from 0° to 90°. Aspect layer has a value expressing orientation in degrees for each cell. 0 represents north and 180 represents south. Now that we have extracted the DSM and building footprints layer, we could do further analysis and extract the solar radiation layer.

2.5   Solar Radiation Layer

Solar radiation layer represents a raster layer that maps how much solar energy reaches rooftop surfaces in our region over the course of a typical year. We used Area Solar Radiation tool which is part of the ArcGIS Spatial Analyst to extract it. It can take only a DSM as input to create a layer for whole region, but we have also provided building footprints layer as input to calculate solar radiation only for the areas with the buildings. In this way we have saved a significant amount of time. This tool calculates radiation

Fig. 6. Digital surface model layer

taking into account the position of the sun throughout the whole year once every hour for each day. In order to consider obstacles such as nearby trees or buildings, also slope and the orientation of the surfaces, this tool checks 32 directions around each cell. The output of the tool is solar radiation raster where each cell has value in watt-hours per square meters [Wh/m^2]. We converted units to kilowatt-hours per square meter [kWh/m^2] since some cells had very big values. Figure 7 shows solar radiation layer. Red and orange colors indicate higher amounts of solar radiation (maximum value is 1262.63 kWh/m^2, while yellow and blue tones indicate lower amounts (minimum value is 0.0045 kWh/m^2. As we can see points that are outside building footprints layer have not been computed.

2.6 Suitable Rooftops

Suitable rooftops should meet 4 criteria:

- Have a slope of 45° or less. Steeper slopes receive less sunlight. We used slope layer to extract rooftops which meet criteria.
- Have at least 800 kW h/m^2 of solar radiation. For the installation of a solar panel to pay off, it must receive a sufficient amount of radiation. We used solar radiation layer to extract suitable rooftops when it comes to this criterion.
- Should not face north because these rooftops receive less sunlight. We used aspect layer to extract rooftop which does not face north.
- Area covered by its suitable cells greater than 30 m^2. We used building footprints layer to extract this information.

Fig. 7. Solar radiation layer. Red and orange colors indicate higher amounts of solar radiation, while yellow and blue tones indicate lower amounts.

When there are only rooftops that meet all the criteria left, we also have a calculation for them (in addition to the area covered by suitable cells for installing solar panels) about the average solar radiation (in kW h/m^2) that these cells receive. With that and with the suitable surface we calculated the amount of solar radiation received per year by each building's usable area. We got this number by multiplying these two values. We divided the obtained values by 1000 to avoid large numbers and get the values in megawatt-hours.

2.7 Converting to Electric Power Potential

Now, when we have an estimate for every building about how much solar radiation, they receive on suitable surfaces we can convert that to electric power production potential. As we mentioned earlier in the text, power production of solar panels depends not only on solar radiation, but also the solar panels' efficiency and the installation's performance ratio. In Introduction (Sect. 1) we said that the solar panels can convert 16% of incoming solar energy into electricity, and then 86% of that electricity is preserved as it goes through installation [13]. Therefore, to obtain electric power potential we multiplied the amount of solar radiation received per year for each building's suitable area with 0.16 ∗ 0.86. The end result is expressed in MWh. Figure 8a shows on the map electric power potential per building and Fig. 8b shows color map for representation.

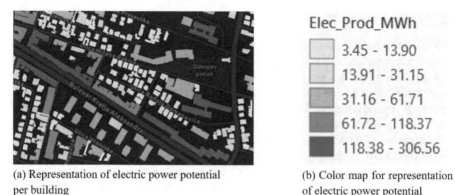

(a) Representation of electric power potential per building

(b) Color map for representation of electric power potential

Fig. 8. Electric power potential

3 Results

The total estimated amount of electrical energy that can be produced in this part of the Zvezdara municipality in Belgrade per year is 19877.5 MWh. Some buildings have more potential and some less. Of course, it depends a lot on the size of the building itself, so larger buildings have greater potential, but also consume more energy. Figure 9 shows the distribution of the buildings with their electrical energy production potential. We can see that most of the buildings have relatively small production potential, but these buildings/houses are small so that production is enough for them. Bigger buildings have larger suitable surfaces for solar panels and can produce more energy. The maximum amount of energy that can be produced by one building from this area is 306.56 MWh.

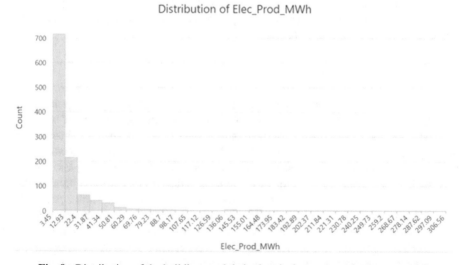

Fig. 9. Distribution of the buildings and their electrical energy production potential.

3.1 Economic Analysis

To do a short economic analysis we used three buildings from our data source. According to our solar potential calculation all three fall into the category 31.16–61.71 MWh of solar potential (Fig. 8). One is a large chain private furniture store, and other two are government buildings, with the roof square meter areas of 525, 620, and 612 m^2. If we use 700 m^2 out of a total of 1700 m^2 (it is not possible to install the panel on the roof ideally in all places), according to the solar calculator web site [20], the cost of commercial solar panel system (equipment and installation) of a capacity of 120kW would be around 100000 EUR. We estimated that the combined monthly electricity consumption for these three buildings is approximately 12 MWh. This translates to an annual expense of approximately 15000 EUR on electricity, given the average daytime electricity rate in Serbia is 110 EUR per 1 MWh. By implementing a 120 kW solar panel system, they have the capacity to generate enough electricity to cover their energy requirements, translating to annual cost savings of 15,000 EUR. With an initial installation cost of 100,000 EUR, the solar systems' payback period spans 7 years. Subsequently, there remains an extended 18-year period during which the electricity generated by the solar panels is essentially free, aligning with the typical 25-year warranty for such systems. Over this lifespan, the cumulative savings amount will be 325,000 EUR. Moreover, this transition to solar power significantly contributes to a substantial reduction in CO_2 emissions. A potential minor challenge lies in Serbian law, which mandates that systems exceeding 50 kW necessitate a construction permit rather than qualifying for a simplified procedure. However, there's an opportunity to address this obstacle. Given the proximity of these three buildings, each approximately 40 m apart, they could collectively pursue a single permit. This collaborative approach not only streamlines administrative processes but also opens the door to government incentives. Furthermore, it might be worth exploring the feasibility of establishing a shared battery storage system, allowing them to generate surplus energy, potentially exceeding their own consumption. This surplus could be advantageous if the current government prosumer deals for selling excess electricity prove to be advantageous and attractive.

4   Conclusions

This work describes a complete method for the estimation of solar power potential of rooftops in an urban landscape. The method considers only the rooftops of buildings and calculates the solar radiation for them throughout the year on an hourly basis. Also, all obstacles that do not allow the sun to reach the rooftop are taken into account. The total energy potential is then calculated based on solar radiation. Complete method was implemented using ArcGIS tools and applied to the part of Zvezdara municipality in Belgrade. However, there is a lack of literature related to the existing systems for predicting solar power potential in Serbia and wider so we can not exactly compare this method with existing ones. The good thing about our method is that all the processing steps can be combined into one model and all that is needed is to insert the LIDAR point cloud data into the model and the solar power potential will be obtained for that data. The results showed that only this part of the city can annually produce almost 20000 MWh. Also in the results, there is data for each building separately and how much electricity it could produce. Additionally, a brief economic analysis is given for three buildings belonging to the high potential range of solar energy production. Therefore, this method and these results can be useful for the development of the city in terms of solar energy production and for further urban planning.

Acknowledgments. This research was supported by the city of Belgrade which provided us with LiDAR point cloud data for part of the Zvezdara municipality for which we performed the estimations.

References

1. Organization, W.H.: The World Health Report 2002: Reducing Risks, Promoting Healthy Life (2002)
2. Stocker, T.: Climate Change 2013: The Physical Science Basis: Working Group I Contribution to the Fifth Assessment Report of the Intergovernmental Panel on Climate Change (2014)
3. Panwar, N.L., Kaushik, S.C., Kothari, S.: Role of renewable energy sources in environmental protection: a review. Renew. Sustain. Energy Rev. **15**(3), 1513–1524 (2011)
4. Lewis, N.S.: Toward cost-effective solar energy use. Science **315**(5813), 798–801 (2007)
5. Kannan, N., Vakeesan, D.: Solar energy for future world: - a review. Renew. Sustain. Energy Rev. **62**, 1092–1105 (2016)
6. Regional Spatial Plan of the Administrative Area of the City of Belgrade (Official Gazette of the City of Belgrade, No. 10/04 and 38/11.). Official Gazette of the City of Belgrade, No. 10/04 and 38/11
7. Hofierka, J., Suri, M., et al.: The solar radiation model for open source gis: implementation and applications. In: Proceedings of the Open Source GIS-GRASS Users Conference, vol. 2002, pp. 51–70 (2002)
8. Lindberg, F., et al.: Urban multi-scale environmental predictor (umep): an integrated tool for city-based climate services. Environ. Model. Softw. **99**, 70–87 (2018)
9. Lindberg, F., Jonsson, P., Honjo, T., Wastberg, D.: Solar energy on building envelopes – 3d modelling in a 2d environment. Sol. Energy **115**, 369–378 (2015)
10. Prieto, I., Izkara, J.L., Bejar, R.: A continuous deployment-based approach for the collaborative creation, maintenance, testing and deployment of citygml models. Int. J. Geogr. Inf. Sci. **32**(2), 282–301 (2018)
11. Bohner, J., McCloy, K., Strobl, J.: SAGA - Analysis and Modelling Applications (2006)
12. Fu, P., Rich, P.M.: Design and implementation of the solar analyst: an arcview extension for modeling solar radiation at landscape scales. In: Proceedings of the Nineteenth Annual ESRI User Conference, San Diego, vol. 1, pp. 1–31 (1999)
13. Green Power Equivalency Calculator - Calculations and References. Environmental Protection Agency. https://www.epa.gov/green-power-markets/green-power-equivalency-calculator-calculations-and-references
14. Ord'onez, J., Jadraque, E., Alegre, J., Martinez, G.: Analysis of the photovoltaic solar energy capacity of residential rooftops in andalusia (Spain). Renew. Sustain. Energy Rev. **14**(7), 2122–2130 (2010)
15. Redweik, P., Catita, C., Brito, M.: Solar energy potential on roofs and facades in an urban landscape. Sol. Energy **97**, 332–341 (2013)
16. Prieto, I., Izkara, J.L., Usobiaga, E.: The application of lidar data for the solar potential analysis based on urban 3d model. Remote Sens. **11**(20), 2348 (2019)

17. Protic, D., Kilibarda Milan, S., Nenkovic-Riznic Marina, D., Nestorov, I.: 3d urban solar potential maps-case study of the i-scope project. Therm. Sci. **22**(1B), 663–673 (2018)
18. Bajat, B., et al.: Space-time high-resolution data of the potential insolation and solarduration for montenegro. Spatium (44) (2020)
19. Abdishakur.What is Lidar Point Cloud Data? Spatial Data Science (2022)
20. (2023). https://solarnikalkulator.rs/

Platform for Efficient Building Operation and Demand Response Flexibility Provision

Lazar Berbakov[1](✉) [iD], Valentina Janev[1] [iD], Marko Jelić[1] [iD], Nikola Tomašević[1] [iD], and Lilia Bouchendouka[2] [iD]

[1] Institute Mihajlo Pupin, Volgina 15, Belgrade, Serbia
`lazar.berbakov@pupin.rs`
[2] Engie, Place Samuel de Champlain 1, 92400 Courbevoie, France

Abstract. This paper describes a SGAM (Smart Grid Architecture Model) compliant collaborative platform including its logical components in terms of functionalities and interfaces and their relationships. It aims at facilitating the service deployments, establishment and management of a Citizen Energy Community (CEC) by stakeholders along the energy value chain (consumers, energy managers, grid operators, service providers). Its foundation is based on standard-enabling technologies and practices and recommendations from EU projects (NEON and SINERGY). Unified Modeling Language (UML) is used for illustrating potential scenarios of using the services by NEON piloting partners. The energy dispatch optimization service developed by Institute Mihajlo Pupin (IMP) has been tested for a CEC from Spain.

Keywords: SGAM Architecture · Interoperability · Services · API · KPIs · Standards

1 Motivation

In the last few years, particularly in Europe, there has been a notable increase in the number of citizen-led energy initiatives focused on producing, distributing, and consuming energy from renewable energy sources (RES) at a local level. This growth is evident when considering the *Inventory of citizen-led energy action*, which includes data from 29 countries and highlights over 10,000 such initiatives [1].

These initiatives, commonly referred to as Citizen Energy Communities (CEC), vary in size, configuration, and capacities in terms of the renewable energy sources involved, such as photovoltaic or wind plants, as well as other devices deployed, including energy storage batteries, energy consumption devices, and green hydrogen production devices, among others. The primary objective shared by these initiatives is to enhance self-consumption of locally produced renewable energy.

Service providers, which may include ICT companies specializing in integrating various energy services, can derive benefits from these initiatives. They may earn service fees based on the contracted share of energy savings and receive payments for providing unlocked flexibility and automated demand response (DR) mechanisms [2] under Energy

© The Author(s), under exclusive license to Springer Nature Switzerland AG 2024
M. Trajanovic et al. (Eds.): ICIST 2023, LNNS 872, pp. 375–383, 2024.
https://doi.org/10.1007/978-3-031-50755-7_35

Performance Contracting (EPC) [3] and Pay-for-Performance (P4P) arrangements [4] established with utilities.

Simultaneously, stakeholders involved in the power grid stand to gain advantages from the rise of citizen-led energy initiatives. These advantages encompass reduced maintenance and operation costs resulting from improved grid stability and lower transmission losses, courtesy of the increased hosting capacity for local renewable energy sources.

Furthermore, the proliferation of citizen-led energy initiatives is not only beneficial in terms of renewable energy production but also contributes to the overall sustainability goals of communities. By promoting the use of clean energy sources and reducing reliance on traditional fossil fuel-based power generation, these initiatives help combat climate change and reduce greenhouse gas emissions. They also foster a sense of local empowerment and engagement, as citizens actively participate in the transition towards a more sustainable energy future. In addition to the environmental and community benefits, citizen-led energy initiatives also have the potential to create new economic opportunities. These initiatives often generate local jobs, ranging from installation and maintenance of renewable energy systems to the development of innovative technologies and services. Moreover, they encourage entrepreneurship and foster a supportive ecosystem for local businesses, such as renewable energy equipment suppliers, energy consultants, and energy efficiency specialists.

To ensure the long-term success and scalability of citizen-led energy initiatives, collaboration and coordination among various stakeholders are crucial. This includes close cooperation between the community members, local authorities, energy service providers, and utility companies. By working together, these stakeholders can develop effective regulatory frameworks, streamline administrative processes, and establish clear guidelines for the integration of citizen-led energy initiatives into the existing energy infrastructure. The increasing number of citizen-led energy initiatives in Europe reflects a growing momentum towards decentralized, renewable energy systems. These initiatives not only contribute to the transition to a cleaner and more sustainable energy sector but also bring about economic opportunities, community engagement, and local empowerment. With continued support and collaboration, citizen-led energy initiatives have the potential to revolutionize the energy landscape, leading to a more resilient, environmentally friendly, and inclusive energy future.

2   Research Questions

In this paper, we present a novel software platform that serves as a solution for integrating diverse data-driven services within Citizen Energy Communities (CEC). The platform not only facilitates seamless connectivity between the physical energy assets but also enables efficient data exchange and processing among different actors and software systems involved in the CEC ecosystem.

By leveraging advanced technologies and data analytics capabilities, the proposed software platform empowers CECs to optimize their energy management strategies and enhance the overall performance of renewable energy assets. It enables real-time monitoring and control of energy generation, consumption, and storage systems, allowing for efficient allocation and utilization of resources. The platform also supports the integration of emerging technologies such as demand response mechanisms, energy forecasting algorithms, and grid optimization tools, enabling CECs to actively participate in grid balancing and provide valuable flexibility services.

The research conducted in this paper is closely aligned with the objectives of the NEON (Next-Generation Integrated Energy Services for Citizen Energy Communities) and SINERGY (Capacity building in Smart and Innovative eNERGY management) projects. The NEON project aims to develop innovative solutions for integrating renewable energy resources, enhancing energy efficiency, and fostering sustainable practices within CECs. In NEON, the platform provides a robust and scalable solution for data-driven energy management in CECs from Italy, France and Spain, while in SINERGY, the platform is adopted for testing services relevant for the IMP campus in Serbia.

Furthermore, we acknowledge the significance of adoption of interoperability standards for future data spaces, tested in the EU project OMEGA-X (Orchestrating an interoperable sovereign federated Multi-vector Energy data space built on open standards and ready for GAia-X) as a relevant initiative in this context. The aim of OMEGA-X is to implement a data space (based on European common standards), including federated infrastructure, data marketplace and service marketplace, involving data sharing between different stakeholders and demonstrating its value for real and concrete Energy use cases and needs, while guaranteeing scalability and interoperability with other data space initiatives, not just for energy but also cross-sector. By aligning with the principles and objectives of the OMEGA-X project, the proposed software platform for CECs aims to promote interoperability and integration between different energy systems, devices, and software applications, fostering a more interconnected and efficient energy ecosystem.

Furthermore, this paper introduces a software platform that addresses the integration challenges within CECs, enabling seamless connectivity, data exchange, and processing among diverse actors and software systems. The platform aligns with the goals of the NEON project and emphasizes the importance of interoperability solutions exemplified by initiatives like OMEGA-X. Through the adoption of this software platform, CECs can unlock the full potential of data-driven energy management, leading to improved energy efficiency, increased renewable energy utilization, and enhanced grid integration capabilities.

3   Methodology

The design of the platform architecture is a result of analysis and consideration of various standard-enabling technologies [5] and practices. These include cloud-based infrastructures, service-oriented architectures, blockchain technology, flexibility and loosely coupled design principles, interoperability, security and privacy by design, and configuration management. By incorporating these elements, the platform architecture aims to create a robust and scalable foundation for the integration of diverse CEC energy services.

One key aspect considered during the design process is the COSMAG (COntextualisation of the Smart Grid Architecture) and SGAM (Smart Grid Architecture Model) specifications. These specifications provide a comprehensive framework for structuring and organizing the various components and functions within smart grid systems.

By aligning with COSMAG and SGAM, the platform architecture ensures compatibility and harmonization with existing smart grid infrastructures, enabling seamless integration and interoperability between CECs and the broader energy grid ecosystem.

Moreover, an "ethics by design" approach is followed to guarantee compliance with the European ethical and legal framework. This approach encompasses adherence to regulations such as the NIS (Network and Information Security) Directive, eIDAS (electronic Identification, Authentication, and Trust Services), and GDPR (General Data Protection Regulation). By integrating ethical considerations from the early stages of design, the platform architecture prioritizes data protection, security, and privacy. This approach ensures that the platform safeguards the personal and sensitive information of individuals while promoting transparency and accountability in data handling processes.

In addition to legal and ethical compliance, the platform architecture emphasizes the importance of configuration management. This aspect involves effectively managing and controlling the various configurations and settings of the platform to ensure optimal performance and adaptability. Through robust configuration management practices, the architecture enables efficient customization and adaptation of the platform to suit the specific needs and requirements of different CECs, while maintaining stability, reliability, and consistency.

By incorporating standard-enabling technologies, following ethical design principles, considering COSMAG and SGAM specifications, and implementing effective configuration management, the platform architecture is designed to empower CECs with a secure, interoperable, and customizable framework for integrated energy services. It sets the stage for the deployment of advanced energy management solutions within CECs, promoting sustainable practices, and enabling the seamless integration of renewable energy resources into the broader energy ecosystem.

In Fig. 1, we present the platform architecture. During the initial year of the NEON project (in 2021 and 2022), the Institute Mihajlo Pupin team dedicated significant effort to conducting a comprehensive analysis of the available components and services that could be utilized within the project. These components and services were carefully examined and subsequently mapped to the interoperability layers specified in the SGAM:

- Business Layer encompasses the applications and dashboards that facilitate the management and visualization of data within the NEON project. This layer focuses on providing user-friendly interfaces and tools for CECs to monitor and control their energy systems effectively.
- Function Layer constitutes a crucial aspect of the NEON project architecture, as it consists of various components and services that perform specific functions. This layer plays a vital role in enabling the desired energy management capabilities and services within CECs.
- Information Layer is responsible for managing the information used and exchanged between different functions, services, and components within the NEON project. It serves as a crucial communication hub, ensuring the seamless flow of data across various aspects of the project.
- Communication Layer focuses on defining the protocols and mechanisms necessary for the interoperable exchange of information between the different components within the NEON project. This layer ensures that the various systems and devices

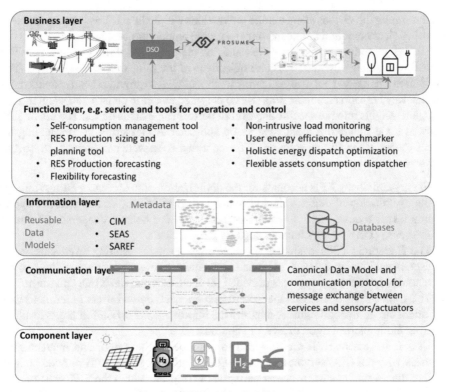

Fig. 1. NEON platform architecture

involved can communicate and share data effectively, promoting interoperability and seamless integration.

- Component Layer pertains to the physical distribution of all the participating components within the smart grid context. This layer encompasses the deployment of hardware and software components across the CECs, enabling the realization of the NEON project's goals in a tangible and practical manner.

By organizing the NEON project's components and services into these interoperability layers, the consortium established a structured and comprehensive framework for the development and implementation of the project. This approach ensured that all necessary aspects, from business applications to physical components, were considered and integrated harmoniously, fostering the successful deployment and operation of the NEON platform within Citizen Energy Communities.

The UML (Unified Modeling Language) modelling language was utilized to elaborate on all aspects of the self-consumption process within the NEON project. This required analysis of the specific requirements and needs of four different Citizen Energy Communities (CECs) - BERCHIDDA in Spain, DOMAINE DE LA SOURCE in France, POLÍGONO INDUSTRIAL LAS CABEZAS in Spain, and STAINS CITY in France. The UML modelling language provides a standardized and comprehensive framework for describing the various elements and interactions within the self-consumption process, ensuring clear communication and understanding among project stakeholders.

To assess and measure the performance of the pilot sites during operation, it was crucial to evaluate how the goals and objectives of the pilot sites were achieved. This evaluation was carried out using scientific methodologies to provide accurate and reliable results. Key Performance Indicators (KPIs) provided means to quantify different metrics and gain insights into the specific and overall performance of the CECs. The use of KPIs allowed for a standardized and systematic approach to measuring and evaluating the effectiveness of the NEON solutions. The identified KPIs were categorized into several key areas:

- Energy Efficiency KPIs accounts for the optimization of users' energy usage through the exploitation of demand flexibility and energy efficiency of multi-carrier opportunities. It focuses on the benefits derived from the holistic cooperative Demand Response (DR) strategy implemented within the CECs.
- The Economic KPIs evaluates the economic savings resulting from changes in user behaviour as a result of their engagement and energy usage following the recommendations and services provided for the CECs and the NEON platform.
- The Comfort KPIs assesses the benefits experienced by end users in terms of their indoor environment. It aims to measure the improvements in comfort levels resulting from the implementation of NEON solutions.
- User Engagement KPIs are designed to describe the behaviour and interaction of users with the CEC services and the NEON platform. These KPIs provide insights into the level of engagement and participation of users within the CEC ecosystem.
- The Social KPIs explores how the required levels of flexibility intersect with social norms and everyday practices, such as routines and family life. It also considers the effects of CECs on health and well-being, emphasizing the social impact of NEON solutions.
- Environmental KPIs evaluate the impact of NEON solutions on the local environment, focusing on aspects such as carbon footprint reduction, greenhouse gas emissions, and other environmental indicators.
- Technical category encompasses KPIs that evaluate different technical characteristics of the CEC services and systems. These KPIs provide insights into the performance, reliability, and functionality of the technical infrastructure supporting the NEON platform.

By defining and measuring these diverse categories of KPIs, one can comprehensively evaluate the performance and impact of the proposed solutions. This allows for evidence-based decision-making, continuous improvement, and the refinement of the platform and services to ensure optimal outcomes within the CECs (Fig. 2).

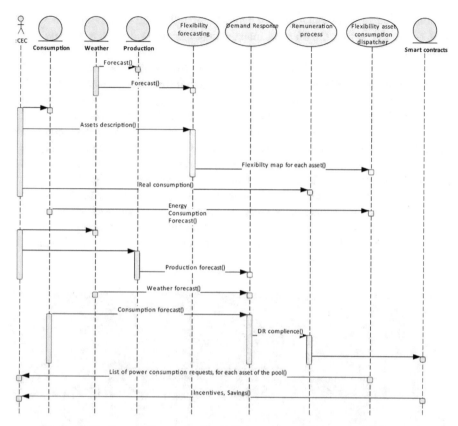

Fig. 2. CEC - Demand Response, Flexibility services and Remuneration Process

4   Solution/Discussion

The platform designed, installed and tested at the Institute Pupin premises in the NEON project framework, has been adopted for the forthcoming activities in SINERGY [6] and OMEGA-X projects [7]. This installation serves as a crucial step in the development and validation of the platform's capabilities. During the testing phase, services for energy dispatch optimization, demand and production forecasting have been put to the test. These services focus on optimizing the dispatch and distribution of energy resources within the platform. By analysing the available data and utilizing advanced algorithms for production and demand forecasting and optimization, the energy dispatch optimization service aims to maximize the efficiency and effectiveness of energy distribution.

The data utilized in the testing process is sourced from Spanish CEC, providing a real-world context for evaluating the performance and functionality of the platform. Overall, the installation of the platform at the Institute Pupin premises and the subsequent testing using data from Spain, along with the components provided by the Institute Pupin, represents a significant milestone in the development and evaluation of the NEON project. These activities contribute to the refinement and enhancement of the platform's capabilities, ensuring its suitability for deployment within Citizen Energy Communities (CECs) and promoting the efficient management and utilization of renewable energy resources.

In Fig. 3, we show Grafana web interface for visualization of results provided by services. The top graph represents the PV production forecasts. In essence, it provides predictions regarding the amount of electricity that will be generated by solar panels over a specific timeframe. The middle one shows the demand forecast. This graph presents projected energy consumption levels during the same period. It effectively communicates when energy usage is expected to peak or decline based on predictive analytics. The graph at the bottom presents the optimal demand which signifies the recommended approach for energy consumption. This recommendation is based on an alignment of energy consumption with the availability of renewable energy generated by the PV plant and other system parameters such as battery capacity, maximum allowed power from/into grid, etc. As it can be seen, the optimization recommends to shift the energy consumption towards the period where renewable energy from the PV plant is available, while also respecting the system constraints.

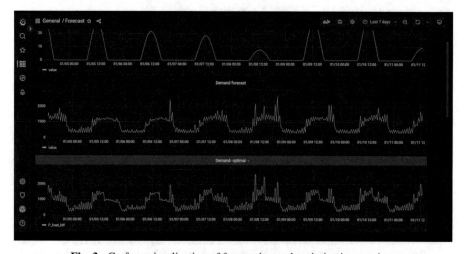

Fig. 3. Grafana visualization of forecasting and optimization services

5 Conclusion

In this paper we have highlighted the diverse scales and configurations of these initiatives, emphasizing the goal of increasing self-consumption and the potential benefits for service providers and grid stakeholders.

In this paper, we have discussed the proposed solution of the NEON platform, which serves as a software platform for integrating data-driven services and connecting physical energy assets within CECs. The design of the NEON platform architecture was elaborated, considering standard-enabling technologies, interoperability solutions, and ethical and legal compliance. The NEON project's progress and accomplishments were presented, including the detailed analysis of components and services mapped to SGAM interoperability layers, as well as the measurement of performance through key performance indicators (KPIs). Finally, we have provided more details on implementation of platform with the focus on forecasting and optimization service which were applied on the data obtained from Spain CEC. As part of the future work, additional short, mid- and long-term planning services will be integrated and tested using data from the IMP campus.

Acknowledgement. This work was supported by the EU H2020 funded projects NEON (Next-Generation Integrated Energy Services fOr Citizen Energy CommuNities, GA No. 101033700); SINERGY (Capacity building in Smart and Innovative eNERGY management, GA No. 952140) and OMEGA-X (Orchestrating an interoperable sovereign federated Multi-vector Energy data space built on open standards and ready for GAia-X, GA No. 101069287).

References

1. Wierling, A., Schwanitz, V.J., Zeiss, J.P., et al.: A Europe-wide inventory of citizen-led energy action with data from 29 countries and over 10000 initiatives. Sci. Data **10**, 9 (2023)
2. Jelić, M., Batić, M., Tomašević, N.: Demand-side flexibility impact on prosumer energy system planning. Energies **14**(21), 7076 (2021)
3. Shang, T., Zhang, K., Liu, P., Chen, Z.: A review of energy performance contracting business models: status and recommendation. Sustain. Cities Soc. **34**, 203–210 (2017). https://doi.org/10.1016/J.SCS.2017.06.018
4. Szinai, J., et al.: Putting Your Money Where Your Meter Is: A Study of Pay-for-Performance Energy Efficiency Programs in the United States (2017)
5. Berbakov, L., Batić, M., Tomašević, N.: Smart energy manager for energy efficient buildings. In: Proceedings of the 18th International Conference on Smart Technologies (IEEE EUROCON 2019), Novi Sad, pp. 1–4 (2019)
6. SINERGY Project Pilot 3. https://project sinergy.org/Pilot-3
7. OMEGA-X Project Pilots. https://omega-x.eu/pilots/

Framework for Optimizing Neural Network Hyper Parameters for Accurate Wind Production Forecasting

Dea Pujić$^{(\boxtimes)}$ ⓘ and Valentina Janev ⓘ

Institute Mihajlo Pupin, University of Belgrade, Volgina 15, 11060 Belgrade, Serbia
dea.pujic@pupin.rs

Abstract. To address the escalating issue of burnt fossil fuels, there has been a persistent rise in the integration of Renewable Energy Sources (RES) within the energy production sector. However, this transition has led to a destabilization of the electrical grid due to an inherent mismatch between the production of renewable energy and the corresponding demand. Consequently, the accurate day-ahead forecasting of RES production has emerged as a pivotal component in energy planning and dispatch. Recognizing the efficacy of neural networks in production forecasting, this paper introduces a comprehensive framework aimed at optimizing the hyperparameters of these networks. The primary objective of this framework is to minimize the time required for model selection while ensuring an automated and generic approach to the modeling process. By fine-tuning the hyperparameters of neural networks, researchers and practitioners can enhance the precision of RES production forecasts. The optimization process allows for the identification of the optimal configuration, resulting in more accurate predictions. This, in turn, assists energy planners and operators in effectively managing the challenges associated with the integration of renewable energy into the electrical grid. The proposed framework offers a systematic and efficient method for determining the hyperparameters of neural networks. By reducing the time needed for model selection, it enables swift decision-making processes and facilitates the adoption of an automatic and generic approach to RES production forecasting. This research endeavors to contribute to the ongoing efforts to address the destabilization of electrical grids caused by the growing penetration of renewable energy sources.

Keywords: wind production forecaster · neural networks · artificial intelligence · grid stability

1 Introduction

In recent years, the introduction of renewable energy sources (RES) such as photovoltaic panels and wind turbines has been instrumental in the global effort to reduce reliance on fossil fuels. However, one of the inherent challenges associated with these sources is their stochastic nature, primarily driven by the strong dependency between their production and prevailing meteorological conditions. This dependency makes the planning and utilization of RES more complex.

M. Trajanovic et al. (Eds.): ICIST 2023, LNNS 872, pp. 384–391, 2024.
https://doi.org/10.1007/978-3-031-50755-7_36

As RES continue to gain prominence and increase their market share in the electricity sector, maintaining a stable grid becomes paramount. The key to achieving grid stability lies in effectively matching the energy production from RES with the overall demand. To accomplish this, accurate predictions of accessible energy become crucial. In light of this, the focus of this paper is on the development of an accurate RES production forecaster, specifically targeting wind turbine production.

The unpredictable and fluctuating nature of wind poses a significant challenge for energy planners and grid operators. Reliable forecasts of wind turbine production enable better management of grid resources and facilitate the integration of wind power into the existing energy infrastructure. By providing accurate predictions, the forecasters contribute to the efficient planning and balancing of the electricity grid, reducing the risk of power imbalances and ensuring a stable energy supply.

This paper recognizes the need for an advanced and precise wind turbine production forecaster as an essential component in the successful integration of wind energy into the electrical grid. Through detailed analysis and modelling techniques, the research aims to enhance the accuracy of wind production forecasts, enabling more informed decision-making processes and promoting the optimal utilization of renewable energy resources.

2 State of the Art Analysis

In the field of RES production forecasting, a wide range of solutions have been proposed, each tailored to the specific renewable energy source and forecasting horizon under consideration. When it comes to short-term production forecasting, probabilistic models such as modified versions of Box-Jenkins models have gained popularity [1]. However, the focus of this paper lies in day-ahead production forecasting for grid balancing purposes.

A study by [2] delved into a similar setup, providing a comprehensive comparison of various machine learning approaches for production forecasting of photovoltaic and solar thermal collectors. It highlights the significance of machine learning (ML) and deep learning models, particularly in scenarios where ample data is available for day-ahead forecasting [3]. Therefore, this research aims to leverage the benefits of hybrid deep neural networks.

The literature offers a diverse range of neural network architectures for production forecasting, including multilayer perceptron [4], radial basis function neural networks [5], convolutional neural networks (CNN) [6], and long short-term memory (LSTM) networks [7]. However, [1] emphasizes that the highest performances are achieved through the use of hybrid neural network configurations. Consequently, the primary focus of this paper revolves around the development of a framework to optimize the hybrid neural network architecture specifically for day-ahead wind production forecasting at an hourly resolution.

3 Methodology

Neural networks (NNs) have proven to be highly effective in solving complex problems with remarkable precision, such as image recognition and natural voice processing. However, achieving optimal estimation performance heavily relies on carefully selecting the network architecture and hyperparameters. The vast number of potential solutions can make hyperparameter optimization a daunting task, necessitating the use of automated tools. With the objective of developing a high-precision solution, this paper presents a Python-based framework designed to optimize the hyperparameters and architecture of NNs specifically for wind production forecasting.

In this framework, the optimization engine aims to identify the most performant architecture and corresponding training parameters, thereby enabling the creation of the most accurate wind production forecasting model. While various parameters can be optimized, this paper focuses on the following targets: the **number** and **types** of **hidden layers**, the **number** of **training epochs**, and the **learning rate**. To explore the architecture optimization space comprehensively, the framework tests numerous combinations of four different types of hidden layers: LSTM, convolutional, dense, and dropout layers.

An evolutionary heuristic approach was used for the model selection. Namely, an initial set of models has been randomly initialized. Afterwards, the models that showed better performances were favored to be used as the base for slightly modified models in the new generation. However, the introduction of newly created models could also happen, in order to overcome the problem of overfitting. Nevertheless, the probability of creating significantly different models in the new generation was decreasing with the generations. Finally, this was repeated until convergence.

All the optimized models shared common input and output configurations. The model was designed to map estimated meteorological parameters, including wind speed, wind direction, and outside temperature, to the expected wind energy production. Leveraging the available historical data, these inputs were fed into the model, and the output was the anticipated energy production. The models were designed with the aim of daily ahead forecasting application, but could be utilized for up to couple of days ahead horizon, in case corresponding meteorological forecast is available.

By optimizing the architecture and hyperparameters through the proposed framework, this research aims to enhance the precision and reliability of wind production forecasting and reduce time for its development. The inclusion of meteorological parameters as inputs enables the model to capture relevant information and provide accurate predictions, facilitating effective energy planning and grid balancing in the context of renewable energy integration.

4 Results

The development and testing process of the proposed framework began with thorough data analysis, as it is a data-driven approach. Historical data from the Krnovo plant in Montenegro was obtained for a period of six months, consisting of both production and meteorological measurements. However, a discrepancy was observed in the sampling frequency of the meteorological and production data, with the former being recorded every 10 min and the latter every hour. To address this issue, the meteorological data was down sampled to match the production data, ensuring compatibility between the two datasets.

Next, a comprehensive data analysis process was conducted to establish the relationship between the meteorological data and the target production variable. This step was crucial for enabling the models to extract relevant information from the input and map it to the desired output. The primary focus was on investigating the influence of meteorological parameters, particularly wind speed, on wind production. To visually demonstrate this relationship, Fig. 1 and Fig. 2 were provided as representative illustrations.

Figure 1 depicted the peaks in production and wind speed occurring at the same time, indicating a strong correlation between the two variables. Conversely, when wind speed was low, the production level was also observed to decrease. This graphical representation served as evidence supporting the hypothesis that wind production is highly dependent on meteorological factors, particularly wind speed. Apart from the visual presentation, actual Prison correlation coefficient was calculated between wind speed and produced energy resulting in high 0.80.

Furthermore, Fig. 2 showcased the relationship between production and wind speed. The left figure demonstrated the actual dependency between these variables obtained from the historical data. The right figure displayed the theoretical dependency obtained from (8). The alignment between the historical data and the theoretical dependency further confirmed that the extracted input features were indeed correlated with the target variable and exhibited the expected distribution.

To ensure the integrity of the data used for model development, the process of removing outliers was undertaken. The data points were separated into bins of 200 W based on the specific production values. Points located in the tails of the corresponding distribution were identified as outliers and subsequently eliminated. This outlier removal step aimed to enhance the quality of the dataset, thereby improving the accuracy and reliability of the subsequent modeling process.

By conducting this comprehensive data analysis, the proposed framework laid a solid foundation for the subsequent development and testing stages. The down sampling of meteorological data enabled compatibility with the production data, ensuring a consistent and coherent dataset. The established correlation between the meteorological parameters and wind production, as illustrated in Fig. 1 and Fig. 2, provided substantial evidence supporting the framework's underlying assumptions. Furthermore, the removal of outliers from the dataset increased the robustness and reliability of the subsequent modeling efforts.

Once the data analysis, clearance, and preprocessing stages were completed, the next step involved developing and testing NN framework for hyperparameter optimization. The goal was to create models that could accurately forecast the production of a wind turbine with a capacity of 2.85 MW, considering the three meteorological parameters mentioned earlier. The same dataset was used for training, validation, and testing across all models. Available data were split into three data sets – four months for training, and a month for validation and testing.

A total of 250 NN models were trained, and the final model was selected based on the lowest root mean square error (RMSE) on the validation data. The optimal NN architecture is presented in Table 1. Notably, all the layers used the same activation function. Since the model's output is constrained by the turbine's capacity, the decision was made to employ the activation function that is limited. In order to reduce the complexity in the search space, it was decided to select one activation function for all layers between the following – *sigmoid*, *tansig* and *tanh*. *Tansig* was selected after couple of experiments, since it showed the best performance of the three. The optimal architecture comprised a total of nine layers, out of which three (33%) were dropout layers used for regularization. By using predefined rules for including dropout layers, the search space was further narrowed down.

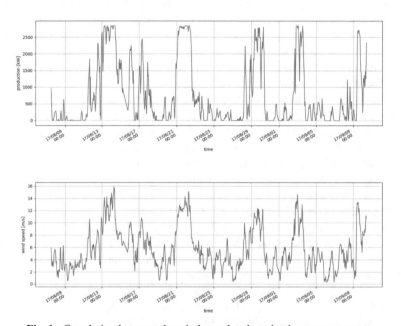

Fig. 1. Correlation between the wind speed and production measurements

For training all NN models, the ADAM optimization method was utilized, with mean square error serving as the criterion function. The optimal model had a learning rate of 0.0001. The training process was carried out for 500 epochs, but the minimal loss on the validation set was achieved by the optimal model in the 387th epoch. The optimal model demonstrated the following RMSE performances: **0.142 MW** on the training data, **0.154 MW** on the validation data, and **0.159 MW** on the testing data.

Aside from the numerical evaluation of the optimal model's performance, Fig. 3 provided an example of the production forecasting model estimation using the testing data. In the upper picture performance on the whole testing set is given, with a month of data sampled on hourly resolution. Therefore, the figure displayed a comparison between the real production and the estimated production of the optimal model per testing sample. It was evident that the model performed with high precision, particularly considering the rapidly changing wind speeds and the fact that the forecast was provided at an hourly time resolution. Additionally, more details could be noticed on the bottom picture, where the results were zoomed form approximately 70th to 160th testing samle.

The accuracy of the optimal model was more than satisfactory, considering the challenges associated with predicting wind production, given the inherent variability of wind speeds. The model's ability to provide accurate forecasts demonstrated its effectiveness in leveraging the available meteorological data to predict the wind turbine's production. This capability is highly valuable for wind energy operators and stakeholders as it allows for better planning, scheduling, and optimization of resources.

Furthermore, the developed NN framework and the optimization process ensured that the final model was selected based on thorough evaluation and validation, providing confidence in its performance. The use of dropout layers for regularization contributed to the model's generalization ability and reduced the risk of overfitting, enhancing its reliability when applied to new and unseen data.

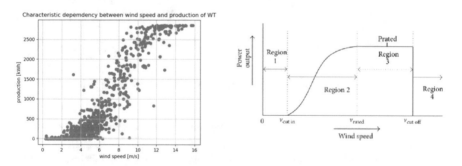

Fig. 2. Correlation between the wind speed and production measurements

In conclusion, the developed framework for optimizing neural network architectures has yielded a highly effective model capable of accurately forecasting wind production based on meteorological conditions. As a result, this framework significantly reduces the workload for future renewable energy production forecasters, as it establishes an automated approach for designing forecasting models. By utilizing the proposed framework, the optimal architecture can be obtained, ensuring precise forecasts and streamlining the forecasting process for renewable energy production.

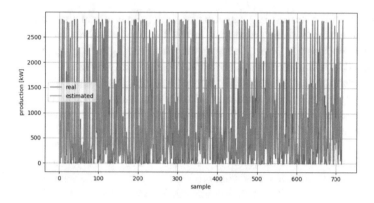

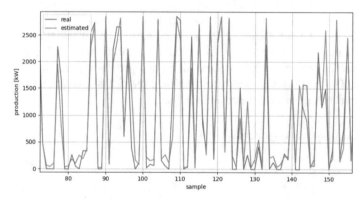

Fig. 3. Example of production forecasting model estimation on the testing data

Table 1. Architecture of the optimal deep neural network for wind production forecasting

Layer	num. of filters	filter size	num. of neurons	activation function	factor
LSTM	–	–	64	tansig	–
LSTM	–	–	128	tansig	–
Dropout	–	–	–	–	0.5
CNN	64	3	–	tansig	–
Dropout	–	–	–	–	0.5
Dense	–	–	300	tansig	–
Dense	–	–	150	tansig	–
Dropout	–	–	–	–	0.5
Dense	–	–	1	tansig	–

5 Conclusion

In this study, a comprehensive framework for optimizing neural network architectures was developed and tested for accurate wind production forecasting based on meteorological conditions. Through thorough data analysis, preprocessing, and model training, the framework successfully generated a highly performant model with precise forecasting capabilities. The proposed approach significantly reduces the workload for future renewable energy production forecasters by establishing an automated methodology for designing forecasting models. By leveraging the optimal architecture obtained through the framework, forecasters can confidently generate reliable predictions, enabling better planning, scheduling, and resource optimization in the wind energy sector. The results of this study demonstrate the effectiveness of the developed framework and its potential to contribute to the advancement of renewable energy forecasting. Further research can focus on expanding the framework to incorporate additional hyperparameters and heuristics for their optimization and exploring its applicability in other renewable energy domains. Overall, the findings of this study provide valuable insights and pave the way for improved forecasting models in the renewable energy industry.

Acknowledgement. The research presented in this paper is partly financed by the European Union (Horizon 2020 PLATOON project GA #872592 and SINERGY project GA #952140) and partly by the Ministry of Education, Science and Technological Development.

References

1. Hanifi, S., Liu, X., Lin, Z., Lotfian, S.: A critical review of wind power forecasting methods—past, present and future. Energies **13**(15), 3764 (2020)
2. Pujić, D., Jelić, M., Tomašević, N.: Comparison between different ML approaches for PV and STC production forecasting using real world data. In: ICIST 2020 Proceedings [Internet]. Society for Information Systems and Computer Networks; 2020 [cited 2020 Nov 26, pp. 94–98. http://www.eventiotic.com/eventiotic/library/paper/592
3. Voyant, C., Notton, G., Kalogirou, S., Nivet, M.L., Paoli, C., Motte, F., et al.: Machine learning methods for solar radiation forecasting: a review. Renew. Energy **1**(105), 569–582 (2017)
4. Bilal, B., Ndongo, M., Adjallah, K.H., Sava, A., Kebe, C.M.F., Ndiaye, P.A., et al.: Wind turbine power output prediction model design based on artificial neural networks and climatic spatiotemporal data. In: 2018 IEEE International Conference on Industrial Technology (ICIT), pp. 1085–1092 (2018)
5. Wu, X., Hong, B., Peng, X., Wen, F., Huang, J.: Radial basis function neural network based short-term wind power forecasting with Grubbs test. In: 2011 4th International Conference on Electric Utility Deregulation and Restructuring and Power Technologies (DRPT), pp. 1879–1882 (2011)
6. Zhu, A., Li, X., Mo, Z., Wu, R.: Wind power prediction based on a convolutional neural network. In: 2017 International Conference on Circuits, Devices and Systems (ICCDS), pp. 131–135 (2017)
7. Shahid, F., Zameer, A., Muneeb, M.: A novel genetic LSTM model for wind power forecast. Energy **15**(223), 120069 (2021)
8. Shilpa, G.N., Sheshadri, G.S.: ARIMAX model for short-term electrical load forecasting. IJRTE **8**(4), 2786–2790 (2019)

Demand-Side Optimization of Hybrid Energy Systems with Heat Pumps

Marko Jelić[(✉)] [iD] and Marko Batić [iD]

Mihajlo Pupin Institute, University of Belgrade, Volgina 15, Belgrade, Serbia
{marko.jelic,marko.batic}@pupin.rs

Abstract. Ensuring sustainability and energy-efficient operation is of utmost importance in the current era as humanity strives to facilitate the transition to green energy. As our energy systems become increasingly complex, incorporating multiple energy carriers and power transformation devices, it is crucial to find innovative solutions to maximize their potential. Heat pumps have emerged as a key enabler of sustainable heating and cooling, while also offering demand-side flexibility. However, effectively harnessing this flexibility poses a significant challenge. This paper presents a comprehensive methodology for optimizing hybrid energy systems in the long term by leveraging various flexibilities. By integrating different energy sources such as renewable energy generation and heat pumps, the system's overall performance and efficiency can be significantly enhanced. The proposed optimization framework aims to determine the optimal operation strategy for hybrid energy systems, considering factors such as energy demand, weather conditions, and electricity prices. It considers the dynamic nature of energy supply and demand, utilizing advanced optimization algorithms and predictive modeling techniques. By intelligently managing the flexibilities inherent in hybrid energy systems, the proposed methodology seeks to minimize operational energy costs, reduce greenhouse gas emissions, and enhance overall system efficiency. The outcomes of this research will contribute to the development of sustainable energy systems that are adaptable, efficient, and capable of accommodating the fluctuating nature of renewable energy sources. The findings can inform policymakers, energy planners, and system operators in making informed decisions to facilitate the transition towards a greener and more sustainable future.

Keywords: hybrid energy system · heat pump · demand-side flexibility · energy dispatch optimization

1 Introduction

In order to determine the feasibility of heat pump (HP) and renewable energy source (RES) installations, various factors need to be taken into consideration. Firstly, the geographical location plays a crucial role in assessing the potential for utilizing RES such as photovoltaics (PVs) and wind turbines. The availability of solar radiation and wind resources in a particular area directly impacts the effectiveness and efficiency of these RES technologies. Moreover, the economic viability of HP and RES installations must be thoroughly examined. This includes considering the initial investment costs, operational and maintenance expenses, as well as potential financial incentives or subsidies available for adopting these technologies. The energy demand profile of the buildings or facilities also plays a significant role in determining the appropriateness of HPs and RES. Assessing the energy consumption patterns and load requirements can help determine the optimal sizing and configuration of the HP and RES systems.

Furthermore, in order to comprehensively analyze RES installations, it is essential to assess their environmental impact. Evaluating the potential reduction in greenhouse gas emissions and the overall sustainability of the energy system is crucial for making informed decisions. Life cycle assessments and environmental impact analyses can provide valuable insights into the environmental benefits and potential drawbacks of implementing HPs and RES technologies. By conducting comprehensive assessments and verifications of the feasibility of HP and RES installations, energy end users can make informed decisions regarding their energy systems. These evaluations can help determine the potential benefits, challenges, and overall viability of integrating HPs and RES technologies into residential and commercial buildings, ultimately contributing to the development of efficient and sustainable energy systems in the 21st century.

With the aim of contributing to the previously mentioned points necessary for assessing RES installations, this paper places a particular focus on operational optimization of hybrid, multi-carrier, energy systems with demand-side flexibility through load modifications.

2 State of the Art

Relevant work in this field approaches this topic from different angles. Various approaches regarding demand-side energy management, particularly oriented towards smart grids, can be found in literature, as presented in [1]. For example, many consider RES and battery-powered microgrids like in [2], but focuses solely on the electric domain. In [3], the authors build upon this idea and also include a portion of the electric demand that is a result of HP operation, but do not explicitly include thermal flexibility. A hybrid system that includes PV and HP is studied in [4], but the approach focuses on low-level operational parameters rather than more abstract efficiency indications. A study from [5] includes a techno-economic analysis with demand-side flexibility aspects, but focuses on higher-scale district heating systems. Some authors apply heuristic approaches for cost reduction in hybrid renewable energy system optimization like in [6]. This paper aims to extend the state of the art by proposing a methodology for modeling these types of system while allowing for the demand-side flexibility to be extended beyond the electric into the thermal domain. It also aims to provide a basis for more complex methodologies to be developed later on as extensions that include more precise flexibility modelling by means of building models.

3 Methodology

The methodology presented in this paper consists of a three-step process that is utilized to assess long-term performances of hybrid energy dispatching systems that can be suitable as a basis for further sizing and economical studies. These steps are:

1. Linearization of complex dynamics within the system;
2. Implementation of a mixed-integer linear programming (MILP) optimization model;
3. Result assessment,

as discussed in greater detail in the following sections. In order to more clearly separate constants from variables in the modeling stages, the former are written in bold letters.

3.1 Resolution of Non-linear Dynamics

Complex and non-linear dynamics in the discussed scenario consist of estimation of variable air-source HP (ASHP) efficiency, maximum heat power output calculation, load and RES production forecasting as well as demand level and demand flexibility forecasting. These processes can be summarized with a set of functions f as given below. The first one depicts a variable HP efficiency estimation depending on the instantaneous source $T_{src}(t)$ and sink $T_{user}(t)$ temperatures as well as a predetermined quality grade parameter $\boldsymbol{\eta_{QG}}$ and an icing penalty factor $\boldsymbol{f_{ice}}(t)$ that defines the performance reduction occurring when ice deposits start to form on the HP's evaporator as summarized in [7] and given by

$$(\forall t) \ \boldsymbol{COP_{ASHP}}(t) = f_1\left(\boldsymbol{T_{src}}(t), \boldsymbol{T_{user}}(t), \boldsymbol{\eta_{QG}}, \boldsymbol{T_{ice}}\right) = \boldsymbol{f_{ice}}(t)\boldsymbol{\eta_{QG}} \frac{T_{user}(t)}{T_{user}(t) - T_{src}(t)}.$$

Once the coefficient of performance (COP) $\boldsymbol{COP_{ASHP}}(t)$ is estimated for each time step, it can be utilized to estimate the total maximum heat power output based on a predefined HP nominal capacity given by $\boldsymbol{\dot{Q}_{ASHP_0}}$ and COP $\boldsymbol{COP_{ASHP_0}}$. This relation can be express using the formula

$$(\forall t) \ \boldsymbol{\dot{Q}_{HP}^{max}}(t) = f_2\left(\boldsymbol{T_{src}}(t), \boldsymbol{T_{user}}(t), \boldsymbol{\eta_{QG}}, \boldsymbol{T_{ice}}\right) = \boldsymbol{\dot{Q}_{ASHP_0}} \frac{COP_{ASHP}(t)}{COP_{ASHP_0}}.$$

For the purposes of RES forecasting, the procedure outlined by [8] is implemented. This procedure allows the PV production to be summarized as a function of several environmental conditions including the ambient outdoor air temperature $\boldsymbol{T_{air}}(t)$, effective solar irradiance $\boldsymbol{E_{rad}}(t)$ and wind speed $\boldsymbol{v_{wind}}(t)$, as given by

$$(\forall t) \ \boldsymbol{P_{PV}^{forecast}}(t) = f_3(\boldsymbol{T_{air}}(t), \boldsymbol{E_{rad}}(t), \boldsymbol{v_{wind}}(t)).$$

On the other hand, demand curves for appliances and heating are calculated based on distributed total yearly consumptions $\boldsymbol{P_{app}^a}$ and $\boldsymbol{\dot{Q}_{demand}^a}$ in accordance with the normalized profiles given in [3], as defined by two power forecasting function

$$(\forall t) \ \boldsymbol{P_{app}^{forecast}}(t) = f_4\left(\boldsymbol{P_{app}^a}, t\right) \text{ and } \boldsymbol{\dot{Q}_{demand}^{forecast}}(t) = f_5(\boldsymbol{\dot{Q}_{demand}^a}, t).$$

On top of these curves, a certain amount of flexibility is assumed. For the electric demand coming from appliances, the ratio between the flexible demand in either direction and the forecasted power is defined as either 5% in inactive hours or 20% in inactive hours, effectively given by a flexibility function

$$\boldsymbol{f_{app}}(t) = \begin{cases} 0.05, & t(\bmod 24) \in [0, 7) \cup [9, 16) \\ 0.2, & \text{otherwise} \end{cases}.$$

Finally, the flexibility ratio for the heating demand is set to depend on the instantaneous ambient outdoor air temperature $T_{air}(t)$ following an effectively continuous case-by-case law given by

$$f_{heat}(t) = \begin{cases} 0.2, & T_{air}(t) \geq 10\,°C \\ 0.074 \log_{10}(T_{air}(k) + 6\,°C), & -5\,°C < T_{air}(t) < 10\,°C \\ 0, & -5\,°C \geq T_{air}(t) \end{cases}$$

effectively restricting heating demand elasticity in cold times.

Pertaining to the electrical power exchange with the grid, it is required to separately implement two variables, one for import and one for export, so that they can be included in the cost-describing criterion function. Namely, in order to ensure that the system cannot import and export electrical power in the same time step, a constraint is defined that links these power variables with two integer variables in the form of

$$(\forall t)\ P_{imp}(t) \leq y_1(t)M \text{ and } P_{exp}(t) \leq y_2(t)M$$

with an additional constraint

$$(\forall t)\ y_1(t) + y_2(t) \leq 1, \quad y_1(t), y_2(t) \in \{0, 1\}$$

that only allows one of these two variables to have a non-zero value at each time step. With $M = 10^{12}$, these constraints in conjunction achieve the desired limitations on the coexistence of imports and exports.

3.2 Mixed-Integer Linear Programming Model Implementation

Once the non-linear dynamics are resolved, the hybrid energy system can be represented as a power network as depicted in Fig. 1. Here, links between the electric and thermal domains are established via two converters, an electric heater (ELH) with a fixed efficiency $\eta_{ELH} = 1$ and a previously mentioned ASHP. Besides these converters, electric energy can be introduced into the system either via imports from the grid depicted by $P_{grid}(t)$ or from local generation modelled as a PV array and a coupled inverter (INV) with a fixed linear efficiency $\eta_{inv} = 0.95$. On the other hand, electric energy consumption can either be attributed to consumption stemming from the input heat pump power $P_{ASHP}(t)$, input electric heater power $P_{elh}(t)$, or to appliance consumption $P_{app}(t)$ and exports back to the grid $P_{exp}(t)$ in case of excess production. As for the thermal bus, it has a supplemental gas power supply given by $\dot{Q}_{gas}(t)$ and a demand sink $\dot{Q}_{demand}(t)$.

The power balances of these busses are implemented as equality constraints

$$(\forall t)\ P_{imp}(t) + \eta_{INV}P_{PV}(t) = P_{ASHP}(t) + P_{app}(t) + P_{elh}(t) + P_{exp}(t),$$

$$(\forall t)\ \dot{Q}_{ASHP}(t) + \dot{Q}_{elh}(t) + \dot{Q}_{gas}(t) = \dot{Q}_{demand}(t).$$

The previously mentioned converters provide a bridge between the two busses as given by

$$(\forall t)\ P_{ASHP}(t) = \frac{1}{COP_{ASHP}(t)} \dot{Q}_{ASHP}(t)$$

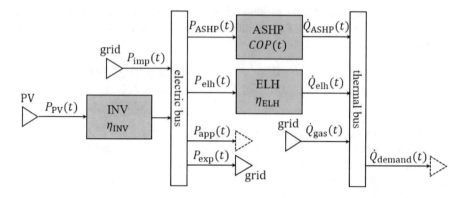

Fig. 1. Network representation of the discussed hybrid energy system

$$(\forall t)\ P_{\text{elh}}(t) = \frac{1}{\eta_{\text{EL}}} Q_{\text{elh}}(t).$$

In order to implement demand-side side flexibility in form of possible positive and negative load modifications when compared to a forecasted baseline, the appliance and heat demand are represented using two sets of bounds

$$(\forall t)\ P_{\text{app}}^{\text{min}}(t) \leq P_{\text{app}}(t) \leq P_{\text{app}}^{\text{max}}(t)\ \text{and}$$

$$\dot{Q}_{\text{demand}}^{\text{min}}(t) \leq \dot{Q}_{\text{demand}}(t) \leq \dot{Q}_{\text{demand}}^{\text{max}}(t)$$

Reflecting a minimum and maximum allowed value. The flexibility results in demand bounds that can be determined as

$$(\forall t)\ P_{\text{app}}^{\text{min}}(t) = \left(1 - f_{\text{app}}(t)\right)P_{\text{app}}^{\text{forecast}}(t)\ \text{and}$$

$$P_{\text{app}}^{\text{max}}(t) = \left(1 + f_{\text{app}}(t)\right)P_{\text{app}}^{\text{forecast}}(t)$$

And likewise, for the heat demand, in line with the previously discussed flexibility factors

$$(\forall t)\ \dot{Q}_{\text{demand}}^{\text{min}}(t) = \left(1 - f_{\text{heat}}(t)\right)\dot{Q}_{\text{demand}}^{\text{forecast}}(t)\ \text{and}$$

$$\dot{Q}_{\text{demand}}^{\text{max}}(t) = \left(1 + f_{\text{heat}}(t)\right)\dot{Q}_{\text{demand}}^{\text{forecast}}(t).$$

Additionally, upwards and downwards load shifts should be balanced out and, assuming this should be true for each day τ_i, two sets of integral constraint can be defined with

$$(\forall i \in \{1, 2, \ldots, 365\})\ \sum_{t \in \tau_i} P_{\text{app}}(t) = \sum_{t \in \tau_i} P_{\text{app}}^{\text{forecast}}(t),$$

$$\sum_{t \in \tau_i} \dot{Q}_{\text{demand}}(t) = \sum_{t \in \tau_i} \dot{Q}_{\text{demand}}^{\text{forecast}}(t).$$

In conjunction with power flow limitations, the given set of equations and corresponding inequalities resulting from variable bounds defines a model that can be solved via mixed-integer linear programming.

Variables depicting unidirectional power flow are limited by a minimum value of zero and a predefined upper bound that corresponds to the power capacity of the corresponding link as given by a set of constraints

$$(\forall t) \ 0 \leq P_{\text{imp}}(t) \leq P_{\text{imp}}^{\text{max}}$$

$$(\forall t) \ 0 \leq P_{\text{exp}}(t) \leq P_{\text{exp}}^{\text{max}}$$

$$(\forall t) \ 0 \leq P_{\text{elh}}(t) \leq P_{\text{elh}}^{\text{max}}$$

$$(\forall t) \ 0 \leq \dot{Q}_{\text{gas}}(t) \leq \dot{Q}_{\text{gas}}^{\text{max}}$$

$$(\forall t) \ 0 \leq \dot{Q}_{\text{ASHP}}(t) \leq \dot{Q}_{\text{HP}}^{\text{max}}(t)$$

In particular, PV power production is assumed to be curtailable, i.e., that its value can be adjusted between zero and a predetermined maximum $P_{\text{PV}}^{\text{forecast}}$ corresponding to the forecasted production as given by

$$(\forall t) \ 0 \leq P_{\text{PV}}(t) \leq P_{\text{PV}}^{\text{forecast}}(t).$$

4 Results and Discussion

The results depicted in the following sections assume a use case of a residential dwelling located in Ferrara, Italy, that is optimized for lowest possible yearly operational cost as given by a criterion

$$J = \min \sum_{t} \left\{ \alpha_{\mathbf{i}}(t) P_{\text{imp}}(t) - \alpha_{\mathbf{e}}(t) P_{\text{exp}} + \beta_{\mathbf{i}}(t) Q_{\text{gas}}(t) \right\}.$$

For ASHP modeling in particular, it is assumed that the user requires a constant temperature of $T_{\text{user}}(t) = 40\,°\text{C}$ in to be supplied while the source temperature equals the ambient outdoor air $T_{\text{air}}(t)$ that can, as with other environmental parameters, be derived from a TMY file [9] for a particular location. Nominal heat pump power values are set with a nominal heat power $\dot{Q}_{\text{ASHP}_0} = 10\,\text{kW}_\text{t}$, electric consumption $P_{\text{ASHP}_0} = 3.25\,\text{kW}_\text{e}$, its quality grade is $\eta_{QG} = 0.4$ while icing is set to occur at $T_{\text{ice}}(t) = 2\,°\text{C}$ with a performance reduction of $f_{\text{ice}}(t) = 0.8$ when applicable.

Yearly consumption for forecasting were assumed to be double the average values for residential consumers [10] i.e., of $P_{\text{app}}^{\text{a}} = 4.65\,\text{MWh}_\text{e}$ and $\dot{Q}_{\text{demand}}^{\text{a}} = 22.3\,\text{MWh}_t$ and are subsequently utilized for hourly demand forecasting using the methodology depicted in [3]. Fixed energy tariffs were assumed with $\alpha_i(t) = 31.15$ cEUR/kWh$_e$, $\alpha_e(t) = 0.11$ cEUR/kWh$_e$ and $\beta_i(t) = 0.18$ cEUR/kWh$_t$. Only heating demand is considered assuming cooling is not needed in summer time. Also, the system is assumed to be equipped with a $P_{PV}^0 = 2$ kW$_e$ capacity PV array.

First, two references use cases are considered. One assumes only resistive heating can be utilized for thermal demand fulfilment. The second assumes that only gas can be utilized. Finally, these are compared against a hybrid system that has the possibility to utilize all three thermal energy flows. The results are as follows:

- Only using (unlimited) resistive heating: $J_1 = 7760.39$ EUR;
- Only using (unlimited) gas supply: $J_2 = 4886.86$ EUR (37% reduction over J_1);
- Using combined ASHP, resistive heating (≤ 10 kW$_e$) and gas (≤ 10 kW$_t$): $J_3 = 3070.66$ EUR (60% reduction over J_1, 37% reduction over J_2).

Once the model is optimized, a selection of 48h-long data is extracted for illustrative visualization purposes. Figure 2 depicts the optimized electric and thermal load profiles. Each graph showcases the forecasted value in the middle with a band of flexibility around it, corresponding to the scale factor discussed previously. Finally, the optimization engine deices when and by how much to utilize this flexibility and the corresponding load upshift and downshift events are illustrated in red and green, respectively. The same approach is utilized for both load types.

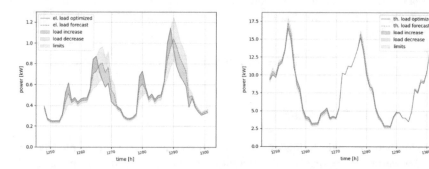

Fig. 2. Optimized electric load (left) and thermal load (right) with indicated flexibility and load modification actions

On the other hand, Fig. 3 depicts the selection of input power from various thermal sources that are utilized to fulfill the optimized thermal load curve. It shows that, due to the set pricing for different carriers and the achievable efficiency of the ASHP, the system chooses to first utilize the ASHP as the most cost-effective means of heating. However, when this output is not sufficient, it is supplemented first by the gas supply, as clearly shown in Fig. 3. However, since the gas supply is never exceeded, the electric heater doesn't turn on in the depicted temporal window.

With the results indicating approximately 1800 EUR/a reduction in energy costs over a gas-powered baseline, this model can be further utilized to run return of investment (ROI) analyses for feasibility assessment of HP installations or be extended with

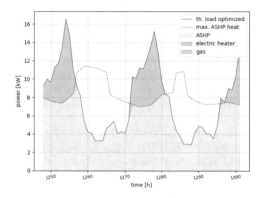

Fig. 3. Optimum utilization of different heat sources

alternative HP variants (e.g., ground-powered with a higher quality grade) to determine which is more suitable for a given use case. It is also worth noting that this type of model can be resolved in only a few tens of seconds which also provides a possibility for multiple configurations to be tested either in parallel or in succession. Also, the model structure is suitable for expansion such as adding electrical energy storages (batteries) and thermal energy storages (buffer tanks, reservoirs) that can also be modelled under the MILP umbrella by means of integral constraints.

5   Conclusion

In summary, this paper presents a methodology based on mixed-integer linear programming suitable for longer-term optimization and assessment of multi-carrier energy dispatching systems with demand-side flexibility. This methodology is illustrated for a use case depicting a mix of electrical and thermal power. The considered system employs locally installed renewable generation in the form of photovoltaic panels as an addition to an existing grid connection while, on the thermal side, an air-source heat pump is utilized in conjunction with gas and electric-powered heating devices to fulfill thermal demand. The utilized model is presented in form of a set of constraints with the results presented and discussed for a year-long data set with different combinations of utilized energy sources and convertors. As the results show, the utilization of heat pumps may have a notable impact on operational costs in these types of cases where the system can dynamically make use of the provided flexibility.

Acknowledgement. The research presented in this chapter is partly financed by the European Union (H2020 IDEAS project, Grant Agreement No.: 815271 and SINERGY project, Grant Agreement No.: 952140) and the Ministry of Education, Science and Technological Development of the Republic of Serbia.

References

1. Bakare, M.S., Abdulkarim, A., Zeeshan, M., Shuaibu, A.N.: A comprehensive overview on demand side energy management towards smart grids: challenges, solutions, and future direction. Energy Inf. **6**(1), 59 (2023). https://doi.org/10.1186/S42162-023-00262-7
2. Atia, R., Yamada, N.: Sizing and analysis of renewable energy and battery systems in residential microgrids. IEEE Trans. Smart Grid **7**, 1204–1213 (2016). https://doi.org/10.1109/TSG.2016.2519541
3. Jelić, M., Batić, M., Tomašević, N.: Demand-side flexibility impact on prosumer energy system planning. Energies **14**, 7076 (2021). https://doi.org/10.3390/EN14217076
4. Bellos, E., Tzivanidis, C., Nikolaou, N.: Investigation and optimization of a solar assisted heat pump driven by nanofluid-based hybrid PV. Energy Convers. Manag. **198**, 111831 (2019). https://doi.org/10.1016/J.ENCONMAN.2019.111831
5. Arnaudo, M., Topel, M., Laumert, B.: Techno-economic analysis of demand side flexibility to enable the integration of distributed heat pumps within a Swedish neighborhood. Energy **195**, 117012 (2020). https://doi.org/10.1016/J.ENERGY.2020.117012
6. Amer, M., Namaane, A., M'Sirdi, N.K.: Optimization of hybrid renewable energy systems (HRES) using PSO for cost reduction. Energy Procedia **42**, 318–327 (2013). https://doi.org/10.1016/J.EGYPRO.2013.11.032
7. Maruf, M.N.I., Morales-España, G., Sijm, J., et al.: Classification, potential role, and modeling of power-to-heat and thermal energy storage in energy systems: a review. Sustain. Energy Technol. Assess. **53**, 102553 (2022). https://doi.org/10.1016/J.SETA.2022.102553
8. Holmgren, W.F., Hansen, C.W., Mikofski, M.A.: pvlib python: a python package for modeling solar energy systems. J. Open Source Softw. **3**, 884 (2018). https://doi.org/10.21105/JOSS.00884
9. Huld, T., Paietta, E., Zangheri, P., Pascua, I.P.: Assembling typical meteorological year data sets for building energy performance using reanalysis and satellite-based data. Atmosphere **9**, 53 (2018). https://doi.org/10.3390/ATMOS9020053
10. Household energy consumption per dwelling by end-use — European Environment Agency. https://www.eea.europa.eu/data-and-maps/daviz/energy-consumption-by-end-uses-1#tab-chart_1. Accessed 10 Feb 2023

Scalable Real-Time Simulation of Electrical Power System Models in the Cloud

Bojana Ivanovic[(✉)] [ID], Vanja Mijatov [ID], and Branko Milosavljevic [ID]

Faculty of Technical Sciences, University of Novi Sad, 21000 Novi Sad, Serbia
{bojana.ivanovic,vanja.mijatov,mbranko}@uns.ac.rs

Abstract. This paper proposes a scalable system for real-time simulation of models of different electrical power systems. The main goal of this system is to simulate electrical energy fluctuation of a complex electrical power system which allows investigation of how different system components behave under different circumstances. Proposed system works with both live and historical data which are collected from the physical site. On the other hand, by using historical data, the system can mimic the behavior of the physical system over a longer period of time and to eventually show how the real system would function in the given period of time. Benefit of this system is that it allows the analysis of how electrical power flows through the system. Moreover, it simulates how the system functions in various scenarios which is achieved by using parametrization of the model. In addition, the observed shortcomings of the system were highlighted with suggestions for possible future work and improvements of the system.

Keywords: electrical power system · model simulation · real-time simulation · digital twin · hardware-in-the-loop · HIL

1 Introduction

Electrical power system is a complex system consisting of different electrical components and subsystems that participate in fluctuation of electrical energy. Due to its complexity, it is difficult to test how the system operates, since malfunctioning of any part of the system can have great consequences. Because of it, computer modeling and simulation of electrical power systems can be very beneficial. Modeling and simulation enable studying how a system performs in different scenarios and circumstances which allows observing how the system components are behaving and identifying and eliminating drawbacks of the system. Possessing detailed knowledge of the system provides an opportunity of creating a thorough model of that system. Having such a model allows simulating an electrical power system as a digital twin.

Hardware-in-the-loop (HIL) as a concept that implies simulating how a device operates in the real time can be useful in simulations of models of electrical power systems. By running simulations of a model using HIL simulator, real data collected from a physical system could be injected into a simulation which would allow the HIL simulator to replicate how electric energy fluctuates in the physical system in the real time. Outputs

© The Author(s), under exclusive license to Springer Nature Switzerland AG 2024
M. Trajanovic et al. (Eds.): ICIST 2023, LNNS 872, pp. 401–409, 2024.
https://doi.org/10.1007/978-3-031-50755-7_38

of the simulations would be data that correspond to different processed and unprocessed signal values that are collected from different components of the simulated model. To simulate an electrical power system using HIL simulator, it is mandatory that digital representation of that system is designed as a model that can be interpreted by software that allows interaction between simulation and the person that initiates and manages it.

HIL is very useful in the early stages when developing power electronic systems since it removes all difficulties regarding setup configuration and focuses on testing the physical system and its controllers. With HIL there are no dependencies with the physical sites, experimental equipment and no need for building different test benches. Moreover, HIL provides a controlled environment convenient for testing system boundaries and performance. Additionally, it can be beneficial once the system is put in operation since it can be used for system behavior prediction and better future management and maintenance [1].

The aim of this paper is to propose a system that allows simulating complex electrical power systems by using live and historical data collected from physical systems. The rest of the paper is organized as follows. Section "Related work" gives an overview of papers with associated topics regarding simulations of models and designing systems for the purpose of simulating the models. Section "System design" describes the proposed solution for simulating electrical power system models in real-time using live and historical data. Section "Results evaluation and discussion" analyzes the results of the proposed system, while in the section "Conclusion" summarizes the paper and emphasizes points of future work.

This research was conducted in co-operation with company Typhoon HIL, which is a partner in a consortium that participated in Horizon 2020 project with title "CREATing cOmmunity eneRgy Systems" or shorter "CREATORS" [2].

2   Related Work

Different systems for simulation of models of electrical power systems have been introduced in the past few years because of the increasing emphasis on green power systems and the electronic power system sector in general. The general idea is to define a model of a power plant, electric vehicle, part of a city or city in whole and simulate behavior of the model on a device in real-time.

Many papers discuss HIL simulations and their results such as paper [3] where the authors described a system which was modeled and simulated using Typhoon HIL software and hardware [4]. Described system is used for permanent magnet synchronous machine characteristics and motor control. It was simulated using a HIL 604 device. Furthermore, authors of the paper [5] describe a grid integrated solar photovoltaic system modeled with Typhoon HIL software and simulated using HIL 402 for the purpose of maximum power point analysis under various conditions.

Complete electric power system that contains several components presented in paper [6] was implemented using MATLAB [7], Simulink [8] and LabVIEW [9] and simulated with a designed Power System Emulator (PSE). The results of the simulated system were verified by comparing MATLAB and HIL simulation results. Authors aimed for PSE instead of a commercial simulation environment because PSE is a low-cost solution and it was more flexible for the authors due to its non-bounding policy to the choice of the hardware and software.

In the field of electrical vehicles (EV) there is an ongoing need for modeling and simulating some of EV components like the battery. One of the papers in this field analyzed two different battery models (BWC, BWOC) as well as two different alternator models (VTVA, CVTA) of the EV in respect of ease-of-use and accuracy and presented their results. The models were created using MATLAB and Simulink with the idea to simulate and analyze the vehicle electric power flow. The authors presented their simulation results and gave a conclusion that BWC battery model is more accurate but has difficulties with identification of its model parameters, unlike BOWC battery model which is more favored because of its simplicity and therefore favored for EV power system model. When it comes to alternator models, the VTVA has shown more accuracy compared to the CVTA model [10].

Authors of the paper [11] discuss the design of a solution for scalable electric power system simulator (SEPSS) as a general solution for electric power simulation tool. MATLAB was used for representation of system components and on a higher level the components were connected through interfaces which enabled the authors scalability. Different types of interfaces allowed the addition of different kinds of system components to the system. By adding new components to the system and using appropriate interface(s) for component communication, the system scales. Authors also mentioned one of the general problems simulators have and that is time. Each system component of their system has its own logical clock and the step function which overcomes the problem of time synchronization. Authors did not address the scenario of having more than one independent simulation.

Authors of the paper [12] also presented a power system simulator, describing challenges and solutions for both hardware and software parts of a system that later became a commercially available simulator. As a part of their discussion about possible solutions and tools that can be used for the implementation of the simulator, the authors highlighted Simulink and MATLAB as some of the possible tools. The fact that was problematic for the authors was that the mentioned toolbox was designed for offline simulations and it is not optimized for hard-real-time simulations and simulations that need to run in parallel.

3   System Design

The aim of this research was to design a system that allows simulating how an electrical power system would behave by using both historical data and live data that is collected constantly in the physical systems. Therefore, we developed two versions of the system that collects data differently and simulates them in different manners. Both versions of the system have architecture that is illustrated in Fig. 1. Core component of the system is a web application with an API that allows choosing which model to simulate and how to parametrize the simulation. User interacts with the system using the UI. When simulation is started, the system starts a Docker container with an environment that is established for running simulations as it is explained in paper [13]. This way each simulation is executed in a clean and isolated environment without interference with other simulations or external factors. It also makes the system scalable since it allows it to run multiple independent simulations at the same time in separate containers. Besides software for running simulations, the container includes an application that allows communicating with the core component via WebSocket protocol. Core component sends commands to the container that behaves as a WebSocket client which processes them and forwards processed content to the software that manages the simulation.

In the version of the designed system that allows user running simulations of models in real-time where live data is being fed to the simulation, HIL devices are being used for simulating model's behavior. In this research, we used HIL 402 devices produced by Typhoon HIL. To start the simulation, the user chooses which model to simulate and how to parametrize the model. Available parameterization options of the model are turning on or off certain components of the system and choosing values of input parameters that certain system's components provide. After user chooses model and its parameters, following steps occur:

1. Core application of the system starts a Docker container with all the necessary software that is required for running a simulation and gives it permissions to communicate with HIL device
2. Chosen parameters are injected into model
3. Model is compiled
4. Model is loaded to HIL device
5. Simulation begins.

Duration of listed steps varies and depends on hardware which is used and complexity of the simulated model.

After simulation begins, the system periodically fetches data from the external system in the cloud that collects live data from the physical power electronic system and stores it. In addition, it periodically reads different signal values from the simulation and uploads it to another external system in the cloud that requires the data for further processing and visualization. The benefit of this version of the designed system is that it allows the user to observe in real-time how the physical system works by studying the behavior of the virtual system. After making certain changes on the physical system, the user can in real-time observe how those changes affect the system by observing results of its virtual counterpart.

In the version of the designed system that allows user running simulations of a model using historical data, specialized software that allows SIL (software-in-the-loop) testing is used to simulate model's behavior. In this research, we used Virtual HIL application that is part of Typhoon HIL Control Center toolchain which is produced by Typhoon HIL. To start the simulation, the user chooses which model to simulate, past time frame for which collected data should be used (e.g., simulating how the system would work with data collected from January 1st until December 31st in the previous year) and how to parametrize the model. Available options for model parameterization are the same as in the first version of the system. After user chooses a model and its parameters, following steps occur:

1. Core component of the developed system retrieves historical data from the external system in the cloud

2. Core component of the developed system starts a Docker container with all the necessary software that is required for running a simulation and gives it permissions to communicate with Virtual HIL
3. Chosen parameters are injected into the model
4. Model is compiled
5. Model is loaded to Virtual HIL
6. Simulation begins.

During the simulation system injects downloaded data to mimic behavior of the physical system. Contrary to simulations by the previous version of the system where simulation speed (i.e., execution rate) is same as real time, simulations of this system are accelerated so that one calendar year can be simulated for significantly less time, depending on the model complexity. The simulations last from a few minutes (for simpler models) to a few hours (for complex models). In addition, it periodically reads different signal values from the simulation and uploads it to another external system in the cloud that requires data for further processing and visualization. In case of simulating one calendar year, exported data is uploaded after each simulated month. The benefit of this version of the designed system is that it allows the user to observe how the physical system worked during a longer period and how it could work if its components were different. With provided parametrization of the model, the user can moderate simulation by modifying model's components which would result in different simulation results. For example, the user can change the size of a certain battery or PV system and after the simulation can evaluate if investing funds into improving these parts of the system would bring larger financial benefits. This way the designed system can be used as a tool in conducting a financial study regarding potential investments in a certain electrical power system.

Data that is collected from the simulations is uploaded to an external system in the cloud whose purpose is to collect and visualize data. Collected data is processed and

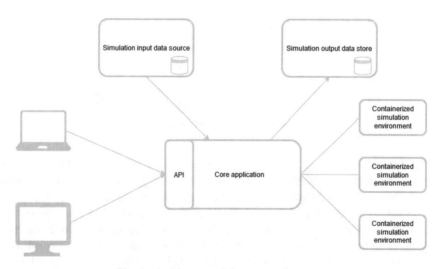

Fig. 1. Architecture of the proposed system.

stored as time series data. Received data is grouped by the version of the system that ran the simulation and identifier of the simulation which simplifies data retrieval.

4   Results Evaluation and Discussion

Proposed system was evaluated by simulating a complex model of Port of Barcelona, Spain [14]. Port of Barcelona is an energy community that consists of multiple objects that are great energy consumers such as fisher houses, an ice factory, restaurants and a couple of buildings. Model of the Port of Barcelona that was simulated is illustrated in Fig. 2. Each of the components including Grid can be turned on or off before the simulation by the user. Component "New fish market" has one PV system and one battery and the component "Casetas/Zona de Redes" has two PV systems and one battery. For batteries user can modify their:

1. Nominal active power (W)
2. Nominal apparent power (VA)
3. Nominal capacity (Ah)
4. Nominal voltage (V)

and for PV system user can modify their:

1. Nominal active power (W)
2. Nominal apparent power (VA).

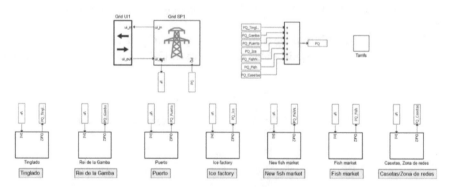

Fig. 2. Port of Barcelona model.

To simulate this model, software for running simulation requires different energy values collected for the model's components and meteorological data such as temperature and irradiance which are required for simulating power generation by PV systems.

In order to evaluate the outputs of the simulations and therefore the efficiency of the proposed system, collected data from simulations were uploaded to an external application in the cloud whose purpose was to collect and visualize collected data. Collected data was stored in the database and was visualized in Grafana [15] dashboard using time series charts. Features of the charts allowed observing results of multiple simulations simultaneously and choosing which signal values to observe and for which time frame.

In case of simulations that used live data, we could observe how electrical energy fluctuated during day. This allows us to investigate how electrical energy fluctuated in different parts of day, how much of physical system's consumption was covered by energy that was produced by its PV systems and how could it be changed in case of modifying physical system's components (such as eliminating certain components or modifying PV systems).

In case of simulations that used historical data, we could observe their outputs for the same period of time but with different parameters. This allows us to investigate how a physical system could work in the long run with modified components and how beneficial would specific improvements to its infrastructure be in the course of time.

Evaluation was conducted by using a developed dashboard which proved that the described system is capable of simulating complex models and to allow the user to modify it in order to simulate the model in different scenarios. In addition, due to its scalability it allows the user to run multiple simulations of different models at the same time which is not a characteristic of most platforms that allow running only one HIL or SIL simulation at the time.

Proposed solution can serve as a useful tool for multiple purposes such as monitoring a virtual clone of a complex electrical power system or for planning the modifications to a complex electrical power system. Figure 3 presents an example of a dashboard where it can be seen how the same signal (simulation output) has different values in different scenarios. Figure shows Grid active power value measured at certain time of day with data collected from October 20th 2022 until October 27th 2022. We can observe Grid active power in following scenarios:

1. Scenario with identifier 53 where model is simulated with following parameters:
   a. All components turned on except for the batteries
   b. New fish market PV has nominal active power 29.9 kW and nominal apparent power 60 kVA
   c. Casetas PV has nominal active power 109 kW and nominal apparent power 300 kVA
   d. Redes PV has nominal active power 29.5 kW and nominal apparent power 100 kVA.
2. Scenario with identifier 54 where model is simulated with following parameters:
   a. All components turned on except for the batteries
   b. New fish market PV has nominal active power 200 kW and nominal apparent power 600 kVA
   c. Casetas PV has nominal active power 500 kW and nominal apparent power 800 kVA
   d. Redes PV has nominal active power 300 kW and nominal apparent power 500 kVA.
3. Scenario with identifier 55 where model is simulated with following parameters:
   a. All components turned on except for the batteries
   b. New fish market PV has nominal active power 700 kW and nominal apparent power 1.2 MVA

c. Casetas PV has nominal active power 1 MW and nominal apparent power 2 MVA
d. Redes PV has nominal active power 800 kW and nominal apparent power 1.5 MVA.

By observing data for these three scenarios, we can conclude that in the case where all three scenarios have the same power consumers but different power producers, as the size of PV was increased, less power was consumed from the grid. Moreover, in the third scenario at certain times the production of power was higher than the consumption which resulted in grid active power being negative meaning that excess power was transferred into the grid which resulted in the system acting as a power plant.

The only drawback of the solution is that in the version of the system whose simulations use historical data, in case of rapid simulations where one calendar year is simulated in less than two hours, output data can not be collected more frequently than a few times per each simulated day due to causing a decrease in simulation speed. This issue does not exist in the version of the system where live data is used due to simulation speed being identical to real-time.

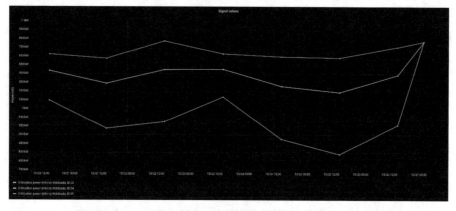

Fig. 3. An example of a dashboard for three simulated scenarios.

5   Conclusion

This paper describes a scalable system for running real-time simulations of the electrical power systems in the cloud which allows users to simulate different models of complex systems and investigate how their physical counterparts work. It is achieved by designing a system with multiple components where the core component allows starting simulations of different models with their parametrization prior to being simulated. Simulations are executed in an isolated environment and are being orchestrated by the core component. Two versions of the system are proposed: one where live data is collected during a simulation and model is simulated using physical HIL devices and the other where historical data is collected before the simulation, that has higher execution speed and the model is simulated by the Virtual HIL. Values of different signals are collected during a simulation and uploaded to the external system that stores signal outputs. Solution was validated using a dashboard that allows observing simulation outputs and comparing outputs of different simulations.

Future work could be directed into making the system even more scalable by decomposing system architecture into multiple subsystems. Proposed system scales in the number of parallel simulations but is limited with the number of resources that a computer that is running the system has. With a more distributed solution where the core component of the system could be orchestrating multiple subsystems that would manage containers, running simulations would result in the system being able to furtherly expand outside of the bounds of a single computer. In addition, addressing the identified drawback with simulation speed in one of the versions of the system would make a proposed solution an ideal tool for any kind of study of an electrical power system's operations.

References

1. Bai, H., Liu, C., Majstorovic, D., Gao, F.: Real-Time Simulation Technology for Modern Power Electronics. Academic Press (2023)
2. https://cordis.europa.eu/project/id/957815
3. Moldovan, T., et al.: Typhoon HIL real-time validation of permanent magnet synchronous motor's control. In: 2021 9th International Conference on Modern Power Systems (MPS). IEEE (2021)
4. https://www.typhoon-hil.com/products/hil-software/
5. Pal, S., Sahay, K.B.: Modeling of solar energy grid integration system using typhoon HIL. In: 2018 International Electrical Engineering Congress (iEECON). IEEE (2018)
6. Parizad, A., et al.: Power system real-time emulation: a practical virtual instrumentation to complete electric power system modeling. IEEE Trans. Indust. Inform. **15**(2), 889–900 (2018)
7. https://www.mathworks.com/products/matlab
8. https://www.mathworks.com/products/simulink
9. https://www.ni.com/en-rs/shop/labview
10. Lee, W., Choi, D., Sunwoo, M.: Modelling and simulation of vehicle electric power system. J. Power Sources **109**(1), 58–66 (2002)
11. Ilić, M.D., Jaddivada, R., Miao, X.: Scalable electric power system simulator. In: 2018 IEEE PES Innovative Smart Grid Technologies Conference Europe (ISGT-Europe). IEEE (2018)
12. Snider, L., Bélanger, J., Nanjundaiah, G.: Today's power system simulation challenge: high-performance, scalable, upgradable and affordable COTS-based real-time digital simulators. In: 2010 Joint International Conference on Power Electronics, Drives and Energy Systems and 2010 Power India. IEEE (2010)
13. Mijatov, V., Ivanović, B., Milosavljević, B.: System for software testing in real-time simulation environment. In: Zdravković, M., Trajanović, M., Konjović, Z. (eds.) ICIST 2022 Proceedings, pp. 22–25 (2022)
14. https://www.creators4you.energy/
15. https://grafana.com/

Advanced Information Systems

Relying on E-Contracting and Smart Contracts to Facilitate Legally Enforceable Conformance Checking in Collaborative Production

Balša Šarenac[(⊠)] [ID], Nikola Todorović [ID], Nenad Todorović [ID], and Goran Sladić [ID]

Faculty of Technical Sciences, University of Novi Sad, Trg Dositeja Obradovića 6, Novi Sad, Serbia
balsasarenac@uns.ac.rs

Abstract. Close collaboration and horizontal integration between value chain participants during collaborative production execution provide them with new possibilities and mutual benefits, resulting in improved operational and business performance. With our previous work, we introduced a novel methodological approach that facilitates runtime monitoring in collaborative networks based on Distributed Ledger Technology. In this approach, smart contracts are used to define and enforce process monitoring and conformance-checking logic for finding commonalities and discrepancies between the contracted and observed behavior during collaborative production execution. However, as smart contracts as tools can not be considered legally binding agreements, the collaborating parties still must go through the bureaucratic hurdle of defining the same conformance-checking logic within textual agreements. In this paper, we present the adaptation of the methodological approach that relies on e-contracting processes and activities to make the logic implemented with smart contracts legally enforceable. The benefits of this adaptation are that it speeds up the establishment of collaborative networks and further promotes trustworthy collaboration. Along with presenting the adaptation of the approach, we describe its implementation.

Keywords: E-Contracting · Collaborative Production Processes · Conformance-checking · Distributed Ledger Technology

1 Introduction

Cross-organizational business processes (CBPs) include multiple independent parties involved in high-level interactions directed at joint endeavors [1]. Close collaboration between involved parties when executing CBPs provides them with new possibilities and mutual benefits and allows for improvements in their operational and business performance. In our work, we aim to investigate and propose improvements in the domain of collaborative product design and development, where multiple parties collaborate to manufacture small batches of highly customized products jointly. Here, the prime concern of collaborating parties is manufacturing innovative products cost-effectively [2].

© The Author(s), under exclusive license to Springer Nature Switzerland AG 2024
M. Trajanovic et al. (Eds.): ICIST 2023, LNNS 872, pp. 413–422, 2024.
https://doi.org/10.1007/978-3-031-50755-7_39

During the execution of CBPs, multiple data sets are generated, owned by the involved parties, and maintained within their respective Information Technology (IT) systems. This process execution data must be shared between the parties in a timely manner to facilitate CBP execution monitoring. Execution monitoring provides opportunities for better coordination of production and logistic activities between involved parties and provides them with a mutual under-standing of the generated records. Distributed Ledger Technology (DLT) can be used as a distribution mechanism for sharing execution data between the parties [3]. By writing execution data to a distributed ledger, each party ensures that they are propagated to other participants who have access to the DLT network. Furthermore, the validation of the distributed data can be performed using a component of enterprise DLT solutions, smart contracts. Smart contracts represent computer programs whose execution is guaranteed by system rules and for which the execution outcome is verifiable and auditable by all network participants. Once validated using smart contracts, distributed data are stored in a distributed ledger in a persistent and immutable way, making it impossible for the parties that caused an issue to hide or alter information that proves their ill behavior.

As a part of our previous research [4], we introduced a novel approach that promotes trustworthy collaborative production execution where production data are distributed between the included parties using DLT and smart contracts. The approach is based on the CE-MultiProLan Domain-Specific Modeling Language (DSML). CE-MultiProLan contains concepts required to allow process designers to model collaborative production processes in multiple levels of detail. The DSML can be used by collaborating parties to design CBP models that specify what data should be shared between them during the production execution and define conformance checks that must be performed on the shared data. Once the models have been designed, the Model-Driven Software Development (MDSD) principles are used to automatically generate smart contracts that implement the defined data-sharing and conformance-checking logic.

In the approach, smart contracts are used only as a tool for implementing a conformance-checking logic. Current research suggests that, in general, smart contracts as tools can not be considered legally binding agreements between contracting parties, as there is not enough clarity on their validity and enforceability [5]. However, making the logic implemented with smart contracts legally binding would allow parties to avoid the bureaucratic hurdle of defining the same logic within textual agreements, thus accelerating the establishment of collaborative networks. Since, in our approach, designers express conformance-checking logic using CBP models, these could be utilized as ancillary artifacts to the textual agreements. By digitally signing CBP models, these could be deemed legally binding. Because of this, we aim to adapt our approach to include e-contracting processes and activities. Apart from making the contracting process more efficient, this adaptation would also further promote trustworthy collaboration between parties, as the inherent properties of digital signatures would also be utilized. That is, the integrity of these artifacts would be guaranteed, and the ownership of the signatures would be verifiable.

Based on the fact that each party involved in the collaboration has to agree on the specification of the collaborative process before the production takes place, we ask the following question: *How can the CBP specification process include the e-contracting activities so that production conformance to the resulting CBP specifications is legally binding?* To address this question, we have first analyzed the existing literature on utilizing e-contracting processes in collaborative networks. Then, we devised an adaptation and implemented a solution to include e-contracting processes and activities to the existing approach.

The remainder of this paper is structured as follows. In Sect. 2, an overview of background and related work is given. The devised adaptation of our approach is presented in Sect. 3, and the implementation of the adaptation is described in Sect. 4. An outlook on the contributions of the paper and a conclusion are presented in Sect. 5.

2 Background and Related Work

We have investigated the literature in the collaborative production domain relevant to monitoring production processes and applying e-contracting processes to this domain. Section 2.1 gives insight into the advantages of utilizing process monitoring techniques for auditing production execution and the benefits of relying on DLT and smart contracts for this purpose. Then, in Sect. 2.2, we present our MDSD approach that can be used for facilitating runtime monitoring of collaborative production. Finally, in Sect. 2.3, we discuss details that must be considered when utilizing e-contracting processes to facilitate legally enforceable conformance checking in collaborative production.

2.1 Process Monitoring of Collaborative Production

The upcoming transformation of industrial production into shared production implies that companies should form value-added cross-company networks for each new order [6]. These cross-company networks then execute CBPs to fulfill the production orders. Here, process monitoring techniques can be used to keep track of the running CBP instances, detect deviations or identify anomalies during process executions [3]. Our work mainly focuses on facilitating runtime production process monitoring techniques. The goal of runtime monitoring is to passively track and collect event logs about the statuses of process steps performed during process executions [7]. The collected data can then be used for conformance checking. Conformance checking aims to find commonalities and discrepancies between the modeled and the observed behavior. Furthermore, conducting conformance checking is relevant for business process alignment and auditing, as the stakeholders can verify that the execution aligns with the process description. For example, collected data can be replayed on top of a process model to find undesirable deviations suggesting fraud or inefficiencies [8].

During the execution of CBPs, the collaborating parties can rely on a Business Process Management System (BPMS) to automate runtime monitoring activities. For this to be possible, the IT systems of the included parties must be integrated to facilitate the establishment of a BPMS that would enable data distribution between actors and production monitoring with conformance checking. Furthermore, the systems must be integrated in a trustworthy and transparent way to facilitate trust between the actors. Here, DLT can be used as a basis for facilitating such an integration [9]. The innovative power of using DLT for IT system integration and process monitoring stems from it allowing parties to collaborate with others they do not fully trust [10]. Besides the distributed ledger, the main components of DLT that enable effective and efficient process monitoring are smart contracts. Smart contracts represent computer programs whose execution is guaranteed by system rules and for which the execution outcome is verifiable and auditable by all network participants [11]. Collaborating parties can use smart contracts to analyze the status of the process enactment, perform transparent conformance checking, and validate that the process steps are executed according to contracted specifications. Once validated, records of process step executions are stored in a distributed ledger in a persistent and immutable way. The record immutability makes it impossible for the parties that caused an issue to hide or alter information that proves their ill behavior.

One of the issues that occur when automating runtime monitoring activities using DLT and smart contracts is that automation may decrease flexibility in a value network [12]. The decrease in flexibility in the network stems from the fact that all the included parties must mutually agree upon the defined process monitoring logic before launching the collaborative production. Furthermore, once production is deployed, it is challenging to introduce changes to the defined logic. Hence, facilitating runtime monitoring means that data-sharing and conformance-checking requirements must be analyzed, negotiated, and defined in great detail before deployment. Here, process models can be used to define the expected behavior of each participant. Then, the defined process models can be used to generate BPMS artifacts automatically - this is one of the approaches offered by the Model-Driven (MD) paradigm and the MDSD methodology.

2.2 MDSD Approach for Runtime Monitoring of Collaborative Production

In [4], we introduced a novel MDSD approach for trustworthy collaborative production planning and execution in non-hierarchical collaborative networks. The approach is centered around the CE-MultiProLan DSML that offers collaborating parties concepts required for negotiating and clearly defining the expected behavior of each participant during collaborative production execution. The expected behavior is defined using different types of process models, as explained below. Once process models are defined using the DSML, they are used for generating code artifacts that implement the defined process monitoring logic.

CE-MultiProLan offers three different process model types for collaborative CBP modeling to allow disclosure of internal process data to collaborating parties: (i) Private Process Models (PPMs) that represent internal processes executed by a single party; (ii) Interface Processes Models (IPMs), i.e., process views that are used to coordinate internal actions with activities of external partners while concealing private data; and (iii) Cross-organizational Process Models (CPMs), used to describe how participants collaborate to produce the end product. As this paper does not cover a discussion related to internal production processes, details about PPMs are omitted from the paper. More details about PPMs can be found in [13]. IPMs represent process views built upon PPMs, used to expose private production data to collaborating parties. In the approach, IPMs are also utilized for defining process monitoring requirements, i.e., defining what data should be shared between parties during production execution and what conformance checks should be performed on the shared data. CPMs are created to coordinate production activities between collaborating parties when executing shared production. Our approach utilizes CPMs for defining process steps that should be executed to produce the end product, enriched with service details like delivery dates and cost.

Once IPMs and CPMs are defined, they are used for generating smart contracts used for conducting runtime process monitoring. IPMs are used to generate smart contracts for monitoring a production process executed by a single participant. Smart contracts generated from IPMs handle data distribution between the parties and perform conformance-checking on that data. Based on CPMs, smart contracts that monitor the enactment of the defined cross-organizational production processes be generated. These smart contracts validate that the defined economic properties have been fulfilled, e.g., delivery dates and set product quantities have been met. After the smart contracts have been generated, they can be deployed to the established DLT network, the production can be initiated, and the shop-floor devices can start sending production data to the DLT-based BPMS.

2.3 Relying on E-Contracting for Facilitating Collaborative Production Monitoring

Many different research efforts are being invested in utilizing DLTs and smart contracts for collaborative process execution and monitoring. However, to the best of our knowledge, researchers rarely focus on the legal aspects of such settings, focusing more on data distribution and monitoring. Since there is still insufficient clarity on the legal validity and enforceability of the logic implemented with smart contracts, parties cannot rely solely on them to create legally binding agreements. We aim to make the logic defined in the introduced CBP models of our approach, which are used to generate smart contracts, legally enforceable through e-contracting processes.

Contract creation, i.e., the definition of legally enforceable agreements in which two or more parties commit to certain obligations in return for certain rights, is one of the stages of creating collaborative networks [14]. The e-contracting process can generally be split into two phases: contract establishment and contract enactment [15]. Figure 1 displays these two phases alongside the activities that belong to the phases. The figure was created based on Fig. 1 - E-contracting process and activities from [15] but was refined to emphasize activities related to defining process monitoring requirements.

The contract establishment phase implies that the identified collaborating parties perform contract negotiation, i.e., negotiation of the terms and conditions of the contract, and contract validation, before signing a final agreement. As a part of the negotiation process, parties should clearly define process monitoring terms and requirements. In the domain of collaborative production, these terms should define what data should be distributed between the collaborating parties during production execution and what conformance checks will be applied to that data. In contrast, the contract enactment phase covers tracking contract fulfillment by relying on contract monitoring, i.e., observing the activities performed by the parties, knowing the state of contract execution, and detecting contract violations. The contract enactment phase aims to ensure that the executed processes follow the signed final agreement. Section 3 describes how we adapted our approach to include contract establishment and contract enactment phases.

Adapting our approach to include contract establishment and contract enactment phases, alongside the accompanying activities, should facilitate legal enforceability of

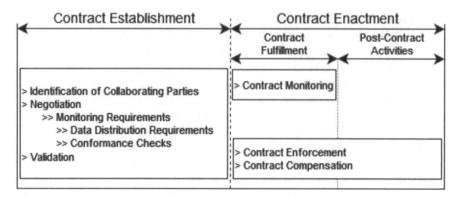

Fig. 1. E-contracting Process and Activities

the defined process monitoring and conformance checking activities. The following section provides details on the devised and implemented adoption of the approach.

3   Methodology

The current approach that facilitates data distribution and conformance-checking based on DLTs and smart contracts was adapted to support e-contracting processes and activities. First, the contract establishment process is performed, where parties use different types of business process models to negotiate different aspects of the contracts. After the final agreement has been achieved, e-signatures are used by the collaborating parties to sign the resulting artifacts. Then, the contract enactment process is performed, where contract fulfillment is tracked through collaborative production monitoring. The outline of the adapted approach is depicted in Fig. 2. Here, newly introduced components and approach steps are marked with a bright red color, while the adapted existing components are marked with a dark red color.

During the contract enactment phase, as step 1 of our approach, collaborating parties use the DSML Tool, a custom modeling tool that implements CE-MultiProLan DSML. This tool was adapted to support e-contracting by facilitating negotiations and the digital signing of two different process models - Interface Process Models (IPMs) and CBP Models (CBPMs). IPMs are created based on a production process specification and used to (i) define what event logs should be distributed between parties during process execution and (ii) define conformance checks that will be performed during contract enactment. Also, the tool should be used by process designers and collaborating parties to create CBP Models (CBPM) for coordinating actions between different participants in a value chain. The contents of these process models should be negotiated and agreed upon by all core partners of the collaboration network.

Once the final agreement has been negotiated, in step 2, parties use the DSML tool to sign the created models digitally. The signed models are then sent to the E-Contract Service, a utility service that collaborating parties can use to keep track of and browse all the digitally signed artifacts. For this to be possible, E-Contract Service relies on the Signed Artifacts Archive (SAA), a database used for archiving all the signed artifacts.

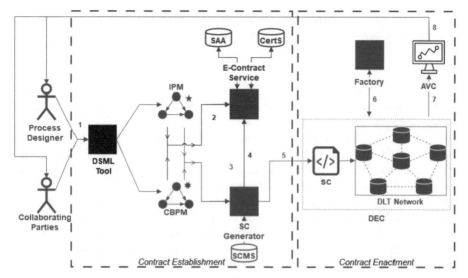

Fig. 2. The outline of the adapted approach

In addition, a digital Certificate Store (CertS) is used for the verification of the signed artifacts, as it stores the digital certificates of all the involved parties.

Next, in step 3, the Smart Contract (SC) Generator component is used to generate smart contracts. While IPMs can be used to generate smart contracts used to monitor processes executed by a single organization, CBPMs can be used to generate smart contracts that monitor the enactment of CBPs. In step 4, SC Generator performs a check by relying on the E-Contract Service to ensure that only approved and signed models are utilized for contract generation. After smart contracts are generated, in step 5, they are stored on the configured DLT Network.

After the production is started, the Contract Enactment phase is initiated. Here, the DLT Network and smart contracts represent a basis for creating the Data Exchange Component (DEC), the system component that is central in tracking contract fulfillment and monitoring. DEC facilitates the exchange of production data between the collaborating parties and enables process monitoring and conformance-checking on the exchanged data. Smart resources available on the Factory shop floor can then send records about production execution events to the DEC in step 6, automatically triggering actions specified as a part of the stored smart contracts. The role of smart contracts, generated based on CBPMs and I-PPMs, is to monitor these records and validate that the production execution is conducted according to the contracted specifications. With two final steps of the approach, steps 7 and 8, collaborating parties can oversee the contract fulfillment by analyzing the state of the immutable DLT store using a set of data visualization tools provided by the Aggregation and Visualization Component (AVC).

As we mainly deal with temporary collaborative networks, the included parties should take steps towards managing regular and premature network dissolvement scenarios. Regular network dissolvement is caused by the cease of the need for further network operations, as the goal of the network has been fulfilled. In contrast, unexpected causes might induce a premature network dissolution, and the parties should define strategies for contract enforcement and compensation for such cases.

4 Solution

In this section, we present the solution that implements the adapted approach and allows collaborating parties to rely on e-contracting to facilitate legally enforceable conformance-checking collaborative production. More specifically, the solution implements a workflow for the digital signing of CBP and IP models to make the logic within them legally binding. The devised and implemented workflow for signing process models is shown in Fig. 3. As step 1 of the workflow, after the contract negotiations are complete, one of the core participants should upload the final version of models, represented as XML artifacts, to the E-Contract service. The E-Contract service was built using the Spring Framework to enable storing and verifying digital signatures. Uploaded XML artifacts are stored by the E-Contract service to the SAA, built using the Postgres relational database. During the upload, in the case of IPM, all sensitive information is removed from the contained XML artifact to avoid potential data leakage. In the case of CBP, there is no private data to be removed, and this step is skipped. In the context of storing XML artifacts, XML-a transformations, specifically XML canonicalization, are employed to ensure the attainment of repeatability.

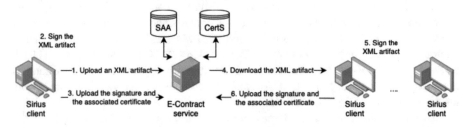

Fig. 3. The workflow of the solution

Next, in step 2, the participant that uploaded an artifact digitally signs it. By doing so, the participant ensures that the integrity and authenticity of the uploaded artifact can always be verified. To support the signing of process models, the DSML tool, developed as a Sirius [16] client application, was extended with options to upload and digitally sign their XML representation. Signing is conducted by creating an XML signature of the model using the party's private key, issued by a trusted certificate authority. XML signatures are generated utilizing the ECDSA (Elliptic Curve Digital Signature Algorithm) cryptographic scheme [17] in conjunction with the SHA-256 (Secure Hash Algorithm 256-bit) hashing algorithm [18]. In this setting, XML signatures are created using the Apache Santuario library.

In order to accommodate multiple signatures of the same document, the signatures are stored in a detached manner. A process is implemented where the signed signatures are extracted from within the document and subsequently stored separately. This approach simplifies the verification procedure, as the signatures are not in any way related nor ordered, and each can be verified separately without additional information. In step 3, the E-Contract service is used to store the resulting XML signature to the SAA, while the party's certificate is stored in CertS, developed using Java KeyStore technology. Storing signed XML artifacts to SAA and parties' certificates to CertS enables involved parties to verify the existing signatures at any moment. For example, other parties can verify the existing signatures before signing and uploading their own signature to ensure that the document they are about to sign is indeed the one uploaded by the proposing party.

After an artifact has been signed and uploaded by the proposing party, all other core virtual organization parties must also sign it. By signing the models, they agree that the model is valid and approve the continuation of the collaborative process deployment. Therefore, each remaining core participant must load the uploaded artifacts into their Sirius environment (step 4) and initiate the signing action (step 5). Once all core participants have signed all the needed XML artifacts, contract enactment can happen. The SC Generator relies on the E-Contract Service to perform verification of existing signatures and proceed with smart contract generation. Then the process deployment can take place.

5 Conclusion

This paper describes an e-contracting approach to facilitating legally enforceable process monitoring in collaborative production. It presents the results of an investigation into how the CBP specifications, defined during the creation of a collaborative network, can be made legally binding.

As the first part of the investigation, we analyzed the existing literature on CBPs, utilizing DLT and smart contracts for process monitoring and making the performed conformance checks legally binding through e-contracting processes in collaborative networks. Based on the analysis, as the second part of the investigation, we adapted our existing MDSD approach for facilitating collaborative production to include e-contracting processes and activities. The adaptation introduced contract establishment and contract enactment phases to the approach, thus enabling parties to utilize CBPs for contract negotiations to define legally binding conformance-checking logic that should be implemented. Once the contract negotiations are complete, CBPs are digitally signed and made ancillary artifacts to the final agreement. Furthermore, the adaptation allows parties to monitor contract fulfillment using the generated smart contracts that implement the conformance-checking logic.

Acknowledgements. This research (paper) has been supported by the Ministry of Science, Technological Development and Innovation through project no. 451-03-47/2023-01/200156 "Innovative scientific and artistic research from the FTS (activity) domain".

References

1. Berre, A.-J., et al.: The ATHENA interoperability framework. In: Gonçalves, R.J., Müller, J.P., Mertins, K., Zelm, M. (eds.) Enterprise Interoperability II, pp. 569–580. Springer, London (2007). https://doi.org/10.1007/978-1-84628-858-6_62

2. Shamsuzzoha, A., Toscano, C., Carneiro, L.M., Kumar, V., Helo, P.: ICT-based solution approach for collaborative delivery of customised products. Product. Plann. Control **27**, 280–298 (2016). https://doi.org/10.1080/09537287.2015.1123322

3. Di Ciccio, C., Meroni, G., Plebani, P.: On the adoption of blockchain for business process monitoring. Softw. Syst. Model. **21**, 915–937 (2022). https://doi.org/10.1007/s10270-021-00959-x

4. Todorović, N., Vještica, M., Todorović, N., Dimitrieski, V., Luković, I.: A novel approach and a language for facilitating collaborative production processes in virtual organizations based on DLT networks. In: Proceedings of the 2nd International Conference on Innovative Intelligent Industrial Production and Logistics, Valletta, Malta, pp. 197–208 (2021). https://doi.org/10.5220/0010720900003062

5. Lipton, A., Levi, S.: An introduction to smart contracts and their potential and inherent limitations. The Harvard Law School Forum on Corporate Governance, 26 May 2018. https://corpgov.law.harvard.edu/2018/05/26/an-introduction-to-smart-contracts-and-their-potential-and-inherent-limitations/. Accessed 30 May 2023

6. Köcher, A., et al.: A reference model for common understanding of capabilities and skills in manufacturing. at - Automatisierungstechnik. **71**, 94–104 (2023). https://doi.org/10.1515/auto-2022-0117

7. Pourmirza, S., Peters, S., Dijkman, R., Grefen, P.: BPMS-RA: a novel reference architecture for business process management systems. ACM Trans. Internet Technol. **19**, 1–23 (2019). https://doi.org/10.1145/3232677

8. van der Aalst, W.: Process Mining. Springer, Heidelberg (2016). https://doi.org/10.1007/978-3-662-49851-4

9. Xu, X., Weber, I., Staples, M.: Architecture for Blockchain Applications. Springer, Cham (2019). https://doi.org/10.1007/978-3-030-03035-3

10. Mendling, J., Weber, I., Aalst, W.V.D., Brocke, J.V., Cabanillas, C., et. al.: Blockchains for business process management - challenges and opportunities. ACM Trans. Manag. Inf. Syst. **9**, 1–16 (2018). https://doi.org/10.1145/3183367

11. Hileman, G., Rauchs, M.: 2017 global blockchain benchmarking study. SSRN J. (2017). https://doi.org/10.2139/ssrn.3040224

12. Agrawal, T.K., Angelis, J., Khilji, W.A., Kalaiarasan, R., Wiktorsson, M.: Demonstration of a blockchain-based framework using smart contracts for supply chain collaboration. Int. J. Product. Res. **61**, 1497–1516 (2023). https://doi.org/10.1080/00207543.2022.2039413

13. Vještica, M., Dimitrieski, V., Pisarić, M.M., Kordić, S., Ristić, S., Luković, I.: Production processes modelling within digital product manufacturing in the context of Industry 4.0. Int. J. Product. Res. 1–20 (2022). https://doi.org/10.1080/00207543.2022.2125593

14. Goodchild, A., Herring, C., Milosevic, Z.: Business Contracts for B2B (2000)

15. Xu, L., Vrieze, P.: Fundaments of virtual organization e-contracting. In: Camarinha-Matos, L.M., Afsarmanesh, H., Novais, P., Analide, C. (eds.) Establishing the Foundation of Collaborative Networks, pp. 209–216. Springer, Boston (2007). https://doi.org/10.1007/978-0-387-73798-0_21

16. Madiot, F., Paganelli, M.: Eclipse sirius demonstration. P&D@ MoDELS **1554**, 9–11 (2015)

17. Johnson, D., Menezes, A., Vanstone, S.: The elliptic curve digital signature algorithm (ECDSA). IJIS **1**, 36–63 (2001). https://doi.org/10.1007/s102070100002

18. Penard, W., van Werkhoven, T.: On the secure hash algorithm family. Cryptogr. Context. 1–18 (2008)

Can Memes Beat the Market? Forecasting Financial Asset Returns by Using Social Media Data and Machine Learning

Milan Zdravković(⊠) ⬤, Pavel Dudko ⬤, and Maxim Kucherov ⬤

ZenPulsar, London, UK
`milan.zdravkovic@zenpulsar.com`

Abstract. The objective of the research presented in this paper is to demonstrate the effect of the social media data on the return on investment in crypto trading. To do that, three methods were used in vectorized back-testing on selected 17 crypto-currencies, namely, Machine Learning-driven strategy, sentiment signal or feature-driven strategies and hybrid strategies. Two sentiment signal strategies are trialed, so-called naïve sentiment momentum and percentile strategy. One hybrid strategy was tested, namely, combined crossover and naïve sentiment momentum strategy. The performance of all methods is assessed by using normalized ROI and Sharpe ratio as indicators. The best results in back-testing on equally weighted portfolio were achieved by using ML-driven strategy, namely normalized ROI of 6.14 and Sharpe ratio of 1.98.

Keywords: Machine Learning · Algorithmic Trading · Market Forecasting

1 Introduction

Financial asset investment return prediction is a difficult problem, typically addressed by the quantitative researchers with using technical and fundamental analysis, as well as other exogenous features analysis, such as news and media data. Technical analysis attempts to uncover patterns in the stock price data by using historical prices and volumes of trading. Fundamental analysis considers financial statements of the industries and businesses and their competitors, as well as overall state of economy data. The predictions made by using technical and/or fundamental analysis cannot consistently produce excessive profits, because historical price information, patterns and all other historical public events are already reflected or embedded in actual stock prices. This statement is a paraphrase of the Efficient Market Hypothesis (EMH).

When a so-called strong EMH (all public or private market information is embedded in an asset's price) is adopted, market price fluctuations are then considered fully random, and distribution of returns is Gaussian or normal. In reality, the distribution of returns is characterized by many outliers, with a total range of distribution significantly larger than the interquartile range. In finance, this phenomenon is called "fat tails" and it implies strong impact of extreme observations on expected future return. This situation can be

M. Trajanovic et al. (Eds.): ICIST 2023, LNNS 872, pp. 423–434, 2024.
https://doi.org/10.1007/978-3-031-50755-7_40

explained by the less-strong forms of EMH. Semi-strong EMH implies that all public (but not private) market information is embedded in the current asset price, meaning that there are exogenous features that can be used as good predictors if we have access to the corresponding data. Weak EMH implies that all historical asset price data is embedded in the current asset price, meaning that while technical analysis is not useful for forecasting asset price, fundamental and other analyses may point out some strong predictors.

Expected return models are widely used in event studies which aim at quantifying the particular event's economic impact in so-called abnormal returns. Abnormal returns are defined for asset i at time t as difference of its actual and return predicted by using some expected return model: $AR_{i,t} = R_{i,t} - PR_{i,t}$. Event studies aim at finding cross-references between abnormal returns and specific types of events, for example, earnings announcements, mergers and others. Machine Learning (ML) techniques, associated with vast and diverse data availability can be used to generalize the problem as defined by the event studies, to identify the exogenous data patterns causing the abnormal return occurrences. While the news articles are proven not to have a consistent impact on the asset returns (EMH confirmed), this is not the case with social media, which provides continuous stream of encoded public and private events data.

Based on the above, the problem of accurate forecasting of financial asset returns may be summarized to forecasting returns or return directions (determining buy-sell orders), by using technical and fundamental analysis data and social media data. The hypotheses of the research were set as follows:

- Orders, produced by ML model trained with sentiment and financial signals, when used in back-test, beat the market, bring positive ROI and produce better ROI than orders produced by ML model trained with financial signals only.
- Social media sentiment signals, when used in feature-driven back-tests, beat the market and/or bring positive ROI. Social media sentiment signals, when reaffirmed with technical analysis trend indicators, beat the market and/or bring positive ROI.

In this paper, we present the evidence to support these hypotheses. First, we provide extensive literature research on the different aspects of correlation and causality between social media activity and market asset prices. Then, based on synthesis of the literature review and some initial experiments, we outline the concept and methodology for the most exhaustive asset return direction forecasting models. Finally, we present the results of the experiments.

2 Background Research

The literature review has provided clear evidence of the existing correlation (and in many cases causality) between the asset returns and relevant social media activity. The research on this topic has been especially dynamic in the past couple of years. Some of the most interesting and relevant findings are highlighted in this section.

Khan et al. (2020) [1] tested different ML methods for market value prediction based on the overall sentiment of social media and financial news on different markets. An important finding was that not all markets are equally sensitive to the effect of social media and financial news. Tweets (with removed spam tweets) and news headlines (from

BusinessInsider) were used in the research. Random forest method was found to produce the best result, with accuracy of 83.22%.

Coelho et al. (2019) [2] explored correlation between sentiment of messages on StockTwits social media for investors and stock market closing prices. Two correlations were confirmed: a) correlation between overall sentiment of the twits for one day and stock price change for that day; b) correlation between overall sentiment of the twits during n days and stock price change for n + 1 day. In another experiment, only those users for whose tweets the correlation between their sentiment and stock price change is confirmed (historically) were taken into account. Accuracy was improved with this approach. Another important conclusion of the authors is that domain specific sentiment analyzer would bring significant improvement over using generic ones. The most exhaustive analysis we have found [3] has determined average sentiment classification accuracy of 61% for general-purpose predictors, with improvement of 11% for domain-specific systems.

Glenski et al. (2019) [4] used social signals from GitHub and Reddit to forecast prices of cryptocurrencies (BTC, ETH, XMR) with Long short-term memory (LSTM) models. Features extracted from subreddits were: 1) volume of comments made each day, 2) 10k-dimensional vectors of word-level daily-normalized statistics that focus on the most frequent unigrams (language), 3) vector representation of the quartiles of Reddit scores (i.e., # upvotes - # downvotes) for comments posted each day (popularity), 4) quartiles for the subjectivity and polarity of comments each day which provides a signal of the distribution of sentiment in discussions. Correlation between social signals and price was evaluated by using Pearson and distance correlation. The best results were achieved with LSTM configuration.

Sousa et al. (2019) [5] tried to improve the accuracy of sentiment predictors, by fine training the pre-trained BERT model with 582 financial news articles manually labeled with sentiments. Such a model was then used for forecasting Dow Jones Industrial Index (DJI). The average sentiment of the news published 5 h before the market opening time was selected as a social media feature and its correlation with the daily direction of DJI (closing-opening) was confirmed in 69% cases.

Sun et al. (2021) [6] developed a model for China Individual Investor Sentiment Index, by using 200M posts from online financial forums and an original sentiment dictionary. It was proven to be more successful than the commonly used BW Investor Sentiment Index. Besides sentiment, each post is associated with an influence factor (by using the number of views and comments of the post). Linear correlation of sentiment index with market value (weekly sentiment with market value of that week, previous and next one) was tested by using Pearson linear correlation and simple regression analysis.

Nguyen et al. (2015) [7] also highlighted that accurate sentiment classification is still a difficult problem. Besides the sentiment towards the actual company/stock, they also introduced the sentiment towards the specific topic related to the company (product, service, dividend, etc.), as a feature. For the social media, Yahoo Finance Message Board messages were used. Several models were made, ranging from the simplistic ones considering only prices, over ones with sentiment to the most complex topic-sentiment models. Support Vector Machine (SVM) was used for predicting the direction. The Latent Dirichlet Allocation (LDA) method was used for topic identification; 50 topics were selected, and the chosen features were average distributions of each topic in the messages at the time t. The averages of the joint probabilities of topic-sentiment pairs in the messages at the time t were used as input features.

Cai et al. (2022) [8] developed The Wall Street Bots – automatic stock trading platform, including NLP analysis of different news and social media content. For NLP, they used FinBERT [9] to analyze sentiment of financial text, built by fine tuning of the BERT language model [10], using a large financial corpus. GME was chosen for the demonstration model as meme stock, assumingly highly sensitive to social media. One of the experiments included an interesting data augmentation approach: augmenting the GME data with price data from different stocks whose returns have a high correlation with GME returns. Different architectures were tested, and the best result was achieved with LSTM.

Zhou et al. (2021) [11] proposed approach to corporate events detection for news-based event-driven trading. Event detection from news is multi-class, multi-label text problem of classifying the events of specific known (hence, multi-class) types (more events can be classified in one news, hence multi-label). Based on event occurrence (namely, acquisitions, clinical trials, new contracts, and others), long/short orders are generated. Only non-periodic events were detected. EDT dataset was used, comprising of 9721 articles, where 2266 have at least one label. The rest of the articles are not related to an event and are kept in the dataset for model to learn to distinguish event from non-event articles. Pre-trained BERT was used as text encoder.

3   Methodology

The research experiments we use to prove the above hypotheses consider social media (extracted from Reddit) and financial data (hourly data series) for the assets in the bucket consisting of ADA, ALGO, AVAX, BTC, IOTA, DOGE, VET, BNB, MATIC, SHIB, SOL, ETH, XRP, THETA, DASH, LTC and DOT in a period of 2017-10-01 18:00:00, to 2022-09-30 20:00:00.

3.1   Social Media Data Acquisition

Selection of asset-relevant social media posts is done via iterative usage of information retrieval methods such as keyword extraction and topic modelling (LDA, BERTopic, etc.). To classify an account posting a specific social media content as a bot or as an authentic user (content trustworthiness), we apply a combination of the following techniques: NLP-based content analysis, Heuristics-based features (speed of posting, statistical characteristics based on NER analysis results, etc.), the format of recent posts from the same user, and analysis of network topology (bots have unique topologies that differ from human accounts). To classify an account posting a specific social media content as either an influencer, a market analyst, or a normal user (content credibility) we apply a combination of the following techniques: NLP-based content analysis, analysis of the account following network characteristics, and number of followers/reddit karma thresholds. For sentiment classification (bullish vs. bearish posts), we utilize transformer-based models (FinBert, CryptoBert, and CryptoRoberta) fine-tuned on our internal datasets.

The model was trained on cryptocurrency and stock data collected from social media, and three categories were produced by the classifiers: bearish, neutral, and bullish.

3.2 Social Media Data and Feature Engineering

Raw sentiment data is a series of counts of 1) bullish and bearish posts, 2) bullish or bearish comments, 3) upvotes/downvotes of bullish or bearish votes, in the past hours, published by bots/non bots, trusted/non trusted users, for the assets in bucket.

The volume of posts and their sentiment is already recognized in literature and practice as the most significant social media feature. It is used in the vast majority of reported works, and it reflects the overall attention of the relevant social media communities (on the specific channels/subreddits). We extend this approach by including social media engagement features (in specific, comments and votes from Reddit), as well as by considering trustworthiness of content by distinguishing between bot and credible user generated content, and between the content produced by trusted and non-trusted users. Besides other benefits, measures of engagement make it possible to better distinguish between trusted and non-trusted content, as the latter is expected to produce less engagement.

Raw data is used to build features by using following steps:

- Data is filtered to include or exclude bots or include or exclude non-trusted users.
- Selected content types (posts, comments, upvotes/downvotes) are included in further processing.
- Data is aggregated (rolling sum) for the given aggregation period.
- Features are constructed by using three different approaches, found in literature: difference, scaled difference and ratio.

$$SMF_{diff} = N_{bull} - N_{bea}, SMF_{diff} = \frac{N_{bull} - N_{bea}}{N_{neu}}, SMF_{ratio} = \frac{1 - N_{bull}}{1 - N_{bea}},$$

where N_{bull} - Number of bullish posts, N_{bea} - Number of bearish posts, N_{neu} - Number of neutral posts.

- Features are postprocessed by using Exponential Moving Average (EMA) and Kaufmann Adaptive Moving Average (KAMA) functions.

Social media feature engineering uses Hyperopt implementation of the Bayesian optimization. It aims to provide the best combinations of parameters for constructing social media features, defined above. The feature engineering is a Hyperopt process of 100 trials on objective function which minimizes Pearson or Spearman coefficient of correlation between social media signal constructed with different combinations of parameters and asset return. The process results with selected 6 features, namely 3 different function types (difference, scaled difference, ratio) on 2 content types (posts, comments, upvotes/downvotes), with optimum scope of social media content (include bots, include non-trusted users), postprocessing type (EMA, KAMA, percentile rank) and its associated numeric parameters. Finally, to diminish potential negative impact of optimization with correlation as objective function, generic features are added, namely 6 raw social media signals (3 function types on 2 content types - posts and comments) and 6 EMA-postprocessed raw signals on default parameters. The figure below presents selected social media sentiment features for BTC asset. The green vertical line demarcates the training data from the test dataset (Fig. 1).

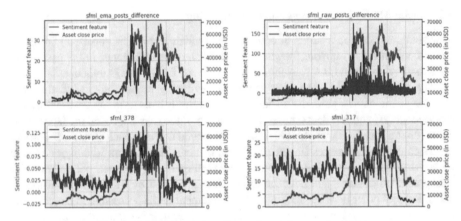

Fig. 1. Selected social media sentiment features for BTC

3.3 Machine Learning

Parametrized experiment is implemented in Python, encompassing financial and social media data preparation and feature engineering. In a summary, the following steps are implemented in financial data preparation: adding Datetime features, adding intra period features, adding Technical Analysis indicators and imputation of the missing values in the financial series by using interpolation. Model is trained to predict direction of the close price/return. Hence the problem is classification; model is expected to predict bullish/bearish market behavior for the assets in the next period.

Extreme Gradient Boosting (XGBoost) algorithm was used (several other algorithms were also tried) with default set of parameters (to facilitate objective performance comparison). Data is scaled before training by using MaxAbs scaling method.

3.4 Sentiment and Boosted Sentiment Signal Strategies

Sentiment signal strategies are using social media sentiment signals to generate long and short orders, according to the pre-defined criteria. Two sentiment signal strategies are introduced: naïve momentum and percentile strategy.

Naïve momentum strategy is implemented on continuous series, namely EMA-smoothed signals. Long order is generated on occurrence of p consecutive timepoints of increased sentiment in the past. Short order is generated on occurrence of q consecutive timepoints of decreased sentiment in the past. Results with default p = 8, q = 8.

Percentile strategy is implemented by applying percentile rank transformation of all three types of signals (see Social media data and feature engineering section for description of different signal types). Long order is generated when sentiment reaches the highest percentile. Short order is generated when sentiment touches the lowest percentile. The floors and ceilings of the signals are parameters, default floor = 0.8, ceil = 0.2 (quintiles).

We also introduce so-called boosted sentiment strategies, in which two or more social media sentiment signals are used to generate long and short orders. In boosted sentiment strategy, long/short orders are generated only when given conditions are met for all used signals. For boosted sentiment strategy to make sense, all signals in the combination must have same post-processing parameters. For example, for EMA, introducing different decay parameters for different signals would introduce reduced responsivity in one of them, resulting with slight shift between two signals. Such shift would diminish if not completely degrade boosting effect.

3.5 Hybrid Trading Strategies

We introduce the concept of hybrid trading strategies, which are using both financial and social media signals (by combining conventional, for example, trend trading strategies with naïve sentiment momentum strategy), to deduct long/short orders under the conditions applied on both signals, defined in advance.

In this report, the results of a trend trading crossover moving average strategy, combined with naïve sentiment momentum strategy will be reported. In this case, conditions for long/short orders are defined as follows:

Long order is generated only when there is occurrence of p consecutive timepoints of increased sentiment in the past, AND short moving average of close price is above long moving average (for the given moving average periods). Short order is generated only when there is occurrence of p consecutive timepoints of decreasing sentiment in the past, AND short moving average of close price is beyond long moving average (for the given moving average periods).

Hybrid strategies are introduced to increase confidence on trading with social media signals, by confirming the forecasts driven by social media signals with forecasts of the conventional, widely recognized and commonly used strategies.

3.6 Validation and Metrics

Validation includes back-testing on the test data, by using trading orders predicted by the trained model. A so-called vectorized back-test is implemented. Trading orders (1 for long, −1 for short), produced by the ML model are shifted one step ahead (as they are forecasts of the return in next time step) and then multiplied with corresponding return - log normalized return at time t.

$$logr_t = \log(\frac{c_t}{c_t - 1})$$

where c_t is close price of the asset at the time t.

While the latter is a return on time t, its cumulative sum is overall log normalized return, from the beginning of testing to time t. Obviously, cumulative sum at the last timepoint of the sample is overall log normalized return on strategy.

$$logR_t = \sum_{i=1}^{t} \log(\frac{c_i}{c_i - 1})$$

Finally, portfolio metrics, namely normalized return and Sharpe are also calculated. Portfolio consists of all crypto currencies in the available dataset. An equally weighted portfolio allocation method was used. Thus, the period returns for the portfolio (at time t) are calculated as averages of the period returns of the traded assets at time t.

Although Sharpe ratio is the most used loss function for optimizing trading strategies, we do not consider it as efficient indicator for ML-driven strategies. Sharpe ratio considers risks by taking into account standard deviation of the log return distribution. As in vectorized back-testing approach, trading is executed in every timestep (short or long signals), it is risky by default, and returns are characterized with very large variance. Introducing risk may diminish the effect of ML model performance. This is confirmed in an optimization experiment which used Sharpe ratio as loss, when great most of the models with 'optimized' set of hyperparameters performed poorly.

4   Results and Discussion

The experiments were carried out on the large datasets. For example, for BTC, more than 1.5M posts, 7M comments and 11M of votes have been considered in this analysis. Other assets with significant representations were DOGE and ETH. The least represented assets were DASH, DOT and AVAX, yet with 10K posts, 50K comments and 50–150K votes. In the initial test, votes were not included, due to lower correlation coefficients and generally worse performance in ML driven methods.

Feature selection on posts and comments was carried out by choosing the features whose parameters performed best in the optimization process, by using Spearman monotonic correlation coefficient as an objective function. To avoid look-ahead bias, the optimization process has been done on the data from ML training dataset. Given very high variance and in some cases varying data representations in training and test dataset, the correlation coefficients of the selected features in the test dataset were not as expected. This is the reason why generic features were introduced, reflecting common knowledge and intuition on the data impact (24 h aggregation period, EMA post-processing with selected decay parameter). Still, some general conclusions on the features correlation can be drawn by looking at the results of the optimization test. Out of 1700 trials (100 per crypto currency) in 500 cases the feature with S > 0.02 was found. Best correlated features had S in range of (0.036–0.04). We highlight that Spearman coefficients usually have lower values than Pearson coefficients in datasets with frequent outliers, which is the case here.

The optimization test shows that the ratio and difference types of the posts and comments, smoothened by using EMA and KAMA are the most superior features.

Multiple ML architectures were tested, including different selections of 'manual' features, loss functions (Pearson or Spearman correlation coefficients) for feature engineering and training, algorithms (XGBoost, Random Forest, KNN and CatBoost), by following the proposed methodology. Final results are presented in a table below (green cells for models with social media data which exceed the performance of those trained with financial features only) (Fig. 2).

	BnH		Without sm features			With sm features		
	log roi	norm roi	log roi	norm roi	sharpe	log roi	norm roi	sharpe
ADA	-1.12	0.33	2.35	10.52	1.28	2.19	8.92	1.34
ALGO	-1.24	0.29	-0.43	0.65	-1.12	0.60	1.81	1.10
AVAX	-1.39	0.25	1.11	3.05	1.10	1.59	4.92	1.04
BTC	-0.53	0.59	1.11	3.03	1.86	0.90	2.45	1.57
IOTA	-0.92	0.40	2.08	7.97	2.11	3.24	25.45	2.23
DOGE	-1.46	0.23	2.37	10.68	1.77	2.80	16.51	1.94
VET	-1.17	0.31	1.38	3.98	0.95	1.05	2.85	1.21
BNB	-0.01	0.99	-0.01	0.99	-0.37	0.28	1.32	1.99
MATIC	-0.65	0.52	0.57	1.78	2.48	1.56	4.77	2.52
SHIB	-0.79	0.45	1.09	2.98	0.94	-0.22	0.80	-1.55
SOL	-1.40	0.25	0.57	1.77	1.21	1.32	3.74	1.29
ETH	-0.33	0.72	0.68	1.98	1.02	-0.24	0.78	0.26
XRP	-0.25	0.78	1.46	4.31	0.87	0.13	1.14	0.50
THETA	-1.83	0.16	2.53	12.59	1.93	4.17	64.64	2.15
DASH	-1.36	0.26	1.34	3.82	0.74	1.36	3.88	0.78
LTC	-0.87	0.42	0.82	2.26	0.71	2.02	7.56	2.53
DOT	-1.01	0.36	1.59	4.88	1.38	1.17	3.24	1.46
PORTFOLIO	-0.82	0.44	1.42	4.14	1.48	1.81	6.14	1.98

Fig. 2. Results of ML model back-test

The above results were achieved with XGBoost classifier with default parameters (for comparison with metrics of the financial model) and with optimization of parameters of features related to posts and comments by using Spearman coefficient as objective function.

There are some common results which can be summarized to the following key conclusions:

- In all approaches, for all cryptocurrencies in the bucket, normalized returns are above BnH and positive, in many cases very high.
- For 9–11 assets from the bucket of 17 cryptocurrencies, ML models trained with social media features perform better than ML models trained only with financial features.
- When looking into all experiments, it appears that social media features perform the best for THETA, DASH, LTC, MATIC and ALGO, as for those assets, the performance of models trained with social media features are consistently better than models trained with financial data only. On the opposite side, models with social media features consistently underperform (when compared to ML models with financial features) for ETH, XRP and DOT.

Variance of results from one experiment setup to another is relatively high, for at least two reasons. First, results vary depending on the adopted test period. Test periods are different because of stripping data with NaN values at the beginning of the overall data period, where the length of the stripped period depend on the (automatically, by

Bayesian optimization) adopted KAMA parameters for the selected best features. Second, Bayesian optimization is stochastic process which does not necessarily produce same set of 'best' features in each run.

In some cases, algorithms perform consistently and expectedly worse, due to misfortunate (for ML training) market dynamics in the period. For SHIB, while there is a very active market in the train period, which is also representative in social media signals, test period is very calm with sparse social media data (see figure below, left). For BNB, the problem is the opposite - very calm market in training period, and high dynamics in test period (see figure above, right). Actually, majority of cryptos show similar behavior, especially THETA, MATIC, BNB and VET (Fig. 3).

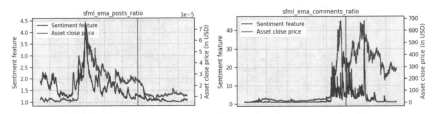

Fig. 3. Example illustrating signal weakening (left)/strengthening (right) in test period

Another reason for occasional unexpectedly low results is due to the nature of correlation coefficients, which sometimes 'favorize' simplistic features in the automatic feature engineering process, with absence or loss of credible predictive information. As a rule, this occurs in cases where there is sparse data on posts or comments.

Finally, the popularity and market cap of crypto is typically proportional to EMH strength. It is expected that social media features show less predictive power in the case of the most popular cryptos, such as BTC and ETH. That hypothesis is weakly confirmed in the experiment - BTC, ETH models with sentiment features, continuously (over different experiments) underperform when compared to ML financial models.

4.1   Back-Testing on Feature-Driven Strategies

In vast majority of cases, use of sentiment features in naïve sentiment momentum strategies brings positive return (normalized ROI > 1). In overall, the least responsive asset to sentiment is BNB, for which only one feature ensures positive return. The most effective feature type, as demonstrated in the back-test is ratio of bullish and bearish content, either posts or comments. On portfolio level, those features bring the highest return. Buy and hold strategy on equally weighted portfolio, produces normalized ROI of 0.44, meaning that market was bearish in the test period. Each of the tested sentiment features beats the market in back-test by using naïve sentiment momentum strategy, with normalized ROI in the range of (1.07, 2.02), with ema_comments_ratio feature showing the best results. The results of percentile strategies were not presented as they were significantly worse.

In the experiment, we tested boosted naïve sentiment momentum by using different combinations of signals. Experiments were performed with same look-back periods for long and short signal adoption, to facilitate comparison with the method implemented with individual features. In all tests, although positive normalized returns were achieved on portfolio, the results were weaker than in implementing naïve sentiment momentum strategy with individual features. For example, for boosted [ema_posts_ratio, ema_comments_ratio] strategy, normalized return on portfolio was 1.46. For boosted strategy where all feature types on posts were used, normalized return on portfolio was 1.49. Other combinations produced lower normalized returns (Fig. 4).

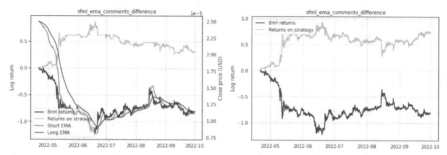

Fig. 4. Comparison of strategy returns between hybrid (left) and naïve momentum (right) back-test approach on SHIB

The hybrid naïve sentiment momentum-ema crossover back-test experiment was carried out with fixed parameters, namely short EMA period of 2 days and long EMA of 7 days. The results on the portfolio did not exceed the results of the naïve momentum back-test with the majority of individual features.

5   Conclusions

The strength of ML techniques in forecasting market returns lies in capability to generalize over vast and diverse data, not on simply fitting the arbitrary sets of input features to buy/sell decisions. Generalization is often synonymous with simplification: two key market forces that ML approach should try to highlight in data are momentum and reversion, where the former is typically long-term, and latter is short-term. EMH makes it almost impossible to highlight those two phenomena based on historical data only. However, there is a huge potential in identifying the points in which the momentum and reversion take place and find causal relationships between the social media bursts and those change points. In this paper, we report on the research aiming to provide forecasting model development methodology that does exactly that. The approach combines vast data coverage with best practices in strengthening the power of that data to represent market dynamics and with the best ML techniques on the market, customized to fit the intended purpose.

In contrast to a dominant way of thinking in the ML community and even in scientific literature, a quest for the best performing asset return forecasting method is not a race for improving accuracy. When different forecasting approaches are evaluated, no single metric should be considered/compared in isolation; the performance of the approach needs to be assessed by looking at the sets of the key performance indicators, where back-testing ROI and Sharpe in different strategies for different sets of parameters are the most informative on the model predictive strength. When using those metrics, our models clearly outperform the market or simple trading strategies, which is strong evidence of the impact social media has in making trading decisions.

In practice, adoption of ML models in real-life scenarios is very low, due to a common lack of trust to so-called black-box models. We consider hybrid strategies as innovative approach to combining predictive strength of social media signals with confident conventional strategies, such as trend following, to mitigate the risks of uncertainty and complexity behind the social media sentiment. In this context, the presented results are proof of concept which is going to be developed in the future.

References

1. Khan, W., Ghazanfar, M.A., Azam, M.A., Karami, A., Alyoubi, K.H., Alfakeeh, A.S.: Stock market prediction using machine learning classifiers and social media, news. J. Ambient. Intell. Humaniz. Comput. (2020). https://doi.org/10.1007/s12652-020-01839-w
2. Coelho, J., D'almeida, D., Coyne, S., Gilkerson, N., Mills, K., Madiraju, P.: Social media and forecasting stock price change. In: 2019 IEEE 43rd Annual Computer Software and Applications Conference (COMPSAC), pp. 195–200 (2019). https://doi.org/10.1109/COMPSAC.2019.10206
3. Zimbra, D., Abbasi, A., Zeng, D., Chen, H.: The state-of-the-art in Twitter sentiment analysis: a review and benchmark evaluation. ACM Trans. Manag. Inf. Syst. 9(2), 1–29 (2018). https://doi.org/10.1145/3185045
4. Glenski, M., Weninger, T., Volkova, S.: Improved Forecasting of Cryptocurrency Price using Social Signals (2019). https://doi.org/10.48550/ARXIV.1907.00558
5. Sousa, M.G., Sakiyama, K., Rodrigues, L.d.S., Moraes, P.H., Fernandes, E.R., Matsubara, E.T.: BERT for stock market sentiment analysis. In: 2019 IEEE 31st International Conference on Tools with Artificial Intelligence (ICTAI), pp. 1597–1601 (2019). https://doi.org/10.1109/ICTAI.2019.00231
6. Sun, Y., Zeng, X., Zhou, S., Zhao, H., Thomas, P., Hu, H.: What investors say is what the market says: measuring China's real investor sentiment. Pers. Ubiquit. Comput. 25(3), 587–599 (2021). https://doi.org/10.1007/s00779-021-01542-3
7. Nguyen, T.H., Shirai, K., Velcin, J.: Sentiment analysis on social media for stock movement prediction. Expert Syst. Appl. 42(24), 9603–9611 (2015). https://doi.org/10.1016/j.eswa.2015.07.052
8. Cai, J., et al.: Wall Street Bots: Building an Automatic Stock Trading Platform based on Artificial Intelligence From Scratch. Medium article (2022)
9. Araci, D.: FinBERT: Financial Sentiment Analysis with Pre-trained Language Models (2019). https://doi.org/10.48550/ARXIV.1908.10063
10. Devlin, J., Chang, M.-W., Lee, K., Toutanova, K.: BERT: Pre-training of Deep Bidirectional Transformers for Language Understanding (2018). https://doi.org/10.48550/ARXIV.1810.04805
11. Zhou, Z., Ma, L., Liu, H.: Trade the Event: Corporate Events Detection for News-Based Event-Driven Trading (2021). https://doi.org/10.48550/ARXIV.2105.12825

Access Control in a Distributed Micro-cloud Environment

Tamara Ranković$^{(\boxtimes)}$ 🄳, Miloš Simić🄳, Milan Stojkov🄳, and Goran Sladić🄳

Faculty of Technical Sciences, Novi Sad, Serbia
tamara.rankovic@uns.ac.rs

Abstract. Proliferation of systems that generate enormous amounts of data and operate in real time has led researchers to rethink the current organization of the cloud. Many proposed solutions consist of a number of small data centers in the vicinity of data sources. That creates a highly complex environment, where strict access control is essential. Recommended access control models frequently belong to the Attribute-Based Access Control (ABAC) family. Flexibility and dynamic nature of these models come at the cost of high policy management complexity. In this paper, we explore whether the administrative overhead can be lowered with resource hierarchies. We propose an ABAC model that incorporates user and object hierarchies. We develop a policy engine that supports the model and present a distributed cloud use case. Findings in this paper suggest that resource hierarchies simplify the administration of ABAC models, which is a necessary step towards their further inclusion in real-world systems.

Keywords: ABAC · Access Control · Cloud Computing · Edge Computing

1 Introduction

The widespread adoption of Internet of Things (IoT) systems in increasingly many industries has introduced novel and more challenging requirements for cloud computing. IoT devices intrinsically generate tremendous amount of raw sensor data and heavily rely on cloud for storage and processing. Some of them even require real-time responses, which is often infeasible due to long distances from data centers [1]. A recurring proposal that addresses these problems is to distribute cloud resources to smaller data centers in places of higher demand [2]. For example, edge computing places an edge layer between the cloud and devices [1].

Edge-IoT collaboration has significant potential to transform and accelerate the growth of many industries [3]. It is estimated there will be billions of edge clients only in the next few years [1]. For these predictions to come to life, it is necessary to rethink all aspects of the traditional cloud and adapt them to the edge environment.

In this paper, we focus on the access control to edge resources. Coarse-grained models like Role-Based Access Control (RBAC) [4] are generally not applicable in cloud-enabled IoT scenarios [1]. Authorization decisions should be based on many contextual conditions, including user, resource, spatiotemporal and relationship context [5]. Access control policies have to be flexible enough to support the following five necessary components: who, when, where, what and how [1]. On the other hand, policy administration has to be fairly straightforward in order to avoid misconfigurations, as a few administrators may manage thousands of devices.

The most prevalent solution, both in literature and in industry, is to opt for some form of ABAC [6] because of its flexible and dynamic nature [7]. ABAC enables declaring fine-grained policies, but that comes at the cost of high policy-management complexity. Extensions of ABAC that address this problem primarily do so by introducing hierarchies that facilitate permission inheritance implicitly through attribute inheritance. Majority of hierarchical ABAC models include user and object group hierarchies [8, 9] or attribute hierarchies [9], but there is little work on direct user and object hierarchies. More research needs to be conducted in this area in order to make ABAC models expressive enough to intuitively emulate widely-adopted constructs like role hierarchies [10]. Only by doing this, ABAC can be fully utilized in domains where its flexibility is essential.

In this paper, we want to explore the implications of user and object hierarchies in ABAC models. Our goal is to determine whether and how hierarchies can simplify policy administration. Additionally, we want to investigate if this extension is useful in edge computing scenarios. To accomplish that, we formally define a Resource Hierarchy ABAC (RH-ABAC) model, a general-purpose model that allows creating user and object hierarchies via dependency relationships. We develop a policy engine that supports RH-ABAC to demonstrate model's feasibility and present a micro-cloud platform use case to display model's usability in the edge computing environment.

Any significant increase in real-world systems that rely on ABAC will be possible only when the administrative overhead is reduced enough so that benefits outweigh the complexity. Our research presents one step toward that goal. Another contribution of our paper lies in the use case we present, as it can help researchers identify requirements for access control in edge computing systems.

The rest of the paper is organized as follows: Sect. 2 presents relevant literature review. In Sect. 3, we present the RH-ABAC model: components, formal definitions and policy conflict resolution strategy. Section 4 outlines the policy engine architecture and implementation details. In Sect. 5 we present a use case that demonstrates RH-ABAC usability in distributed micro clouds. Section 6 concludes the paper and discusses directions of future work.

2   Related Work

User groups simplify policy administration as users inherit permissions from their groups [11]. RBAC extended this mechanism with roles, which group users and permissions simultaneously. Permissions are added and removed from roles, which affects all users assigned those roles. In RBAC$_1$ model [4], role hierarchies additionally reduce redundant permission assignments, as they enable roles to inherit permissions from their junior roles.

While user-role and role-role relationships clearly express inheritance in RBAC, permissions are vaguely described as operations on objects, without any guidance on how to structure and reason about them. This has led to additional development of RBAC models. For example, authors of [7] formalize the access control model employed by Google Cloud Platform (GCP), where permissions can be operations on objects or object types. If someone is allowed to perform an operation on an object type, they are also allowed to do it on all objects of that type.

Straightforward RBAC management comes at the cost of inflexible policies that are insufficient in many cases, when hybrid models taking attributes into account or pure ABAC models are seen as a preferable option [7, 10]. ABACα [6] is a formal model containing only a minimal set of ABAC elements: users, subjects, objects, user attributes, subject attributes, object attributes, policies based on attributes and constraints. Even though ABACα policies can be highly expressive and direct relationship representation via attributes is trivial, it is not easy to represent hierarchy structures, proven to be very useful for inheritance [10].

Hierarchical Group and Attribute-Based Access Control (HGABAC) model [8] attempts to solve this problem via user and object groups. Users can be assigned to groups from which they obtain attributes. If an attribute is assigned to a group, all members of that group inherit it. Additionally, groups form a group graph, which enables attribute propagation from parent groups to children groups. Object groups and attribute inheritance work identically. Authors concluded that HGABAC requires less operations to express certain access control rules when compared to ABACα, while for others it has an administrative overhead. Another potential problem is the authorization function, requiring only one policy to evaluate to true for a permission to be granted, which isn't necessarily desirable behavior. If one policy states that a group should have access to a resource and the other one denies access to a user from that group, the user would still have access because of the first policy. Merging the two policies solves the problem, but it leads to unmaintainable policies as exceptions stack up.

A slight variation of HGABAC, restricted HGABAC (rHGABAC), is proposed in [9]. Authors altered the policy format in order to implement a policy engine backed by rHGABAC. Groups and group hierarchies remained unchanged. To further simplify the attribute management process, rHGABAC model introduced attribute hierarchies. Attribute values form a partially ordered set. If a user is assigned a senior attributed value, it is also assigned all junior attribute values. For example, role and resource hierarchies can be represented in this way. Role hierarchy can be formed in the attribute value set and by assigning one value from the set, which is a role, all junior roles to that role would also be assigned. In this way, all users with senior roles would satisfy policies requiring junior roles.

Hierarchical ABAC models have also been developed in the area of IoT and edge computing. Access control model for a smart cars use case utilizing attribute inheritance is presented in [12]. Multiple objects, usually sensors and actuators, form clustered objects, for example smart cars. Clustered objects can be direct members of one group and groups can be organized hierarchically. Aside from attributes an object was directly assigned, it inherits attributes of its clustered object. Clustered objects inherit additional attributes from their direct group and a group inherits attributes from its senior groups.

Another model, developed for cloud-enabled industrial IoT, was introduced in [3]. The group concept is divided into static and dynamic groups. In the case of static groups, members are explicitly declared, while members of dynamic groups are objects that satisfy the query set by an administrator. Aside from attribute inheritance, groups enable policy inheritance. Policy attached to a group is propagated to all children groups and object members.

3   RH-ABAC Model

This section provides an overview of the RH-ABAC model and its core components, including sets, relations and authorization function. Figure 1 shows core relations between the sets, while formal model definitions can be found in Table 1.

Resources (R) is a set consisting of all entities present in the system. It is a union of two sets with no overlapping values: **Users (U)** and **Objects (O)**. Users are entities that can perform operations, while objects are entities on which those operations are performed. Operations are not directly performed by users, but rather by their representations in the **Subjects (S)** set. Each subject is created by only one user and it executes actions on their behalf, but there can be multiple subjects that correspond to the same user. *User* function maps every subject to its creator. It is worth noting that subjects are equivalent to sessions in the RBAC model.

All resources, both users and objects, are assigned a set of attributes. They represent the basis on which fine-grained access control policies can be defined. Every attribute is a *(name, value)* pair that belongs to the **Resource Attributes (A_R)** set. Attribute assignments are defined by the **Resource Attribute Assignment ($A_R A$)** relation. A resource is constrained to have up to one attribute with the same name. Subjects obtain attributes from users by the *SubAttr* mapping. A subject can hold all attributes of its creator or only a subset of them. This enables the enforcement of the least privilege principle.

In addition to attributes inherited from the user, subjects can be assigned one or more attributes from the **Request Attributes (A_{REQ})** set. Assignments are represented by the **Request Attribute Assignment ($A_{REQ}A$)** relation. Their purpose is to describe conditions at the time of the subject request arrival more closely. Conditions can be related to the environment (e.g. current time), user connection (e.g. the IP address the request originated from) or something else. Since attribute values can be different for each request a subject makes, they should be updated accordingly.

Operations that can be performed in the system are elements of the **Operations (OP)** set. An operation paired with a value from the **Permission Types (PT)** set forms an element in the **Permissions (PERM)** set. This allows for positive and negative permissions. In other words, it is possible to explicitly allow or deny access to a certain resource or group of resources. Permissions are combined with elements of the **Conditions (C)** set to create **Policies (POL)**. Each condition is a function that returns a Boolean value based on object's attributes and subject's resource and request attributes.

Subjects will be granted a permission p only if there is a policy that contains p and its condition evaluates to *true* for provided attribute values. Conditions should be specified in some policy language, which is outside the scope of this paper.

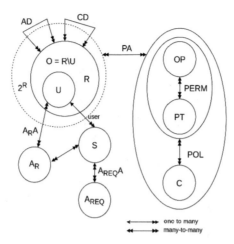

Fig. 1. RH-ABAC conceptual model

Aggregation Dependencies (AD) and **Composition Dependencies (CD)** are strict partial order relations on resources. They both express *part-of* relationships between resources. The difference between them is that the child's life depends on the parent's life cycle in the case of composition, but not in the case of aggregation. **Dependencies (D)** relation, also a strict partial order relation, is the union of aggregation and composition dependencies extended by pairs required to satisfy transitivity. Resources and dependencies form a **Resource Dependency Graph (RDG)**, a directed acyclic graph. These dependencies can be viewed as additional resource attributes, represented in an alternative way. For example, to emulate group memberships, there can be an aggregation dependency from the group to the user. It is equivalent to assigning the user a *group* attribute.

Since dependencies can't be referenced in policy conditions, as they are not regular attributes, there has to be a mechanism for accessing them. The problem is solved by introducing user and object scopes. They are additional conditions assigned to policies through **Policy Assignments (PA)**. Every element of the PA set is a *(policy, subject scope, object scope)* triple. Both the subject and object scope are non-empty sets of resources. For a permission to be granted, aside from the condition that must return true, the subject must be in the subject scope and the object must be in the object scope. A subject is in the subject scope if all resources from the scope are subject's parents, defined by the *SubParents* function, or they are equal to the user who created the subject. The same principle is applied to objects, with the difference that object's parents are determined by the *ResParents* mapping. Multiple resources in the scope correspond to the *AND* logical operator. For example, subject scope *{group: g1, group: g2}* states that a subject must be a member of both the group *g1* and group *g2*.

Table 1. RH-ABAC sets, relations and functions

Sets and relations

- $R, U(U \subseteq R), O(O = R \setminus U), S, A_R, A_{REQ}$ - sets of resources, users, objects, subjects, resource attributes and request attributes
- $OP, PT = \{allow, deny\}, PERM = OP \times PT$ - sets of operations, permission types, and permissions
- C - the set of conditions, where $\forall c \in C$, c is $f: 2^{A_R} \times 2^{A_R} \times 2^{A_{REQ}} \longrightarrow \{true, false\}$
- $POL \subseteq PERM \times C, AD = \{allowed, denied, undefined\}$ - sets of policies and authorization decisions
- $A_R A \subseteq R \times A_R$ - resource attribute assignment (many-to-many relation)
- $A_{REQ} A \subseteq S \times A_{REQ}$ - request attribute assignment (many-to-many relation)
- $PA \subseteq POL \times 2^R \times 2^R$ - policy assignment (ternary relation), where the combination of the operation, permission type, user scope and object scope must be unique, $\forall pa \in PA: pa = (op \in OP, pt \in PT, c \in C, user\_scope \in 2^R, object\_scope \in 2^R)$
- $\preceq_{AD} \subseteq R \times R$ - aggregation dependency, a strict partial order relation on R
- $\preceq_{CD} \subseteq R \times R$ - composition dependency, a strict partial order relation on R
- $\preceq_{AD} \cap \preceq_{CD} = \emptyset \bullet r_1 \preceq_{AD} r_2 \Rightarrow r_2 \npreceq_{CD} r_1$
- $root \in R: (\forall r \in R)(r \preceq_{CD} root)$
- $\preceq_D = \preceq_{AD} \cup \preceq_{CD} \cup \{(r_1, r_3) | (r_1, r_2) \in (\preceq_{AD} \cup \preceq_{CD}) \wedge (r_2, r_3) \in (\preceq_{AD} \cup \preceq_{CD})\}$
- $RDG = (R, \preceq_D)$ - Resource Dependency Graph, a directed acyclic graph

Functions

- $User: S \longrightarrow U$ - mapping a subject to the corresponding user
- $ResAttr: R \longrightarrow 2^{A_R}, ResAttr(r) = \{a \in A_R | (r, a) \in A_R A\}$ - mapping a resource to its attributes
- $SubAttr: S \longrightarrow 2^{A_R}, SubAttr(s) \subseteq ResAttr(User(s))$ - mapping a subject to its attributes
- $ReqAttr: S \longrightarrow 2^{A_{REQ}}, ReqAttr(s) = \{a \in A_{REQ} | (s, a) \in A_{REQ} A\}$ - mapping a subject to request attributes
- $ResParents: R \longrightarrow 2^R, ResParents(r) = \{r_1 \in R | r \preceq_D r_1\}$ - mapping a resource to its parent resources
- $SubParents: S \longrightarrow 2^R, SubParents(s) \subseteq ResParents(User(s))\}$ - mapping a subject to its parent resources
- $ResChildren_C: R \longrightarrow 2^R, ResChildren_C(r) = \{r_1 \in R | r_1 \preceq_{CD} r\}$ - mapping a resource to its children in composition relation
- $TransitiveReduction: 2^R \times 2^{R \times R} \longrightarrow 2^R \times 2^{R \times R}$ - mapping a graph to its transitive reduction

Policy evaluation

$PolEval: POL \times 2^{A_R} \times 2^{A_R} \times 2^{A_{REQ}} \longrightarrow AD, \ PolEval(pol, SubAttrs, objAttrs, ReqAttrs) =$

$$\begin{cases} allowed & pol_{(2)}(SubAttrs, objAttrs, ReqAttrs) = true \wedge pol_{(1)} = allow \\ denied & pol_{(2)}(SubAttrs, objAttrs, ReqAttrs) = true \wedge pol_{(1)} = deny \\ undefined & otherwise \end{cases}$$

$EffectivePolicies: S \times R \times OP \longrightarrow 2^{PA}, \ EffectivePolicies(sub, obj, op) = \{pa = (pol, US, OS) \in PA | (\forall us \in US)(us \in (SubParents(sub) \cup User(sub)) \wedge (\forall os \in OS)(os \in (ResParents(obj) \cup obj)) \wedge pol_{(0)} = op\}$

$PolPriorityBySub: S \times PA \longrightarrow Z, \ PolPriorityBySub(sub, pa) = max(\{p = -Dist(User(s), userScope) | userScope \in pa_{(3)}\})$

(*continued*)

Table 1. (*continued*)

$PolPriorityByObj: O \times PA \rightarrow Z, PolPriorityByObj(obj, pa) = max(\{p = -Dist(o, objScope) | objScope \in pa_{(4)}\})$

- *Dist* calculates distance between two nodes in the $TransitiveReduction(RDG)$ graph

---

$MaxPriorityBySub: S \times O \times 2^{PA} \rightarrow 2^{PA},$

$MaxPriorityBySub(sub, obj, pas) = \{pa$
$\in pas | Eval(pa_{(0,1,2)}, SubAttr(sub), ResAttr(obj), ReqAttr(s))$
$\neq undefined \wedge \neg(\exists pa_1 \in pas)(pa_1$
$\neq pa \wedge Eval(pa_{1(0,1,2)}, SubAttr(sub), ResAttr(obj), ReqAttr(s))$
$\neq undefined \wedge PolPriorityBySub(sub, pa_1)$
$> PolPriorityBySub(sub, pa))\}$

$MaxPriorityByObj: S \times O \times 2^{PA} \rightarrow 2^{PA},$

$MaxPriorityByObj(sub, obj, pas) = \{pa$
$\in pas | Eval(pa_{(0,1,2)}, SubAttr(sub), ResAttr(obj), ReqAttr(s))$
$\neq undefined \wedge \neg(\exists pa_1 \in pas)(pa_1$
$\neq pa \wedge Eval(pa_{1(0,1,2)}, SubAttr(sub), ResAttr(obj), ReqAttr(s))$
$\neq undefined \wedge PolPriorityByObj(obj, pa_1)$
$> PolPriorityByObj(obj, pa))\}$

$Eval: S \times O \times 2^{PA} \rightarrow AD,$

$Eval(sub, obj, pas)$

$= \begin{cases} undefined & pas = \emptyset \\ denied & \left(\exists(pol, us, os) \in pas\right)\left(PolEval\begin{pmatrix} pol, SubAttr(sub), \\ ResAttr(obj), \\ ReqAttr(sub) \end{pmatrix} = denied\right) \\ allowed & otherwise \end{cases}$

$Authorize: S \times O \times OP \rightarrow AD,$

$Authorize(s, o, op)$

$= Eval\left(MaxPriorityByObj\begin{pmatrix} s, o, \\ MaxPriorityBySub\begin{pmatrix} s, o, \\ EffectivePolicies(s, o, op) \end{pmatrix} \end{pmatrix}\right)$

Only policies present in the PA set are evaluated when an access control request is being processed. That poses the question of how to represent policies that should be applicable to all subjects or objects. The R set includes a *root* resource that is connected to all other resources by the composition relationship. When a policy should be evaluated for every subject, the policy's subject scope should contain only the root resource. The same strategy goes for objects.

Every authorization request contains a subject, object and operation and returns a value from the **Authorization Decisions (AD)** set. *EffectivePols* calculates policies that should be evaluated for a request, based on the operation and subject and object scopes. Selected policies can make conflicting decisions when evaluated by the *PolEval* function. For that reason, we present a conflict resolution strategy. Each policy has a subject priority, defined by *PolPriorityBySub* and object priority, defined by *PolPriorityByObj*. The greater the distance of the subject's creator from the subject scope in RDG, the smaller the policy's subject priority. Object priority is determined in the same way. Out of all applicable policies, only the ones that have a defined authorization decision and the maximum subject priority should be kept. From them, only those with the maximum object priority are selected. If there is any policy in the resulting set that evaluates to the *Denied* decision, the operation is denied. If the set is empty, the decision is undefined. In other cases, the operation is allowed. When the decision is undefined, system administrators should decide if the operation is allowed or denied by default.

The model supports operations for creating and deleting users, objects, subjects, attributes, policies and dependency relationships. While most of them only insert or remove one element from the corresponding sets, operations that delete resources also rely on composition dependency relationships. In addition to the specified resource and its attribute and policy assignments and dependency relationships, this operation also deletes all resources that are a part of that resource. *ResChildren$_C$* function returns all resources that should be additionally deleted in one request.

4   Policy Engine

In this section, we present the policy engine that supports RH-ABAC. It provides operations corresponding to model's administrative operations and authorization function. We developed it in order to show our model can be integrated with real-world systems. We optimized it for different use cases, so that a wider range of requirements is met. All optimizations can be turned on and off via configuration to achieve the desired behavior.

The policy engine is written in Go programming language. Clients can invoke any operation synchronously, using gRPC, or asynchronously, by sending requests to the NATS message broker. Asynchronous communication was enabled in order to decouple the engine from the rest of the system more easily. This property is especially important in microservice applications, which often use sagas [13] for transactive data updates in multiple services. To make data changes and response publishing atomic, we implemented the outbox pattern [14]. The message relay is implemented as a Go agent that periodically polls for outbox messages.

All data is currently stored in the Neo4j database, as it has a native graph data model facilitating the RDG traversal. Resources, attributes and policies are nodes in the graph, while dependency relationships and policy scopes are edges. There are two strategies for storing policy assignments. The first one persists only direct assignments. It creates edges from resources in the user scope to policies and from policies to resources in the object scope. For each authorization request, RDG must be traversed to find parents and applicable policies and policy priorities must be calculated. The other strategy removes this overhead as it caches policies and their priorities. It does so by directly connecting policies with all users and objects for which that policy is applicable. The edge from user to policy stores the subject priority and the one from policy to object stores object priority. With this, authorization checks don't need to search potentially long paths to find parents or inherited policies. However, this introduces additional computation in administrative operations. The second strategy is recommended for systems with predominantly static resources and policy assignments, while systems with frequent updates may prefer the first strategy, as policy caching can degrade the engine's performance.

When fast authorization is the first priority, the caching system can be configured. The engine uses multi-level cache where each engine instance has its own local cache, and they all share a global Redis cache. The cache stores users, objects, their attributes and applicable policies for a particular operation. In the case of a cache hit, only the policy conditions have to be evaluated for the request. As the cache can retain stale data for some time, it should be used with caution.

5 Micro-cloud Use Case

Constellations (c12s)[1] is a proof-of-concept implementation of a model for dynamic formation of micro data centers, presented in [15, 16]. The model organizes nodes into disposable data centers in proximity to data sources, with the objective of offering edge computing as a service, like any other service in the cloud. To ensure resiliency and latency, the model consists of multiple layers: clusters, regions and topologies. Multiple nodes are organized into clusters. One or more clusters form a region. A topology comprises one or multiple regions.

The implementation supports the following operations:

- **Mutate** creates, updates and deletes topologies, regions and clusters;
- **List** displays the state of resources the user controls. They can request a high-level overview of all topologies for example, or a detailed view of clusters in a single region;
- **Query** shows free nodes that satisfy requirements set by the user;
- **Logs** returns user's activity history in the form of logs and traces;

---

[1] https://github.com/c12s.

To demonstrate feasibility of the RH-ABAC model, we extended c12s with access control support. When a new user registers, an organization in which he is granted all permissions is also created. Users can add other users to their organization. Every topology in the system belongs to one organization, created by some organization member. To facilitate policy administration, users can be assigned to groups.

Figure 2 shows the transitive reduction of one possible RDG in the system. Users are nodes with the *u* prefix, while all other nodes are objects. Root resource is represented with the *root* node. Every other resource is dependent on it by composition relation. In this example, the system has one free node, *fnode:1*, and one organization, *org:o1*. Users *u:u1* and *u:u2* are members of *org:o1*. User *u:u1* is a member of the group *g:g1* and user *u:u2* is a member of groups *g:g1* and *g:g2*. The organization owns one topology, *top:t1*, with two regions: *reg:r1* and *reg:r2*. Nodes *c:c1, c:c2, c:c3* and *c:c4* are clusters that form those regions, and they consist of nodes *node:1, node:2, node:3* and *node:4*. Policy assignments are arrows whose start marks subject scopes and at their ends are object scopes. Policy assignments of policies *p1, p2* and *p3* and their properties are displayed in Table 2. Policy conditions are omitted since they function identically to the ones of other ABAC models.

The policy assignment of *p1* grants all users in the system the permission to query free nodes in order to form clusters, regions and topologies from them. This is because the subject scope is *root*, which is the parent of every user. Therefore, any user is in this subject scope. This is equivalent to other ABAC models, in which all policies are applicable to all subjects and objects.

The policy *p2* assignment allows all members of the *org:o1* to view information of every node in that organization. The mechanism of narrowing the subject and object scope, as is the case with this assignment, allows us to filter-out possibly many subjects and objects that are out of scope. For example, if we want to ask what users are granted this permission on *node:1*, we can instantly discard all users that are not members of the organization. This is of course true if there are no other policy assignments with the same permission and *node:1* in the object scope.

Authorization request for the subject of *u:u2*, object *node:1* and operation *node.get* evaluates two policies with conflicting decisions, *p2* and *p3*. While *p2* states that all members of the organization can view *node:1* information, policy *p3* denies the execution of that operation for users that are members of both the group *g:g1* and *g:g2*. In this case, conflict resolution strategy must be enforced. Firstly, we must calculate the subject priority for both *p2* and *p3*, which are −2 and −1, respectively. Since *p3* has higher subject priority and a defined authorization decision, it will be kept, and *p2* won't. As there is only one policy left, we can skip other steps and return a response that operation execution is denied. However, if there were multiple policies left after the filtering, we would also have to calculate object priorities.

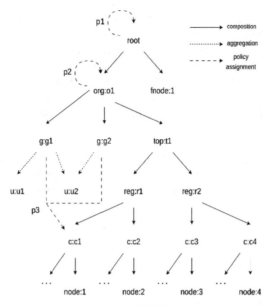

Fig. 2. Transitive reduction of the RDG for the micro-cloud use case

In this example, if a user issued a command to delete *top:t1* resource, apart from it, the regions *reg:r1* and *reg:r2* and all of their children resources would also be deleted. When a resource is being deleted, all of its attributed and dependency relationships should be removed, as well as policy assignments where that resource is a part of the subject or object scope, since they are not valid anymore. This cascading removal of resources requires only one operation for the entire hierarchy of resources, and it doesn't expect the client to know the structure of the hierarchy.

Table 2. Policy assignment properties

Policy	Operation	PermType	SubScope	ObjScope
p1	*freenode.list*	*allow*	*{root}*	*{root}*
p2	*node.get*	*allow*	*{org:o1}*	*{org:o1}*
p3	*node.get*	*deny*	*{g:g1, g:g2}*	*{c:c1}*

6 Conclusion

In this paper, we demonstrated how user and object hierarchies lower the administrative overhead inherent to ABAC and proved hierarchical ABAC usability in distributed cloud environment. We achieved this by formally defining RH-ABAC model and implementing it in a policy engine. We outlined core RH-ABAC characteristics through a use case in micro clouds.

In order to have a thorough understanding of our model's strengths and weaknesses and explore its potential, we want to evaluate it on two levels. The first one involves comparing RH-ABAC expressive power and administration complexity to other hierarchical ABAC models, which is invaluable information when selecting an access control model. The second level is concerned with policy engine efficiency, as it is a major requirement for its integration with real-world systems. We want to quantify the trade-offs made by the two policy storing strategies, so that the most suitable one can be employed for a particular system.

Acknowledgements. █████Funded by the European Union (TaRDIS, 101093006). Views and opinions expressed are however those of the author(s) only and do not necessarily reflect those of the European Union. Neither the European Union nor the granting authority can be held responsible for them.

References

1. Xiao, Y., Jia, Y., Liu, C., Cheng, X., Yu, J., Lv, W.: Edge computing security: state of the art and challenges. Proc. IEEE **107**(8), 1608–1631 (2019)
2. Ferrer, A.J., Marquès, J.M., Jorba, J.: Towards the decentralised cloud: Survey on approaches and challenges for mobile, ad hoc, and edge computing. ACM Computing Surveys (CSUR) **51**(6), 1–36 (2019)
3. Bhatt, S., Pham, T.K., Gupta, M., Benson, J., Park, J., Sandhu, R.: Attribute-based access control for AWS internet of things and secure industries of the future. IEEE Access **9**, 107:200–107:223 (2021)
4. Sandhu, R., Coyne, E., Feinstein, H., Youman, C.: Role-based access control models. Computer **29**(2), 38–47 (1996)
5. Kayes, A., et al.: A survey of context-aware access control mechanisms for cloud and fog networks: taxonomy and open research issues. Sensors **20**(9), 2464 (2020)
6. Jin, X., Krishnan, R., Sandhu, R.: A unified attribute-based access control model covering dac, mac and rbac. In: Cuppens-Boulahia, N., Cuppens, F., Garcia-Alfaro, J. (eds.) DBSec. LNCS, vol. 7371, pp. 41–55. Springer, Heidelberg (2012). https://doi.org/10.1007/978-3-642-31540-4_4
7. Gupta, D., Bhatt, M.G., Kayode, O., Tosun, A.S.: Access control model for google cloud IoT. In: 2020 IEEE 6th International Conference on Big Data Security on Cloud (BigDataSecurity), IEEE International Conference on High Performance and Smart Computing (HPSC) and IEEE International Conference on Intelligent Data and Security (IDS), pp. 198–208. IEEE (2020)
8. Servos, D., Osborn, S.L.: Hgabac: towards a formal model of hierarchical attribute-based access control. In: Cuppens, F., Garcia-Alfaro, J., Zincir Heywood, N., Fong, P.W.L. (eds.) FPS 2014. LNCS, vol. 8930, pp. 187–204. Springer, Cham (2015). https://doi.org/10.1007/978-3-319-17040-4_12
9. Bhatt, S., Patwa, F., Sandhu, R.: Abac with group attributes and attribute hierarchies utilizing the policy machine. In: Proceedings of the 2nd ACM Workshop on Attribute-Based Access Control, pp. 17–28 (2017)
10. Servos, D., Osborn, S.L.: Current research and open problems in attribute-based access control. ACM Comput. Surv. (CSUR) **49**(4), 1–45 (2017)
11. Sandhu, R.S., Samarati, P.: Access control: principle and practice. IEEE Commun. Mag. **32**(9), 40–48 (1994)

12. Gupta, M., Benson, J., Patwa, F., Sandhu, R.: Dynamic groups and attribute-based access control for next-generation smart cars. In: Proceedings of the Ninth ACM Conference on Data and Application Security and Privacy, pp. 61–72 (2019)
13. Garcia-Molina, H., Salem, K.: Sagas. ACM SIGMOD Rec. **16**(3), 249–259 (1987)
14. Richardson, C.: Microservices Patterns: With examples in Java. Manning (2018). https://books.google.rs/books?id=UeK1swEACAAJ
15. Simić, M., Prokić, I., Dedeić, J., Sladić, G., Milosavljević, B.: Towards edge computing as a service: dynamic formation of the micro data-centers. IEEE Access **9**, 114:468–114:484 (2021)
16. Simić, M., Sladić, G., Zarić, M., Markoski, B.: Infrastructure as software in micro clouds at the edge. Sensors **21**(21), 7001 (2021)

E-Government Requirements Specification Based on BPMN

Sonja Dimitrijević[(✉)] [iD], Milovan Marić [iD], and Natalija Trtica [iD]

Mihailo Pupin Institute, Volgina Street 15, 11050 Belgrade, Serbia
`sonja.dimitrijevic@pupin.rs`

Abstract. Software requirements for e-government systems are not easy to specify. Among other reasons, e-government is characterized by complex, and often cross-organizational business processes. In search of effective approaches, BPMN has been seen as a prospectively beneficial aid. This paper presents an approach for e-government requirements specification based on BPMN. In the main, software requirements specification is conducted as the extension or elaboration of BPMN models. The approach is iterative, incremental, and highly flexible, so it can be adjusted to a required or chosen project management and/or development methodology. In addition, the paper critically analyzes the acceptability and effectiveness of the approach based on project artifacts and reflections of participants collected in several projects over time. More specifically, acceptability was found satisfactory based on the ratio of positive and negative comments per participant. Furthermore, the paper identifies and discusses the main factors influencing the effective use of the approach: model quality, familiarity of participants with BPMN, adequate collaboration between participants, and software support. In a nutshell, for the approach to be effective and acceptable, participants in an e-government project should be adequately familiarized with BPMN models, modeling principles, and guidelines that are followed. To this end, collaborative modeling and careful positioning of the models in the requirements specification are highly recommended.

Keywords: e-Government · BPMN · software requirements

1 Introduction

Different definitions of e-government emphasize its different aspects. It can be broadly defined as the use of information and communication technologies (ICT) in combination with organizational changes to improve public services [1]. The main goal of the e-government is to increase the quality and efficiency of government services while reducing costs [2].

Business processes in e-government are characterized by mandatory compliance with multiple regulations and laws, and by public audits. In addition, e-government business processes are primarily inter and intra-collaborative and may spread within several organizational units and/or organizations. At the same time, they should be

© The Author(s), under exclusive license to Springer Nature Switzerland AG 2024
M. Trajanovic et al. (Eds.): ICIST 2023, LNNS 872, pp. 448–458, 2024.
https://doi.org/10.1007/978-3-031-50755-7_42

transparent to citizens. Consequently, e-government business processes are not easy to define, monitor, and evaluate [3].

E-government's collaborative processes involve organizations (employees, technologies), partners (providers, consumers), and users (legal entities, citizens, foreigners). They are based on different e-government models comprising complex interactions and various technologies [4]. Consequently, it is very demanding to conduct the business analysis of e-government processes, even more so to specify requirements when developing or improving e-government systems. In search of effective approaches, BPMN has been seen as a prospectively beneficial aid.

Business Process Model and Notation (BPMN) is a standard notation that OMG [5] developed to support Business Process Modelling (BPM). BPM comprises activities of systematically defining, representing, analyzing, and improving business processes using methods, techniques, and software. The improvement of business processes primarily addresses their quality, efficiency, and facilitation, often with the help of innovations [6, 7].

Therefore, BPMN is intended to be comprehensible to business users and yet flexible enough for technical users to represent complex process semantics [5]. More specifically, BPMN provides a notation that should be relatively clear to different users [1]:

- business analysts who model the processes;
- technical developers who implement the technology applied in the redesigned processes;
- process owners, i.e., persons responsible for managing and monitoring the processes.

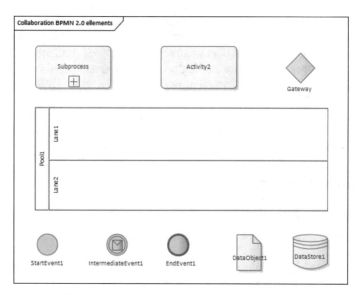

Fig. 1. BPMN 2.0 Subset of collaboration elements

The notation consists of graphical symbols (see Fig. 1) to represent an action, flow, or conduct of a process (e.g., orchestration, collaboration, etc.) [1, 5].

Diagrams based on BPMN were initially used for modeling an organization's business processes. Gradually, they began to be used for requirements engineering [8–10], and consequently for functional size or complexity measurement [8, 11–13]. With the development of Business Process Management Systems (BPMS), BPMN became the foundation for the implementation of BPMS solutions [14].

At the early stages of software development, software requirements are typically elicited in an unstructured form of Functional User Requirements (FUR) [10] and Non-Functional Requirements (NFR) [15]. In the business analysis phase, functional requirements are elaborated and formalized as a Software Requirements Specification (SRS). Requirements are usually specified as structured statements or use cases (with or without use case diagrams), i.e., in textual, natural language form [10, 16, 17].

The comprehension of requirements modeled either with BPMN, use case diagrams (UCD), or both were compared in [14]. BPMN models were found more effective in understanding the requirements that involve a sequence of activities.

There are studies that have investigated the application of BPMN in e-government business analysis [1, 18, 19]. Likewise, there are studies that investigate the application of BPMN in requirements engineering in general [10, 17, 20, 21]. However, studies with a focus on requirements elicitation with the help of BPMN in e-government projects are not very common. Moreover, they typically address the implementation of a specific technology or modeling of a specific type of requirement. One of them, for instance, applies BPMN in the development of smart contracts for e-government [22]. Another one proposed the application of BPMN in compliance requirements modeling in e-government [3].

Therefore, this paper aims to contribute to this research niche by presenting an original approach for e-government requirements specification based on BPMN. Furthermore, the paper analyzes the acceptability and effectiveness of the approach based on project artifacts and reflections of participants collected in several projects over time. In addition, the paper shows software engineers and business people working in the e-government domain that specification of requirements based on BPMN for different e-government systems is a feasible approach.

The rest of the paper is organized as follows. Section 2 defines the research questions. After the introduction of the methodology in Sect. 3, the results are presented and discussed in Sect. 4. Finally, Sect. 5 provides the paper's conclusion.

2   Research Questions

The objective of the paper is twofold:

- First, to present an approach for e-government requirements specification based on BPMN;
- Second, to critically analyze the effectiveness and acceptability of the approach and to analyze and discuss what influences its effective use.

The approach in question was proposed, tailored, and applied in several e-government projects to address the complexity of software requirements related to e-government system implementation.

Therefore, the paper presents the approach and addresses the following questions:

- RQ1. Can the approach for e-government requirements specification based on BPMN be acceptable for different participants in e-government projects?
- RQ2. What are the main factors influencing the effective use of the approach for e-government requirements specification based on BPMN?

3 Methodology

First, the main components and characteristics of the approach and its applications over the years in several e-government projects were summarized. The e-government projects are:

- Sustainable Information System - MAEP SIS (The Ministry of Agriculture and Environmental Protection - MAEP, the Republic of Serbia; 2016–17),
- e-Registration (The Serbian Business Registers Agency; 2016–17),
- e-Register of financial leasing (The Serbian Business Registers Agency; 2017),
- National Measures 2022 - NM22 (The Directorate for Agrarian Payments, the Republic of Serbia; 2022).

The projects were selected based on the following criteria:

- complexity of business processes,
- complexity of collaborations, and
- variety of participants.

The main components of the approach include activities, participants, outputs, and software support. Some of the main characteristics are flexibility, incrementality, and iterative cycles.

Then, the critical analysis of the approach's acceptability and effectiveness was conducted based on the project artifacts and reflections of participants in the roles of business analysts, software engineers, ICT domain experts (the client's side), and business domain experts (the client's side). The reflections of the participants were collected over projects' time in two ways:

- they were written down by the researcher (business analyst) during the working meetings;
- they were provided in the form of comments in draft SRS documents.

At least one participant in each role participated in the phases of modeling business process, development of SRS, and/or validation of SRS (e.g., NM22). MAEP SIS, as the largest project, brought together the most participants. Two teams of three business analysts analyzed business processes in nine MAEP organizational units, each of which has a dedicated lead domain expert/decision maker. A couple of ICT experts (from the client's side) were shared.

Acceptability was assessed with the ratio of positive and negative comments towards the approach per participant in a specific role. In the absence of any participant comments, the responsible researcher would be engaged to elicit the participant's feedback regarding familiarity with the models, their understanding of BPMN, as well as their attitude towards BPMN and the applied approach.

Factors influencing the effective use of the approach were mainly elicited from the participants' feedback and subsequently categorized. In addition, the project artifacts (plans, business process models, SRSs) were analyzed regarding the number of revisions from the draft to the final version, adherence to delivery deadlines, and completeness.

Finally, the key findings were discussed in accordance with the research questions.

4   Results and Discussion

The results were presented and discussed as follows. First, the approach for requirements specification based on BPMN was presented. Then, the acceptability and effectiveness of the approach were summarized and discussed.

4.1   The Approach for Requirements Specification Based on BPMN

Our approach for e-government requirements elicitation based on BPMN comprises the following key activities:

- identification of business processes or groups of business processes;
- planning of the participation of business process owners and relevant stakeholders;
- planning of data collection methods;
- data collection and initial analysis;
- modeling of business processes;
- development of SRS;
- validation of SRS.

Planning the participation of business process owners and relevant stakeholders is critical. This activity primarily concerns identifying the business process owners and relevant stakeholders and ensuring their participation in the project. If the process is collaborative, relevant stakeholders from all organizational units and/or organizations should be involved.

Since process owners (domain experts) and stakeholders have limited time to devote to the project, careful and mutual planning of their participation and data collection is of utmost importance.

Data collection, to which an initial analysis is inherent, may comprise different methods and techniques depending on the context of the project. Interviews, focus groups, workshops, observation of business process activities, sampling business process inputs and outputs, extraction of relevant regulations, and working procedures are the frequent ones.

As for the modeling of business processes, this activity initiates the in-depth analysis of the requirements. At the same time, it should facilitate communication among the participants. In addition, it should allow ambiguities to be clarified and details to be added in the as-is analysis, as well as in the to-be design. However, an assumption that BPMN is easy to apply, read, and understand per-se is greatly exaggerated. BPMN does offer a rich set of symbols for process modeling, but this can be a double-edged sword. To ensure that process models are comprehensible, consistent, and comparable, it is necessary to apply modeling principles and guidelines. This is also true for the effective use of the proposed approach.

The development of SRS is conducted as the extension or elaboration of models. It is flexible so it may be in compliance with the formalities imposed by the software/service purchaser. For instance, use cases were used in some e-government projects. On some other projects, less formal descriptions of requirements were provided.

Validation of SRS implies that SRS is validated by process owners/decision makers, end-users, and stakeholders at the end of the analysis phase, in the implementation phase, or even in the support phase of an e-government project. The main reasons behind the validation of SRS in the implementation and support phases are the dynamic nature of e-government requirements due to changes in legislation, as well as the complexity of e-government projects, and the involvement of a large number of participants in different roles.

The approach is incremental, which means that business processes can be analyzed one by one leading to the gradual development of SRS. It is also iterative meaning that

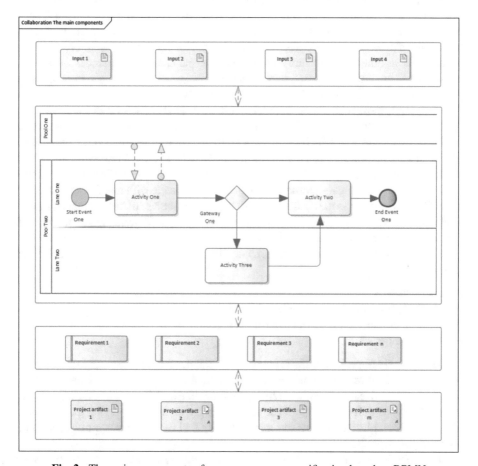

Fig. 2. The main components of an e-government specification based on BPMN

some steps can be repeated to elicit and elaborate the requirements to a satisfactory extent within the project scope.

The outputs of the described activities can be deduced in most cases. The outputs of the modeling of business processes however can be, in addition to models, a report, diagrams, and a part of SRS. One of the main characteristics of the approach is its flexibility, which is necessary due to the particularities of different e-government projects. Software support is also one of the factors that can influence both the acceptability and effectiveness of the approach. Software that allows for collaborative modeling, review, and requirements specification is highly recommended.

Figure 2 shows a simplified view of the main components of an e-government specification based on BPMN.

4.2   Acceptability and Effectiveness of the Approach

The ratio of positive and negative comments towards the approach per business analyst ranged from 3 to 1 in the first project to 5 to 1 or 6 to 1 in later projects. The ratio of positive and negative comments per domain expert was similar and ranged from 4 to 1 in the first project to 6 to 1 in later projects. Misinterpretation of the models in the comments was also considered negative. Insufficient comments were collected for participants in other roles to find such ratios. The increase in the number of positive comments over time is understandable considering the experience gained in the first project and the application of that experience in later ones.

In general, most participants appreciated the use of BPMN for requirements elicitation. This is particularly true for business analysts and business domain experts who worked most closely to clarify business processes and elicit software requirements. Some of the participants were neutral toward the use of BPMN, at least in the beginning. These participants either had an engineering background and were more familiar with other modeling and requirements approaches, or had no experience with modeling approaches at all. However, the approach is flexible and encourages the use of complementary modeling approaches, if necessary. For instance, UML sequence diagrams were found useful for modeling complex exchanges of messages among a larger number of systems.

The following main factors influencing the effective use of the approach were identified:

- model quality,
- familiarity of participants with BPMN,
- adequate collaboration between participants, and
- software support.

The analysis of project artifacts showed that the number of revisions from the draft version to the final version, adherence to delivery deadlines, and completeness are satisfactory. It was noticeable though that such results are mostly influenced by the complexity of the project, as well as the experience of the business analyst in applying the approach. The project complexity impacts the model quality and requires stronger collaboration between participants.

The effective use of the approach is closely related to the model quality, primarily to the simplicity and comprehensiveness of the BPMN models. To this end, it should be ensured that the semantics of the BPMN elements are clear to project participants, especially if the semantics are complex as in the case of events. The types of events were found confusing for some participants regardless of their professional background. Whereas technically oriented participants better differentiated the types of events, they were more likely to lose sight of the broader business meaning of an event. For instance, a message event can have a broader meaning than simply calling a Web service, which was a misinterpretation by some project participants.

An important question that arose when applying the approach was whether to limit the set of BPMN elements. In search of an answer, modeling principles were consulted. The general principles in achieving good-quality models are correctness, relevance, economical efficiency, clarity, comparability, and systematic design [23]. The correctness principle refers to both syntactic and semantic correctness. Syntactic correctness is not difficult to achieve, and modern modeling tools can help to that end. On the other hand, satisfactory cooperation with both business domain experts and software engineers is crucial for semantic correctness, which depends on the context.

To be relevant, models need to capture an appropriate amount of information considering their purpose [19]. Model relevance also depends on the context and has to be validated by project participants.

Predefined model building blocks were found useful for achieving economical design when applying the proposed approach. Moreover, the application of modeling patterns in creating building blocks was encouraged. Typically, modeling environments/tools provide their own ways to extract complex activities or subprocesses and treat them as building blocks. Still modeling collaborative business processes from different abstract levels is usually perceived as complicated. It largely depends on the knowledge and skills of the analyst. Based on the experience from the projects addressed in this paper, the provision of guidance to the less experienced analysts in this regard is highly recommended.

It is of utmost importance to find the right balance between clarity and comprehensiveness. Too much information and the models become cluttered with elements and hard to read. On top of that, the clarity criterion is subjective. To this end, the lack of ontological completeness in BPMN [24] was also found disadvantageous when applying the proposed approach. Construct deficit and construct overload can especially impair clarity. For instance, BPMN does not provide separate elements for systems and subsystems. On the other hand, the BPMN constructs Pool and Lane are used to depict a whole range of concepts such as a process, a specific organizational entity, an application system, a set of entities, and so on. There were several occasions when extra effort was required in the projects to make project participants familiar with the meaning these constructs have in each model.

The recommended way to deal with construct redundancy is to select one of the redundant constructs and use it consistently in models. All analysts should agree on a selected subset of constructs to achieve model comparability.

As for systematic design, there are modeling guidelines that can be helpful. Due to criticism that modeling principles are too theoretical and impractical, some guidelines were also proposed such as [25]:

- Use as few elements in the model as possible;
- Minimize the routing paths per element;

- Use one start and one end event;
- Avoid OR routing elements;
- Use verb-object activity labels;
- Decompose the model if it has more than 50 elements.

Although the given guidelines were proposed with a focus on the Event-driven Process Chain (EPC) notation, most of them are applicable to BPMN models.

More recently, Corradini et al. [26] provided a set of fifty modeling guidelines that can help business analysts improve the understandability of their models. The given guidelines were found quite practical in developing good-quality models when applying the proposed approach.

Adequate familiarization of project participants with BPMN, modeling principles, and guidelines proved to be very useful in ensuring both acceptability and effective use of the e-government requirements specification approach based on BPMN. Therefore, the approach comprised recommendations for providing a BPMN user guide and, prospectively, BPMN training sessions that are tailored to the needs of a specific e-government project.

Moreover, the approach promotes, improves, and benefits from collaboration between project participants. Collaborative working sessions contributed to the quality of the models and at the same time to their acceptability and effective use.

If feasible, the use of collaborative modeling environments is very beneficial. It supports remote collaborative working sessions with the purpose of the development, review, and improvement of the models.

5 Conclusion

The paper presents an approach for e-government requirements specification based on BPMN. The approach was applied in several e-government projects. In the main, software requirements specification is conducted as the extension or elaboration of BPMN models. The approach is iterative, incremental, and highly flexible, so it can be adjusted to a required or chosen project management and/or development methodology.

The paper also analyzes acceptability and factors of effective use of the approach. Based on the ratio of positive and negative comments towards the approach per project participant, the approach was found acceptable by participants in different roles.

In addition, the paper identifies the main factors influencing the effective use of the approach: model quality, familiarity of participants with BPMN, adequate collaboration between participants, and software support.

The model quality is ensured by the experience of business analysts. The provision of high-quality training and guidance to less experienced business analysts contribute to this end. Furthermore, the application of agreed modeling principles and guidelines is of utmost importance for ensuring the model quality. In addition. The model quality is also affected by the other three factors influencing the effective use of the approach.

The familiarity of project participants with BPMN, in general, plays an important role in both the acceptability and effectiveness of the approach. Therefore, a proper introduction of project participants such as domain experts to BPMN and modeling principles, as well as mutual agreement on guidelines to be followed, have a positive impact on the application of the approach. Even more so, the collaborative development, review, and improvement of the models contribute to the effective use of the approach. If feasible, the use of collaborative modeling environments is very beneficial.

Acknowledgment. The research described in this paper was partially funded by the Ministry of Education, Science and Technological Development of the Republic of Serbia.

References

1. Di Martino, B., Marino, A., Rak, M., Pariso, P.: Optimization and validation of eGovernment business processes with support of semantic techniques. In: Barolli, L., Hussain, F.K., Ikeda, M. (eds.) CISIS 2019. AISC, vol. 993, pp. 827–836. Springer, Cham (2020). https://doi.org/10.1007/978-3-030-22354-0_76
2. Al-Mushayt, O.S.: Automating E-government services with artificial intelligence. IEEE Access **7**, 146821–146829 (2019)
3. González, L., Delgado, A.: Towards compliance requirements modeling and evaluation of e-government inter-organizational collaborative business processes. Bus. Process Manag. (BPM) **1**(2), 2079–2088 (2021)
4. Delgado, A., Montarnal, A., Astudillo, H.: Introduction to the digital government and business process management (BPM). In: Proceedings of the Hawaii International Conference on System Sciences (HICSS 2021), pp. 2067–2068, Hal Open Science, France (2021)
5. OMG: Graphical notations for business processes (2022).https://www.omg.org/bpmn/. Accessed 30 May 2023
6. Ko, R.K., Lee, S.S., Wah Lee, E.: Business process management (BPM) standards: a survey. Bus. Process. Manag. J. **15**(5), 744–791 (2009)
7. Santana, F., Nagata, D., Cursino, M., Barberato, C., Leal, S.: Using BPMN-based business processes in requirements engineering: the case study of sustainable design. In: Proceedings of the International Conference on e-Learning, e-Business, Enterprise Information Systems, and e-Government (EEE), pp. 79–85. CSREA Press, Providence, RI (2016)
8. Monsalve, C., Abran, A., April, A.: Measuring software functional size from business process models. Int. J. Softw. Eng. Knowl. Eng. **21**(03), 311–338 (2011)
9. Przybylek, A.: A business-oriented approach to requirements elicitation. In: Proceedings of 9th International Conference on Evaluation of Novel Approaches to Software Engineering (ENASE), pp. 1–12. IEEE, New York City (2014)
10. Sholiq, S., Sarno, R., Astuti, E.S.: Generating BPMN diagram from textual requirements. J. King Saud Univ.-Comput. Inf. Sci. **34**(10), 10079–10093 (2022)
11. Marín, B., Quinteros, J.: A COSMIC measurement procedure for BPMN diagrams. In: Proceedings of 26th International Conference on Software Engineering and Knowledge Engineering (SEKE 2014), pp. 408–411. KSI Research Inc., Pittsburgh, PA (2014)
12. Khlif, W., Haoues, M., Sellami, A., Ben-Abdallah, H.: Analyzing functional changes in BPMN models using COSMIC. In: Proceedings of the 12th International Conference on Software Technologies (ICSOFT), pp. 265–274. SCITEPRESS, Setúbal, Portugal (2017)
13. Yaqin, M.A., Sarno, R., Rochimah, S.: Measuring scalable business process model complexity based on basic control structure. Int. J. Intell. Eng. Syst. **13**(6), 52–65 (2020)
14. Vega-Márquez, O.L., Chavarriaga, J., Linares-Vásquez, M., Sánchez, M.: Requirements comprehension using BPMN: an empirical study. In: Lübke, D., Pautasso, C. (eds.) Empirical Studies on the Development of Executable Business Processes, pp. 85–111. Springer, Heidelberg (2019). https://doi.org/10.1007/978-3-030-17666-2_5

15. Meridji, K., Al-Sarayreh, K.T., Abran, A., Trudel, S.: System security requirements: a framework for early identification, specification and measurement of related software requirements. Comput. Stand. Interfaces **66**, 103346 (2019)
16. Ahmed, A., Prasad, B.: Foundations of Software Engineering. Auerbach Publications, Boca Raton, Fla. (2016)
17. Dalpiaz, F., Ferrari, A., Franch, X., Palomares, C.: Natural language processing for requirements engineering: the best is yet to come. IEEE Softw. **35**(5), 115–119 (2018)
18. Kasemsap, K.: The roles of business process modeling and business process reengineering in e-government. In: Open Government: Concepts, Methodologies, Tools, and Applications, pp. 2236–2267). IGI Global, Hershey, PA (2020)
19. Pantelić, S.D., Dimitrijević, S., Kostić, P., Radović, S., Babović, M.: Using BPMN for modeling business processes in e-government–case study. In: Proceedings of the 1st International Conference on Information Society, Technology and Management (ICIST). The Association for Information systems and Computer networks, Belgrade, Serbia (2011)
20. Zhao, L., et al.: Natural language processing for requirements engineering: a systematic mapping study. ACM Comput. Surv. (CSUR) **54**(3), 1–41 (2021)
21. Zareen, S., Akram, A., Ahmad Khan, S.: Security requirements engineering framework with BPMN 2.0. 2 extension model for development of information systems. Appl. Sci. **10**(14), 4981 (2020)
22. Gómez, C., Pérez Blanco, F.J., Vara, J.M., De Castro, V., Marcos, E.: Design and development of smart contracts for e-government through value and business process modeling. In: Proceedings of the 54th Hawaii International Conference on System Sciences (HICSS-54), pp. 2069–2078. The Association for Information Systems (AIS), Atlanta, GE (2021)
23. Becker, J., Rosemann, M., von Uthmann, C.: Guidelines of business process modeling. In: van der Aalst, W., Desel, J., Oberweis, A. (eds.) Business Process Management. LNCS, vol. 1806, pp. 30–49. Springer, Heidelberg (2000). https://doi.org/10.1007/3-540-45594-9_3
24. Recker, J., Indulska, M., Rosemann, M., Green, P.: How good is BPMN really? Insights from theory and practice. In: Proceedings of the 14th European Conference on Information Systems, pp. 1–12. IT University of Goteborg (2006)
25. Mendling, J., Reijers, H.A., van der Aalst, W.M.: Seven process modeling guidelines (7PMG). Inf. Softw. Technol. **52**(2), 127–136 (2010)
26. Corradini, F., et al.: A guidelines framework for understandable BPMN models. Data Knowl. Eng. **113**, 129–154 (2018)

Towards a Formal Specification and Automatic Execution of ETLs in Cross-organizational Business Processes

Miroslav Tomić[(✉)] [ID], Nikola Todorović[ID], Marko Vještica[ID], Slavica Kordić[ID], and Vladimir Dimitrieski[ID]

Faculty of Technical Sciences, University of Novi Sad, Trg Dositeja Obradovića 6, 21000 Novi Sad, Serbia
{tmiroslav,nikola.todorovic,marko.vjestica,slavica, dimitrieski}@uns.ac.rs

Abstract. Small and medium-sized enterprises seize business opportunities jointly to be more concurrent on the market. They form virtual organizations with the goal of competing with larger companies for new business opportunities. Their collaboration in the domain of Industry 4.0 usually requires production process contracts to be created. To easily handle bureaucratic tasks of creating such contracts and ensure transparent and automatic activation of contract clauses, companies can utilize the CE-MultiProLan solution. The solution contains concepts required for the specification of cross-organizational business processes and utilizes the distributed ledger technology and smart contracts to ensure trust between collaborating parties through transparency and automation of contractual clause fulfilment. Each collaborating party can define which private data from their internal systems should be shared with other collaborating parties for contract conformance-checking and monitoring of production processes. To avoid potential fraudulent behavior, in which a company manually falsifies the shared data, CE-MultiProLan solution needs to be extended with new concepts and components to support automated data sharing. For such purposes, Extract, Transform and Load (ETL) processes could be used to automatically transform data from the shopfloor into the shared data. However, manual development of ETLs is often time-consuming and error prone. In this paper, we examine common ways to formally define the ETL processes, especially in Industry 4.0, and lay ground for future extension of CE-MultiProLan that would allow for faster and easier ETL specification.

Keywords: Extract Transform and Load · Cross-organizational Business Process · Smart Contract · Domain-Specific Language · Industry 4.0

1 Introduction

Nowadays, Small and Medium-sized Enterprises (SMEs) often find themselves in a subordinate position to large companies when trying to acquire new business opportunities. However, technological foundations for more accessible and better connectivity,

© The Author(s), under exclusive license to Springer Nature Switzerland AG 2024
M. Trajanovic et al. (Eds.): ICIST 2023, LNNS 872, pp. 459–470, 2024.
https://doi.org/10.1007/978-3-031-50755-7_43

together with an increasing readiness of other companies to collaborate, multiple SMEs can form a Virtual Organization (VO), which allows them to compete on the market jointly. VO represents a temporary alliance for integrating competencies and resources from several independent collaborative companies to satisfy customers' requirements or seize business opportunities by jointly developing complex products [1].

Although beneficial from the business perspective, such collaboration exposes new challenges. Most notably, there is an issue of trust where some VOs might be comprised of SMEs that have never collaborated before. Although they jointly undertake a business venture, each company wants to protect its know-how and intellectual property. This also might have a negative impact on trust as it could prevent all other companies from monitoring every step of the production process in detail, which they are used to when running the production by themselves. Another issue stems from increasing the number of participants in the value chain as it increases the bureaucratic overhead at all levels: from bureaucracy required to exchange materials and products to creating and dissolving a VO. One of the main bureaucratic tasks is the creation of Cross-organizational Business Processes (CBPs), which represent business processes taking place between multiple independent parties in a VO. The CBP specification needs to hold all the relevant contractual data, processes to be executed, and responsibilities to be attributed. Therefore, it is a critical task to be done right, but often it is done in an informal and non-automated way, causing much delay in VO formation.

In our previous work [2], we introduced a novel methodological approach that could help collaborating parties facilitate the easier creation of CBPs and automate the setup of all the required infrastructure to get a VO running. The approach allows a high-level specification of the private production processes of each company while also enabling parties to define the production process data to be shared between collaborating parties, allowing for global process monitoring without compromising privacy. This approach utilizes Distributed Ledger Technology (DLT) with smart contracts. Each company writes its production data to an immutable and transparent log, hosted in a DLT. This way, transparency is achieved as all the participants have an insight into the state of the global production process execution while also having the means to precisely attribute liability in case something goes wrong. Furthermore, smart contracts are automatically created from the parts of the more detailed CBP specification. These smart contracts are used for monitoring the CBP execution and tracking contractual fulfillment. Smart contracts are used for performing conformance checks on the shared data, i.e., analyzing the discrepancies between the contracted and the executed production processes. This increases the automation degree of the entire collaboration, which in addition to transparency, increases the overall trust between the participants.

Data required for conformance-checking and contractual fulfillment checks mainly refers to data such as status changes of production process steps, required performance indicator values, or measurement values taken as a part of the quality control checks. Although the contractual obligation monitoring is automated through smart contracts, it is still based on the production data shared by each company. Much fraudulent behavior can still occur inside a company's private space, resulting in misleading production data being shared. To prevent such behavior, the process of making private production data available to collaborating parties should be automated as well.

In addition to purely automating the publishing of some private data, an issue arises that smart factories do not necessarily have data for conformance-checking readily available as they mostly gather data generated by the shop-floor resources. Data from shop-floor resources represent real-time data about the resource state and operations, such as robot movement and gripper activities. Such data is too granular for conformance-checking, as smart contracts primarily require values aggregated over multiple resources and time windows. Data transformations should be applied to bridge the discrepancy between available and required data. Bridging the discrepancy could be accomplished with Extract, Transform, and Load (ETL) processes. Here, ETL processes could be used to extract data from different sources, perform data transformations, i.e., filtering and aggregation of data, and load the output data into the target DLT network, thus allowing for the conformance-checking process to take place. Automating the distribution of private production data between collaborating parties would require expanding the approach with the automatic creation and execution of ETLs based on CBP specifications.

The goal of this paper is to explore the possibility of fully automating data sharing between collaborating parties during CBP execution. Exploration is split into two phases. First, we investigate existing approaches and solutions that enable automatic creation and execution of ETLs by defining them at a higher level of abstraction using different modeling techniques. Then, we analyze the prospect of extending the methodological approach introduced in [2] with additional steps to enable the full automation of data sharing within a VO.

This paper is organized as follows. Apart from Introduction, in Sect. 2, an overview of theoretical background on ETLs and related work are given. An outline of the approach to ETL process modeling in the domain of CBPs is described in Sect. 3. The conclusion and future research directions are presented in Sect. 4.

2 Related Work

As the main theme of this paper is to discuss the modeling and usage of ETLs in automating the process of sharing private data in Industry 4.0 and CBP domains, thus mitigating the possibility of fraudulent behavior, we have surveyed research work that roughly falls into two categories:

- **ETL process modeling** – research papers describing the process of modeling ETLs at a higher abstraction level than the level of program code, mostly covering the domain of Data Warehouses (DWs); and
- **ETL process usage in Industry 4.0** – applications of ETLs in the domain of Industry 4.0, which is also the domain in which we aim to apply our approach. However, as far as we are aware, there are no papers covering the application of ETLs in creating and monitoring CBPs.

Next, we discuss the surveyed papers in two subsections, one for each category. At the end of this section, we give a short summary of all the findings.

2.1   ETL Process Modeling

The idea of modeling an ETL process and then generating ETL code, instead of directly writing it, is not new. Many approaches to modeling the ETL processes emerged since the early 2000s [3–11]. Most of them came from the DW domain as during the DW development, one of the most time-consuming and complex tasks is the development of its ETL processes [3]. There are several Domain-Specific Languages (DSLs) that are tailored for the DW domain [3–5]. Also, there are many modeling approaches that use General-Purpose Modeling Languages (GPMLs), such are Unified Modeling Language (UML) [6–9] and Business Process Model and Notation (BPMN) [10, 11], to make conceptual models of ETL processes.

One of the first approaches aimed to formally define an ETL was introduced in 2002 by Vassiliadis et al. [3], thus a formal foundation for ETL conceptual modeling was created with the creation of DSL. The authors paid special attention to the modeling of transformation operators in ETLs and *divided all transformation operators into six groups*: filters, unary transformations, binary transformations, transfer operations, file operations, and composite transformations. Afterwards, the authors improved their app-roach with the introduction of a formal description of attribute mapping [4]. Although the described research work was intended solely for the DW domain, it represents an abstraction that can be used in other domains as well.

An additional family of DSLs for the specification and automation of ETL processes in the DW domain was introduced by Petrović et al. [5]. The authors created several DSLs, each covering concepts relevant for a specific aspect of an ETL process. They created *a language for the specification of data operations* i.e., data flow (ETL-O), and a *language for the specification of data operation execution order* i.e., control flow (ETL-P). Also, three supplementary languages are provided: a *language for the specification of various logical and arithmetic expressions* (ETL-E), a *language for the specification of transformation operation templates* (ETL-T), and a *language for the specification of source and target data models* (ETL-D). Even if these languages are meant to be used in the DW domain, they can be used in other domains as well.

Trujillo and Luján-Mora presented an approach that is based on UML, allowing conceptual modeling of ETL processes [6]. Additionally, they provided mechanisms for the specification of common operations in a way that conforms to the presented approach. The approach they presented comprises six main tasks:

- *Selection of sources for extraction* – often different heterogeneous data sources;
- *Source transformations* – data extracted from sources can be transformed or new data can be derived. Some common transformation operations that the authors iden-tified are the following: filtering data, converting codes, calculating derived values, transforming different data formats, and automatically generating sequence numbers;
- *Joining sources* – in addition to pure one-to-one transformations that produce the value from a single source variable, various sources can be joined to produce new, often enriched data;
- *Selection of targets load* – serve as the destination storage of the transformed data; and
- *Load data* – the selected targets are populated with the transformed data.

Based on the activities outlined by Trujillo and Luján-Mora [6], Muñoz et al. [7] created an approach that proposes the usage of UML activity diagrams. Similar to approach presented by Vassiliadis et al. [3, 4], Muñoz et al. *divided transformation operators into ten groups*: (i) aggregation – aggregates data based on specified criteria; (ii) conversion – changes data type and format or derives new data; (iii) filter – selects and verifies data; (iv) incorrection – redirects incorrect data; (v) join – joins two data sources related to each other with some attributes; (vi) load – loads data into the target of an ETL process; (vii) log – logs activity of an ETL mechanism; (viii) merge – integrates two or more data sources with compatible attributes; (ix) surrogate – generates unique surrogate keys; and (x) wrapper – transforms a native data source into a record based data source. Muñoz et al. then extended their approach to satisfy the main principles of Model-Driven Architecture (MDA) with the goal to allow the automatic code generation of ETL processes [8]. This approach *indicates that it is possible to describe ETL processes as Platform-Independent Models* (*PIMs*) and to create foundations for generating executable ETL code.

Luján-Mora et al. [9] also presented a framework for the design of ETL processes. The key observation of their work is that the task of developing ETLs, either manually or by modeling it at a lower level of abstraction, involves dealing with the specificities of information at low levels of granularity, including transformation rules at the attribute level. The presented approach dives into details about concepts for describing attributes and attribute transformations.

In addition to the UML-based modeling languages, the BPMN modeling language was also used. El Akkaoui et al. [10, 11] proposed a framework for Model-Driven Development (MDD) of ETL processes that is based on BPMN. The authors argue [10] that there are two main benefits of their framework: (i) the usage of vendor-independent models for unified design of ETL processes that is based on BPMN allows the model reuse and portability; and (ii) the automatic transformation of BPMN models into the required vendor-specific code to execute ETL processes reduces development time and mitigates common development-related errors. In the following research paper [11], the authors presented a BPMN-based meta-model for conceptual modeling of ETL processes. The presented meta-model is based on a classification of ETL objects resulting from a study of the most used commercial and open source ETL tools in that moment.

2.2 ETL Process Usage in Industry 4.0

Of all possible applications of ETLs in the domain of Industry 4.0, the most interesting to our cause are the ones applied in conjunction with smart contracts. Data needed for the conformance-checking of smart contracts is stored in a DLT which, in turn, is inserted there from some shop-floor databases often in near real-time. To gather the required data from sources, transform it in an appropriate way, and store it in the DLT, ETL processes are needed.

Lu et al. [12, 13] proposed using smart contracts for grid enterprises doing electricity transactions and charge settlements based on the blockchain technology. Their design of smart contracts enables the usage of blockchain for conformance-checking. The authors used ETL processes to bridge discrepancy between source data and the blockchain. They used general-purpose ETL tools to implement the ETL processes and found out that additional efforts are needed to implement such processes. These additional efforts stem from the fact that a developer must think in terms of two separate domains: ETL development framework and problem domain.

In the case of smart factories, which automate the entire production mainly based on the gathered process data, ETL processes used for data movement and integration have the big impact on the resulting codebase. Santos et al. [14] proposed the Big Data Analytics architecture and stated that ETL processes are required to enable data extraction from various sources, their transformation in different formats, and loading it into the desired targets. An example of the proposed Big Data Analytics architecture is presented in their following work [15] where the proposed platform is implemented in Bosch Car Multimedia – Braga. In their architecture, ETL processes are a part of the data preparation layer and are a foundational part of the architecture without which all other parts would not function.

2.3 Summary

During the literature review, we did not find any approach for modeling and utilizing ETL processes in the domain of CBPs. Most dominantly, ETL papers we found are from the domain of DW development. We found some examples of ETL process applications in Industry 4.0 too, out of which the most relevant ones are presented in this Section.

Although not from the needed domain, most of the modeling concepts described in this paper can be used in our approach as well. These concepts are: the data extraction from source, data filtering, and transformations, such as aggregation and data loading to desired targets. Furthermore, the main sentiment in these papers seems to be that the general-purpose ETLs, although flexible enough to fit any problem domain, introduce an overhead when it comes to the implementation. They require both the experts in a particular ETL framework, who are knowledgeable in the language concepts, and domain experts who know the semantics of data inside out. Additionally, usage of general-purpose tools requires additional effort and time to integrate them into existing tools and technology landscape.

3   ETL Component to Support Automated Data Sharing

In this section, we examine the prospect of extending the approach introduced in [2], with additional steps to enable the full automation of data sharing within a VO. We first give a short overview of all necessary concepts from the existing approach, after which we discuss details of the proposed extensions related to the ETL process modeling. Components of the extended architecture are depicted in Fig. 1, with newly introduced components enclosed by a dashed box with a gray background.

In our previous research, we first introduced a novel methodological approach and a software solution in which the Model-Driven (MD) principles and Domain-Specific Modeling Languages (DSMLs) facilitate a formal specification and automatic execution of production processes within a single collaborating party [16]. The DSML we introduced is called Multi-Level Production Process Modeling Language (MultiProLan) [17] and can be used by process designers to model production processes at two levels of abstraction. At a higher level, there is a Master-Level Production Process Model (MasL-PPM), representing a production process model independent of any concrete production system or resources which are to execute the production process. Such models allow process designers to focus only on value-adding process steps and not to worry about where their process is being executed. On the other hand, if a production process is to be executed, various production system details are required, such as resources, storage, configuration steps and logistics steps. By enriching a MasL-PPM with such production details, a lower-abstraction level model named Detail-Level Production Process Model (DetL-PPM) is created. A DetL-PPM is dependent on a specific production system in which the process is to be executed. Such a process model can be created manually or using an orchestrator software system, a system that automatically performs resource matching and production scheduling.

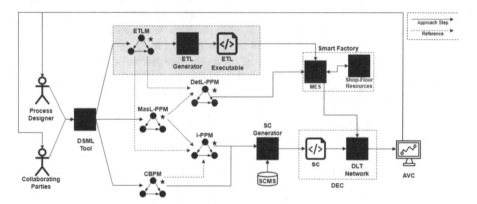

Fig. 1. Components of the extended CE-MultiProLan architecture

The resources that are to be used to execute production processes are described with a specific language for production system modeling, as presented in [18]. All the necessary resource details, such as their interfaces, semantics, constraints, and skills, are described by using such a language. Based on the DetL-PMM process specification and the detailed specification of shop-floor resources, executable production process commands can be generated and forwarded to a Manufacturing Execution System (MES), which handles the production on the shop-floor.

To support collaboration between SMEs, and not only internal production process specifications, we extended MultiProLan with CBP modeling concepts. Such a language extension is a DSML as well, and we named it Collaborative Extension of MultiProLan (CE-MultiProLan) [2]. This language contains modeling concepts utilized by collaborating parties to create a CBP Model (CBPM) to coordinate actions between different participants in a value chain. Such actions require coordinating private production processes, specified in MultiProLan, based only on shared production data. This shared data is defined using an Interface Production Process Model (I-PPM), which references relevant parts of private process specifications in the form of MasL-PPMs. An I-PPM is used by each collaborating party, configuring what data can be shared between the parties and monitored. Collaborating parties need to specify: (i) which production steps from MasL-PPM should be traced during production; and (ii) what data should be persisted alongside the step traces. This data is often denoted as a monitoring metric and is available in a shared repository where a smart contract can access it. Both CBPM and I-PPM can be used by Smart Contract (SC) Generator to generate smart contracts. More precisely, an I-PPM is used to generate smart contracts for monitoring the execution of a process performed by a single organization. In contrast, a CBPM is used to generate smart contracts for monitoring the enactment of CBPs. After smart contracts are generated, they are stored on configured DLT Network together with the monitoring metrics on top of which they choose action to execute. DLT Network and smart contracts represent a basis for implementing Data Exchange Component (DEC). DEC is the component that facilitates the exchange of production data between the collaborating parties and enables process monitoring and conformance-checking on the exchanged data.

Data collected from the shop-floor resources and emitted by the factory information systems may serve multiple purposes. This off-chain data can be used to monitor the performance and health of modeled resources, communicate resource state and availability to collaborating parties, and help monitor the execution of implemented processes. The two latter purposes require a careful specification of a transformation processes, which transforms the private shop-floor resource data of a single collaborating party into publicly available monitoring metric read by other collaborating parties. Such a specification can be seen as an ETL process.

Currently, implementing such ETL processes often requires time-consuming and error-prone manual development. Frequently used general-purpose ETL frameworks and languages are not well suited for the CBP domain, as discussed in Sect. 2. They often require deep knowledge of both the programming framework and the context in which the collaborating parties produce and exchange data. As such, it is hard to assemble teams with an appropriate skillset to implement the ETL process itself.

Our goal is to ease the creation of ETL processes in the CBP domain and allow for it to be specified at the initial, negotiating phase of the collaboration enactment by the production domain experts rather than software developers or system integrators throughout the later stages of the collaboration lifecycle. Therefore, the description of such ETL processes should be made by CBPM modelers or process designers based on: (i) a known DetL-PPM with all the details about concrete shop-floor resources, serving as sources; and (ii) a defined I-PPM which provides a description of data required for the conformance checking, serving as the target.

To specify these ETL processes, we propose an extension of our existing MD approach addressing the need for faster and less error-prone ETL process development. Thus, we propose an extension of DSML Tool, to allow the specification of ETL Models (ETLMs). These ETLMs are to be specified by each CBPM modeler or process designer involved in the collaboration enactment in an easily understandable and easy-to-use way. ETLMs represent models at a higher level of abstraction, without implementational and execution-platform-dependent details. They represent ETL models of transforming private data into publicly monitoring metrics that will be used for conformance-checking. Inspired by the literature survey and commonly encountered steps in the ETL process development, our approach to the ETL model specification comprises at least the following steps:

- **Specification of data sources** – choosing a concrete set of resources from a DetL-PPM enriched process specification such are a robot, machine or a human worker. As data sources are modeled with their interfaces, semantics, and constraints the ETL process designer should choose relevant properties for each resource.
- **Specification of data targets** – choosing a concrete set of relevant monitoring metrics from a DLT as specified within an I-PPM. As data targets are modeled with their interfaces, semantics, and constraints, for each monitoring metric, relevant properties should be chosen by the ETL process designer.
- **Specification of data transformations** – using a predefined set of transformation operations, such are filtering, aggregation, splitting, and conversion between different data formats. Although these transformation types are based on the ones discussed in Sect. 2, they must support any high-level production-process-related variables to be used as operands. For example, there is often a need to combine inputs from: a machine creating a product, an external sensor checking the machine, and a human inspection worker checking the product quality characteristics. These inputs should be combined based on the order identification number and a predefined time frame to output a single public monitoring metric on the production process step.

Once an ETLM is specified, ETL Generator should use it as a blueprint to generate executable ETL process code. ETL Generator comprises model-to-text transformation rules and ETL templates. Based on various use cases, ETL executable code should be able to process incoming data in batches or in a near-real-time fashion from data streams. As a plethora of technologies for such an ETL process execution exist, we will choose the most robust and resilient ones when used in an industrial use case. We will leave possibilities for manual interventions on the generated ETL code. Special care should be taken to enable future regenerations of code and merging with the manually written code fragments.

After the ETL code is generated, an execution platform, like an MES, should gather production process execution data sent by the shop-floor resources, execute the defined ETL processes to transform such data in the format required by smart contracts, and load the output data to DEC. Finally, customers and other value chain participants should be able to oversee the state of production and contract fulfillment by overviewing the monitoring metrics in the DLT through Aggregation and Visualization Component (AVC).

With such an extension, CE-MultiProLan would support the complete automation of creating, executing, and monitoring of CBPs. The main characteristics of such an approach are simplicity and the fast time to execution-ready collaborative solution.

4   Conclusion

In this paper, we examined the prospect of extending the MD approach based on the CE-MultiProLan DSML [2] with an ETL component to support automated data sharing between collaboration parties. Since CE-MultiProLan cannot be used to model data sharing, the complete automation of creating, executing, and monitoring CBPs cannot be achieved.

During the literature survey, we identified that existing ETL process modeling solutions are not directly applicable outside of the domain they are intended for, but that the main concepts and steps to create them are indeed reusable across the domains. Therefore, although applicable, general-purpose ETL tools cannot be used efficiently and at the appropriate level of abstraction in the CBP domain. Even if they are used to model CBPs and data sharing, domain experts would struggle utilizing such tools efficiently and integrating them into their own tool and technology landscape.

We concluded that a novel modeling language needs to be created on top of the identified, general ETL concepts, and by increasing the level of abstraction when applying these concepts. In this paper, we provided foundations for a language used to model ETL processes in the CBP domain. Such a language should be used by domain experts, i.e., collaboration parties or process designers, allowing them to specify: (i) data sources; (ii) data targets; and (iii) data transformations by using high-level, domain concepts available in other specifications created using CE-MultiProLan. This novel language should comprise modeling concepts that are close to the domain experts, enabling them to specify how private corporate data should be transformed into data that could be shared between them. Creating the language in such a way would decrease the time needed for the execution-ready collaborative solution.

Based on the foundation for the formal description of ETL processes presented in this paper, we plan to create a novel methodological approach. Such an approach and the novel ETLM language, as the main part of it, will be specified in detail and created to enable collaborating parties to model ETL processes.

In addition, we plan to design a case study to demonstrate the usage of both the approach and the ETLM language. Our case study would be an extension of the one presented by Todorović et al. [2] where VO participants collaborate to produce a decorative wooden wine box. Resources that execute the process modeled in DetL-PPM of the wooden wine box production generate data that represents the source data for ETLM. The specified I-PPM of the wooden wine box defines the data required for conformance checking, i.e., target data. The data can be used by collaborating parties to track Key Performance Indicators (KPIs) and monitor the production process of a wooden wine box. Since source data differs from target data, a set of transformations should be applied. We plan to utilize the ETLM proposed in this paper to bridge such a discrepancy and to demonstrate its usage. To evaluate the performance of ETLM we will measure the time needed to specify the required set of transformations and compare it with the time needed to implement those transformations with existing general-purpose tools and programming languages. Also, we plan for collaborative parties to evaluate ETLM's quality characteristics, such as usability and productivity, by applying the Framework for Qualitative Assessment of Domain-specific languages (FQAD) [19].

Acknowledgements. This research (paper) has been supported by the Ministry of Science, Technological Development and Innovation through project no. 451-03-47/2023-01/200156 "Innovative scientific and artistic research from the FTS (activity) domain".

References

1. Priego-Roche, L.-M., Rieu, D., Front, A.: A 360° vision for virtual organizations characterization and modelling: two intentional level aspects. In: Godart, C., Gronau, N., Sharma, S., Canals, G. (eds.) I3E 2009. IAICT, vol. 305, pp. 427–442. Springer, Heidelberg (2009). https://doi.org/10.1007/978-3-642-04280-5_33
2. Todorović, N., Vještica, M., Todorović, N., Dimitrieski, V., Luković, I.: A novel approach and a language for facilitating collaborative production processes in virtual organizations based on DLT networks: In: Proceedings of the 2nd International Conference on Innovative Intelligent Industrial Production and Logistics, pp. 197–208. SciTePress, Valletta, Malta (2021). https://doi.org/10.5220/0010720900003062
3. Vassiliadis, P., Simitsis, A., Skiadopoulos, S.: Conceptual modeling for ETL processes. In: Proceedings of the 5th ACM International Workshop on Data Warehousing and OLAP, pp. 14–21. ACM, NY, USA (2002). https://doi.org/10.1145/583890.583893
4. Simitsis, A., Vassiliadis, P.: A methodology for the conceptual modeling of ETL processes. In: Proceedings of the 15th Conference on Advanced Information Systems Engineering (CAiSE '03), pp. 305–316, Klagenfurt/Velden, Austria (2003)
5. Petrović, M., Vučković, M., Turajlić, N., Babarogić, S., Aničić, N., Marjanović, Z.: Automating ETL processes using the domain-specific modeling approach. Inf. Syst. E-Bus. Manag. **15**, 425–460 (2017). https://doi.org/10.1007/s10257-016-0325-8
6. Trujillo, J., Luján-Mora, S.: A UML based approach for modeling ETL processes in data warehouses. In: Song, I.-Y., Liddle, S.W., Ling, T.-W., Scheuermann, P. (eds.) ER 2003, LNCS, vol. 2813, pp. 307–320. Springer, Heidelberg (2003). https://doi.org/10.1007/978-3-540-39648-2_25
7. Muñoz, L., Mazón, J.-N., Pardillo, J., Trujillo, J.: Modelling ETL processes of data warehouses with UML activity diagrams. In: Meersman, R., Tari, Z., Herrero, P. (eds.) OTM 2008, vol. 5333, pp. 44–53. Springer, Heidelberg (2008). https://doi.org/10.1007/978-3-540-88875-8_21
8. Muñoz, L., Mazón, J.-N., Trujillo, J.: Automatic generation of ETL processes from conceptual models. In: Proceedings of the ACM Twelfth International Workshop on Data Warehousing and OLAP, pp. 33–40. ACM, NY, USA (2009). https://doi.org/10.1145/1651291.1651298
9. Luján-Mora, S., Vassiliadis, P., Trujillo, J.: Data mapping diagrams for data warehouse design with UML. In: Atzeni, P., Chu, W., Lu, H., Zhou, S., Ling, T.-W. (eds.) ER 2004, LNCS, vol. 3288, pp. 191–204. Springer, Heidelberg (2004). https://doi.org/10.1007/978-3-540-30464-7_16
10. El Akkaoui, Z., Zimànyi, E., Mazón, J.-N., Trujillo, J.: A model-driven framework for ETL process development. In: Proceedings of the ACM 14th International Workshop on Data Warehousing and OLAP, pp. 45–52. ACM, NY, USA (2011). https://doi.org/10.1145/2064676.2064685
11. El Akkaoui, Z., Mazón, J.-N., Vaisman, A., Zimányi, E.: BPMN-based conceptual modeling of ETL processes. In: Cuzzocrea, A., Dayal, U. (eds.) DaWaK 2012. LNCS, vol. 7448, pp. 1–14. Springer, Heidelberg (2012). https://doi.org/10.1007/978-3-642-32584-7_1
12. Lu, J., Wu, S., Cheng, H., Xiang, Z.: Smart contract for distributed energy trading in virtual power plants based on blockchain. Comput. Intell. **37**, 1445–1455 (2021). https://doi.org/10.1111/coin.12388
13. Lu, J., Wu, S., Cheng, H., Song, B., Xiang, Z.: Smart contract for electricity transactions and charge settlements using blockchain. Appl. Stoch. Models Bus. Ind. **37**, 442–453 (2021). https://doi.org/10.1002/asmb.2570
14. Santos, M.Y., et al.: A big data analytics architecture for industry 4.0. In: Rocha, Á., Correia, A.M., Adeli, H., Reis, L.P., Costanzo, S. (eds.) WorldCIST 2017. AISC, vol. 570, pp. 175–184. Springer, Cham (2017). https://doi.org/10.1007/978-3-319-56538-5_19

15. Santos, M.Y., et al.: A Big Data system supporting Bosch Braga Industry 4.0 strategy. Int. J. Inf. Manag. **37**, 750–760 (2017). https://doi.org/10.1016/j.ijinfomgt.2017.07.012

16. Vještica, M., Dimitrieski, V., Pisarić, M.M., Kordić, S., Ristić, S., Luković, I.: Production processes modelling within digital product manufacturing in the context of Industry 4.0. Int. J. Prod. Res. **61**, 6271–6290 (2023). https://doi.org/10.1080/00207543.2022.2125593

17. Vještica, M., Dimitrieski, V., Pisarić, M., Kordić, S., Ristić, S., Luković, I.: Multi-level production process modeling language. J. Comput. Lang. **66**, 101053 (2021). https://doi.org/10.1016/j.cola.2021.101053

18. Pisarić, M., Dimitrieski, V., Vještica, M., Krajoski, G., Kapetina, M.: Towards a flexible smart factory with a dynamic resource orchestration. Appl. Sci. **11**, 7956 (2021). https://doi.org/10.3390/app11177956

19. Kahraman, G., Bilgen, S.: A framework for qualitative assessment of domain-specific languages. Softw. Syst. Model. **14**, 1505–1526 (2015). https://doi.org/10.1007/s10270-013-0387-8

Task Queue Implementation for Edge Computing Platform

Veljko Maksimović$^{(\boxtimes)}$ ⓘ, Miloš Simić ⓘ, Milan Stojkov ⓘ, and Miroslav Zarić ⓘ

Faculty of Technical Sciences, University of Novi Sad, Novi Sad, Serbia
veljko.maksimovic@uns.ac.rs

Abstract. With the rising number of distributed computer systems, from microservice web applications to IoT platforms, the question of reliable communication between different parts of the aforementioned systems is becoming increasingly important. As part of this paper, a task queue, which facilitates reliable asynchronous communication between different services, will be implemented. In order to control the flow of tasks through the queue and limit the load on downstream components, we are going to explore ways of efficiently restricting throughput, and defining priority queues within the task queue service. The research will also take a look at how different aspects of the task queue, such as the underlying persistence layer, affect their performance, reliability, and resource usage. This task queue will be implemented as a component in the already existing platform for managing clusters, *constellations* (https://github.com/c12s).

Keywords: Task queues · Asynchronous communication · Distributed systems · Edge computing

1 Introduction

1.1 Cloud Computing

At the beginning of the 1960s, a new paradigm for developing and running software applications emerged - cloud computing. The idea was to share computer resources, such as RAM, CPU, and hard disk space. Developers would not have to worry about assembling, monitoring, and repairing hardware components, and could focus on software. Later on, the distribution of entire platforms such as operating systems, web servers, and databases was introduced [1].

Running applications in these cloud environments has a few drawbacks. Data centers where applications are hosted can be distant from end users, introducing delays due to network communication [2]. There are also scenarios, depending on the countries where the application is being used, and the domain, in which user data can't leave state borders [3]. In this case, applications need to be hosted in each state individually, which could eliminate some cloud providers who have their data centers in only a handful of locations.

1.2 Edge Computing

"Edge computing is a new computing model that deploys computing and storage resources (such as cloudlets, micro data centers, or fog nodes, etc.) at the edge of the network closer to mobile devices or sensors." [4] It adds another layer of infrastructure to modern information systems that is geographically closer to end users. The goal is the reduction of traffic between clients and data centers by offloading some of the tasks to the edge nodes. Internet service providers (ISPs) have been using servers on the edge of their networks for caching web pages so they wouldn't have to route all of the requests to their final destination, therefore reducing communication costs [5].

A similar approach could be applied to cloud computing: The entire infrastructure can be expanded by adding smaller clusters of computer nodes near every bigger population center. Developers would be able to run processes for aggregating or anonymizing data on those nodes before passing it on to the central cluster for further processing and storage. These small clusters positioned geographically near users are called *micro-clouds* [6].

The goal of the *constellations*[1] project, explained in detail in *Towards Edge Computing as a Service: Dynamic Formation of the Micro Data-Centers* is to create a control unit for provisioning and managing *micro-clouds*. Further implementation of the task queue, which this paper follows, will be done as a part of the above-mentioned project.

1.3 Current Research

Edge computing introduces a new way to structure applications, but it also introduces some new challenges. Since the hardware resources that are available at the edge aren't as powerful as those in the cloud, some of the modern security best practices like complex cryptography algorithms aret feasible. Also, this is a computing mode that integrates multiple trust domains with authorized entities as trust centers, traditional data encryption and sharing strategies are no longer applicable [7]. Currently, research into security and privacy protection for edge computing is still in the early stages, with limited findings available.

Another area of research is computation offloading. Given the constraints in computational power mentioned above, it's critical to use these resources efficiently. Different strategies for offloading are being researched, where decisions are made on whether to process data locally or send it to a more powerful node or the cloud. This takes into account aspects like workloads, required response times, and the state of the network [8]. Machine learning and AI are also being employed to predict and adapt to varying edge environments, ensuring optimal performance and minimal resource wastage.

2 Communication in Distributed Systems

This chapter will briefly go over different ways of implementing communication over the network between different services.

---

[1] https://github.com/c12s.

2.1 Synchronous Communication

Synchronous communication between software components implies that the component that initiated communication is waiting for the response in the idle state, before executing any other commands. This isn't a problem if both components are part of the same process, or different processes on the same operating system, as the time to pass the message from one component to the other is rarely a bottleneck.

However, if we are talking about distributed systems where communication is carried over the network, this can become a problem. The time it takes to propagate the request and later response through the network can be greater than the time needed for the rest of the processing, therefore leaving the process in the idle state for most of the time (Fig. 1), and delaying the response to the end user.

The second issue is the fact that downtime in one component can lead to downtime in other components: in case of component B not being reachable, component A will never receive the response, and therefore won't be able to finish processing any requests, making it unreachable as well.

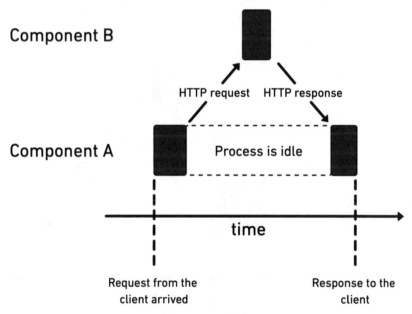

Fig. 1. A graph depicting synchronous communication between the components

2.2 Asynchronous Communication

If component A can finish its task without relying on component B to complete its functionality, we can use asynchronous communication. This consists of component A sending a message to component B and proceeding with subsequent commands without waiting for the response (Fig. 2). This approach will solve the problem of overall response time, as the client will get the response regardless of how long it takes to propagate messages between components. It will also avoid the propagation of errors from component B to component A, as processing will be complete regardless of whether component B is reachable.

If the task that component B performs isn't crucial, and system behavior isn't affected even though communication between two components might fail occasionally, this way of async communication is suitable.

What if component B has to correctly process each message it receives? How to keep track of messages that weren't processed? We would also need to develop some sort of a retry mechanism so that unprocessed messages can be handled.

These problems are not within the scope of any specific component in our system and are outside of the domain as well. Therefore, it would be useful to have a separate component whose sole task would be the propagation of messages between services in our application. This component needs to be fault tolerant, decoupled from implementation of any specific service that is using it, and configurable. Users should be able to define how many messages can pass through this component in any given time frame, how frequently it should try to resend messages that weren't processed, and how many times to retry before declaring that a message can not be processed.

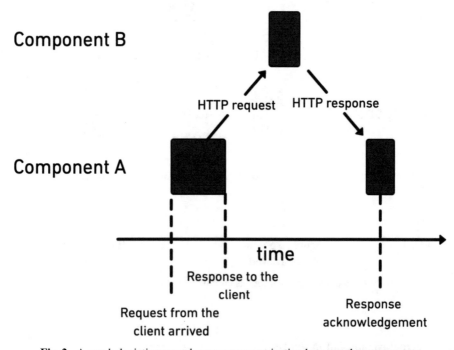

Fig. 2. A graph depicting asynchronous communication between the components

2.3 Message Queues

Message queues are components utilized in distributed systems to handle asynchronous communication. Their purpose is to make communication between services fault tolerant. If the destination service is not reachable, the queue will persist the message, and retry delivering it after the specified time. This simplifies the implementation of distributed systems, such as microservice applications, as developers don't have to focus on the implementation of reliable communication.

The example used in chapters 2.1 and 2.2 to illustrate communication between components shows *point-to-point* communication [9], and as mentioned above, queues make that type of communication easier to implement. However, message queues also open up a new way of propagating information throughout the system, where a single message is delivered to multiple services, and we can dynamically change which components need to receive those messages. This model of communication is called *publisher/subscriber,* or *pub/sub* for short, and it allows for easier scaling of distributed systems [10].

2.4 Different Characteristics of Message Queues

Depending on system requirements, we can use different types of message queues.

The first thing to take into consideration is the tradeoff between message durability and system efficiency. Is it important that each message reaches its destination, even if the message queue component itself crashes unexpectedly? If so, all incoming data should be stored in the persistence layer such as the file system. Which would slow down message processing but would give the queue ability to pick up where it left off after it is restarted. If we need to protect our message queue even from hardware failure, file replicas stored on different machines or a separate database need to be used, which would slow down message processing even further as it requires network calls after each message is received.

Will the queue deliver messages following the *push* or *pull* principle? Push queues are sending messages to recipients as soon as they arrive. This approach is good if we care about how quickly messages are delivered to consumers. On the other hand, this can overwhelm or even crash services if they start receiving messages faster than they can process them. Pull queues work in the opposite direction, where services that are processing data periodically ping the queue for new messages, ensuring they have enough capacity to process them.

There are a few other things that can define how message queue works, such as sending messages one by one, or in batches, and message encryption, but those fall out of the scope of this paper.

3   Implementation

3.1   Solution Requirements

This paper follows the implementation of the task queue service within the constellations project. The source code for this component is open source and can be found at https://github.com/c12s/blackhole/tree/vm/implementation_2.0

Task queues are a subgroup of message queues focused on starting task execution on different components within the system. Requirements for the task queue service were the ability to create multiple different queues where each can be assigned a different priority, the throughput of each queue can be controlled to not overwhelm downstream services, and the number of retries and layoff period between retries can be configured.

Implementation was done in the programming language *Go, gRPC* over HTTP/2 was used for communication between services, and *protocol buffers* were used to define the Application Programming Interface (API) of task queue. All of these technologies and standards are open source.

The queue delivered messages using the *push* principle and expects acknowledgment from the recipient service that the task was successfully executed in the form of the HTTP response with status code 2xx.

3.2   Definition of a Task

A *task* is a data structure based on which the queue knows which service it should communicate with, and how to handle if the communication or task execution is not successful. In this implementation, the task contains the following data: Unique identification, name, identification of the queue which the task should be written into, the destination of the service which should execute the task, in the form of a URL address, and which HTTP method should be used when sending the task to its destination. Additional data that can be provided with each task, if needed, includes retry policies: how many times should the queue try to execute the same task, what is the layoff period before retrying to send the same task and how long should the queue wait for the confirmation from the destination service that the task was successfully processed. Each task goes through different states, as it is being handled by the queue (Fig. 3).

3.3   Definition of a Task Queue

A task queue is a structure used for grouping similar tasks. Up until this point the term "task queue" was used to describe the entire component in the system which handles remote execution of tasks between different services. The task queue service can have multiple different queues within it, where each of them can be used to propagate tasks (Fig. 4). This chapter refers to the task queue structure within the task queue service. Metadata for each task queue includes unique UUID, unique name, the maximum number of tokens, and frequency of token regeneration (these terms are explained in chapter 3.4.).

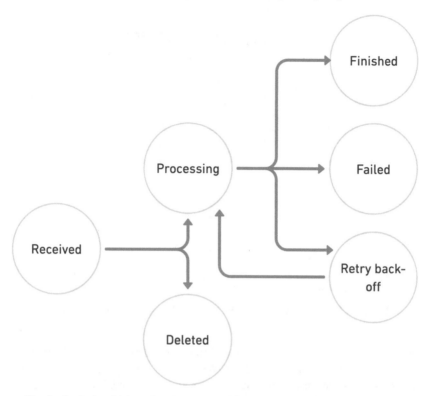

Fig. 3. States in which each task can be as it is being processed by the task queue

The persistence layer used for storing tasks can be changed depending on the needs of the queue itself because this component is defined in the form of an interface. Individual queues are managed by the component named "TaskQueueManager" which is responsible for writing tasks in the corresponding queue, and forwarding them later to the workers for further processing.

3.4 Token Bucket Algorithm

In systems that are used for transferring data, including message and task queues, there is a need for controlling and limiting the amount that can pass through in a certain timeframe. This is important in order to prevent other components from getting more data than they can process and exceeding their capacity. Users need a way to configure how often can tasks go from the queue to their destination to be processed.

The token bucket algorithm was developed for restricting the number of packets that can pass through the telecommunication network [8]. In this paper, we will try to implement the same login for restricting the throughput of any specific queue.

Each queue is assigned a token bucked upon its creation, which consists of two parameters, the maximum number of tokens it can hold, and the frequency of generating new tokens.

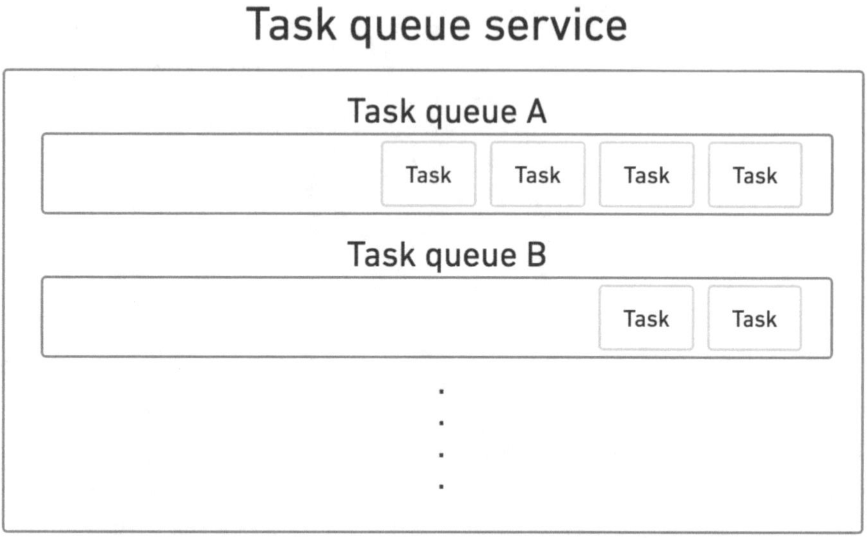

Task queue service

Fig. 4. Difference between the task queue service and individual task queues

When tasks reach their designated queue, they can't proceed with further processing until they are assigned a token from the bucket. With the maximum number of tokens a bucket can hold we can define how many tasks can pass at once. This parameter should reflect how many jobs can be processed in parallel by the consumer. The frequency for generating new tokens should reflect the time it takes for a single job to be processed.

Besides restricting data flow, this algorithm can be used to implicitly define the priority of each individual queue. If the frequency of new tokens in one queue is much higher than in any other, it will have the same effect as assigning that queue a higher priority.

3.5 Task Execution

After obtaining the token, only task execution is left. Given that this operation can take some time, a standalone component within the task queue service should be implemented, so as not to burden the queue manager. This component is called a worker. It should extract the information that is contained within the task structure, construct an HTTP request based on it, and send it to the service for execution. If it fails to send the request, or the service doesn't confirm the successful task execution in the predefined timeframe, the worker checks whether it should retry sending the task, or not. If the response is no, it moves the task into the "execution failed" state, and if the answer is yes, it spends the interval predefined for waiting before attempting to send it again. These tasks don't need to wait for the token again. If we would enforce a rule where tasks need to get a token before they are sent again to their end destination, it would not be possible to guarantee the backoff time between retries that is specified within the task definition.

To achieve parallelism between the component that manages the task queue and the workers, they are executed in separate routines. Instead of creating and destroying new routines for each incoming job, it is more efficient to have a set of routines that will work non-stop - a worker pool, and to send them tasks for execution. Based on the needs of the system and available resources, we can change the number of workers that will be available.

4   Results and Conclusion

In this paper we implemented a dedicated software component that makes communication between distributed services reliable, fault-tolerant, and easily configurable. We defined the scope of the problem and explained the context in which the task queue would be used, as a separate service in the existing microservice platform for managing computer resources. After implementing and testing the solution we came to the conclusion that the token bucket algorithm can be used to efficiently limit the amount of messages going through the queue and also define queue priorities.

In order to see how efficient the queue is, and how implementation changes might affect it, we created a testing version of the task queue. It creates a set number of queues prefilled with generated tasks, and we measured how long it takes before all tasks have transitioned to the "Finished" state.

After defining a framework for measuring the throughput of tasks through the task queue, we experimented with different persistence layers to see how will they affect it. The conclusion we reached during these tests is that the limiting factor in task queue efficiency is the number of parallel processes/routines that can work on processing individual tasks. Writing data to persistent storage like an SSD takes considerably more time (35–40 microseconds) than writing to DRAM (100 ns) [11], but none of those tasks are bottlenecks when each worker has to deliver the message over the network and wait for the response. Choosing a technology that can easily create new processes which are efficient and lightweight, like the Go routines of Erlang processes is more important than optimizing for the time it takes to store data about each task.

Acknowledgment. ▨Funded by the European Union (TaRDIS, 101093006). Views and opinions expressed are however those of the author(s) only and do not necessarily reflect those of the European Union. Neither the European Union nor the granting authority can be held responsible for them.

References

1. Sheth, A., et al.: Research paper on cloud computing. Contemporary research in India (ISSN 2231–2137), no. April, 2021
2. Rogier, B.: What is network latency? How it works and how to reduce it ? Kadiska, 10 November 2020. https://kadiska.com/what-is-network-latency-how-it-works-and-how-to-reduce-it/. Accessed 4 Oct 2022

3. Cory, N.: Cross-Border Data Flows: Where Are the Barriers, and What Do They Cost? Information Technology and Innovation Foundation (ITIF), 1 May 2017. https://itif.org/pub lications/2017/05/01/cross-border-data-flows-where-are-barriers-and-what-do-they-cost/. Accessed 4 Oct 2022

4. Satyanarayanan, M.: The emergence of edge computing. Computer **50**(1), 30–39 (2017)

5. Desertot, M., et al.: Towards an autonomic approach for edge computing. Concurrency and Computation: Practice and Experience, vol. 19 (2007)

6. Bh, D., et al.: Exploring Micro-Services for Enhancing Internet QoS. Trans. Emerging Telecommun. Technol. 29 (2018)

7. Cao, K., Liu, Y., Meng, G., Sun, Q.: An overview on edge computing research. IEEE Access **8**, 85714–85728 (2020). https://doi.org/10.1109/access.2020.299173

8. Cao, J., Zhang, Q., Shi, W.: Challenges and Opportunities in Edge Computing. SpringerBriefs in Computer Science, 59–70 (2018). doi:https://doi.org/10.1007/978-3-030-02083-5

9. Särelä, M., et al.: RTFM: Publish/Subscribe Internetworking Architecture. In: ICT-MobileSummit 2008 Conference Proceedings Paul Cunningham and Miriam Cunningham (2008)

10. Ramasamy, K., Medhi, D.: Network Routing: Algorithms, Protocols, and Architectures. Elsevier Science (2017)

11. Popescu, D.A.: Latency-driven performance in data centers. Technical report, University of Cambridge, vol. 937 (2019)

Authentication and Identity Management Based on Zero Trust Security Model in Micro-cloud Environment

Ivana Kovacevic$^{(\boxtimes)}$ ⬥, Milan Stojkov ⬥, and Milos Simic ⬥

Faculty of Technical Sciences, 21000 Novi Sad , Republic of Serbia
`kovacevic.ivana@uns.ac.rs`

Abstract. The abilities of traditional perimeter-based security architectures are rapidly decreasing as more enterprise assets are moved toward the cloud environment. From a security viewpoint, the Zero Trust framework can better track and block external attackers while limiting security breaches resulting from insider attacks in the cloud paradigm. Furthermore, Zero Trust can better accomplish access privileges for users and devices across cloud environments to enable the secure sharing of resources. Moreover, the concept of zero trust architecture in cloud computing requires the integration of complex practices on multiple layers of system architecture, as well as a combination of a variety of existing technologies. This paper focuses on authentication mechanisms, calculation of trust score, and generation of policies in order to establish required access control to resources. The main objective is to incorporate an unbiased trust score as a part of policy expressions while preserving the configurability and adaptiveness of parameters of interest. Finally, the proof-of-concept is demonstrated on a micro-cloud platform solution.

Keywords: Distributed systems · cloud computing · authentication · IAM service · zero trust security

1 Introduction

In the digital era, IoT and mobile devices interconnect using heterogeneous protocols in both human-centric and machine-centric networks. Taking into consideration the significant number of connected devices and massive data traffic, it became inevitable to incorporate cloud services in many domains, from smart cities and self-driving vehicles to social networks and the gaming industry. Nevertheless, the increased amount of raw data being generated in real-time created a bottleneck in meeting the required quality of services due to these devices' computational, storage, and bandwidth constraints. Therefore, there is a strive to move data centers to the edge, closer to data resources, in order to decrease latency and increase on-demand network access, introducing the concept of edge computing. The basic idea of edge computing is to employ a hierarchy of edge servers with increasing computation capabilities to handle mobile and heterogeneous computation tasks offloaded by the low-end IoT and mobile devices, namely,

© The Author(s), under exclusive license to Springer Nature Switzerland AG 2024
M. Trajanovic et al. (Eds.): ICIST 2023, LNNS 872, pp. 481–489, 2024.
https://doi.org/10.1007/978-3-031-50755-7_45

edge devices [1]. Such advantages over cloud computing have caused edge computing to grow rapidly in recent years. As the general interest and number of users of cloud platforms increase, the security of cloud services is slowly becoming a requirement of uttermost importance.

Along with data migration, new consideration was born. Since customers have a high demand for privacy and restricted access control over their assets, many security challenges have arisen. In this context, there is a need to develop a flexible security service that could be deployed with other cloud services to provide them with mandatory security features. The abilities of traditional perimeter-based security architectures are rapidly decreasing as more enterprise assets are moved toward the cloud environment. From a security viewpoint, the zero trust framework can better track and block external attackers while limiting security breaches resulting from insider attacks in the cloud paradigm. Furthermore, Zero Trust can better accomplish access privileges for users and devices across cloud environments to enable the secure sharing of resources.

Multiple trust domains coexist in the cloud computing environment, and numerous user entities participate in communication and interaction. Therefore, implementing authentication of the application system is crucial [3]. User authentication in cloud computing is the process of validating the identity of the user to ensure that it is legitimate to access cloud resources [4]. Authentication is a critical aspect of security given that it protects from the impersonification of entities, preventing attackers from illegally accessing the system. Authentication in cloud computing (PCC) is classified into six types: username and password authentication (password-based authentication), multi-factor authentication, mobile trusted, Single Sign On, Public Key Infrastructure, and biometric authentication. Despite the growing number of innovative ways to authenticate users, password-based authentication is still one of the most popular methods [5]. Besides, from the aforementioned, authentication is generally performed between devices and servers. If an attacker intends to access protected servers or devices directly, it would be blocked by the authentication system.

In this paper, we proposed a novel micro-service Zero Trust architecture based on multiple technologies to enhance the security of the system. Our proposed architecture incorporates a multi-factor authentication mechanism for authenticating users, as well as certificate management performed by PKI component. Furthermore, we employ an attribute-based authorization. The key aspect of our approach is the utilization of a trust score in conjunction with access rules to facilitate access control. By incorporating a trust score as an additional factor, we enhance the decision-making process for granting access.

The remainder of the paper is organized as follows: Sect. 2 provides a brief overview of the related work considering zero trust architecture. In Sect. 3, the proposed zero trust model is explained, emphasizing the services for user and device authentication, the definition of policies using the ABAC model, and the trust score calculation. Subsequently, in Sect. 4, the conclusion is presented along with the future directions for research.

2   Related Work

Many solutions have been put forth with the objective of effectively integrating or mitigating the cloud environment's security in order to meet the requirements of a zero-trust
architecture. Syed et al. [6] gave a detailed overview of the Zero Trust (ZT) security
paradigm. The article employs a descriptive approach to present the core principles
of ZTA, predominantly focusing on the role of authentication and access control. The
authors contributed an in-depth discussion of cutting-edge authentication and access
control techniques in diverse scenarios. This paper impacted our work by explaining
various methods for calculating the level of trust, along with an overview of relevant
markup languages used for defining access control policies in attribute-based access
control models.

Another concise overview of the various measures that need to be taken into account
when implementing Zero-Trust Architecture (ZTA) is given in [7]. The paper elaborates
on the essential steps required to achieve continuous authentication and authorization for
all network participants. Additionally, the authors introduce a trust engine component
that dynamically calculates the overall trust of a user, device, or application within a
specific network, assigning it a trust score.

Dimitrakos et al. [8] introduced a new model for trust-aware continuous authorization and a novel technology implementing this model that is scalable, efficient, and
lightweight enough to be effective in a consumer IoT setting. Their implementation
focuses on trust in authentication and follows a trust fusion approach where information
from many sources is combined. Such an approach is formalized in Subjective Logic
(SL). Generally, in cases with multiple sources of belief, it can be assumed that a collection of different beliefs can reflect the ground truth better than each belief independently.
A formal description of multi-source fusion is presented in [9] and can be referenced for
a formal description when deciding on parameter weights in the proposed trust algorithm
in this paper.

3   Zero Trust Security Model

Zero Trust encompasses a set of concepts aimed at reducing or ideally eliminating
the ambiguity associated with enforcing precise access decisions for each request by
considering the network as compromised. Zero Trust Architecture (ZTA), in turn, refers
to the specific system design intended to facilitate this objective. ZTA involves the
implementation of a wide range of fundamental principles to secure enterprise assets,
including data, devices, users, and infrastructure components. The key principles for
achieving ZTA are authentication and access control, as these are the means by which
the user's identity is established [6]. ZTA is based on the concepts of least privilege,
granular access control, and dynamic and strict policy enforcement, ensuring that no
user or device is implicitly trusted, regardless of status or location.

The Zero Trust model is based on continuous or context-based authentication. Communication is secure regardless of network location, meaning all messages transferred
between nodes are encrypted. Popular authentication methods include symmetric key
authentication, lightweight public key infrastructure (PKI), and Open Authorization 2.0

[6]. In asymmetric key authentication, digital certificates can be utilized to prove the identity of a device before communication is established.

3.1 Proposed Zero Trust Architecture

Our approach is based on multiple technologies, including multi-factor authentication, public key infrastructure, and attribute-based access control (ABAC) expressed by XACML [10] policy language. The proposed model consists of several interconnected components responsible for implementing different security requirements. The responsibility of the authentication service is to handle requests from both users and devices. Its implementation is designed to support various protocols, specifically HTTPS and gRPC secure, ensuring secure communication. NIST suggests using software-defined perimeters as one of the key strategies in ensuring the effective implementation of ZTA [11]. The authentication service represents the entry point into the software-defined perimeter (SDP) zone, where each request is intercepted and verified.

All relevant information from requests is collected and transferred to the logging component, which is responsible for integration with the security information and event management (SIEM) system. The segregation of logging activities into a distinct service is undertaken with the objective of ensuring enhanced scalability and performance within the system architecture. By separating the logging functionality, the primary components are freed from the responsibility of performing logging operations, thereby enabling them to concentrate on their core features. To achieve loose coupling and mitigate potential blocking or delayed responses, the communication between the logging service and other components within the system is designed to be asynchronous through the utilization of a message queue. By implementing asynchronous communication through a message queue, the logging service can operate independently and interact with other components in a decoupled manner. That enables IAM services to perform tasks without causing delays or impacting the overall system's responsiveness.

Upon successful authentication, entities are redirected to the policy enforcement point, a separate service that aggregates relevant logs regarding the entities' historical behavior. These logs are then prepared and made available to the policy engine (PE). The policy engine consists of functions for defining and managing access policies, using domain-specific language for writing policies. Trust score is calculated in a separate function within the PE, based on aggregated inputs from PEP. Subsequently, the trust score is compared against an empirically established predefined threshold. Policy engine, authorization entry point (AEP), authentication service, and PKI service are all part of the IAM system, which can be described as the framework of policies and technologies that ensure authorized people within an organization have the appropriate access to network resources [14]. That is a micro-service-oriented architecture utilizing in-line context enrichment and publish-subscribe protocols to optimize concurrency among various components such as policy parsing, attribute value retrieval, trust level determination, and policy evaluation (Fig. 1).

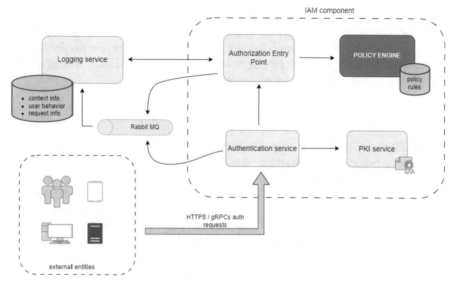

Fig. 1. Proposed zero trust architecture

3.2   Authentication of Users

Authentication enables the various entities participating in a distributed environment to confirm each other's identities before initiating actual communication. For user authentication, a multi-factor authentication scheme is proposed. The trust in authenticity increases exponentially when more factors are involved in the verification process [12]. The initial stage of user authentication verifies credentials, namely username and password. However, additional information apart from credentials validation is necessary during the authentication process. The authentication service is responsible for the extraction of contextual details such as the user's geographical location, the IP address associated with the request, and the timestamp. This context-specific information is then transmitted to a logging service and stored in a database. It can serve as a factor within an implicit trust mechanism, working in conjunction with other parameters like the time duration since the last successful login.

Relying solely on contextual information for implicit authentication is not advisable for a cloud environment, given the diversity of the users, devices, and resources involved. Rather than introducing expensive or vendor-specific solutions, we have chosen to utilize one-time software tokens. As a second factor, a soft token is generated through an authenticator application. This approach does not require deployment complexity and comes with a low cost making it appropriate for cloud environments.

3.3   Authentication of Devices

Node authentication and verification are implemented through the Zero Trust security model. Nodes provide certificates upon joining the cloud computing environment. Trusted certificates are stored centrally and managed by PKI service as a part of the IAM component. The IAM component is developed as an identity-based certificate management system. It enables securing, storing, and controlling secrets, passwords, certificates, and encryption keys for protecting users and sensitive data. PKI service also implements mechanisms for signing and verifying digital signatures. Sensitive information, including encryption keys and secrets, is stored relying on Vault [13].

3.4 Definition of Attribute-Based Access Control Policies

The ABAC can be defined as an access control model where subjects' requests to perform operations on objects are evaluated by considering the attributes of the subject, object, and environment, along with the policies defined around these attributes and conditions. Extensible Access Control Markup Language (XACML) supports and implements the ABAC model. XACML defines policies by using logical formulae involving attributes. The attribute concept in ABAC offers a flexible and extensible abstraction for capturing characteristics of things and keeping access rules abstract enough to be applicable to heterogeneous resources [8].

XACML comes with a robust architecture consisting of multiple software components responsible for creating and managing policies, rules, advice, and obligations. Policy administration point (PAP) creates a policy or set of policies, and it is designed to manage the lifecycle of a policy. The policy decision point (PDP) component is responsible for taking authorization requests as input and returning one of the following responses: PERMIT, DENY, or INDETERMINATE. It is important to note that access decision functions are fed from two different sources. Some user data is stored within the Policy Information Point (PIP). The PIP serves as a repository for granted or denied permits, making it accessible for policy evaluation and decision-making when needed. This separation allows efficient and controlled access to user information within the XACML architecture. To prevent the unnecessary transfer of a large volume of contextual data with each authorization request in XACML, the PIP service is periodically updated with aggregated user data collected in the logging service. PIP also stores records of unsuccessful access attempts that resulted in denial, as well as records of initially granted access to a resource. The latter information is essential, especially because in the proposed ZTM, distinct policy rules are applied based on whether users have an existing authorization history or are new to the system.

Our solution employs a hybrid authorization model, combining criteria and a score-based trust algorithm (TA) along with applying policy enforcement. The criteria-based algorithm is used on users without relevant behavior history, requiring all factors to be fulfilled before evaluating policy rules. This approach enforces strict policies specifically for newly registered users or those requesting access to a particular resource for the first time. In contrast, the score-based trust algorithm calculates a score by assigning weights to various attributes and compares this score against a predetermined threshold value. By continuously successfully passing the authorization mechanism, the user can be awarded for his behavior by transiting to score-based evaluation. The authorization is performed in cycles, solely determined by users' past behavior and current location within a predefined period. Subsequently, the entire process is reinitiated, ensuring a continuous evaluation and adaptation based on the user's behavior history. An example of a condition for enabling an organization member to READ resources if he is in a perimeter of 100 km from the resource from the same organization is shown in Fig. 2.

```
<Condition>
    <Apply FunctionId=
"urn:oasis:names:tc:xacml:3.0:function:string-equal">
        <Apply FunctionId=
"urn:oasis:names:tc:xacml:3.0:function:map-distance">
        <Apply FunctionId=
"urn:oasis:names:tc:xacml:1.0:function:double-bag">
            <Apply FunctionId=
  "urn:oasis:names:tc:xacml:1.0:function:string-bag">
            <AtributeValue>
               {user_lat},{user_long}
            </AttributeValue>
        </Apply>
            <Apply FunctionId=
  "urn:oasis:names:tc:xacml:1.0:function:string-bag">
            <AttributeValue>
                {resource_lat},{resource_long}
            </AttributeValue>
        </Apply>
        </Apply>
        </Apply>
        <AttributeValue>100</AttributeValue>
    </Apply>
</Condition>
```

Fig. 2. Access rule condition example

3.5 Calculation of Trust Score

Trust assessment plays a crucial role in trust-based access control for cloud computing. Nonetheless, a significant challenge lies in assigning reasonable and unbiased weights to various trust factors or parameters involved in the assessment process. The algorithm for calculating trust can be thought of as a function with inputs being relevant attributes such as access request type, the previous behavior of the requesting entity, resource usage history, prior history of penalties, current IAM policies, trust of the group to which the entity belongs to and present threat scenario [15]. Our approach considers the following factors as TA inputs: request metadata (time of request, current details of service or application being used to make the request, and IP address), number of previous requests made to the same resource, resource usage history, previous history of penalties and request geo-location. Aggregated data comes from AEP to the policy engine for each authorization request. The final decision is made considering the result from XACML PDP and the calculated trust score compared against the specified threshold. If

PDP returns the PERMIT value and the trust score passes the threshold boundary, access to the resource is granted. Table 1 shows the possible outcomes of the evaluation.

Table 1. Possible outcomes from trust evaluation

	Trust score >= threshold	Trust score < threshold
PERMIT	allowed	denied
DENY	denied	denied
INDETERMINATE	re-evaluate rules	denied

4 Discussion

This section discusses and compares all techniques mentioned above with other possible solutions. Multifactor authentication is an effective method for establishing a robust authentication mechanism without compromising the user experience or relying on dedicated hardware components to safeguard sensitive information. It eliminates the need for distributing and managing physical tokens, resulting in potential cost savings. On the other side, certificates provide a strong security mechanism for device authentication. They utilize public-key cryptography to establish trust between the device and the authentication server. Since all computational work is transferred and performed in a cloud environment, there is no need to use lightweight encryption algorithms. Although the initial setup required effort, certificates with external key storage using a vault scaled effectively on a micro-cloud platform. The use of certificates simplifies the management of device identities, and the vault handles the storage and retrieval of keys efficiently, enabling secure authentication for a large number of devices. To enable the granularity of access policy rules, it is argued that more than role-based access control is needed. In that sense, RBAC can be seen as a special case of ABAC. XACML involves multiple complex features for condition matching, resolution of conflicts, and computation of rules, which enables a strong base for extending and adapting policies in the future. Separating the trust score from access rules allows greater flexibility in defining and adjusting authorization decisions. A separate trust score also enables adaptive authorization by allowing the system to dynamically adjust access privileges based on the current trust level of the user or entity. Given that we already gather and log various contextual information and event histories, it is possible to utilize this data as input for a machine learning (ML) algorithm as a part of future work. By leveraging ML techniques, we can fine-tune the trust assessment process and improve the accuracy of the trust score calculation.

5 Conclusion

The implementation of the proposed zero trust approach involves a distinct IAM component comprising multiple services. This component is integrated with an existing micro-cloud platform for evaluation. For user authentication, MFA is suggested as an additional layer of security. In contrast, device authentication is accomplished through a two-way SSL authentication process, wherein digital certificates are provided and verified. The system incorporates fine-grained attribute-based access policies that are integrated with the calculated trust score to achieve adaptiveness and continuous verification. The authorization process operates continuously, adhering to the principle of "never trust - always verify" as prescribed by the NIST report on ZTA [11]. Even though the proposed proof of concept promises essential safeguarding of a cloud computing environment, there are certain improvements to be considered, namely regarding performance and coverage of other aspects zero-trust maturity model (ZTMM), including intrusion detection systems, continuous event monitoring, and network segmentation.

Acknowledgment. Funded by the European Union (TaRDIS, 101093006). Views and opinions expressed are however those of the author(s) only and do not necessarily reflect those of the European Union. Neither the European Union nor the granting authority can be held responsible for them.

References

1. Xiao, Y., et al.: Edge computing security: state of the art and challenges. Proc. IEEE **107**(8), 1608–1631 (2019)
2. Furfaro, A., Garro, A., Tundis, A.: Towards security as a service (secaas): on the modeling of security services for cloud computing. In: 2014 International Carnahan Conference on Security Technology (ICCST). IEEE (2014)
3. Li, X., et al.: Smart applications in edge computing: overview on authentication and data security. IEEE Internet Things J. **8**(6), 4063–4080 (2020)
4. Eldow, A., et al.: Literature review of authentication layer for public cloud computing: a meta-analysis. J. Theoretical Appl. Inf. Technol. **97**, 12 (2006)
5. Shen, C., et al.: User practice in password security: an empirical study of real-life passwords in the wild. Comput. Secur. **61**, 130–141 (2016)
6. Syed, N.F., et al.: Zero trust architecture (zta): A comprehensive survey. IEEE Access (2022)
7. Mehraj, S., Banday, M.T.: Establishing a zero trust strategy in cloud computing environment. In: 2020 International Conference on Computer Communication and Informatics (ICCCI), Coimbatore, India, 2020, pp. 1–6 (2020). https://doi.org/10.1109/ICCCI48352.2020.9104214
8. Dimitrakos, T., et al.: Trust aware continuous authorization for zero trust in consumer internet of things. In: 2020 IEEE 19th International Conference on Trust, Security and Privacy in Computing and Communications (TrustCom). IEEE (2020)
9. Wang, D., Zhang, J.: Multi-source fusion in subjective logic. In: 2017 20th International Conference on Information Fusion (Fusion). IEEE (2017)
10. OASIS XACML Homepage. https://www.oasis-open.org/committees/tc_home.php?wg_abbrev=xacml. Accessed 23 May 2023
11. Rose, S., et al.: Zero trust architecture. No. NIST Special Publication (SP) 800-207. National Institute of Standards and Technology (2020)
12. Babaeizadeh, M., Bakhtiari, M., Mohammed, A.M.: Authentication methods in cloud computing: a survey. Res. J. Appl. Sci. Eng. Technol. **9**(8), 655–664 (2015)
13. Vault Homepage. https://www.vaultproject.io/. Accessed 23 May 2023
14. Carnley, P.R., Kettani, H.: Identity and access management for the internet of things. Int. J. Future Comput. Commun. **8**(4), 129–133 (2019)
15. Sarkar, S., et al.: Security of zero trust networks in cloud computing: a comparative review. Sustainability **14**(18), 11213 (2022)

Supporting Integrative Code Generation with Traceability Links and Code Fragment Integrity Checks

Nenad Todorović[(✉)] [ID], Bojana Dragaš[ID], and Gordana Milosavljević[ID]

Faculty of Technical Sciences, University of Novi Sad, Trg Dositeja Obradovića 6, Novi Sad, Serbia
{nenadtod,bojana.zoranovic,grist}@uns.ac.rs

Abstract. It is not always desirable or even possible to separate generated from handwritten code. However, trying to have generated and handwritten code intertwined and integrated means that the problem of persisting manual changes has to be solved in another way. This paper proposes a general workflow based on incremental code generation and API-based generators to allow integrative code generation. We combine traceability links with API-based generation to simplify tracking previously generated code elements. Additionally, we automate detecting manually introduced changes with code fragment integrity checks. To facilitate implementing this workflow, we introduce our solution. The solution includes the extension of RoseLib, our library that allows API-based code generation in C# language. The extension includes support for CSPath, a new language for selecting C# code elements that can be used for tracing, and functionalities that allow performing integrity checks. Finally, we showcase an implementation of the workflow with a proof-of-concept example.

Keywords: Model-Driven Development · Incremental code generation · API-based generators · Traceability

1 Introduction

Model-driven development (MDD) [3] is an approach to software development where models play a core role as primary development artifacts. The advantage of raising the abstraction level and relying on models is that only relevant problem aspects must be considered during development, facilitating better understanding and discussion between stakeholders. Additionally, models can serve as a guide for the automated implementation of solutions based on code generation. Code generation can shorten the production time but also provides other upsides, such as a reduced number of bugs and lower complexity of resulting solutions. Significant contributions have already been made to leverage abstraction and automation in many domains where models are key to success in modern software engineering processes, such as railway systems, automotive, business process engineering, and embedded systems [4].

© The Author(s), under exclusive license to Springer Nature Switzerland AG 2024
M. Trajanovic et al. (Eds.): ICIST 2023, LNNS 872, pp. 490–501, 2024.
https://doi.org/10.1007/978-3-031-50755-7_46

In most scenarios where code generation is used, generated code is kept separate from the hand-written code by using one of the available separation mechanisms [8]. This way, code generators are prevented from overwriting the manual changes in future generation cycles. Unfortunately, there are scenarios where clear separation is not desirable or even possible [1]. For example, some separation mechanisms are available only for specific technologies, such as partial classes and methods, aspects, or inheritance. Furthermore, implementing separation mechanisms usually requires adjusting the code structure or the architecture of a generated solution (e.g., generation gap mechanism). These drawbacks could hinder the adoption of the MDD approach. This problem is especially evident in cases where maintenance and further development of an existing solution could benefit from automation, but an adequate architecture for a straightforward MDD implementation does not exist. Current research shows stakeholders can doubt whether the transition to MDD would improve their situation. They can be hesitant to accept the introduction of significant changes to their solutions [18]. Thus, it would be better to adjust the MDD approach to fit the existing solution in these cases.

To support cases where hand-written and generated code have to be intertwined and integrated, we must find a way to preserve manual changes between generation cycles. Without a clear separation, there is a high risk that code generators will make unwanted changes to the code, introducing bugs or even breaking the solution completely. Manipulating the code on the level of syntax trees using API-based generators [19] provides the ability to perform precise changes only on specific parts of the code while preserving syntax correctness. In this approach, having as few code changes as possible per each generation cycle would be preferable. Only the relevant changes to the input model should affect the corresponding parts of the code, which is the main premise behind the incremental code generation approach [13].

As a part of our research, we investigate how to introduce MDD into a software development lifecycle of manually built solutions. This paper explores how to facilitate integrative code generation where manual changes on the same artifacts are still possible. We base our approach on incremental code generation and API-based generators. More concretely, we analyze a workflow that would enable such a code generation approach and showcase our solution that enables implementing such workflow. To provide the solution, we extend our open-source library named RoseLib [15], which provides API-based generators for manipulating C# files with support for traceability links in the form of expressions written in CSPath, our new language for selecting C# code elements. We also extended RoseLib with functionalities to ensure that previously generated code was not changed manually between generation cycles using integrity checks. The remainder of this paper is structured as follows: Section 2 further explains the theoretical background that provides a basis for this paper and surveys the relevant literature. Section 3 explains the general workflow for integrative code generation. Section 4 provides an overview of a solution that enables implementing such a workflow. Section 5 presents a proof-of-concept to showcase how the approach can be implemented. Section 6 concludes the paper.

2   Background and Related Work

Bernaschina et al. [2] have a similar perspective on code generation, as they consider explicit separation of hand-written code unnecessary and somewhat limiting. In their approach, code generators regenerate the whole code with each generation cycle. However, records of previously introduced hand-written changes are maintained, and hand-written changes are merged into generated code using a version control system. This approach has the downside of needing to manually resolve emerging conflicts. In order to avoid manual resolving of conflicts, we resort to the automated integration of hand-written and generated code. There are a few questions that we have to address for the successful automatic integration:

– What would a workflow that allows integrative code generation look like?
– How should generated code be tracked through code evolution?

In order to address the first question, we took inspiration from the advancements in the field of round-trip engineering [7, 12] (RTE). The main objective of round-trip engineering is keeping the input model and the output (target) code or model in synchronization, as they can diverge due to software evolution. The key difference to our work is that we only want to synchronize code with model changes. We do not aim to update the model according to code changes because we expect that a model does not contain all the implementation-level details. Nevertheless, to derive a workflow that would enable integrative code generation, we took into consideration a common RTE workflow and contained processes [1].

To enable tracking generated code through code evolution, we rely on traceability. Traceability implies keeping track of the relationships between the different artifacts involved in any development process [16] with a wide range of possible applications. In MDD, introducing traceability improves the assessment of which parts of the code are affected by each model change and each generation cycle. This leads to easier integration of hand-written and generated code. Kahani et al. [11] conducted a survey on model transformation tools and languages and analyzed several aspects of transformation, including traceability. Most tools included in the survey support some form of traceability. Traceability is supported either through a manual definition of links or through traces automatically generated by a transformation tool.

In the available literature, traceability is applied to different technical domains. Guana et al. in [10] and previous work described ChainTracker, a framework that aids developers in maintaining and synchronizing code-generation environments with a focus on collecting and visualizing traceability information. This tool tracks two types of traceability links - explicit and implicit. The explicit links are extracted from the transformation itself. On the other hand, the implicit ones are gathered from the dependencies between metamodel elements and associations used to query, select, or navigate the source metamodel, often extracted from OCL expressions. Westfechtel and Greiner in [20] and later Greiner et al. in [9] apply traces in the domain of software product lines to synchronize variability annotations.

Ogunyomi et al. in [13] seek to improve incremental model-to-text transformation by reducing its scale. They propose a runtime analysis to determine how the changed source model affects generated textual artifacts. The output of this analysis is property access traces, structures that aid in making the transformation more efficient. Similarly to our solution, the authors capture property access traces upon the first transformation and store them in a non-volatile storage. In subsequent generation cycles, traces are used to determine whether the source model has changed, allowing the transformation to be partially re-executed. While investigating the relevant literature, we have not found an example where traceability is used in combination with API-based code generators.

3   Methodology

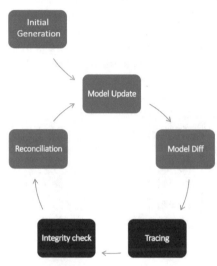

Fig. 1. The integrative workflow

Let us examine how a general workflow for incremental code generation aware of possible code changes could look (Fig. 1). After the initial version of the model is submitted and the first code is generated, every following generation cycle has the next main phases. The cycle starts when the model gets updated, and previously generated code gets out of sync. Once the next generation process is run, relevant model changes need to be calculated to find the differences between the model versions. It is important to find the differences because we want to limit the impact on the code as much as possible. Every unnecessary change made by a code generator increases the chances of a conflict with hand-written changes.

Propagating model changes to corresponding code so that they become synchronized, also called reconciliation, depends on the ability to relate model elements with code fragments. Instead of writing custom logic to link them for each separate code generator, we can generalize the issue and rely on traceability links. We are then able to unify the logic behind the linking. Once a related code fragment is found, we need to ensure that it was not changed between the generation cycles. One way to perform the comparison would be to regenerate the previously generated fragment and compare it to the current code version. However, we can skip this process if we calculate the hash of the code fragment right after the generation takes place and save it for future comparison. That way, we only have to perform an integrity check by comparing the hash value of the current fragment with the previously saved one to conclude if a change has happened. If a change is detected, the workflow is stopped, and the reconciliation must be done manually. If not, the workflow can continue, and the automatic reconciliation can be performed safely. Both traceability links and fragment code hashes need to be recorded between the generation cycles, so we also need to introduce additional data structures associated with the model and maintained by concrete generators.

4   Solution

A solution that can be used to implement the described workflow is presented in Fig. 2, and we will explain all the included elements. To allow domain modeling in a standardized manner, we integrated PyECore, an implementation of ECore [14] for Python. It provides an API to handle meta-models and models, which serve as input for code generators. As explained previously, to support incremental code generation, code generators also need information about model changes. We introduced the Difference component (annotated with the letter delta) to prepare it. The Difference component compares the version of the model at the moment of the latest code generation with the current version to determine the difference. The last input for code generators is stored in the Traceability link and Hash Storage (TaHS). Data from TaHS contains records of traceability links and hash values associated with model elements. Stored records are results from previous code generation cycles, and each code generator maintains its own set.

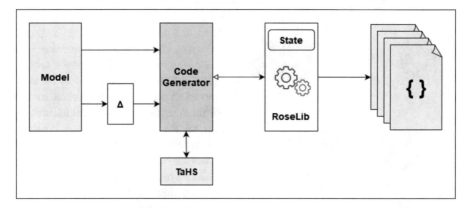

Fig. 2. The solution which enables the workflow implementation

We rely on our open-source library named RoseLib[1] to build code generators that make precise and syntactically correct changes. RoseLib library provides an API designed to enable code generation by manipulating C# syntax trees. The API allows basic syntax tree-level operations - selection, addition, removal, and editing of code related to concepts of the C# programming language. To support implementing the desired workflow and building code generators that utilize traceability links and code fragment integrity checks, we have extended RoseLib with the needed functionalities. Traceability links are based on our new language for addressing parts of a C# document that we call CSPath. Expressions written in CSPath contain concepts related to the C# language, such as namespaces, classes, structs, etc. RoseLib can now generate CSPath expressions based on the selection methods used to reach the current code fragment (state). We also extended RoseLib with an engine that can interpret expressions written in CSPath. Expressions provide directions that the engine uses to navigate through a

---

[1] https://github.com/nenadTod/RoseLib

syntax tree, find the addressed code fragment, and update the state accordingly. Lastly, to support integrity checks, RoseLib now also contains functionalities for normalizing a selected code fragment and calculating its hash value, which can be used for deducing if the selected fragment has been changed afterward. In the following subsections, we will clarify the details of this extension of RoseLib.

4.1 The CSPath Grammar

The CSPath grammar is inspired by the XPath language [5] because it serves a similar purpose, and expressions written using it have a similar form. CSPath was implemented using TextX meta-language [6]. The grammar is omitted from the paper because of the constraints, but a corresponding class diagram is shown in Fig. 3, together with representative expressions. As is shown, a path is composed of multiple parts. Each part denotes a descent towards an instance of some concept. Descent is always relative in CSPath, meaning that the descendant syntax node of interest does not need to be a direct descendant of the current one found in the state. CSPath concepts are mapped to concepts of the C# language. Most structural C# concepts are supported by RoseLib, such as namespaces, classes, and their members, etc. Every descent starts at the syntax node corresponding to the *Compilation unit* concept of C#. This concept is not explicitly defined as a CSPath concept but is automatically implied and found in the state. CSPath expressions can also contain predicates associated with concepts, which serve to find the instance of a concept (syntax node) with the appropriate attribute value. Possible attributes are dependent on the concrete concept the predicate is applied to and are not defined by the grammar but by the RoseLib library instead. These attributes correspond to the parameters of RoseLib's API methods for navigating syntax trees.

Examples of expressions shown in Fig. 3 show the following:

- A descent towards a first namespace with a name matching the regular expression.
- A descent towards a first method with *TestMethod* as a name. From the expression itself, it is not known what the parent concept of such a method is; it could be a class, an interface, or some other C# concept.
- A descent towards a method named *TestOut* with a class as an ancestor.

4.2 Syntax Tree Navigation and Generation of CSPath Expressions

A simplified class diagram of the syntax tree traversal part of RosLib's API is shown in Fig. 4. Classes central to traversing trees are *navigators*. Concrete navigators are related to C# language concepts, hence their names. The starting point for the traversal using a concrete navigator is a selected syntax node of the corresponding syntax kind found in the state. Traversal is implemented through selection methods organized in different *selector* interfaces, which enable selecting descending nodes at an arbitrary depth. One navigator class can implement multiple selectors. Selection methods are designed to allow chaining so that usage follows the language rules (stemming from the C# grammar). Navigator methods' return value types reflect this but are omitted from the diagram for clarity. Each navigator inherits *StatefulVisitor* class to track the visited

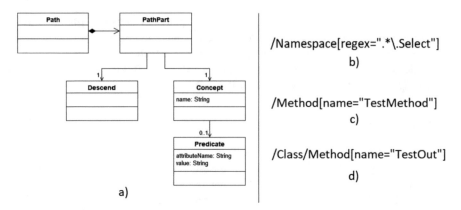

/Namespace[regex=".*\.Select"]

b)

/Method[name="TestMethod"]

c)

/Class/Method[name="TestOut"]

d)

Fig. 3. Showing a) the CSPath model and b–d) example expressions

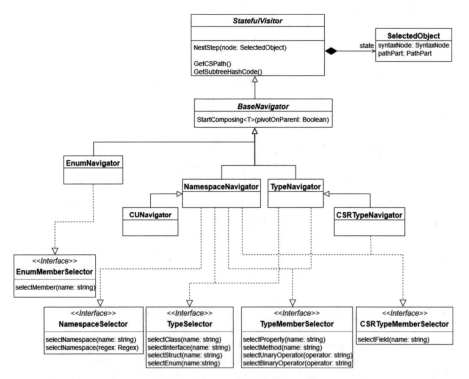

Fig. 4. The tree traversal part of the RoseLib's API

nodes of a tree, which contains the previously mentioned state as a set of *SelectedObject* instances.

Upon successful selection, each of these selection methods has to create an appropriate instance of *SelectedObject* class, constructing an instance of the *PathPart* class with values that can serve to re-run the selection. It must also update the state with this *SelectedObject* instance using the *NextStep* method. When selection is finished, a complete CSPath expression can be obtained using the *GetCSPath* method. RoseLib's classes that implement tree manipulation methods have a reference to instances of *StateVisitor* subclasses and navigate through trees using selection methods. These, too, can then be used to obtain assembled expressions.

4.3 CSPath Engine

Now that we have explained how CSPath expressions are automatically assembled by the selection methods, we can explain how the engine interprets them. A diagram with the engine's design is shown in Fig. 5. The engine relies on TextX to create an instance of the CSPath model based on a given expression. The processing of path parts is done by the logic implemented using the Chain of Responsibility pattern [17]. The chain comprises handlers inheriting the *BaseHandler* class, each with reference to the successor handler, except the last one, and the *Handle* method.

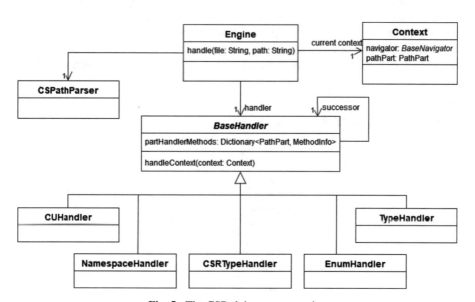

Fig. 5. The CSPath interpreter engine

For interpretation to occur, the engine instantiates a context, combining *CUNavigator* instance with a path part first in line for processing. This initial context is then passed through the chain. If a handler can handle a specific navigator and path part pair, the descent is performed by applying the appropriate selection method. The selection method returns a navigator appropriate for the selected syntax tree node, and the processing of the current context is finished. If none of the handlers can handle the context, the path cannot be processed. Path parts are iteratively combined with succeeding navigators to form new context instances, which get passed through the chain. The handling procedure is repeated until all the parts have been processed and the descent is over. The last navigator returned by the handle method contains an appropriate state and can be used for further descent or to start syntax tree modification.

4.4 Integrity Checks

Once a node has been selected, it is possible to calculate a hash value for the code snippet corresponding to the syntax tree fragment rooted at this node. The way this node got selected is not significant, and it could be either directly using selection methods or indirectly using methods for manipulating syntax trees. Both navigator and tree manipulation classes have the *GetSubtreeHashCode* method, which performs the calculation. Before hashing occurs, we need to ensure that secondary notation does not influence the value, which is important because insignificant changes to the code snippet would impact the resulting value, making the integrity checks less useful. For example, changes in code indentation or adding code comments would cause the integrity check to fail, even though the semantics of the code did not change. To prevent this from happening, all the comments are temporarily removed from the code snippet, and indentation is normalized. Then, we can calculate the hash value of the code snippet. To perform an integrity check later in time, code generators must invoke the calculation on the same tree fragment once again and compare the newly derived value with the one from the TaHS storage. If these are not identical, a change did happen between the generation cycles.

5 Proof-Of-Concept Example

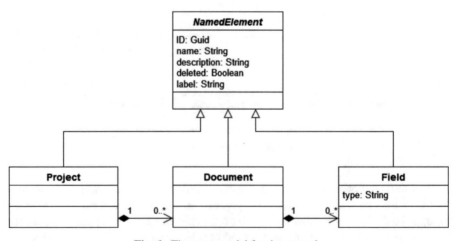

Fig. 6. The meta-model for the example

To showcase how the workflow can be implemented, we present an example based on the meta-model shown in Fig. 6. In this example, an instance of a document is mapped to a C# class with the same name, and document fields are mapped to the corresponding fields of the class. Let us now define a simple model that conforms to the presented meta-model, where we have a *Work order* as a document, with *Order ID* as an Integer and *Order date* as a DateTime as its two fields. Lastly, suppose that such a class has been generated in the initial generation cycle. Now, we will describe how an example code generator can handle a change of the model, where the *Order ID* field is modified to be of the String type.

A sample code of a code generator that can handle such change is presented in Fig. 7. To change the type of the field in the code, the generator first needs to ensure that this field is in the navigator state. As is shown, the generator does not need to implement any case-specific logic for descending through the syntax tree, as it can rely on *CSPathEngine* to interpret the CSpath expression and perform the descent, providing the resulting navigator as an output. To check that the field has not been changed between the cycles, the generator can calculate its current hash value and compare it to the value saved in the previous generation cycle. If the check is unsuccessful, the code generation cycle is stopped. The integrity check logic is also not case-specific. The example finishes with changing the field's type and obtaining the resulting code.

```
CPathEngine cSPathEngine = new CPathEngine();
var navigator = cSPathEngine.Evaluate(
    reader,
    "/Class[name='WorkOrder]/Field[name='OrderId']'"
);

var hashValue = navigator.GetSubtreeHashCode();
if (!hashValue.Equals(previousHashValue))
{
    throw Exception("Code was changed!")
}

var output = navigator.StartComposing<FieldComposer>()
.SetType("String")
.GetCode();
```

Fig. 7. A code generator that can update the type of a field

6   Conclusion

In this paper, we explained why integrative code generation is preferable in some code generation scenarios and how it could be achieved by relying on the presented workflow. Furthermore, we described our solution, which allows workflow implementation and the technologies used to create it. The solution includes the extension of the RoseLib library that provides API for manipulating C# syntax trees, allowing for precise code changes. The extension introduces support for CSPath, a newly created language for addressing parts of C# code. Additionally, it allows checking the integrity of code fragments in subsequent code generation cycles. Finally, we provided a proof-of-concept example to showcase the workflow implementation.

Acknowledgements. This research (paper) has been supported by the Ministry of Science, Technological Development and Innovation through project no. 451-03-47/2023-01/200156 "Innovative scientific and artistic research from the FTS (activity) domain".

References

1. Antkiewicz, M., Czarnecki, K.: Framework-specific modeling languages with round-trip engineering. In: Model Driven Engineering Languages and Systems: 9th International Conference, MoDELS 2006, Genova, Italy, October 1–6, 2006. Proceedings 9, pp. 692–706. Springer (2006)
2. Bernaschina, C., Falzone, E., Fraternali, P., Gonzalez, S.L.H.: The virtual developer: Integrating code generation and manual development with conflict resolution. ACM Trans. Softw. Eng. Methodol. (TOSEM) **28**(4), 1–38 (2019)
3. Brambilla, M., Cabot, J., Wimmer, M.: Model-driven software engineering in practice. Synthesis Lectures Softw. Eng. **3**(1), 1–207 (2017)
4. Bucchiarone, A., Cabot, J., Paige, R.F., Pierantonio, A.: Grand challenges in model-driven engineering: an analysis of the state of the research. Softw. Syst. Model.. Syst. Model. **19**(1), 5–13 (2020)
5. Clark, J., DeRose, S., et al.: Xml path language (xpath) (1999)
6. Dejanović, I., Vaderna, R., Milosavljević, G., Vuković, Ž: Textx: a python tool for domain-specific languages implementation. Knowl.-Based Syst..-Based Syst. **115**, 1–4 (2017)
7. Eden, A.H., Gasparis, E., Nicholson, J., Kazman, R.: Round-trip engineering with the two-tier programming toolkit. Software Qual. J. **26**, 249–271 (2018)
8. Greifenberg, T., et al.: A comparison of mechanisms for integrating handwritten and generated code for object-oriented programming languages. In: 2015 3rd International Conference on Model-Driven Engineering and Software Development (MOD- ELSWARD), pp. 74–85. IEEE (2015)
9. Greiner, S., Nieke, M., Seidl, C.: Towards trace-based synchronization of variability annotations in evolving model-driven product lines. In: Proceedings of the 16th International Working Conference on Variability Modelling of Software-Intensive Systems, pp. 1–10 (2022)
10. Guana, V., Stroulia, E.: End-to-end model-transformation comprehension through fine-grained traceability information. Softw. Syst. Model.. Syst. Model. **18**, 1305–1344 (2019)
11. Kahani, N., Bagherzadeh, M., Cordy, J.R., Dingel, J., Varró, D.: Survey and classification of model transformation tools. Softw. Syst. Model.. Syst. Model. **18**, 2361–2397 (2019)
12. Marah, H., Kardas, G., Challenger, M.: Model-driven round-trip engineering for tinyos-based wsn applications. J. Comput. Lang. **65**, 101051 (2021)
13. Ogunyomi, B., Rose, L.M., Kolovos, D.S.: Incremental execution of model-to-text transformations using property access traces. Softw. Syst. Model.. Syst. Model. **18**, 367–383 (2019)
14. Steinberg, D., Budinsky, F., Merks, E., Paternostro, M.: EMF: eclipse modeling framework. Pearson Education (2008)
15. Todorović, N., Lukić, A., Zoranović, B., Vaderna, R., Vuković, Z., Stoja, S.: Roselib: a library for simplifying .net compiler platform usage. In: Konjović, Z., Zdravković, M., Trajanović, M. (eds.) ICIST 2018 Proceedings, vol. 1, pp. 216–221, pp. 216–221 (2018)
16. Vara, J.M., Bollati, V.A., Jiménez, Á., Marcos, E.: Dealing with traceability in the mdd of model transformations. IEEE Trans. Softw. Eng.Softw. Eng. **40**(6), 555–583 (2014)
17. Vinoski, S.: Chain of responsibility. IEEE Internet Comput.Comput. **6**(6), 80–83 (2002)

18. Vogelsang, A., Amorim, T., Pudlitz, F., Gersing, P., Philipps, J.: Should i stay or should i go? on forces that drive and prevent mbse adoption in the embedded systems industry. In: Product-Focused Software Process Improvement: 18th International Conference, PROFES 2017, Innsbruck, Austria, November 29–December 1, 2017, Proceedings 18, pp. 182–198. Springer (2017)
19. Völter, M.: A catalog of patterns for program generation. In: EuroPLoP, pp. 285–320 (2003)
20. Westfechtel, B., Greiner, S.: Extending single-to multi-variant model transformations by trace-based propagation of variability annotations. Softw. Syst. Model.. Syst. Model. **19**, 853–888 (2020)

Model for Evaluating Points-to Analysis in GraalVM Native Image Using Instrumentation-Based Profiling

Petar Đekanović[1,2](✉) ⓘ, Maja Vukasović[1,2] ⓘ, and Dragan Bojić[1] ⓘ

[1] School of Electrical Engineering, University of Belgrade, Belgrade, Serbia
{maja.vukasovic,bojic}@etf.bg.ac.rs
[2] Oracle Labs, Belgrade, Serbia
petar.dekanovic@oracle.com

Abstract. Static analysis is an essential part of modern ahead-of-time compilers. The level of optimisation and the execution performance of a program directly depends on the analysis results. This paper presents a methodology for empirically evaluating the quality of static analysis implemented in the state-of-the-art Java AOT compiler – GraalVM Native Image. The model combines the data from the points-to analysis with the data from the instrumentation-based profiler to evaluate a generated binary. Finally, we present the initial results of our prototype on exemplary Java programs.

Keywords: Points-to Analysis · Instrumentation-based Profiling · Ahead-Of-Time Compilation · Java · GraalVM

1 Introduction

There are two general approaches for implementing an execution system for a programming language: interpretation and compilation. An interpreter executes the program directly by performing the effects of higher-level operations using their machine-level equivalents. On the other hand, a compiler first translates that same program that contains various levels of abstractions (e.g., classes) into something more understandable to machines, e.g., assembly code.

Modern object-oriented programming languages such as Java and C# combine these two approaches to get the benefits from both – their runtime system called a *virtual machine* (VM) interprets the program but also uses a *Just-In-Time* (JIT) compiler to optimize the frequently used code paths (also known as *hot* code paths) during the execution of the program [1]. Despite all the positive aspects of such hybrid systems, there are certain disadvantages (especially in the performance area). First, the execution of the code is slow at the beginning as the program is first interpreted (in fact, substantially slower compared to the equivalent machine code). Second, there is a significant overhead of compilation of hot code paths and its reverse operation of returning to interpretation (case where control is transferred from the hot path to so-called *cold* code, i.e., *deoptimization*) [2].

M. Trajanovic et al. (Eds.): ICIST 2023, LNNS 872, pp. 502–512, 2024.
https://doi.org/10.1007/978-3-031-50755-7_47

Traditional *Ahead-Of-Time* (AOT) compiler statically translates the program into machine instructions before its execution. One of the main advantages of such compilation is that it can immediately run the produced machine code as efficiently as possible compared to interpretation. However, JIT compilation has the opportunity to employ more precise code optimizations due to the information about the current execution gathered during that same program run.

The AOT compiler heavily relies on some static code analysis. In the case of *points-to analysis*, its purpose is to find all the code parts the application can potentially be used in some execution scenario. Such a program fragment is called *reachable* code (more precisely defined in Sect. 3). The quality of the analysis directly affects the size of the produced binary in terms of the quantity of the code that can be optimized during the compilation. Since the analysis is performed before the actual execution of the program, it must be *conservative* and include all the code for which it cannot tell with certainty that it is not reachable, which can be considered a deficiency [3].

In contrast, there are also dynamic code analyses. These are performed during the program's execution to collect information about its performance that can be useful for further improvement. For the data to be accurate, this type of analysis has to be repeated enough times to cover the majority of possible program behaviours.

One way of measuring a program's performance is by using profiling techniques. These techniques are categorized as instrumentation-based or sampling-based. Instrumentation-based profiling assumes injecting the profiling-code snippets, which are responsible for tracking the information of interest, into an original program. One downside of this approach is that the introduced code extensions may impose significant overhead on the program's performance. On the other hand, sampling-based profiling does not alter the program, which makes it less disruptive than instrumentation-based techniques. It is mainly used in the online profile collection, where it is crucial to reduce the overhead of the collection itself. However, most of the sampling tools are implemented using a *safepoint* mechanism, which makes them less accurate than the instrumentation-based profiling schemes [4].

Various kinds of metrics can be tracked when profiling a program. These metrics are used afterwards to perform more complex program analysis or to aid compiler optimizations, e.g., loop unrolling or method inlining [5]. Our work focuses on *code coverage* – a function that shows a relation between the executed code and code emitted into the binary (from a particular program execution). The code coverage can be measured using different attributes (e.g., the number of methods or the package code size) and at varying levels of granularity (e.g., coarse class level or finer line level). We will focus mainly on the code coverage using the number of the methods, i.e., *method-based code coverage*, since our model later uses it.

The main contribution of this paper is to propose a methodology for measuring the efficiency of static analysis by comparing its results with the factual data gathered by the instrumentation-based profiler. The model is designed with GraalVM Native Image Java AOT compiler in mind, for which we evaluate its state-of-the-art points-to analysis relying on its existing profiler. We also show the motivating results for a simple *HelloWorld* example and a more complex Java benchmark.

The remainder of this paper is organized as follows. Section 2 provides the necessary background for static analysis (points-to analysis in particular) and profiling (focusing on method-level instrumentation-based profiling). Section 3 defines the proposed model for GraalVM Native Image's points-to analysis evaluation. The initial results for the exemplary Java programs are shown and discussed in Sect. 4. Section 5 presents previous work related to our paper. The last section concludes the paper with an overview and the proposed future work.

2   Background and Related Work

2.1   Static Analysis

An analysis performed on a piece of code without its execution is called a static analysis. AOT compilers use these analyses to statically model and predict the behaviour of a program over all possible execution scenarios. We will focus on a fundamental static analysis called *pointer* or *points-to analysis*. It is used for determining sets of all possible type values pointer variables may have, particularly object references and object fields in the case of Java programming language. This kind of analysis uses object allocation sites as sources of object types. These types are then propagated throughout the rest of the program via points of their usage - direct reference assignments, reads and writes of a particular object's field or method invocation sites. The analysis converges when no new types are discovered, i.e., all type sets are stable. These sets are then considered as *reachable types* for a particular reference. The analysis result, in the form of all these type sets, represents at the same time the static model of an application's runtime memory, i.e., heap. This information is crucial as it is used by various code optimisations, such as virtual call resolution [6].

The two principal dimensions, in terms of analysis design in general, are accuracy and performance. Considering more and more information about a potential program run improves the quality of produced type sets and consequently makes the analysis more precise. The literature describes this as *sensitivity* to specific information, e.g., a particular context or flow. Context-sensitive analyses recognise different method invocations as different contexts for that same method. The most used context-sensitive analysis types are *call-site* sensitive [7] and *object-sensitive* analyses [8]. Alternatively, there are flow-sensitive analyses; they consider the control-flow points in a particular method and produce separate sets for each point, but we will not concentrate on them in this paper.

It has been previously shown that analysis precision has a dramatic impact on its efficiency. In fact, an analysis rapidly becomes practically unfeasible due to the increased collected contextual information. Concretely, only a depth of one can be considered doable in the case of call-site sensitivity [8]. As for the object sensitivity, the conclusion is similar, where the limit is reached at a depth of 2 (with a head sensitivity of 1) [9]. This can also explain why the foundational *context-insensitive* and *flow-insensitive* algorithms are still relevant (e.g., Steensgaard's algorithm variation as a part of LLVM [10]).

GraalVM Native Image uses a custom implementation of *context-insensitive* points-to analysis. It is worth noting that the initial version of the analysis was designed as context-sensitive with added "sensitivity" for the object's fields and arrays. It was concluded that the contexts must be deep for analysis to show practical precision gains in a concrete case of Java language. This led to the eventual abandonment of context sensitivity overall and shifted the focus on optimising the analysis for production. Additional refinements are *saturated* type states and method inlining for small methods before the analysis, which empirically demonstrated the performance gains [11].

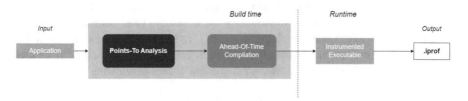

Fig. 1. Image instrumentation pipeline

2.2 Profiling

In contrast to points-to-analysis, which tracks the information about a program statically, profiling captures the program's behaviour dynamically - during its execution. The result of profiling, i.e., profiles, contains different metrics relevant to the analysis of a program, such as the frequency of a method and branch execution. A single profile entry contains an execution number and a call stack, which corresponds to a location of a measured event, and is referred to as a calling context. Calling context contains at least one code location, in which case it is a context-insensitive profile, and at most all locations on the program's call stack up to a program's starting point. Such profiles are fully context-sensitive.

The primary classification of the profiling tools is based on the technique they use to collect profiles. With that said, profilers can be instrumentation-based or sampling-based. The instrumentation technique injects dedicated code snippets into the program's intermediate representation. While it provides precise metrics, this technique can be time consuming, which is why it is dominantly employed in the offline environments. On the other hand, sampling takes the snapshots of a program's call stacks periodically, at pre-determined intervals. It is considered a less time-consuming technique, but it may lead to biased samples and thus contains less precise information [4].

In GraalVM Native Image, the profiling is performed and used by building and running two separate binaries in the following manner. First, the program is compiled with the additional code snippets into a dedicated profiling binary, as shown in Fig. 1. This is implemented by injecting the profiling counters in the graph IRs of a program. Upon running this binary, the profiles are stored in an *iprof* file, which is afterwards consumed during the second compilation of an optimized code. While the profiles in GraalVM Native Image contain information about various event types, this paper focuses on the method execution counters, also referred to as method *hotness*.

3   Model and Methodology

To define the model for analysis evaluation, we will first precisely define all the key concepts. The root abstraction is a program, and the domain of all the possible programs is denoted as P. In our case, a Java program can be abstracted as a collection of user-defined classes. These classes are provided as a program's classpath. Additionally, every Java program needs a runtime infrastructure to be executed and uses additional tools available as a part of the Java Developer Kit (JDK). The runtime can be similarly viewed as a program – a collection of internal classes JDK provides. The domain of all of these classes together is denoted as C. We formally define these as following:

$$classpath :: P \rightarrow C \tag{1}$$

$$runtime \subset C \tag{2}$$

A points-to analysis performed on a given program discovers all the object types that can be instantiated in some execution scenario. We will focus on methods rather than classes as they are later used for the model's primary evaluation metric - method count. With that in mind, we can also say that the result of the analysis can indirectly be seen as a set of reachable methods (each originating from its corresponding reachable type):

$$analysis :: P \rightarrow M \tag{3}$$

We can now define our first method set of interest for a program p – **reachable methods** as:

$$RM\,(p) = analysis(p) \tag{4}$$

The user-provided code never relies on the entire runtime implementation; therefore, we can also say that the result of the analysis is always a strict subset of the program's classpath (defined in Eq. 1) and runtime (defined in Eq. 2):

$$RM\,(p) \subset classpath(p) \cup runtime | \forall p \in P \tag{5}$$

After the analysis, the next step in building a native image is the actual compilation. It consists of various complex code optimizations that make the final machine code efficient as possible. In the case of GraalVM Native Image, the ahead-of-time compilation starts from the so-called *entry points* of the program, i.e., the main method and any other starting method of other threads. These methods are compiled while all the encountered method invocations schedule those methods for the next round of compilation. This process iteratively repeats until a fixed point is reached – there are no newly discovered non-compiled methods. As the model does not benefit from defining each step more precisely, we define the compilation as a process of mapping the input set of reachable methods into the set of methods that will be compiled and emitted into the final executable:

$$compilation :: M \rightarrow M \tag{6}$$

A set of **emitted methods** of a particular program p can now be defined as following:

$$EM(p) = compilation(RM(p)) = (compilation \circ analysis)(p) \qquad (7)$$

In Eq. 6, we say that the analysis result, in the form of the set of reachable methods, is the input for the compilation process, and the set of the emitted methods is its output. In practice, these sets almost always differ due to the employed optimizations. This can be formally described using the following relation:

$$EM(p) \subseteq RM(p) | \forall p \in P \qquad (8)$$

Each execution is characterized by a set of values provided as an input and it maps the emitted methods into the methods that are executed:

$$execution :: M \rightarrow I \rightarrow M \qquad (9)$$

such that I is a domain of all possible input tuples for a particular program p. This set of methods we call the **executed methods** of a program p for a given input i:

$$XM(p, i) = execution(EM(p), i)$$

$$= (execution \circ (compilation \circ analysis)(p))(i) \qquad (10)$$

The next relation is a direct consequence of previous definition:

$$XM(p, i) \subseteq EM(p) | \forall p \in P, \forall i \in I \qquad (11)$$

In an ideal situation, these three sets would be identical. As this is not technically possible due to the limitation of static program modelling in general, we define the *efficiency* of the analysis as the magnitude of the difference or simply a ratio between these sets:

$$efficiency :: M \rightarrow I \rightarrow \mathbb{R}$$

$$efficiency = \frac{|XM(p, i)|}{|RM(p)|} | \forall p \in P, \forall i \in I \qquad (12)$$

4 Motivating Examples and Discussion

To get the initial results, we created a prototype tool, which extends GraalVM Native Image and extracts the set of reachable methods (defined in Eq. 4) from the analysis and the set of methods that end up in the image (defined in Eq. 7). We implemented the prototype using the hooks provided by Native Image's public *Feature* interface [12]. The runtime data is collected by running a dedicated image build for profiling purposes (so-called instrumentation build) and executing the produced instrumented binary afterwards. Our tool then extracts the data on which methods were executed (defined in Eq. 10) based on hotness counters from previously generated profile. Finally, it visualizes all the information using a simple sunburst chart (i.e., hierarchical pie chart).

```
1    class HelloWorld {
2      public static void main(String[] args) {
3        System.out.println("Hello world!");
4      }
5    }
```

Fig. 2. *HelloWorld* example

The example we used to test the analysis performance is a small *HelloWorld* Java program shown in Fig. 2. Although this program is straightforward, to compile this piece of code into a standalone binary, the significant part of JDK and Java runtime must be compiled as part of the final application. To prove this, we compared the size of a regular *classfile* generated by a standard Java programming language compiler (i.e., *javac*) with the size of a binary produced by the GraalVM Native Image compiler. Classfile, which only consists of the user program's bytecode, contains less than 500 B, while in the latter case, a native executable of the same program contains around 8 MB. Even though machine code is usually larger than the corresponding bytecode, and binaries include more than just executable code (e.g., read-only data), this enormous difference in size evidently indicates that most of the final assembly code must indeed come from internal JDK classes. Consequently, this example provides a valuable preview of what we can expect from the analysis. The result is visualized in Fig. 3.

Fig. 3. Method-based code coverage for *HelloWorld*

As we can first observe, the number of methods of all three sets is quite different. The analysis statically models the application; therefore, it must be conservative and include all methods that might be executed, especially in the case of virtual method invocations. Although not entirely equal, the set of reachable and emitted are similar in size (almost 90% of reachable methods were emitted into the executable). That was expected, as the GraalVM Native Image compiler directly uses the analysis reachability model to decide what to compile. Also, the slight decrease in size could be explained by additional code pruning by optimizations performed after the analysis (e.g., function inlining).

The relationship between emitted and executed methods is more interesting. Only 10% of the generated binary is recorded as executed. Although surprising at first, this phenomenon can also be explained. As previously said, our example looks uncompli-cated from an end user's perspective. However, to run this as an independent binary, a considerable amount of internal Java runtime code must also be included within the final executable. As an analysis's natural limitation, the prediction of what code can and will be used when executing the application is imperfect, and the constructed reachability model can get unnecessarily inflated. One good example of this can be Java's mem-ory model implementation (with its non-neglectable *garbage collection*), which Native Image must include in the binary, even though it is most likely not used by this example.

Although we propose a theoretical model for measuring the quality of points-to analysis in this work, thus not focusing on performing the extensive evaluation, we additionally present the results for another, more complex, Java program. We picked one of the benchmarks from the *Renaissance* benchmark suite [13]. This popular suite that contains various kinds of modern applications made for JVM. In particular, we used *fj-kmeans* – a benchmark that runs the k-means algorithm using the *Fork/Join* framework in Java. Figure 4 displays the sunburst visualization for this application.

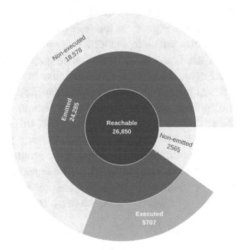

Fig. 4. Method-based code coverage for *fj-kmeans*

As we expected, the visualized results for the two applications show a lot of simi-larities. We can clearly see that the ratio between the number of reachable and emitted methods is stable (around 90%), which is understandable. An increase in the number of executed methods, from 10 to 20%, also aligns with our previous hypothesis. First, the latter application contains more user code (compared to only one *main* method in the first example), and second, more of that code, along with the longer running times, increases the chance of "covering" more execution paths, which also entails more considerable usage of emitted JDK code overall.

5   Related Work

The work done by Liang et al. [14] aimed to adapt existing points-to analyses to work with Java programs. They primarily worked on extending flow-insensitive, context-insensitive analysis implementations – namely, Steensgaard's [15] and Andersen's [16] algorithms. Although the work handled many Java OOP language features (e.g., virtual method calls, and object fields), their implementation did not cover complete Java specification (unlike GraalVM Native Image, which provides a reflection mechanism for example). The authors also provided a detailed evaluation, concluding that making these algorithms both efficient and precise is possible. Still, the authors stated that they only compared their results with predicted worst-case scenarios, not actual execution data.

The subsequent paper by Liang et al. [17] complements the former work by focusing more on analysis evaluation. The proposed approach used dynamic information about the program execution and compared it with the results from points-to analyses. They also based their evaluation on comparing the number of methods found by the analysis and recorded as executed. Since the extended work also relies on Andersen's algorithm, their approach still lacks full Java support. More importantly, their comparison is conducted on user code only, thus leaving out the Java infrastructure (and any other external library) needed for a proper program execution.

The more recent work by Stancu et al. [18] shares a similar motivation with the work presented in this paper and the work conducted by the previous authors [17]. Again, the mutual goal is to gain insights into analysis performance by comparing static and dynamic (i.e., runtime collected) information. This work is most related to ours as both compilation processes use Graal IR as intermediate graph representation. The first significant difference is the actual compilers – we specialise in production-tested Native Image compiler when designing the evaluation model, while they used their custom Java AOT compilation solution (which (also) lacks full feature support like reflection and dynamic class loading). Second, we collect dynamic data differently – Native Image builds and runs a dedicated instrumented image, which directly produces an *iprof* file. In contrast, they collected the profiling data by running the program on the *HotSpot VM* using its built-in profiling infrastructure. More so, we designed the model primarily to assess the quality of produced binaries by evaluating the precision of the analysis itself. Last, unlike any of previously discussed works, we use our model to create a program visualisation tool prototype.

6   Conclusion and Future Work

This paper introduces a novel model for evaluating state-of-the-art points-to analysis in GraalVM Native Image. The model combines the data from the instrumentation-based profiler with the reachability data from the analysis to give a programmer an insight into the general efficiency of the produced executable. We used two exemplary Java programs to get the initial insights.

The future work assumes a complete model evaluation using a standard set of Java benchmarks. The essential difference between a simple *HelloWorld* and a more complex benchmark application the tool can visualise is much larger method sets in the latter case.

Another thing we expect is that the larger applications would have a bigger ratio of the user-defined code compared to runtime code. However, in order to confirm that, we need to localise the methods precisely. Method classification by their origin (e.g., on a package- or class-level) is a prerequisite for a more extensive evaluation and represents an additional work expansion.

Another research direction is adding support for new evaluation metrics, such as the machine-code size of a method, which could provide more precise code coverage information. Additionally, finer-grained branch or line-level profiling could further increase the precision but may require significant effort.

Acknowledgments. Oracle and Java are registered trademarks of Oracle and/or its affiliates. Other names may be trademarks of their respective owners.

References

1. Singer, J.: JVM versus CLR: a comparative study. In: Proceedings of the 2nd International Conference on Principles and Practice of Programming in Java, pp. 167–169. Computer Science Press, Inc., New York (2003)
2. Duboscq, G., Würthinger, T., Stadler, L., Wimmer, C., Simon, D., Mössenböck, H.: An intermediate representation for speculative optimizations in a dynamic compiler. In: Proceedings of the 7th ACM workshop on Virtual machines and intermediate languages, pp. 1–10. Association for Computing Machinery, New York (2013)
3. Native Image Compatibility and Optimization Guide, Oracle GraalVM Enterprise Edition 20 Guide. https://docs.oracle.com/en/graalvm/enterprise/20/docs/reference-manual/native-image/Limitations/. Accessed 24 May 2023
4. Hofer, P., Gnedt, D., Mössenböck, H.: Lightweight Java profiling with partial safepoints and incremental stack tracing. In: Proceedings of the 6th ACM/SPEC International Conference on Performance Engineering, pp. 75–86. Association for Computing Machinery, New York (2015)
5. Samples, A.D.: Profile-Driven Compilation. Technical Report. University of California, Berkeley (1991)
6. Smaragdakis, Y., Balatsouras, G.: Pointer analysis. Found. Trends® Program. Lang. 2(1), 1–69 (2015)
7. Allen, F.E.: Control flow analysis. ACM Sigplan Notices 5(7), 1–19 (1970)
8. Milanova, A., Rountev, A., Ryder, B.G.: Parameterized object sensitivity for points-to analysis for Java. ACM Trans. Softw. Eng. Methodol. (TOSEM) 14(1), 1–41 (2005)
9. Smaragdakis, Y., Bravenboer, M., Lhoták, O.: Pick your contexts well: understanding object-sensitivity. In: Proceedings of the 38th annual ACM SIGPLAN-SIGACT symposium on Principles of programming languages, pp. 17–30. Association for Computing Machinery, New York (2011)
10. Available *AliasAnalysis* implementations, LLVM Alias Analysis Infrastructure. https://llvm.org/docs/AliasAnalysis.html#the-steens-aa-pass. Accessed 24 May 2023
11. Wimmer, C.: GraalVM native image: large-scale static analysis for Java (keynote). In: Proceedings of the 13th ACM SIGPLAN International Workshop on Virtual Machines and Intermediate Languages, p. 3. Association for Computing Machinery, New York (2021)
12. Feature (GraalVM SDK Java API Reference), GraalVM SDK Java API Reference 20. https://docs.oracle.com/en/graalvm/enterprise/20/sdk/org/graalvm/nativeimage/hosted/Feature.html. Accessed 24 May 2023

13. Prokopec, A., et al.: Renaissance: Benchmarking suite for parallel applications on the JVM. In: Proceedings of the 40th ACM SIGPLAN Conference on Programming Language Design and Implementation, pp. 31–47. Association for Computing Machinery, New York (2019)

14. Liang, D., Pennings, M., Harrold, M. J.: Extending and evaluating flow-insensitive and context-insensitive points-to analyses for java. In: Proceedings of the 2001 ACM SIGPLAN-SIGSOFT Workshop on Program analysis for Software Tools and Engineering, pp. 73–79. Association for Computing Machinery, New York (2001)

15. Steensgaard, B.: Points-to analysis in almost linear time. In: Conference Record of the 23rd ACM Symposium on Principles of Programming Languages, pp 32–41. Association for Computing Machinery, New York (1996)

16. Andersen, L.: Program Analysis and Specialization for the C Programming Language. Technical Report. University of Copenhagen, Copenhagen (1994)

17. Liang, D., Pennings, M., Harrold, M.J.: Evaluating the precision of static reference analysis using profiling. In: Proceedings of the 2002 ACM SIGSOFT international Symposium on Software Testing and Analysis, pp. 22–32. Association for Computing Machinery, New York (2002)

18. Stancu, C., Wimmer, C., Brunthaler, S., Larsen, P., Franz, M.: Comparing points-to static analysis with runtime recorded profiling data. In: Proceedings of the 2014 International Conference on Principles and Practices of Programming on the Java platform on Virtual machines, Languages, and Tools, pp. 157–168, Association for Computing Machinery, New York (2014)

A Reference Architecture for Secure IoT

Marija Jankovic[1]([📧]) [iD], Nikos Kefalakis[2] [iD], Sofia Tsekeridou[2] [iD],
and Dionysios Kehagias[2] [iD]

[1] Centre for Research and Technology Hellas, Thessaloniki, Greece
jankovicm@iti.gr
[2] Netcompany-Intrasoft, Luxembourg City, Luxembourg

Abstract. This paper addresses the urgent need for a secure reference architecture in the rapidly evolving field of Internet of Things (IoT). We present a comprehensive strategy for IoT security, which extends the ISO/IEC 30141 IoT Reference Architecture (RA). This approach, reflected in a unique domain-based RA, serves as a robust foundation for different IoT sectors, balancing specific security needs with unified, standard-based practices. The paper details the IoTAC RA's design, its application in a smart home case study, and future directions for its validation and standardization, underscoring its potential to fortify the IoT security landscape.

Keywords: IoT · Reference Architecture · Security

1 Introduction

With the rapid expansion of the Internet of Things (IoT), an estimated 41.6 billion devices, from household appliances to industrial machinery, will be interconnected by 2025, generating approximately 79.4 ZB of data [1]. This vast network of interconnections, as projected by Globaldots, will revolutionize nearly every sector. However, with this digital transformation come significant challenges, particularly concerning security. As the IoT ecosystem continues to evolve, it becomes increasingly vulnerable to security threats [2]. A botnet attack such as Mirai serves as a stark reminder of the vulnerabilities within the IoT environment [3], underscoring the importance of robust security frameworks and strategies.

Designing a security strategy for the IoT, however, is not a trivial endeavour. The heterogeneous nature of IoT devices, their vast numbers, and the complexity of the IoT ecosystem amplify the challenge. IoT devices and systems are frequently susceptible to attacks due to a lack of fundamental security controls, such as encryption, authentication, and access control. Retrofitting these security controls to devices and systems not originally designed with security in mind is a significant hurdle [4]. Thus, as IoT systems become more powerful and versatile, it is crucial to prioritize security in their development and to implement robust security controls within their architecture [4]. There is a clear need for a secure reference architecture (RA) that can serve as a standard for the development and deployment of IoT systems. Traditional security approaches

© The Author(s), under exclusive license to Springer Nature Switzerland AG 2024
M. Trajanovic et al. (Eds.): ICIST 2023, LNNS 872, pp. 513–524, 2024.
https://doi.org/10.1007/978-3-031-50755-7_48

and reference architectures often prove inadequate in addressing the unique needs and challenges of the IoT [5].

In response to this need, this paper presents our contribution - the IoTAC Reference Architecture (RA). Developed within the IoTAC project funded under Horizon 2020 framework, the IoTAC RA extends the ISO/IEC 30141 standard, providing a comprehensive strategy for IoT security [6]. The IoTAC RA's strength lies in its capacity to serve as a solid foundation for different IoT domains, allowing them to build their specific security requirements while ensuring adherence to a unified, standards-based approach.

The rest of this paper is structured as follows: Sect. 2 covers related work, Sect. 3 details the proposed ISO/IEC 30141 extension approach, Sect. 4 discusses the IoTAC RA and its security components, Sect. 5 presents a Smart Home case study, Sect. 6 outlines future directions for IoTAC RA validation and potential standardization, and Sect. 7 concludes the paper.

2   Related Work

This section compares various IoT Reference Architectures from existing standards, articles, and vendor white papers, focusing on their security provisions and challenges.

Standard IoT RAs. The *ISO/IEC 30141* standard outlines an IoT Reference Architecture, encompassing system characteristics, conceptual and reference models, and architectural views [7]. While it identifies security as a part of the vertical, cross-domain trustworthiness dimension, it does not provide specific implementation details. Though it emphasizes security risk management, it lacks guidance on threat modeling and concrete solutions for mitigating security risks. Additionally, the standard outlines selected IoT security challenges, like the need for efficient and secure communication, but it does not fully cover the entire spectrum. The *ETSI M2M Reference Architecture* establishes a foundational standard for M2M and IoT technologies, encompassing Device and Gateway, and Network Domains [8]. It outlines security solutions including key hierarchy, mutual authentication, and access control, but lacks details on specific security threats or countermeasures. The *ITU-T Y.4460* document stresses the importance of information security in IoT devices, urging the inclusion of security best practices in the architectural models. However it does not explicitly detail any security aspects [9]. While it recommends risk assessment and appropriate safeguards, it does not specify particular IoT threats or countermeasures.

Vendor IoT RAs. The *Intel IoT Reference Architecture* tackles security challenges using a layered approach across endpoint devices, networks, and the cloud [10]. It focuses on device integrity, network traffic security, and end-to-end cloud protection, advocating for multiple defence mechanisms via robust hardware and software protection to ensure a secure IoT ecosystem. The *Azure IoT Reference Architecture* addresses security by implementing measures across all domains. It utilizes Azure IoT Hub for secure device communication, Azure Cosmos DB and Blob Storage for encrypted data storage, and Azure Active Directory for authentication, authorization, and auditing [11]. The Azure IoT framework underscores the importance of security in IoT setups, advocating for sophisticated and reliable security alternatives, but refrains from elaborating on distinct countermeasures.

IoT RAs Proposed in the Literature. The IoT-A reference architecture addresses both general and a large number of specific security challenges, analysing threats and offering guidelines, while integrating security measures and best practices in IoT system design [12]. It emphasizes recommended IoT architecture, principles, and subsystems, highlighting security's importance but lacking details on specific countermeasures. Alaba et al. [13] present a streamlined reference architecture segmented into Application, Perception, and Network layers. Their primary focus is on identifying the security challenges, threats, and vulnerabilities prevalent in the IoT environment. Additionally, they propose specific solutions and recommendations such as trust models and privacy policies to mitigate these risks. Yan et al. [14] propose a reference architecture for IoT security, encompassing access control and key management, although it is noted that not all security objectives are met. They provide a thorough analysis of security issues in the IoT domain and offer insights into security countermeasures and future directions. Researchers in [15–17] delve into various IoT challenges, shedding light on the features of emerging technologies and potential countermeasures.

Table 1. Comparison between related IoT Reference Architectures

Ref.	RA Type	Security Aspects	Challenges/Threats	Countermeasures
[7]	Standard RA	Partial	Partial	No
[8]	Standard RA	Yes	No	No
[9]	Standard RA	No	No	No
[10]	Vendor RA	Yes	Yes	Yes
[11]	Vendor RA	Yes	Yes	No
[12]	Proposed RA	Yes	Yes	No
[13]	Proposed RA	Yes	Yes	Partial
[14]	Proposed RA	Yes	Yes	Partial
[15]	Proposed RA	Yes	Yes	Partial
[16]	Proposed RA	Yes	Yes	Partial
[17]	Proposed RA	Yes	Yes	Partial

The analysis outlined in Table 1 suggests a common trend among the reviewed studies; they either demonstrate limited coverage or only partially address the security challenges and countermeasures inherent to IoT security.

At the core of IoT security challenges are device vulnerabilities, a lack of standardization, and inadequate authentication mechanisms. These deficiencies make devices susceptible to unauthorized access and exploitation. A specific issue is the inability of many IoT devices to receive over-the-air updates, which leaves them vulnerable to known exploits. Additionally, the vast interconnectedness of these devices provides a conducive environment for launching Distributed Denial of Service (DDoS) attacks. Securely delivering updates to IoT devices is complex, and there is a notable lack of user awareness regarding necessary security practices.

Key countermeasures aimed at enhancing IoT security include AI-powered anomaly detection for real-time threat identification and proactive mitigation, Zero Trust Architectures for continuous verification to limit attacker movement within networks, and Federated Identity Management for secure, decentralized authentication and authorization. These measures aim to simplify user management while reducing the attack surface, illustrating strides toward addressing the multifaceted security challenges inherent in IoT environments. The identification of key challenges and emerging solutions provided valuable insights for conceptualizing the extension approach for ISO/IEC 30141 and establishing a suitable threat identification methodology, as reported in [18].

3 Methodology

The ISO/IEC 30141 IoT RA, being a flexible and comprehensive framework, is well-suited for various IoT domains. However, it is evident that the standard lacks in-depth coverage on security aspects. Thus, our proposal to extend ISO/IEC 30141 RA concentrates on its security enhancements. We employed a six-step interconnected methodology to execute the proposed extension, as outlined below.

Step 1: *Derivation of the baseline security requirements of the IoTAC framework by performing threat modeling using STRIDE method.* During the first step, the foundational security requirements are established for the IoTAC framework and various domains, such as smart home. STRIDE (Spoofing, Tampering, Repudiation, Information Disclosure, Denial of Service, and Elevation of Privilege) method is used to identify possible threats to an organization's security systems.

Step 2: *Analyzing the ISO/IEC 30141 RA and selecting the Domain-based reference model as a suitable extension base.* This step analyses the ISO/IEC 30141 RA document, with the explicit aim of extending the Reference Model (RM) by following the ISO/IEC 30141 approach. Due to the functional organization across various areas, Domain View is selected for the extension. Using the Domain-based RM within the RA, we align our efforts with an established structure that supports the integration of a vast array of heterogeneous IoT devices and operations.

Step 3: *Specifying the individual IoTAC run-time components to address the identified shortcomings.* As ISO/IEC 30141 RA lacks detail in areas such as implementation guidance, threat modeling, and specific security control measures, this step involves developing specific IoTAC run-time components. These components are designed to address these identified gaps, enhancing the security and resilience of the IoT system.

Step 4: *Mapping the IoTAC components to the ISO/IEC 30141 Six-Domain Reference Model, thereby introducing new components to enhance security.* Following the definition of core functions, the approach proceeds with mapping the identified IoTAC functional components to the corresponding ISO/IEC 30141 Domains. Through mapping our newly developed IoTAC components to this model, we introduce enhanced security measures across the domains, strengthening the overall security profile of the system.

Step 5: *Elaborating on the detailed IoTAC reference model, indicating the overall information flow between the components.* This final step involves elaborating on the relationships and interactions among the various IoTAC components. By detailing the information flow between components, we can provide a comprehensive view of the system's operation. This will not only enable a better understanding of the system, but also help to identify any potential security weaknesses or points of failure.

The emphasis of our approach is on enhancing the cross-domain security and trustworthiness aspects of the Six-Domain Reference Model, to provide a more comprehensive, standardized, and secure framework for IoT design and implementation. By extending the ISO/IEC 30141 Reference Architecture in this manner, we aim to create a more secure IoT environment, addressing the evolving nature of IoT security threats, and making it more adaptable and future-proof.

4 Proposed Reference Architecture

In this section, we describe the resulting IoTAC Reference Architecture (RA), illustrated in Fig. 1 [19]. As previously explained, the newly introduced IoTAC security components are systematically integrated within the framework's existing domains. After providing a brief overview of each domain's functionality, as described in ISO/IEC 30141, we discuss the roles and significance of the newly added components.

*The **Physical Entity Domain (PED)*** includes all physical objects that are part of the IoT system, such as sensors, actuators, and devices as illustrated in Fig. 1. *The **Sensing and Controlling Domain (SCD)*** is responsible for collecting data from sensors and controlling actuators based on that data. IoTAC's RA Model relies heavily on SCD, which comprises most of its groundbreaking features. These include the advanced *Attack Detection* and *Honeypot* components, integrated with the *IoT Security Gateway*. *The **Resource Access and Interchange Domain (RAID)*** provides access to resources such as data storage, processing power, and communication networks. The *Front-end Access Management (FEAM)* is a novel component that implements a delegated capability-based access control, adhering to the Zero Trust concept. It closely collaborates with the *FEAM Gateway* which is logically placed in the SCD. *The **Application and Service Domain (ASD)*** implements application and service logic for service providers in the IoT system, including various services and data stores. The *Data Bus* serves as a communication channel for all real-time data, supporting publish-subscribe functionality. The *Security Configuration (SecCM) Repository* stores data about all IoT system assets and security-related data monitored by the IoTAC platform. The *Observational Repository* offers permanent storage of monitored or processed data from the IoTAC platform. The *Attack Detection Repository* hosts both the offline-trained and online-trained versions of the Attack Detection model for parameter storage and performance evaluation, respectively. *The **Operation and Management Domain (OMD)*** is responsible for managing the overall operation of the IoT system, including monitoring performance, managing security, and handling faults. *The Run-time Monitoring System (RMS)* collects and analyses security-related data from monitored IoT system components or applications for abnormal behavior detection. Finally, the ***User Domain (UD)*** includes all users who interact with the IoT system through various interfaces. Examining Fig. 1 reveals that all individual IoTAC runtime components generate outcomes aligned with a Threat Reporting messaging scheme. This Threat Report is then conveyed to the Data Bus, situated within the ASD, utilizing a publish/subscribe function. As a result, a reporting dashboard or any third-party application may subscribe to these messages, thereby enabling the display of Threat Reports to the end-user or facilitating their further processing.

The Front-End Access Management (FEAM), as an integral part of the IoTAC RA, strengthens IoT security. It manages user authentication and transaction authorization, using secure elements to store access credentials, cryptographic keys, and certificates. The chip card, acting as a secure element, is used by the end user. This method decentralizes the OAuth process and allows for secure transactions to be carried out offline. FEAM satisfies the most important Zero Trust requirements by implementing multi-factor authentication, providing fine-grained privilege management, and adopting to the least and just in time privileges at the session level. It also enhances security by separating access privilege delegation from authentication and authorization processes, which minimizes the possibility of unauthorized access to sensitive data.

The Run-Time Monitoring System (RMS) is a central component of the IoTAC Reference Architecture that is designed to enable real-time services for data collection, transformation, filtering, routing, and management, leveraging security-related data gathered from monitored IoT system components and applications. This architecture supports various data sources, aiming to enhance data accessibility for consumers. The collected data can be used to generate threat reports for measurements that falls outside of nominal thresholds or can be used to drive analytics algorithms that detect patterns of abnormal behavior. The RMS features lightweight monitoring probes that may be used for the data collection and publishing to the monitoring platform. The RMS plays a crucial role in enhancing the security of IoT systems by providing continuous monitoring and detection capabilities as well as valuable insights into security postures of IoT systems, enabling continuous improvement and optimization of security measures, and reducing the risk of cyber-attacks and data breaches.

The IoTAC Security Gateway enhances the security of IoT ecosystems by safeguarding data transmission between IoT devices and enterprise networks. It ensures that network access is granted only to trusted devices with valid credentials, thereby protecting the network from unauthorized access and impersonation attacks. Advanced intrusion detection and prevention systems actively monitor and respond to potential cyber threats, decreasing the likelihood of a breach and maintaining network integrity. The gateway is equipped with robust encryption techniques to safeguard sensitive data and prevent unauthorized access. Additionally, it enforces security policies and controls data flows to minimize attack surfaces, which enhances system security. As illustrated in Fig. 1, the IoTAC Security Gateway also forwards network packets to other components, improving the overall security architecture.

The Attack Detection (AD) Module, utilizing the Dense Random Neural Network (DRNN) model and novel network metrics derived from online network traffic measurements, plays a pivotal role in ensuring IoT security. It learns the communication patterns between IoT devices during normal network operation, enabling it to detect malicious activities based on deviations from these patterns. Additionally, the AD module can be trained offline using normal traffic data collected during the cold-start phase, allowing it to recognize and identify malicious traffic even without prior knowledge of specific attacks. This adaptive approach strengthens the overall defence against emerging threats in the IoT architecture. The Attack Detection component is linked to the IoT Security Gateway, and if malicious activity is identified, it sends Threat Notification messages.

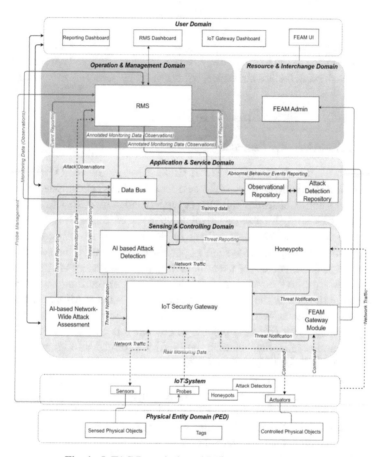

Fig. 1. IoTAC Domain-based Reference Architecture

The AI-Driven Network-Wide Attack Assessment (NW-AA) meticulously assesses each IoT device's security, providing a holistic network security evaluation. It incorporates features such as individual device assessments for vulnerabilities, continually updating parameters and weights for device connections, enhancing network resilience against evolving threats. NW-AA can identify compromised devices by analysing their decision-making pattern changes, implying infections or attacks. Regular training with attack decision data from local detection systems ensures the AI system's adaptability to new threat patterns, providing a robust, adaptable security solution for IoT networks.

Honeypots are integral to IoTAC RA, serving as passive network participants that monitor and analyse network traffic to identify threats. They utilize classic detection techniques to recognize and counter common attacks like Portscan, Login Hacking, Denial of Service (DoS), and malware infections. Furthermore, they feature advanced detection mechanisms, utilizing a distributed learning approach across multiple nodes. This facilitates early identification and prevention of potential threats, even before attackers complete their network scans or exploit vulnerabilities. Housed within an efficient, lightweight architecture, IoTAC honeypots optimize resource usage, contributing to the streamlined operation of the IoT network. Overall, this dual-pronged approach, blending classic and advanced detection strategies, significantly bolsters IoT network security.

5   Case Study - Secure Smart Home

5.1   Smart Home Security Concept

The selected case study is part of the Smart Home Pilot, led by CERTH/ITI within the EU-funded IoTAC project. The Smart Home, located in CERTH campus in Greece, is a cutting-edge two-floor residential building, equipped with a myriad of sensors and smart appliances. The case study focuses on two distinctive use case scenarios derived from the Smart Home system's operations, as outlined in Fig. 2. These carefully chosen scenarios offer a deep understanding of the system's operations and its essential security requirements. The first, UC1, involves the user interacting with the system via the Smart Home Dashboard, operating IoT devices, and the IoTAC system securing the infrastructure against potential threats. In the UC1, we simulate a Man-in-the-Middle (MitM) attack within the Smart Home system. We utilize the IoTAC honeypot and the IoTAC Front End Access Management (FEAM) module to counter this attack. While the honeypot lures attackers and reveals their tactics, the FEAM scrutinizes all communications to prevent unauthorized access and mitigate risks.

The second, UC2, simulates an attack scenario where an external malicious actor attempts to gain unauthorized access to the Smart Home Energy Management system using Distributed Denial of Service (DDoS) attacks such as packet sniffing, port scanning, SYN flood, and UDP flood. This system incorporates various IoT assets like dimmers, sensors, and energy meters. The IoTAC system, showcasing modules like the Honeypot, Attack Detection (AD), Data Bus, Observation Repository, Runtime Monitoring System (RMS), and Dashboard, adeptly identifies and reacts to these simulated attacks, demonstrating its effectiveness in countering DDoS threats [20].

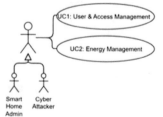

Fig. 2. Smart Home Pilot Scope

In the Smart Home case study, we focus on two critical security dimensions: (1) unauthorized access to the system's data, and (2) false sensor measurements reported to the system's administrator. Use case scenarios UC1 and UC2 correspond to these dimensions respectively. The Smart Home System currently employs only basic authentication and role-based access control, which, despite being effective, could be improved. Furthermore, there is a lack of policy to confirm the integrity of measurements reported by Smart Home Assets, a crucial requirement for maintaining the system's trustworthiness. The detailed threat analysis has allowed us to identify 16 misuse cases specific to the energy domain (UC2). These misuse cases were evaluated and classified based on their impact, with 5 characterized as low priority, 11 as medium priority, and 1 as high priority [21]. To counter identified misuse cases, we have created a security baseline of 64 security requirements, demonstrating our commitment to safeguarding the Smart Home system from all threats [18].

5.2 Smart Home Pilot Logical Architecture

This section outlines a comprehensive description of the Smart Home logical architecture, illustrating the role of various IoTAC modules and how they bolster its operational security. The IoTAC RA essentially serves as an architectural blueprint that guides the Smart Home system's overall structure and operational mechanisms. This strategic guideline ensures the system's robust security and operational effectiveness.

Smart Home Platform (as-is). The unified ITI Smart Home platform follows a centralized client-server implementation approach [22]. In the original configuration, the Smart Home system incorporates an array of *Smart Home Assets* such as sensors and intelligent appliances as illustrated in Fig. 3. These assets relay their measurements to the *Backend Server* using various communication protocols. This server, housing an internal database, persistently stores these inputs. A *Smart Home RESTful API* is integrated to distribute resources to external systems and end users. Interaction with the system is facilitated via a *Smart Home Gateway* and a user-friendly web application known as the *Smart Home Dashboard*. End users are

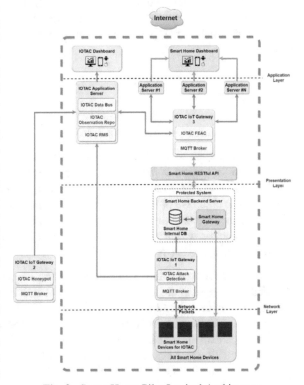

Fig. 3. Smart Home Pilot Logical Architecture

authenticated through a simple scheme and are capable of monitoring and responding to system anomalies. However, despite the advanced functionalities, the system has vulnerabilities to local area network (LAN) and Internet-based attacks, and its process of user authentication needs improvement.

Smart Home Platform (to-be). The integration of the IoTAC introduced additional components into the Smart Home system significantly enhances its security. These newly incorporated IoTAC components, which have further bolstered system security, are marked in blue in Fig. 3. One of these additions is the *IoTAC IoT Gateway - 1*, situated within the Smart Home, equipped with an *MQTT broker* and the *IoTAC Attack Detection (AD)* module. This setup effectively bolsters security against malicious attacks originating from the LAN. The IoTAC *IoT Gateway - 2*, positioned outside the Smart Home, hosts an *MQTT broker* and an instance of the *IoTAC Honeypot module*. Serving as a replica of the protected system, this gateway is designed to lure potential attacks from the Internet, thus increasing protection against external threats. Within the Smart Home, the *IoTAC IoT Gateway - 3* is installed, featuring components of the IoTAC Front-End Access Management (FEAM) module. This configuration improves the authentication and authorization of end users accessing the Smart Home resources. The IoTAC Application Server, which hosts the *IoTAC Data Bus*, the *IoTAC Observation Repo*, and the *IoTAC Runtime Monitoring System (RMS)*, is also included. These components aid in the collection and persistent storage of data generated by the operation of the IoTAC components deployed on the gateways. Finally, the *IoTAC Dashboard* is provided to users, offering them a comprehensive, real-time view of the security status of the Smart Home System.

6   Future Work

The intended course for future work falls into two distinct but interrelated directions.

Real-Life Scenario Validation: A crucial next step in our research agenda involves the practical validation of the IoT RA instantiated reference architecture. We intend to execute this validation within a real-world context, leveraging the CERTH Smart Home scenario. The objective of this initiative is to evaluate the effectiveness and applicability of the proposed architecture in an authentic IoT setting. The testing and validation of the pilot architectures will provide valuable feedback and insights, allowing us to enhance and refine our approach as required.

Standardization Activities: Concurrently, we plan to disseminate our findings and contribute to ongoing standardization efforts at ETSI and ISO. Our proposed IoTAC RA has already introduced two new work items (NWI) to the ETSI Technical Committee "Methods for Testing and Specification" (TC MTS), namely: 'IoT security module testing' (technical specification) and 'IoT security architecture conformity' (technical report). Our active participation in these standardization bodies facilitates the exchange of knowledge and feedback, enabling us to further enhance and standardize the IoTAC RA. We are particularly pleased to report that comment DE39, suggesting an extension approach to the 2nd Edition of the ISO/IEC 30141, has been accepted. This amendment is scheduled for inclusion in the forthcoming version of the standard under clause 10.2.

These two directions will serve to not only affirm the effectiveness of our architecture in the practical field but also ensure our efforts contribute to the broader standardization activities, driving the development of a more secure and resilient IoT ecosystem.

7 Conclusion

To summarize, the accelerating growth of the IoT presents both interesting opportunities and significant security challenges. This paper introduced the IoTAC Reference Architecture, a promising solution to the complex security challenges presented by the IoT environment. By extending the ISO/IEC 30141 standard, the IoTAC RA offers a comprehensive, standard-based security strategy for IoT systems. It serves as a solid foundation upon which different IoT domains can build their specific security needs, thus ensuring a unified approach to IoT security. Through our Smart Home case study, we demonstrated the practical applicability and efficacy of the IoTAC RA. This case study illustrated how the IoTAC RA can be implemented in a real-world IoT system, showcasing its versatility and adaptability in handling the various security needs of the IoT environment.

Looking forward, we aim to validate and potentially standardize the IoTAC RA, emphasizing its potential to strengthen the IoT security landscape. It is our hope that the IoTAC RA will be adopted by more organizations and IoT domains, driving the development of a more secure and resilient IoT ecosystem. The advancement of IoT security is a dynamic pursuit, necessitating perpetual research, refinement, and multisector collaboration. To keep pace with IoT's rapid evolution, it is paramount that we sustain this dialogue and exploration, ensuring the maximization of IoT benefits while mitigating associated risks and vulnerabilities.

Acknowledgement. This work is funded by the European Union's Horizon 2020 Research and Innovation Programme through IoTAC project under Grant Agreement No. 952684.

References

1. Globaldots, A.: 41.6 Billion IoT Devices will be Generating 79.4 Zettabytes of Data in 2025, GlobalDots. https://www.globaldots.com/resources/blog/41-6-billion-iot-devices-will-be-generating-79-4-zettabytes-of-data-in-2025/. Accessed 22 May 2023
2. Kolias, C., Kambourakis, G., Stavrou, A., Voas, J.: DDoS in the IoT: Mirai and other Botnets. Computer **50**(7), 80–84 (2017). https://doi.org/10.1109/MC.2017.201
3. Antonakakis, M., et al.: Understanding the Mirai Botnet. In: The 26th USENIX Security Symposium (USENIX Security 2017), pp. 1093–1110 (2017). Accessed 22 May 2023
4. El Bekkali, A., Essaaidi, M., Boulmalf, M., el Majdoubi, D.: Systematic literature review of Internet of Things (IoT) security. Adv. Dyn. Syst. Appl. **16**, 1671–1692 (2022)
5. Sadeghi, A.-R., Wachsmann, C., Waidner, M.: Security and privacy challenges in industrial internet of things. In: Proceedings of the 52nd Annual Design Automation Conference, in DAC 2015, New York, NY, USA. Association for Computing Machinery, pp. 1–6, June 2015. https://doi.org/10.1145/2744769.2747942

6. IoTAC (Security By Design IoT Development and Certificate Framework with Front-end Access Control). https://iotac.eu/

7. ISO/IEC 30141 Internet of Things (IoT) - Reference Architecture Edition 1.0 2018-08 (2018)

8. ETSI TS 102 690 V2.1.1 (2013-10), Machine-to-machine Communications (M2M); Functional Architecture (2013)

9. ITU-T Telecommunication Standardization Sector of ITU, Y.4460 (06/2019). Architectural reference models of devices for internet of things applications (2019)

10. The Intel IoT Platform, Architecture Specification White paper Internet of Things (IoT), April 2015. https://d885pvmm0z6oe.cloudfront.net/hubs/intel_80616/assets/downloads/general/Architecture_Specification_Of_An_IOT_Platform.pdf

11. Microsoft Azure IoT Reference Architecture, Version 2.1, September 2018

12. D1.2 Initial Architectural Model for IoT (IoT-A) (257521), June 2011. https://cocoa.ethz.ch/downloads/2014/01/1360_D1%202_Initial_architectural_reference_model_for_IoT.pdf

13. Alaba, F.A., Othman, M., Hashem, I.A.T., Alotaibi, F.: Internet of Things security: a survey. J. Netw. Comput. Appl. **88**, 10–28 (2017). https://doi.org/10.1016/j.jnca.2017.04.002

14. Yan, Z., Zhang, P., Vasilakos, A.V.: A survey on trust management for Internet of Things. J. Netw. Comput. Appl. **42**, 120–134 (2014). https://doi.org/10.1016/j.jnca.2014.01.014

15. Caiza, G., Saeteros, M., Oñate, W., Garcia, M.V.: Fog computing at industrial level, architecture, latency, energy, and security: a review. Heliyon **6**(4), e03706 (2020). https://doi.org/10.1016/j.heliyon.2020.e03706

16. Sengupta, J., Ruj, S., Das Bit, S.: A comprehensive survey on attacks, security issues and blockchain solutions for IoT and IIoT. J. Netw. Comput. Appl. **149**, 102481 (2020). https://doi.org/10.1016/j.jnca.2019.102481

17. Kaur, M., Khan, M.Z., Gupta, S., Alsaeedi, A.: Adoption of blockchain with 5G networks for industrial IoT: recent advances, challenges, and potential solutions. IEEE Access **10**, 981–997 (2022). https://doi.org/10.1109/ACCESS.2021.3138754

18. IoTAC Project, D2.4 Security Baseline of the IoTAC Architecture, 31 August 2021

19. IoTAC Project (952684), D2.3 Architecture Design Document', 28 February 2022

20. Gelenbe, E., Nakıp, M.: Real-time cyberattack detection with offline and online learning. In: IEEE International Symposium Local Metropolitan Area Networks 2023 IEEE LANMAN 2023, 10–11 July 2023, London, UK (2023)

21. IoTAC Project (952684), D2.2 Requirements and Use-cases Specification, 31 August 2021

22. CERTH/ITI, ITI Smart Home Digital Innovation Hub. https://smarthome.iti.gr/

Author Index

Printed in the United States
by Baker & Taylor Publisher Services